T0360709

Noble Metal
Noble Value

Ru-, Rh-, Pd-catalyzed
Heterocycle Synthesis

CATALYTIC SCIENCE SERIES

Series Editor: Graham J. Hutchings *(Cardiff University)*

*To view the complete list of the published volumes in the series, please visit:
http://www.worldscientific.com/series/css

CATALYTIC SCIENCE SERIES — VOL. 15

Series Editor: Graham J. Hutchings

Noble Metal Noble Value

Ru-, Rh-, Pd-catalyzed Heterocycle Synthesis

edited by

Xiao-Feng Wu

Zhejiang Sci-Tech University, China &
Leibniz-Institut für Katalyse e.V. an der Universität Rostock, Germany

Imperial College Press

ICP

Published by

Imperial College Press
57 Shelton Street
Covent Garden
London WC2H 9HE

Distributed by

World Scientific Publishing Co. Pte. Ltd.
5 Toh Tuck Link, Singapore 596224
USA office: 27 Warren Street, Suite 401-402, Hackensack, NJ 07601
UK office: 57 Shelton Street, Covent Garden, London WC2H 9HE

Library of Congress Cataloging-in-Publication Data
Names: Wu, Xiao-Feng, 1985–
Title: Noble metal, noble value : Ru-, Rh-, Pd-catalyzed heterocycle synthesis /
    Xiao-Feng Wu (Zhejiang Sci-Tech University, China &
    Leibniz-Institut für Katalyse e.V. an der Universität Rostock, Germany).
Description: New Jersey : Imperial College Press, 2016. | Series: Catalytic science series ; volume 15
Identifiers: LCCN 2015049362 | ISBN 9781783269235 (hc : alk. paper)
Subjects: LCSH: Heterocyclic compounds--Synthesis. | Heterocyclic chemistry. | Transition metal
    catalysts. | Ruthenium catalysts. | Rhodium catalysts. | Palladium catalysts.
Classification: LCC QD400.5.S95 W865 2015 | DDC 547/.59045--dc23
LC record available at http://lccn.loc.gov/2015049362

British Library Cataloguing-in-Publication Data
A catalogue record for this book is available from the British Library.

Desk Editors: Anthony Alexander/Mary Simpson

Typeset by Stallion Press
Email: enquiries@stallionpress.com

Printed in Singapore

Dedicated to my wife and children
Qing-Yuan Wei, Nuo-Yu Wu, and Nuo-Lin Wu
Thanks for their understanding and support!

Acknowledgments

I wish to take this chance to thank Prof. Pierre H. Dixneuf who gave me a chance to study in Rennes 1 University for my Master and become his grandson in Chemistry. Prof. Christophe Darcel must be thanked as my Master supervisor who taught me hand by hand how to play with organometallic and catalyst. The support from Prof. Matthias Beller (my PhD supervisor) should never be forgotten, without his support, there is no chance for me to edit this book.

The assistance and efforts from Merlin Fox, Catharina Weijman and Mary Simpson are appreciated!

Xiao-Feng Wu
2015-09-01

Contents

CHAPTER 1

Ru-Catalyzed Five-Membered Heterocycles Synthesis

Bin Li

State Key Laboratory of Element-Organic Chemistry,
College of Chemistry, Nankai University, Tianjin 300071, China

Ruthenium-catalyzed protocols play an important role in the synthesis of five-membered heterocycles. An overview of reported examples in this field will be discussed in this chapter, including mechanism presentations for some typical transformations.

Keywords: Catalysis, Cyclization, Five-Membered Heterocycles, Ruthenium

1. Introduction

Heterocycles are some of the most important structural motifs found in naturally occurring compounds and pharmaceuticals. The synthesis of heterocycles has attracted considerable attention because of their important physiological and biological activities.[1]

Heterocyclic synthesis involving transition metal complexes has become a common use over the past decade because a transition metal-catalyzed reaction can directly build complicated molecules from readily accessible starting materials under mild conditions. Especially, rapid progress in transition metal-catalyzed C–H bond functionalization in recent years has provided significant benefits toward the synthesis of a variety of heterocycles. Among these reports, ruthenium-catalyzed protocols play an important role.[2] Herein, this chapter is intended to provide an overview of reported examples for ruthenium-catalyzed five-membered heterocycles

1

synthesis till the end of 2014, including mechanism presentations for some typical transformations.[3]

2. Ru-Catalyzed Synthesis of Five-Membered N-Heterocycles

2.1. Synthesis of indoles and their derivatives

In 1986, Jones and co-workers reported the formation of 7-methyl indoles from ruthenium-catalyzed $C(sp^3)$–H bond activation of 2,6-xylylisocyanide (Eq. 1).[4] The catalytic reaction proceeds by using ruthenium (II) complex $Ru(dmpe)_2H_2$ [dmpe: 1,2-bis (dimethylphosphino)ethane] or $Ru(dmpe)_2$(2-naphthyl)H as catalyst (20 mol%) in C_6D_6 at high temperature (140°C). Further studies indicate that the substrate scope is restricted to 2,6-disubstituted (2,6-dimethyl, 2-ethyl-6-methyl, 2,6-diethyl) isocyanides, since less substituted isocyanides only produced stoichiometric indole N–H oxidative addition adducts with $[Ru(dmpe)_2]$.[4(b)]

$$Ru(dmpe_2)H_2 \ (20 \ mol\%)$$
$$C_6D_6, \ 140°C, \ 94 \ h$$

70%   (Eq. 1)

A proposed reaction mechanism is shown in Scheme 1. After coordination of isocyanide to $[Ru(dmpe)_2]$ and dissociation of one end of a dmpe ligand, oxidative addition of a benzylic C–H bond gives rise to a six-membered metallocycle **1**. Then insertion of the isocyanide into the Ru–CH_2 bond and closure of the dmpe chelation generates intermediate **2**. Followed by tautomerism of the methylene hydrogen to the nitrogen and isomerization, the resulting intermediate *cis*-**3** undergoes reductive elimination to give the final product. The isolation and characterization of intermediate *trans*-**3**, and its isomerization to *cis*-**3** were realized under less severe reaction conditions.

Watanabe and co-workers demonstrated that indole derivatives were readily obtained from 2-aminophenethyl alcohols *via* $RuCl_2$ $(PPh_3)_3$-catalyzed intramolecular cyclization in refluxing toluene

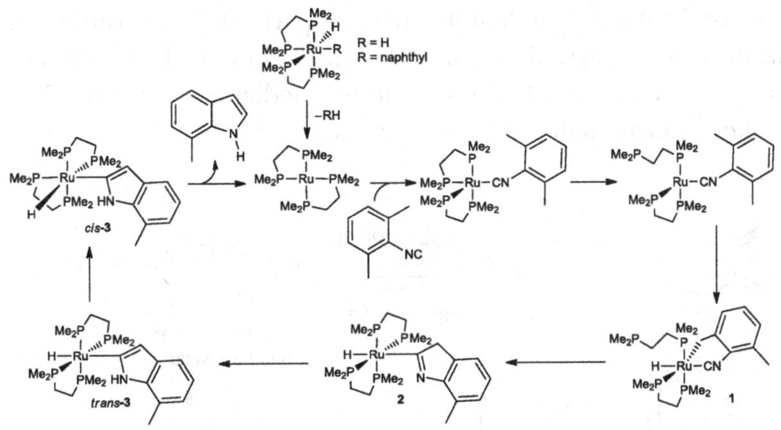

Scheme 1. Proposed mechanism for ruthenium-catalyzed indole synthesis from 2,6-xylylisocyanides

(Eq. 2).[5] A stoichiometric amount of hydrogen was spontaneously released during the reaction. With a heterogeneous and homogeneous binary catalyst system [Rh/C-RuCl$_2$(PPh$_3$)$_3$], indoles could be afforded in one-pot from 2-nitrophenethyl alcohols under a hydrogen atmosphere (Eq. 3).[5(b)]

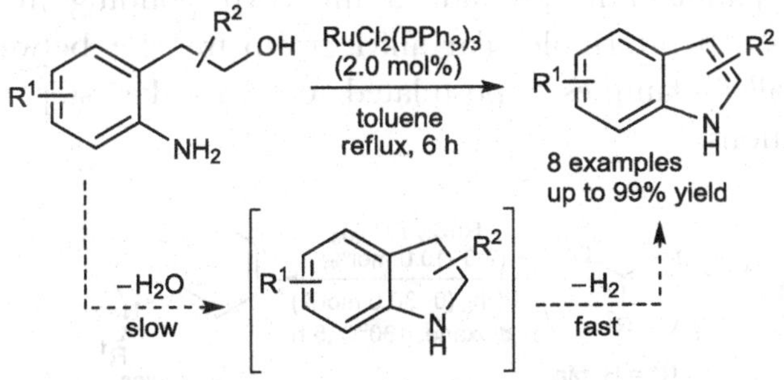

(Eq. 2)

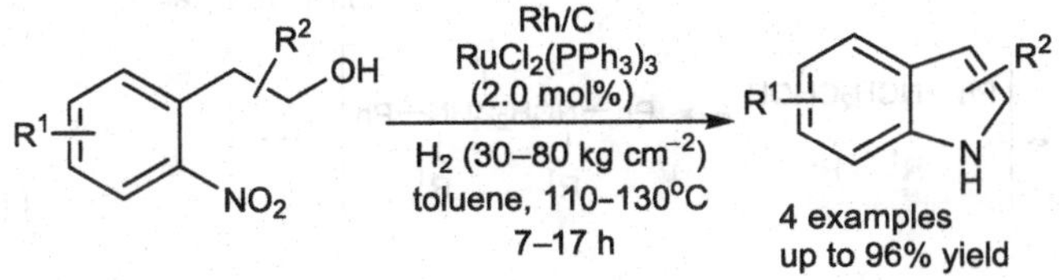

(Eq. 3)

Almost the same time, they also found that *N*-substituted anilines react with glycols in the presence of a catalytic amount of RuCl$_2$(PPh)$_3$

to give *N*-substituted indole derivatives (Eq. 4).[6] This intermolecular cyclizations were carried out at 180°C in dioxane with spontaneous hydrogen evolution, and *N,N'*-diarylethylenediamine was postulated as the key intermediates of the reactions.

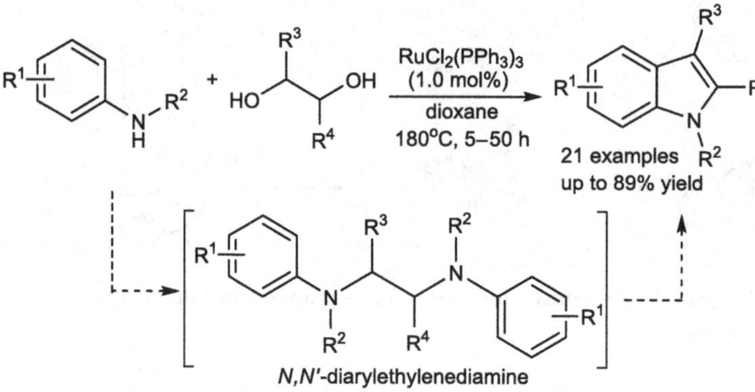

N,N'-diarylethylenediamine   (Eq. 4)

Then in 1996, Shim and co-workers reported that triethanolamine cyclization took place with *N*-substituted anilines under the same catalytic system, leading to the formation of 1-substituted indoles (Eq. 5).[7] Replacement of triethanolamine by *N*-benzyldienthanolamine or triisopropanolamine also afford the corresponding indoles in good yields. An intermolecular alkyl group transfer between anilines and alkanolamines is postulated to be the key step of these transformations.

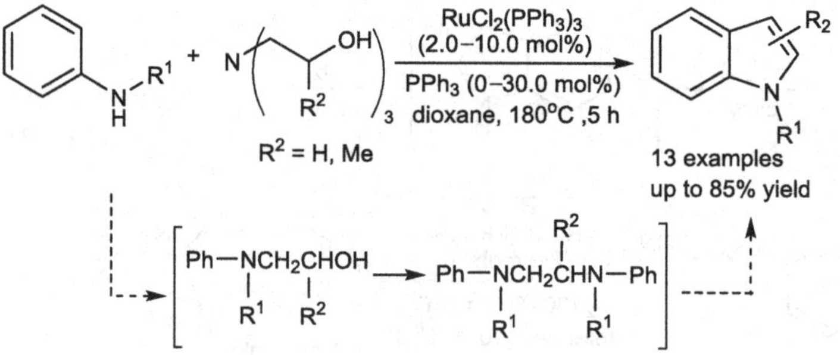

                                                                       (Eq. 5)

Primary aromatic amines were then used as cyclization counter-parts to react with trialkanolamines in dioxane using $RuCl_2(PPh)_3$ as catalyst together with tin(II) chloride dehydrate (Eq. 6).[8] As a result, the corresponding *N*-free indoles are obtained in moderate to good yields. It is postulated that the reaction pathway initiates from the formation of 2-anilinoethanols by amine exchange reactions between anilines and trialkanolamines, then it proceeds a similar catalytic cycle to that which has already been proposed in ruthenium-catalyzed synthesis of indoles from anilines and ethylene glycols.[5]

$$R^1 \!-\! \overset{\text{NH}_2}{\bigotimes} \; + \; N\!\left(\!\overset{\text{OH}}{\underset{R^2}{\bigvee}}\!\right)_{\!3} \quad \overset{\text{RuCl}_3 \, n\text{H}_2\text{O (7.0 mol\%)}}{\underset{\substack{\text{SnCl}_2 \cdot 2\text{H}_2\text{O (1 equiv)} \\ \text{dioxane, 180°C, 20 h}}}{\xrightarrow{\text{PPh}_3 \text{ (2.0 mol\%)}}}} \quad R^1 \!-\! \overset{}{\underset{\text{H}}{\bigotimes}}\!\!\overset{R^2}{N}$$

R² = H, Me

17 examples
up to 99% yield

(Eq. 6)

Based on the above work, a regioselective synthesis of 2-substituted indoles *via* ruthenium-and SnCl₂-catalyzed ring opening of epoxides by anilines was disclosed by the same group (Eq. 7).[9] Recently, Beller and co-workers established a new ruthenium catalytic system for this transformation (Eq. 8)[10]: in the presence of [Ru₃(CO)₁₂] (1.0 mol%) and 1,1'-Bis(diphenylphosphino)ferrocene (dppf) (3.0 mol%), a variety of indole derivatives were obtained in good to high yields at 150°C in dioxane.

$$R^1 \!-\! \overset{\text{NH}_2}{\bigotimes} \; + \; \overset{R^2}{\underset{O}{\triangle}} \quad \overset{\text{RuCl}_3 \, n\text{H}_2\text{O (5.0 mol\%)}}{\underset{\substack{\text{SnCl}_2 \cdot 2\text{H}_2\text{O (1 equiv)} \\ \text{dioxane, 180°C, 20 h}}}{\xrightarrow{\text{PPh}_3 \text{ (15.0 mol\%)}}}} \quad R^1 \!-\! \overset{}{\underset{\text{H}}{\bigotimes}}\!\!\overset{}{N}\!-\! R^2$$

14 examples
up to 98% yield

(Eq. 7)

$$R^1 \!-\! \overset{\text{NHR}^2}{\bigotimes} \; + \; \overset{R^4}{\underset{R^3}{O\!\!\triangleleft}} \quad \overset{\substack{\text{1) Zn(OTf)}_2 \text{ (2.0 mol\%)} \\ \text{neat, rt, 1 h}}}{\underset{\substack{\text{2) [Ru}_3\text{(CO)}_{12}\text{] (1.0 mol\%)} \\ \text{dppf (3.0 mol\%)} \\ p\text{-TsOH (10 mol\%)} \\ \text{1,4-dioxane, 150°C}}}{\xrightarrow{}} \quad R^1 \!-\! \overset{R^4}{\underset{R^2}{\bigotimes}}\!\!\overset{}{N}\!-\! R^3$$

33 examples
up to 89% yield

(Eq. 8)

The Nicholas group described the annulation reactions of nitroaromatics ($ArNO_2$) with alkynes catalyzed by $[Cp*Ru(CO)_2]_2$ in 2002. The reactions are performed under carbon monoxide and regioselectively produced indole derivatives (Eq. 9).[11]

R^1 —⦀— NO_2 + ⦀ $\begin{smallmatrix}R^2\\R^3\end{smallmatrix}$ $\xrightarrow[\substack{750\ psi\ CO\\benzene,\ 170°C\\48\ h}]{\substack{[Cp*Ru_2(CO)_2]_2\\(5\ mol\%)}}$ R^1 —⦀— indole $\begin{smallmatrix}R^2\\R^3\end{smallmatrix}$

10 examples
up to 53% yield

(Eq. 9)

They also showed that indoles could also be formed from $[Cp*Ru(CO)_2]_2$-catalyzed reductive annulation of nitrosoaromatics (ArNO) with alkynes (Eq. 10).[12] Although only moderate yields were observed for both methods, the transformations were conducted under neutral reaction conditions and exhibited high regioselectivity.

R^1 —⦀— NO + ⦀ $\begin{smallmatrix}R^2\\R^3\end{smallmatrix}$ $\xrightarrow[\substack{750\ psi\ CO\\benzene,\ 170°C\\24–72\ h}]{\substack{[Cp*Ru_2(CO)_2]_2\\(5\ mol\%)}}$ R^1 —⦀— indole $\begin{smallmatrix}R^2\\R^3\end{smallmatrix}$

10 examples
up to 64% yield

(Eq. 10)

Ruthenium complex Cp*Ru(COD)Cl (COD = cycloocta-1,5-diene) was found to be an active catalyst for the cyclization of an azabenzonorbornadiene with a propargylic alcohol to produce the dihydrobenzo[g]indole as a single regio and stereoisomer in good yield. For other alkynes, selective formation of the dihydrobenz[g]indole is more favorable by using a cationic complex $[Cp*Ru(CH_3CN)_3]PF_6$ as catalyst (Eq. 11).[13] This transformation involves the cleavage of one C–N bond of the bicyclic alkene and formation of two (C–C and C–N) bonds at the acetylenic carbons. However, the reaction scope was limited to electron-poor alkynes. Such a problem could be overcome by introducing $CpRuI(PPh_3)_2$ [in situ generated from $CpRuCl(PPh_3)_2$ and MeI] as a catalyst for this transformation.[14]

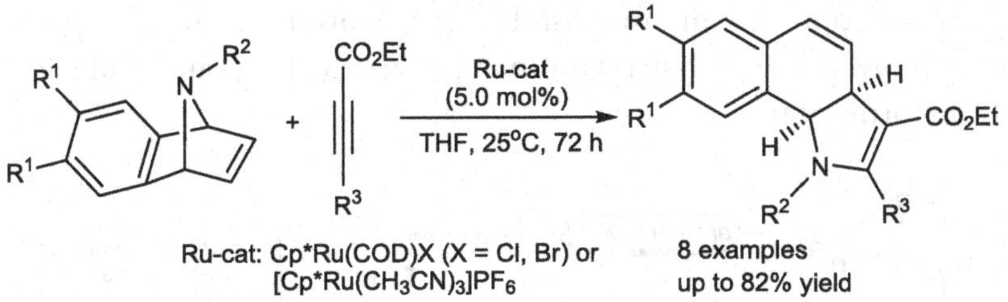

R1
R2
N
R1
+
CO2Et
R3
Ru-cat
(5.0 mol%)
THF, 25°C, 72 h
R1
R1
H
CO2Et
H
N
R2
R3
Ru-cat: Cp*Ru(COD)X (X = Cl, Br) or
[Cp*Ru(CH3CN)3]PF6
8 examples
up to 82% yield

(Eq. 11)

An intramolecular C–H amination reactions of ortho-aryl phenylazides, 1-azido-2-arylvinylazides, and 1-azido-1,3-butadienes were promoted by readily available and relatively cheap metal salt $RuCl_3$ to give carbazoles, indoles, and pyrroles, respectively (Eq. 12).[15] According to computational and experimental results, a two-step process involving formal electrocyclization is involved in the catalytic reaction.

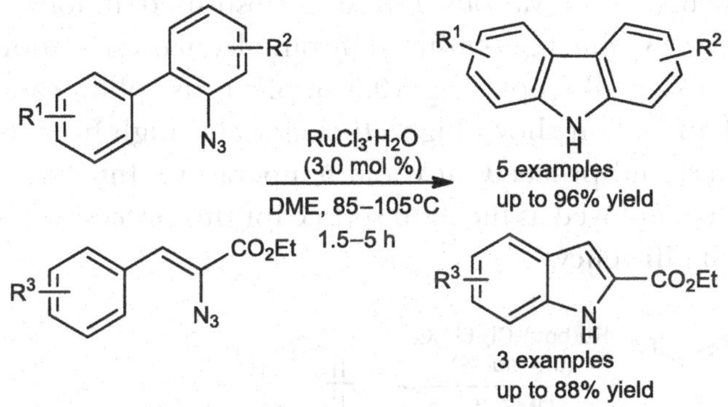

R2
R1
N3
RuCl3·H2O
(3.0 mol %)
DME, 85–105°C
1.5–5 h
R1
R2
N
H
5 examples
up to 96% yield
CO2Et
R3
N3
R3
CO2Et
N
H
3 examples
up to 88% yield

(Eq. 12)

Maity and Zheng developed a visible-light-mediated photocatalytic oxidative C–N bond formation/aromatization of styryl anilines for the preparation of *N*-arylindoles (Eq. 13).[16] With 4 mol% of $[Ru(bpz)_3](PF_6)_2$ (bpz = 2,2′-bipyrazine) as photocatalyst, the *in situ* generated nitrogen-centered radical cation can undergo electrophilic addition to the tethered alkene, then a cascade reaction occurs involving either aromatization or C–C bond migration followed by aromatization to form the final products. The present approach can be conducted under mild aerobic oxidation conditions: ambient temperature, open to air, and visible light. Moreover, only the *para*-alkoxy phenyl-protected styrylanilines proved to be compatible for this transformation.

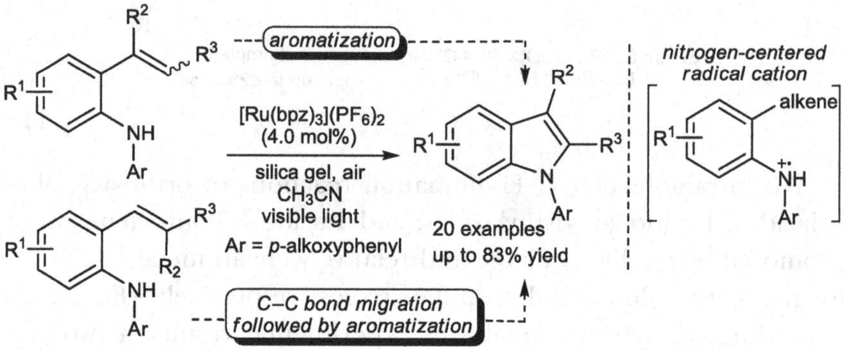

(Eq. 13)

The construction of various *N*-free 2-substituted indoles has been realized by a visible light-induced intramolecular cyclization of styryl azides using $Ru(bpy)_3Cl_2$ (0.5 mol%) as photocatalyst (Eq. 14).[17] The reaction shows high efficiency and high functional group tolerance, and proceeds at room temperature. Interestingly, sunlight can be employed as the light source for this process without loss of reaction efficiency.

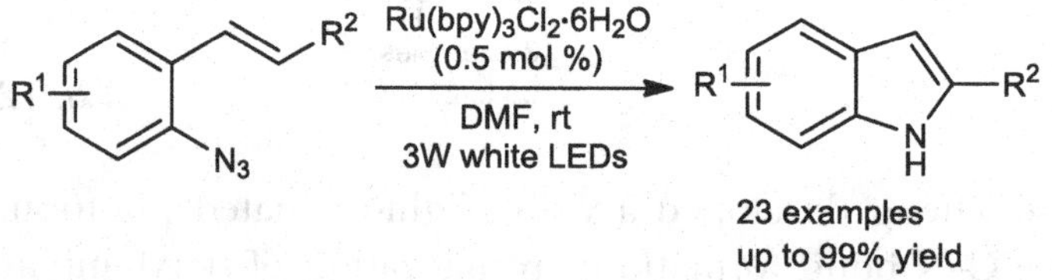

23 examples
up to 99% yield   (Eq. 14)

A novel intermolecular cyclization of aryl azides with alkynes was reported recently (Eq. 15).[18] Interestingly, with a Ru porphyrin catalyst $Ru(TPP)(NAr)_2$ [TPP = dianion of tetraphenylporphyrin, Ar = $3,5\text{-}(CF_3)_2C_6H_3$], the reaction afforded substituted indoles instead of triazoles. The substituted C3-functionalized indoles, which bear electron withdrawing group (EWG) on the fragment coming from the azide, were produced in high yield with high regioselectivity. No oxidant and/or additives were required for this transformation, and many important functional groups are tolerant.

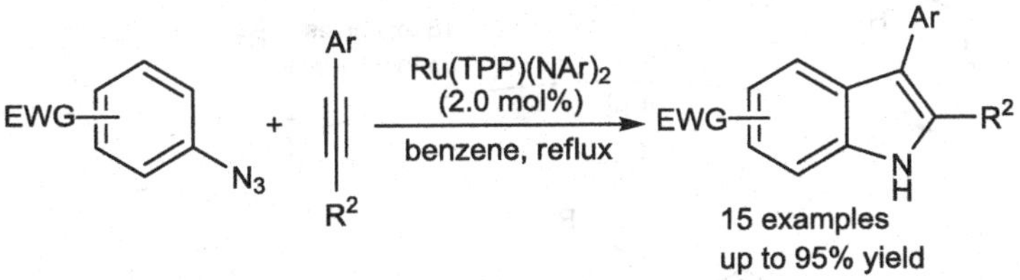

(Eq. 15)

A tandem Ru-catalyzed hydrogen-transfer Fischer indole synthesis from alcohols and phenylhydrazines using microwave irradiation has been developed by Porcheddu and co-workers (Eq. 16).[19] The present strategy features operational simplicity, large selection of reagents, and higher stability by the use of alcohols instead of aldehydes or ketones as starting materials.

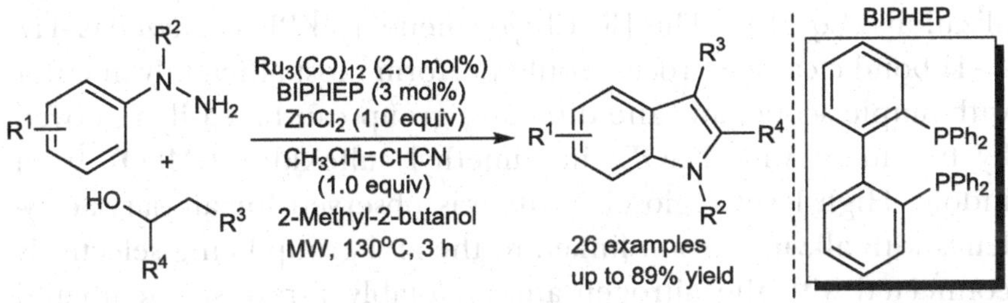

(Eq. 16)

The synthesis of 1*H*-indole-3-carbaldehydes by Ru-catalyzed intramolecular annulation of alkynes with amides involving a formyl translocation process was developed by Li and co-workers (Eq. 17).[20] The reaction process involves coordination of active Ru species to the alkyne motif, followed by oxidative addition into the *N*–CHO bond and nucleophilic cyclization, and final reductive elimination to give the desired product. Notably, acetyl-migration also takes place leading to the corresponding 1*H*-indole-3-ethanone under the same conditions with relative lower yield.

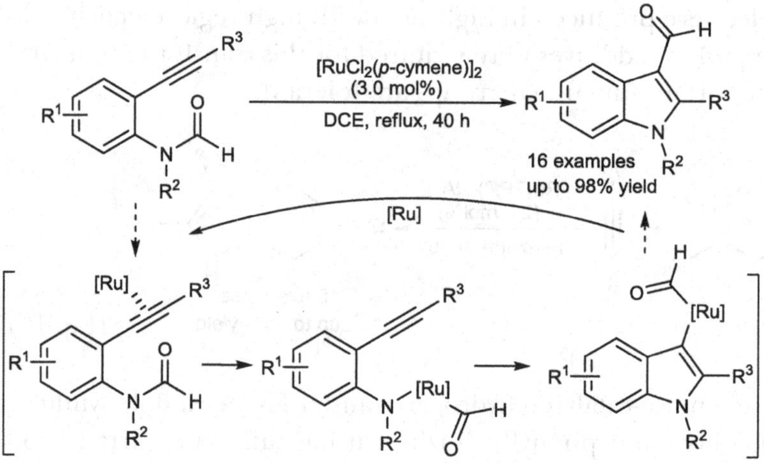

(Eq. 17)

N-2-pyrimidyl (or pyridyl)-substituted anilines reacted with internal alkynes to generate indole derivatives under the conditions of cationic ruthenium(II) complexes enabled C–H bond functionalizations (Eq. 18).[21] This [RuCl$_2$(*p*-cymene)]$_2$/KPF$_6$-catalyzed C–H/ N–H bond cleavage process could perform most efficiently in water with ample scope, and the directing group can be easily removed by treatment with NaOEt in dimethyl sulfoxide (DMSO) from indole. High level regioselectivity was observed for alkylarylacetylenes with alkyne carbon linked to the aryl group being selectively connected with the nitrogen atom. Notably, a rare six-membered ruthenacycle was proposed as a key intermediate for the cyclometalation step.

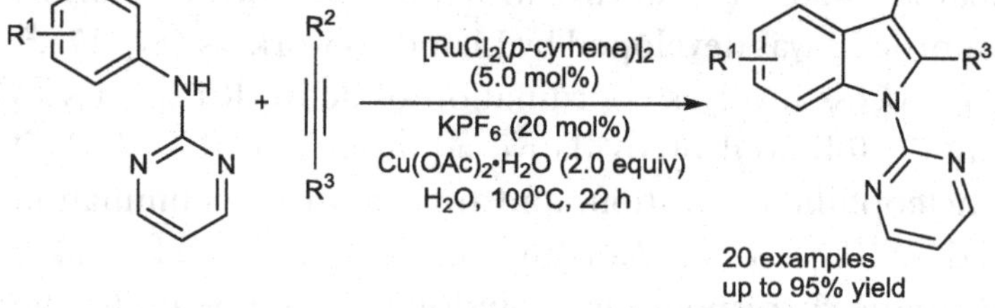

(Eq. 18)

A ruthenium-catalyzed redox-neutral C–H activation reaction *via* cleavage of N–N bond was reported by Huang and co-workers (Eq. 19).[22] In this context, 3-(1*H*-indol-1-yl)propanamide derivatives are obtained in high yields by using pyrazolidin-3-one as an oxidative directing group. A broad substrate scope for both phenylpyrazolidin-3-ones and alkynes are well compatible with this catalytic transformation, especially the latter including commonly incompatible terminal alkynes. Meanwhile, excellent regioselectivity are demonstrated with single regioisomers of the indole products obtained for unsymmetrical alkynes (2-aryl-3-alkylindoles for alkylarylacetylenes and 2-substituted indoles for terminal alkynes). Of note, the internal cleavage of the directing group results in installation of a pharmacologically significant *N*-propanamide functionality.

30 examples
up to 94% yield   (Eq. 19)

A Ru(II)–Ru(IV)–Ru(II) catalytic cycle is proposed (Scheme 2). After carboxylate-assisted cyclometalation and alkyne insertion, a Ru(IV) species **4** is formed by the cleavage of the cyclic N–N bond. Subsequent reductive elimination of **4** gives rise to the resulting indole product and regenerates the catalytic active Ru(II) acetate species.

2.2. Synthesis of pyrroles and their derivatives

With ruthenium catalysis, Watanabe and co-workers reported that *N*-substituted pyrroles and pyrrolidines were synthesized from readily available 1,4-diols and primary amines (Eq. 20).[23] 2-Butyne-1,4-diols react with aliphatic amines in the presence of a catalytic amount of [RuCl$_2$(PPh$_3$)$_3$] to form *N*-alkyl pyrroles in good yields. In the case of

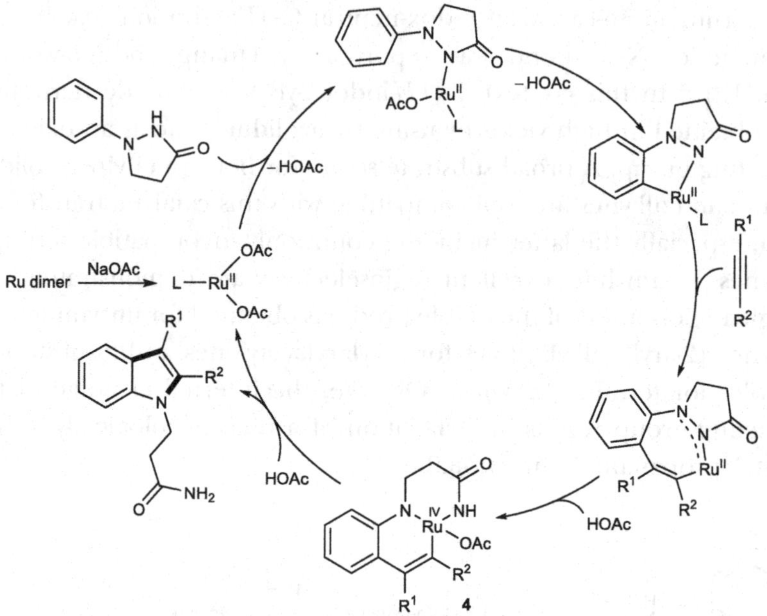

Scheme 2. Proposed mechanism for ruthenium-catalyzed 3-(1H-indol-1-yl) propanamides synthesis from phenylpyrazolidin-3-ones

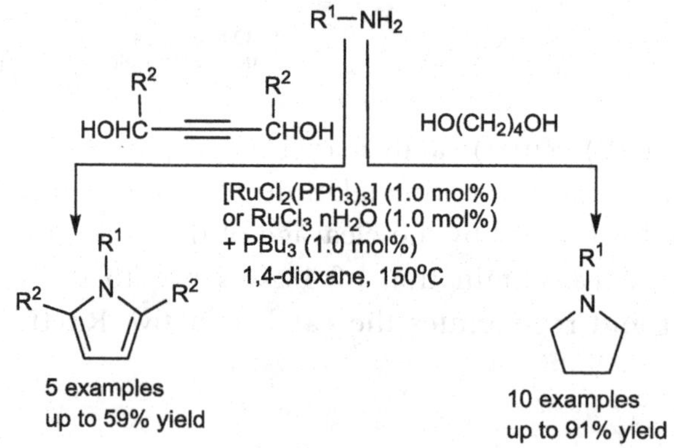

(Eq. 20)

1,4-butanediol, pyrrolidines are obtained $RuCl_2(PPh_3)_3$ and $RuCl_3 \cdot nH_2O/PBu_3$ show highest catalytic reactivity for aromatic and aliphatic amines in these transformations, respectively. When 2-butene-1,4-diol is used as a coupling counterpart, 1:1 mixture of N-substituted pyrroles and pyrrolidines were produced in high yields.

A ruthenium-catalyzed reductive *N*-heterocyclization of α-nitrocarbonyl compounds by carbon monoxide was realized by the same group (Eq. 21).[24] By using a $Ru_3(CO)_{12}$/1,10-phenanthroline catalytic system, this reaction provided a useful method for the synthesis of 1-pyrroline derivatives.

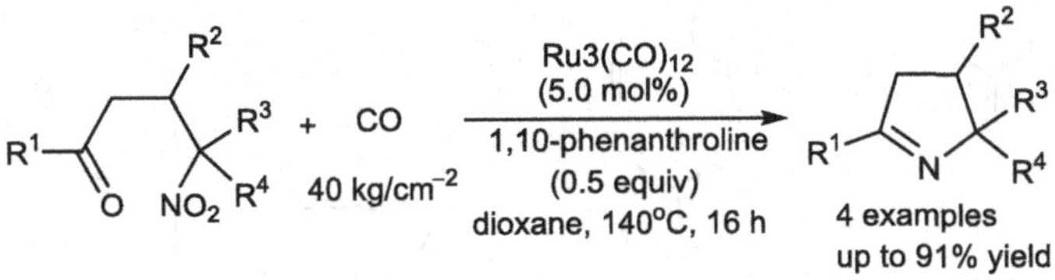

4 examples
up to 91% yield   (Eq. 21)

Trost and co-workers demonstrated that the coupling of *N*-benzylhexa-4,5-dienamines with vinyl ketones using [CpRu$_2$ (NCCH$_3$)$_3$]PF$_6$ as catalyst and TiCl$_4$ as co-catalyst efficiently led to the formation of pyrrolidines in an atom-economical fashion (Eq. 22).[25] The catalytic system shows compatibility with basic amines and good tolerance of other functional groups.

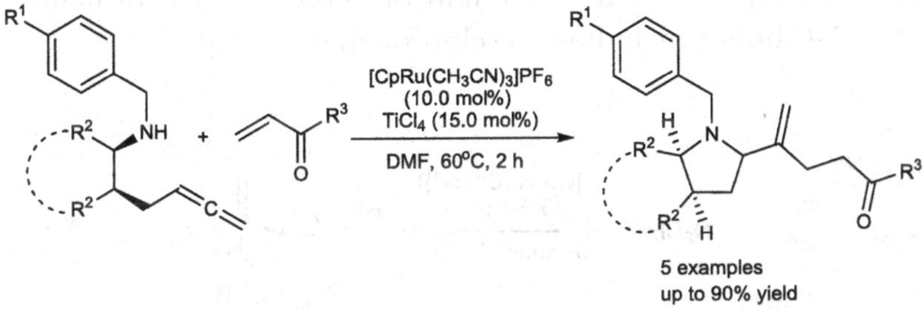

5 examples
up to 90% yield

(Eq. 22)

Two mechanistic pathways were proposed (Scheme 3). One possibility involves allylruthenium intermediates *via* coordination of both the allene and the enone to form ruthenacycle. Another possibility involves initiating the reaction by forming a vinyl ruthenium species from allene, which is then inserted into the enone.

Dixneuf and co-workers disclosed that the reaction of sulfonamide tethered enynes with diazoalkanes catalyzed by Cp*RuCl (COD) led to the formation of pyrrolidine-based 1-alkenyl bicycle

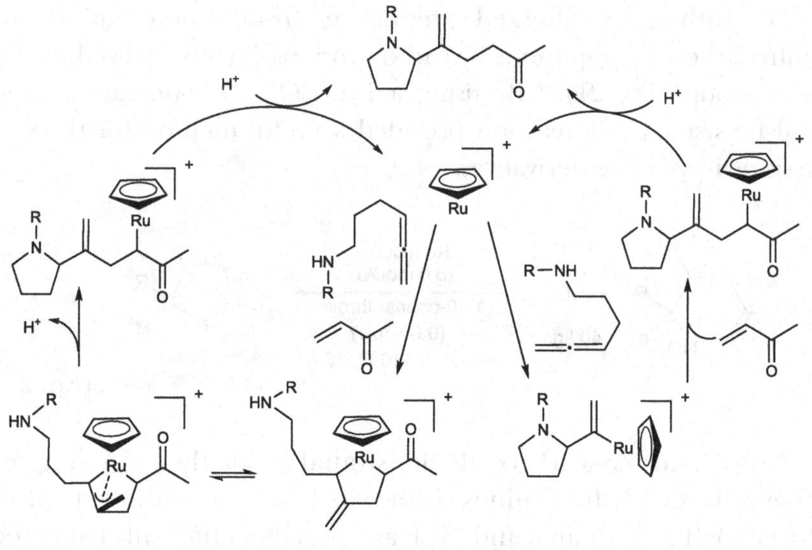

Scheme 3. Proposed mechanism for ruthenium-catalyzed pyrrolidines synthesis from *N*-benzylhexa-4,5-dienamines

compounds (Eq. 23).[26] The reaction involves the selective formation of three C–C bonds including a cyclopropanation step.

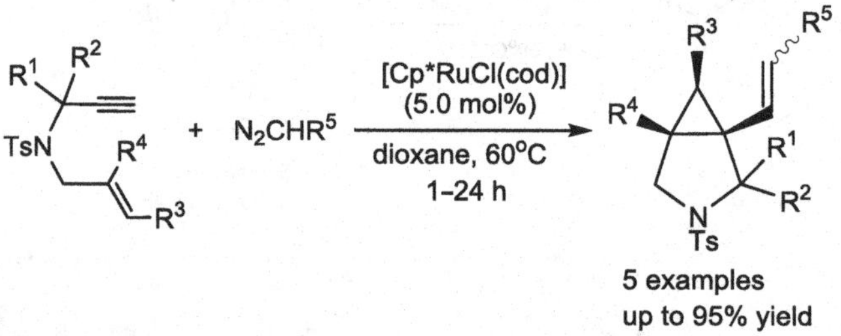

5 examples
up to 95% yield   (Eq. 23)

Cyclization of non-terminal diynes in the presence of ruthenium catalyst $[CpRu(CH_3CN)_3]PF_6$ and water to synthesize α,β-unsaturated ketones was reported (Eq. 24).[27] In the cases of sulfonamide tethered diynes, the corresponding products with pyrrolidine motifs were formed in good to excellent yields. Besides terminal diynes, electron-deficient diynes also did not lead to expected products.

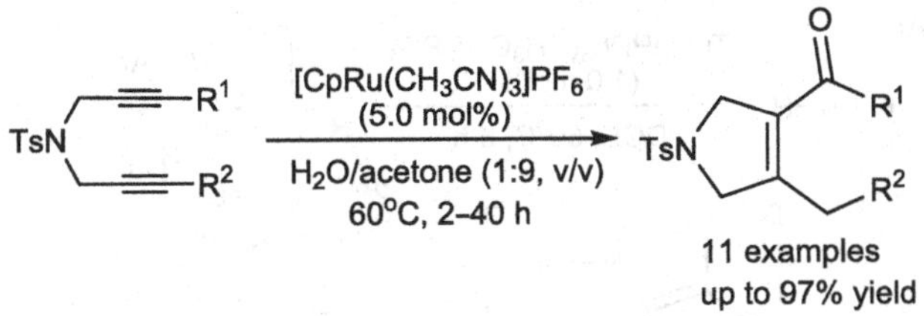

11 examples
up to 97% yield

(Eq. 24)

Ruthenium porphyrin [Ru^{II}(TDCPP)(CO)] proved to be excellent catalyst for three-component coupling reactions of α-diazo esters with a series of *N*-benzylidene imines and alkenes to form functionalized pyrrolidines in excellent diastereoselectivities (Eq. 25).[28] The reaction proceeds *via* a reactive ruthenium-carbene intermediate, and its subsequent reaction with imine to generate azomethine ylide, which reacts with alkenes through 1,3-dipolar cycloaddition. Furthermore, dimethyl acetylenedicarboxylate is also suitable as a dipolarophile to form the desired cycloadducts 2,5-dihydro-1*H*-pyrroles in good to excellent yields.

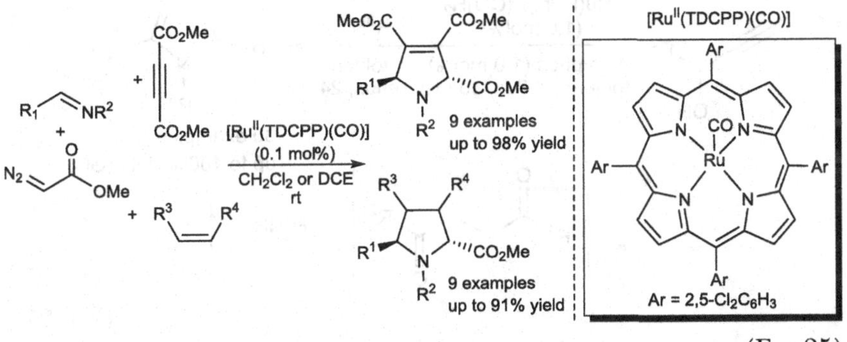

9 examples
up to 98% yield

9 examples
up to 91% yield

Ar = 2,5-Cl$_2$C$_6$H$_3$

(Eq. 25)

Cyclization of 3-en-1-ynyl imines with nucleophiles under ruthenium complex catalysis afford functionalized pyrroles in good yields (Eq. 26).[29] A low loading (1 mol%) of TpRuPPh$_3$(CH$_3$CN)$_2$PF$_6$ [Tp = tris(1-pyrazolyl)borate] catalyst promoted the catalytic cyclization without the formation of by-products. This process is proposed to proceed *via* (2-pyrrolyl)carbenoid intermediates derived from 5-*exo-dig* cyclization.

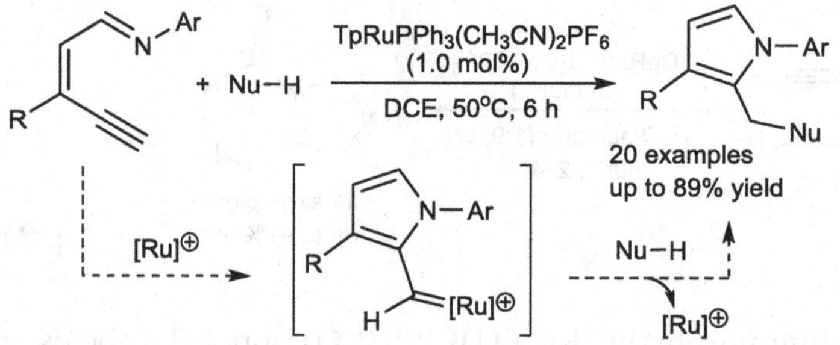

(Eq. 26)

The conversion of 1,4-alkynediols into 1,2,5-substituted pyrroles has been achieved with excellent selectivities using a $Ru(PPh_3)_3(CO)$ H_2 and 4,5-Bis-(diphenylphosphino)-9,9-dimethylxanthene (Xantphos) catalytic system (Eq. 27).[30] The reaction involves a ruthenium-catalyzed isomerization of the 1,4-alkynediol into a 1,4-diketone and subsequent Paal–Knorr cyclization into the corresponding pyrrole in the presence of amine.

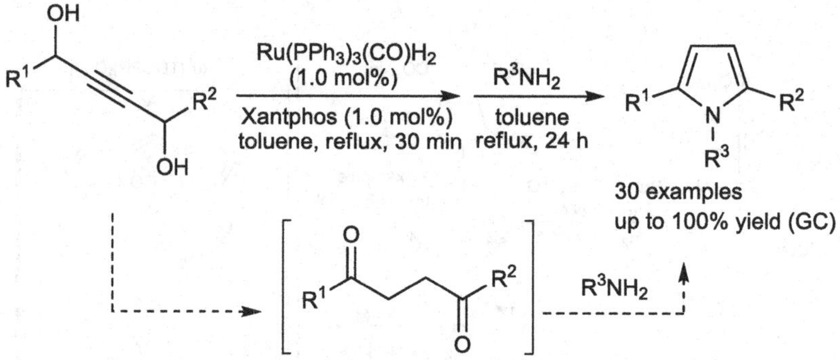

(Eq. 27)

A facile method for the synthesis of pyrroles by the condensation of 2,5-hexadione with primary amines in the presence of a catalytic amount of ruthenium(III) chloride was introduced by De (Eq. 28).[31] The reactions are conducted under solvent-free conditions. This ruthenium promoted Paal–Knorr reaction provided an easy, rapid, and efficient access for the synthesis N-substituted pyrrole derivatives.

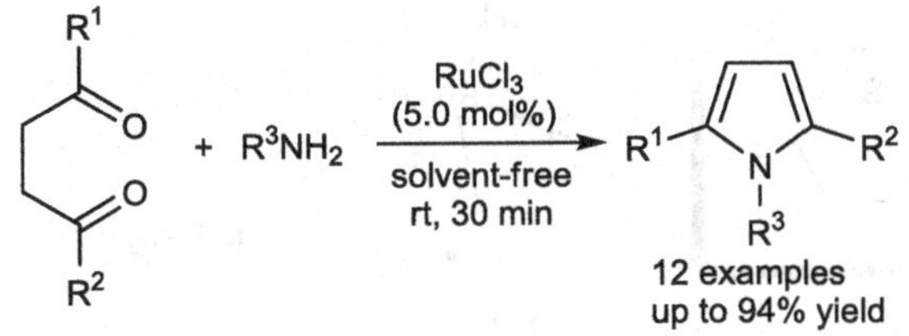

12 examples
up to 94% yield

(Eq. 28)

Reactions of 6-aminohex-2-yn-1-ols with catalytic amounts of the ruthenium complex [IndRu(PPh$_3$)$_2$Cl] (Ind: indenyl), in the presence of indium triflate and camphorsulfonic acid (CSA), provided pyrrolidine in high yield (Eq. 29).[32] Both sulfonamides and carbamates are compatible with this atom-economical, catalytic domino reaction.

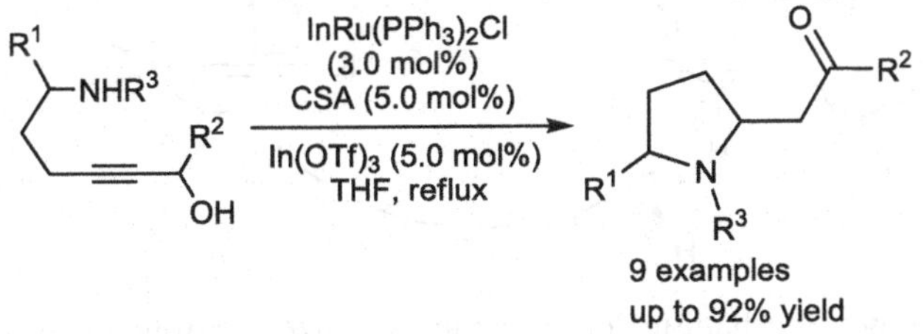

9 examples
up to 92% yield

(Eq. 29)

The whole reaction process is proposed to proceed in a reaction cascade composed of redox isomerization and intramolecular cyclization (Scheme 4).

Aryl- and aliphatic-substituted 3-hydroxyprolines were synthesized with high diastereoselectivities (*cis*, *cis*-) *via* a [RuCl$_2$(*p*-cymene)]$_2$-catalyzed one-pot intramolecular carbenoid N–H insertion reaction (Eq. 30).[33] Notably, the catalytic reactions are insensitive towards air and moisture.

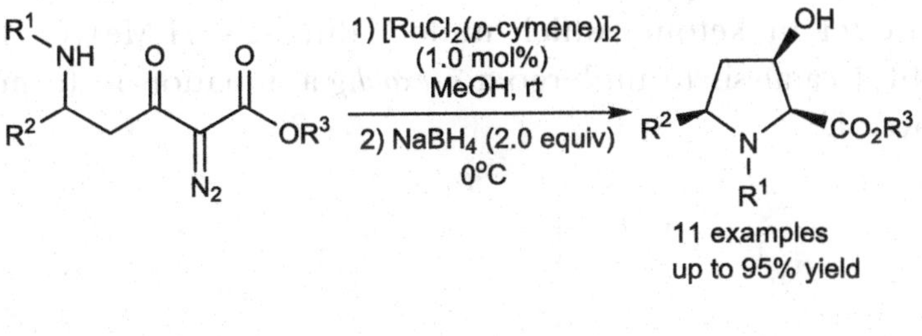

11 examples
up to 95% yield

(Eq. 30)

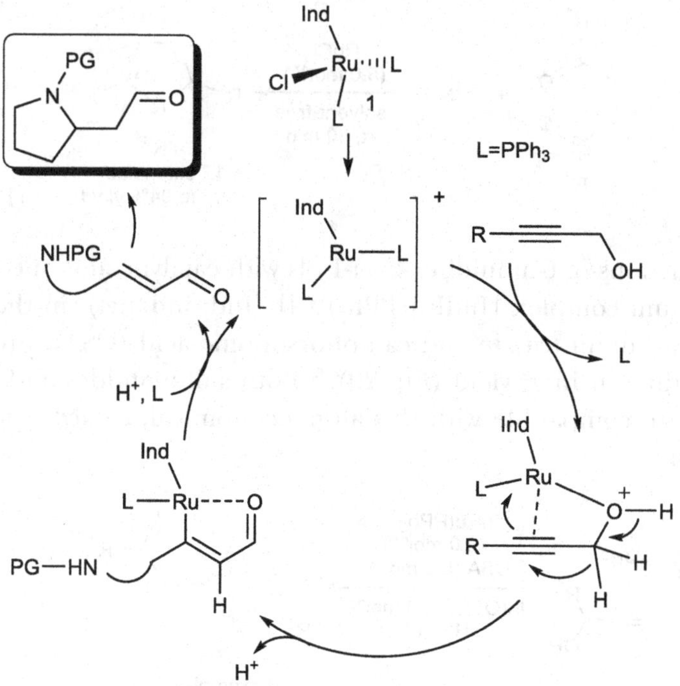

Scheme 4. Proposed mechanism for ruthenium-catalyzed pyrrolidine synthesis from 6-aminohex-2-yn-1-ol

Starting from readily accessible secondary propargylic alcohols, 1,3-dicarbonyl compounds and primary amines, a ruthenium-catalyzed one-pot multicomponent reaction to form fully substituted pyrroles was developed (Eq. 31).[34] This method involves initial propargylation of the 1,3-dicarbonyl compound promoted by CF_3CO_2H and subsequent condensation between the resulting γ-keto alkyne and the primary amine to afford a propargylated β-enamino ester or ketone, which under $[Ru(\eta^3\text{-}2\text{-}C_3H_4Me)(CO)(dppf)][SbF_6]$ catalysis to undergo a 5-*exo-dig* annulation to form the final pyrrole.

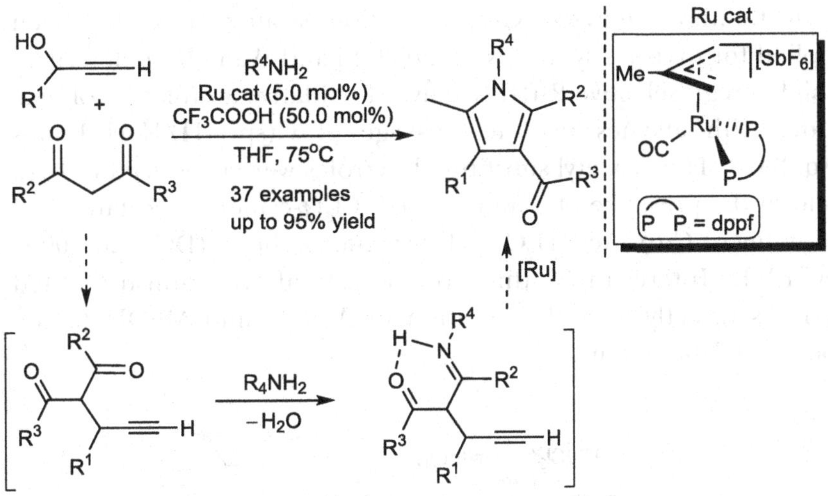

(Eq. 31)

It is described that ruthenium-catalyzed reactions of 1-en-4-yn-3-ols with primary amines to yielded highly substituted pyrroles (Eq. 32).[35] Water was the only waste product of this process. The allylation/cycloisomerization sequence is catalyzed by a single ruthenium(0) complex that contains a redox-coupled dienone ligand. In the cases that secondary allylamines were used, the sequence can be extended by a [3,3] rearrangement.

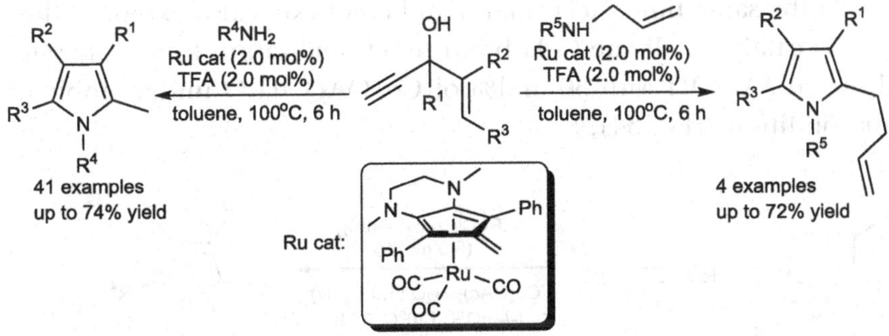

(Eq. 32)

Ruthenium-catalyzed C–H activation strategy has also been applied for pyrroles synthesis. In 2013, Li and Wang have described a highly regioselective Ru(II)-catalyzed oxidative annulation of enamides with alkynes *via* the cleavage of C(sp^2)–H/N–H bonds (Eq. 33).[36] The *N*-acetyl substituted pyrroles were produced in high yield in the presence of 5 mol% of [RuCl$_2$(p-cymene)]$_2$ catalyst and 0.5 equiv of Cu(OAc)$_2$·H$_2$O in 1,2-Dichloroethene (DCE) at 100°C for 12 h. Interestingly, this process can afford *N*-unsubstituted pyrroles directly with the addition of AgSbF$_6$ and MeOH to the above catalytic system.

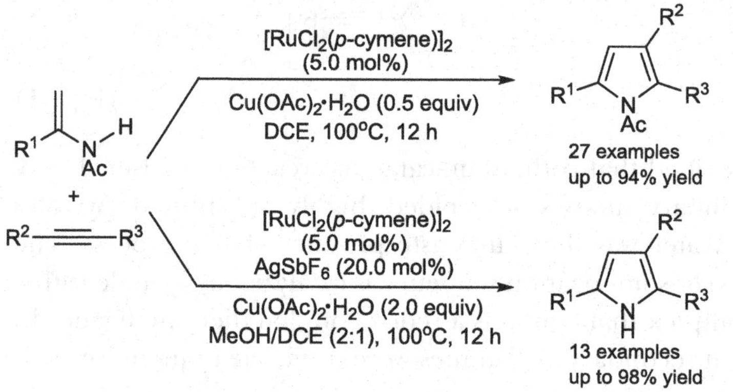

$$\text{(Eq. 33)}$$

At the same time, Ackermann and co-workers also explored this Ru(II)-catalyzed alkene C–H bond functionalization to pyrrole synthesis in *t*-AmOH with 30 mol% of Cu(OAc)$_2$·H$_2$O under ambient air conditions (Eq. 34).[37]

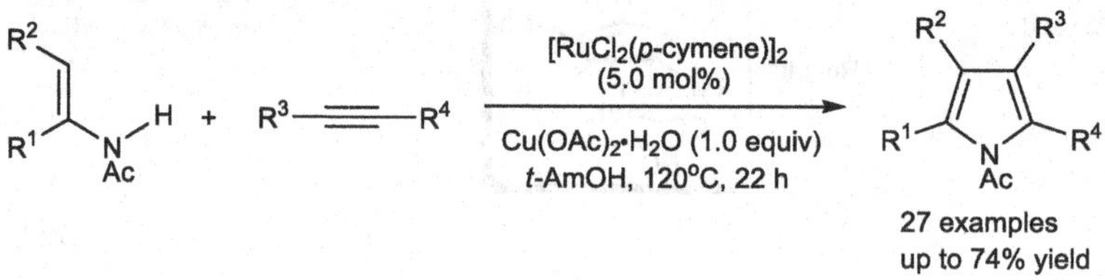

$$\text{(Eq. 34)}$$

By employing $[Ru_3(CO)_{12}]$/Xantphos catalyst system, the Beller group has achieved the synthesis of a variety of classes of multiply substituted pyrroles (1,3- and 2,3-disubstituted, 2,3,5- and 1,2,3-trisubstituted, 1,2,3,5- and 2,3,4,5-tetrasubstituted, as well as pentasubstituted derivatives) from easily available benzylic ketones, vicinal diols, and amines including primary anilines and alkyl amines as well as ammonia (Eq. 35).[38] This atom-economical synthetic protocol proceeds efficiently with high regioselectivity. Furthermore, they found that the present process can also be realized by commercially available $[Ru(p\text{-cymene})Cl_2]_2$/Xantphos/t-BuOK catalyst system in high efficiency.[39(a)]

Reaction conditions A: $[Ru_3(CO)_{12}]$ (1.0 mol %), xantphos (3.0 mol %) K_2CO_3 (20 mol %), t-amyl alcohol, 130°C, 36 examples, up to 90% yield

Reaction conditions B: $[RuCl_2(p\text{-cymene})]_2$ (1.0 mol %), xantphos (3.0 mol %) t-BuOK (20 mol %) t-amyl alcohol, 130°C, 49 examples, up to 92% yield

(Eq. 35)

The reaction initiates the formation of enamine intermediate from *in situ* condensation of ketone and amine. Then, depending on the reaction position of the enamine intermediate with the carbonyl group arising from the metal-induced double or mono dehydrogenation of the diol, four possible reaction pathways are proposed (Scheme 5): hydrogen-transferring N–H or C–H alkylation pathways and tautomerization N–H or C–H alkylation pathways. The pyrrole products are released from the thermodynamic favorable dehydration.

Milstein and co-workers present a dehydrogenative coupling of β-Amino alcohols with secondary alcohols catalyzed by well-defined pincer ruthenium complexes with H_2 and H_2O as waste products (Eq. 36).[40] This transformation provides an environmentally

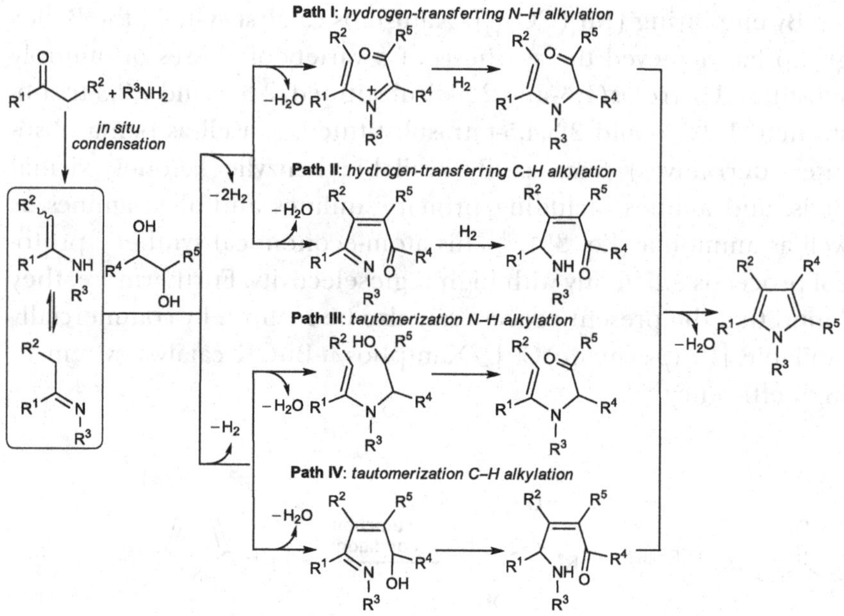

Scheme 5. Proposed mechanism for ruthenium-catalyzed pyrrole synthesis from benzylic ketones, vicinal diols, and amines

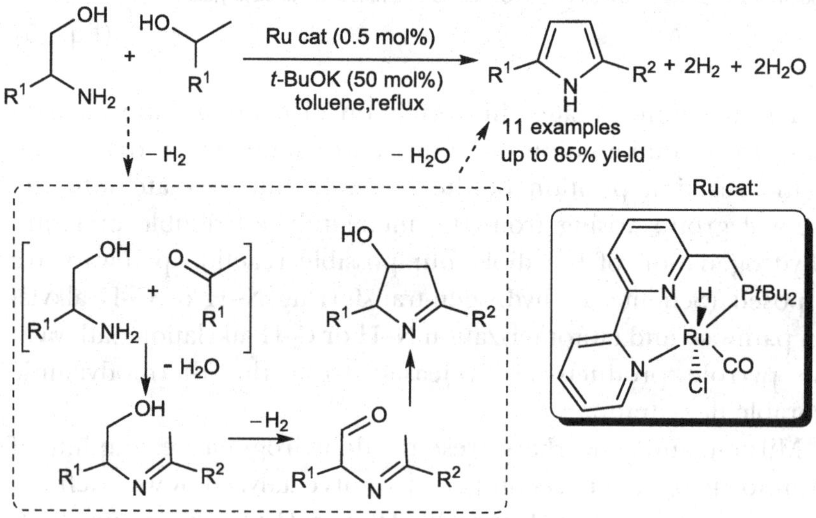

(Eq. 36)

benign, atom-economical efficient, one-step methodology for pyrrole synthesis from readily available starting materials under mild conditions.

A new Ru complex dichlorobis(dicyclohexylphosphinomethylpy ridine)-ruthenium was designed for the condensation of valinols and acetones to form N-unsubstituted pyrroles (Eq. 37).[41] The reactions were conducted under solvent-free conditions in a high atom and step economy manner. A fully unmasked α-amino aldehyde intermediate was generated by the dual effects of a catalytic ruthenium complex and an alkali metal base.

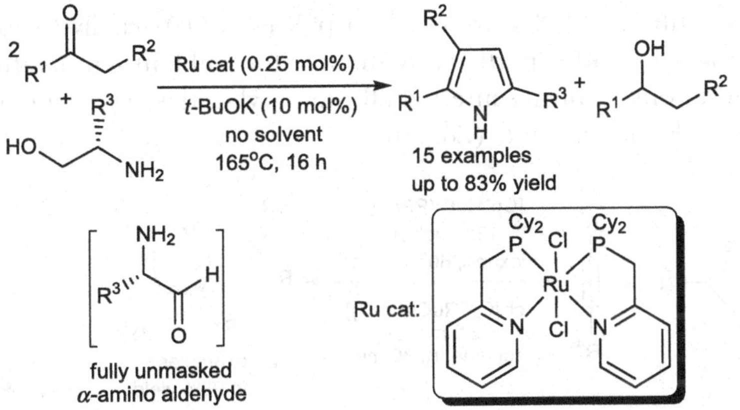

(Eq. 37)

2.3. Synthesis of triazoles and their derivatives

In 2005, Jia and co-workers developed the first ruthenium-catalyzed azide–alkyne cycloaddition reaction (RuAAC). In contrast to the successful Cu(I)-catalyzed azide–alkyne cycloaddition (CuAAC), which promotes the cycloaddition of terminal alkynes to regiospecifically provide 1,4-disubstituted 1,2,3-triazoles, the present RuAAC selectively produces 1,5-disubstituted and 1,4,5-trisubstituted 1,2,3-triazoles from organic azides, terminal and internal alkynes, respectively (Eq. 38).[42] Thus, the RuAAC process together with the CuAAC reaction provides ready access to all 1H-1,2,3-triazole regioisomers.

$$R^1 \overset{}{\underset{n}{\frown}} N_3 \; + \; \overset{R^2}{\underset{R^3}{\big|\big|\big|}} \quad \xrightarrow[\substack{\text{benzene, 80°C} \\ 2\text{–}12\text{ h} \\ (n = 0, 1, 2)}]{\substack{\text{Cp*RuCl(PPh}_3)_2 \\ (1.0\text{ mol\%})}} \quad R^1 \overset{}{\underset{n}{\frown}} \overset{N \approx N}{\underset{R^2 \quad R^3}{\big\langle\big\rangle}}$$

12 examples
up to 94% yield   (Eq. 38)

Extensive investigation indicates that [Cp*RuCl] compounds are the most efficient and regioselective catalysts for RuAAC: in the presence of Cp*RuCl(PPh$_3$)$_2$ catalyst, primary, and secondary azides react with various terminal alkynes containing a range of functionalities, usually at elevated temperature, to selectively produce 1,5-disubstituted 1,2,3-triazoles; the Cp*RuCl(COD) catalyst shows higher activity and is particularly suitable for ambient temperature cycloadditions involving internal alkynes, aryl azides, and generally thermally labile reactants (Eq. 39).[43]

$$\overset{R^2}{\underset{R^1}{\big\rangle}}\!-N_3 \; + \; \overset{R^3}{\underset{R^4}{\big|\big|\big|}} \quad \xrightarrow[\substack{\text{or [Cp*RuCl(COD)]} \\ (2.0\text{ mol\%}) \\ \text{toluene, rt, 30 min}}]{\substack{[\text{Cp*RuCl(PPh}_3)_2] \\ (2.0\text{ mol\%}) \\ \text{dioxane, 60°C, 12 h}}} \quad \overset{R^2}{\underset{R^1}{\big\rangle}}\!\overset{N \approx N}{\underset{R^3 \quad R^4}{\big\langle\big\rangle}}$$

21 examples
up to 95% yield   (Eq. 39)

A proposed mechanism is shown in Scheme 6. The reaction might proceed *via* oxidative coupling of the azide and alkyne reactants to give a six-membered ruthenacycle intermediate, in which the first new C–N bond is formed between the more electronegative carbon of the alkyne and the terminal nitrogen of the azide. Then followed by reductive elimination, the triazole product is formed. Density functional theory (DFT) calculations support such a mechanistic proposal and indicate that the reductive elimination step is rate determining.

The discovery of the ruthenium catalyst allows the use of ynamides [with Cp*RuCl(PPh$_3$)$_2$ as catalyst] and trifluoromethylated propargylic alcohols [with (Cp*RuCl$_2$)$_n$ as catalyst] to undergo

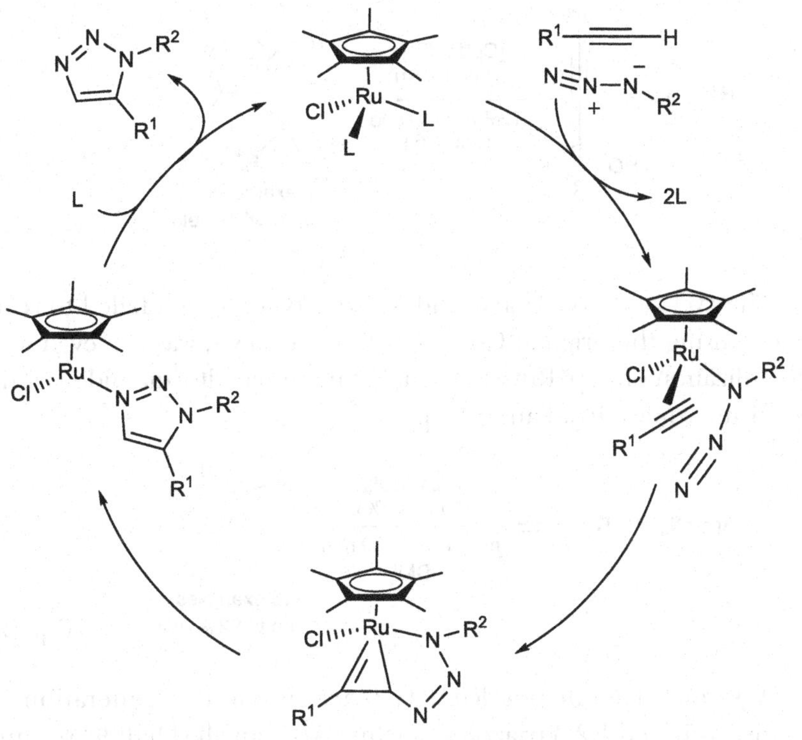

Scheme 6. Proposed mechanism for ruthenium-catalyzed 1,5-disubstituted 1,2,3-triazoles synthesis from azides and alkynes

1,3-dipolar cycloaddition with azides to yield 1-protected 5-amido 1,2,3-triazoles (Eq. 40)[44] and 4-trifluoromethyl-1,4,5-trisubstituted-1,2,3-triazoles (Eq. 41)[45] in high yields, respectively.

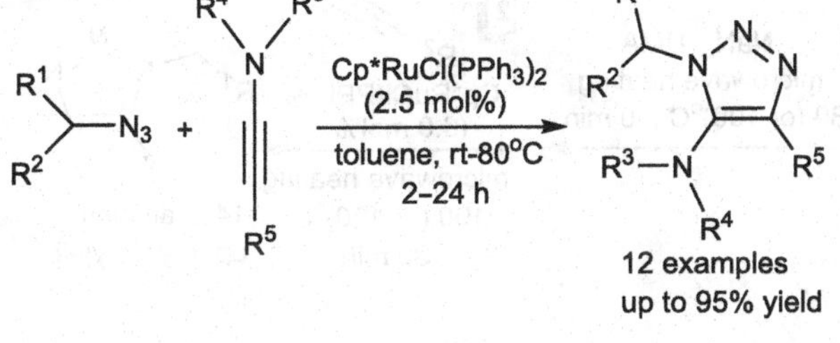

12 examples
up to 95% yield   (Eq. 40)

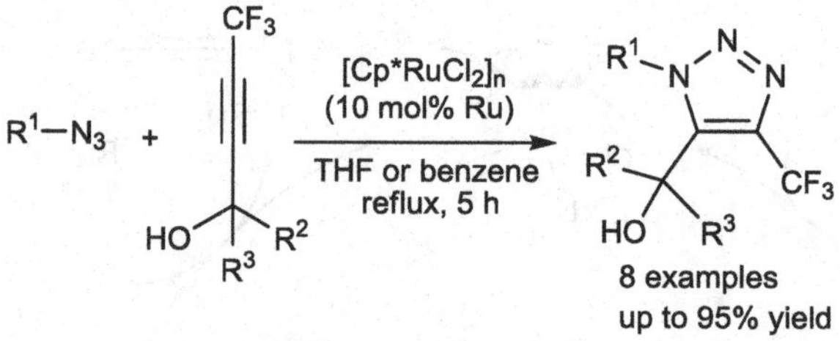

8 examples
up to 95% yield   (Eq. 41)

The cycloaddition of aryl azides and alkynes, which failed to react cleanly using the original $Cp*RuCl(PPh_3)_2$ catalyst, was achieved with the utilization of $[Cp*RuCl]_4$ as catalyst in dimethylformamide (DMF) under microwave irradiation (Eq. 42).[46]

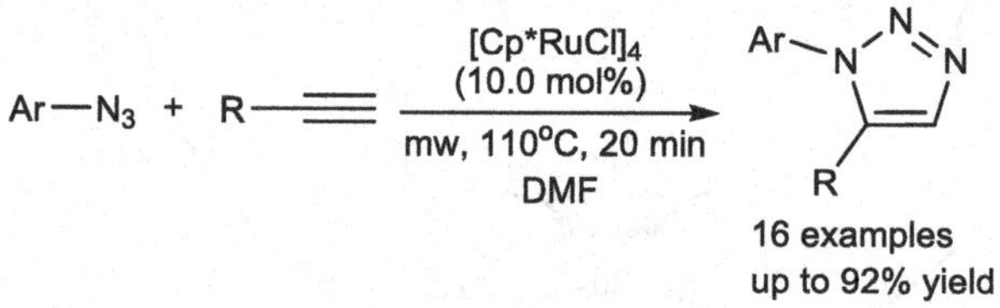

16 examples
up to 92% yield   (Eq. 42)

A sequential one-pot RuAAC reaction for the generation of 1,5-disubstituted 1,2,3-triazoles, starting from an alkyl halide, sodium azide, and an alkyne, has been developed by Johansson and co-workers (Eq. 43).[47] The reaction of primary alkyl halide with sodium azide in dimethylacetamide (DMA) under microwave heating affords the organic azide *in situ*. After addition of $[Cp*RuCl(PPh_3)_2]$ and the alkyne, the desired cycloaddition product was formed in good to excellent yields after further microwave irradiation.

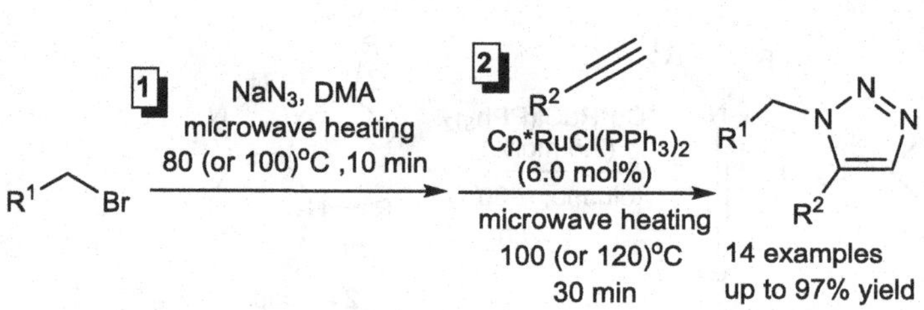

14 examples
up to 97% yield   (Eq. 43)

In 2012, Liu and co-workers have reported the first catalytic system of non-copper metal complexes for the cycloaddition of azides with terminal alkynes to selectively produce 1,4-disubstituted 1,2,3-triazoles (Eq. 44).[48] With ruthenium hydride complex $RuH_2(CO)(PPh_3)_3$ as catalyst, the reactions underwent with complete selectivity and provided the desired products in moderate to excellent yields.

$$R^1{-}N_3 \ + \ \overset{}{\underset{R^2}{\diagup\!\!\!\equiv}} \quad \xrightarrow[\text{THF, 80°C, 2 h}]{\substack{RuH_2(CO)(PPh_3)_3 \\ (5.0 \text{ mol\%})}} \quad R^1{-}N\overset{\displaystyle N}{\underset{N}{\diagdown\!\!\diagup}}{\Bigg\backslash}{=}\!\overset{R^2}{}$$

19 examples
up to 87% yield   (Eq. 44)

Soon after this work, they also showed that ruthenium complexes $RuH(\eta^2\text{-}BH_4)(CO)(PCy_3)_2$ is effective catalyst for the cycloaddition of terminal alkynes and azides to give selectively 1,4-disubstituted 1,2,3-triazoles (Eq. 45).[49] The reaction proceeds well with primary and secondary azides and a range of terminal alkynes containing various functionalities. The diacetylide complex $Ru(C{\equiv}CPh)_2(CO)$ $(PCy_3)_2$ exhibits a similar catalytic activity to that of $RuH(\eta^2\text{-}BH_4)(CO)$ $(PCy_3)_2$, which is as expected since $RuH(\eta^2\text{-}BH_4)(CO)(PCy_3)_2$ can react with terminal alkyne to give the diacetylide complex.

$$\underset{R^1}{\overset{R^2}{\diagdown\!{\diagup}}}{N_3} \ + \ \overset{}{\underset{R^3}{\diagup\!\!\!\equiv}} \quad \xrightarrow[\substack{THF \\ 80°C, 1.5 \text{ h}}]{\substack{Ru\text{-cat} \\ (2\text{--}2.5 \text{ mol\%})}} \quad R^1\underset{}{\overset{R^2}{\diagdown}}{-}N\!\!\diagdown\!\!N\!\!\diagup\!\!N$$

15 examples R^3
up to 93% yield

| Ru-cat | Ru-cat |

(Eq. 45)

Based on experimental and computational studies, the reaction mechanism (Scheme 7) is proposed to involve Ru-acetylide species as the key intermediate, which undergoes formal cycloaddition with azide to give a Ru(triazolyl) complex **5**. The resulting ruthenium

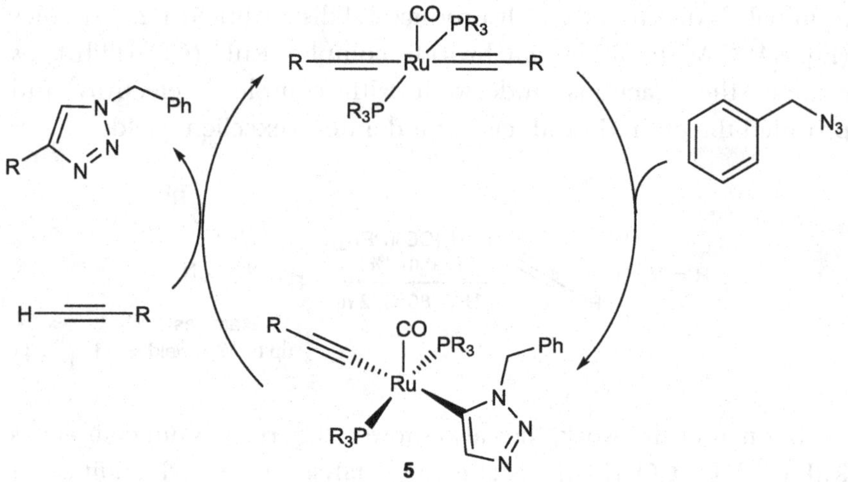

Scheme 7. Proposed mechanism for ruthenium-catalyzed 1,4-disubstituted 1,2,3-triazoles synthesis from azides and terminal alkynes

triazolide complex **5** undergoes a metathesis reaction with a terminal alkyne *via* a four-centered transition state to give the final product and to regenerate the catalyst precursor complex. The metathesis reaction appears to be the rate-determining step.

Very recently, Fokin and co-workers reported a CpRuCl(COD)-catalyzed cycloadditions of organic azides with electronically deficient 1-choro-, 1-bromo-, and 1-iodoalkynes, resulting in the formation of 5-halo-1,2,3-triazoles (Eq. 46).[50] Reactive 1-haloalkynes include propiolic amides, esters, ketones, and phosphonates. Halogenated azole products can be further functionalized by using palladium-catalyzed cross-coupling reactions. In this transformation, the low catalytic activity with [Cp*RuCl(COD)] is attributed to steric demands of the Cp*(η^5-C$_5$Me$_5$) ligand in comparison to the parent Cp(η^5-C$_5$H$_5$).

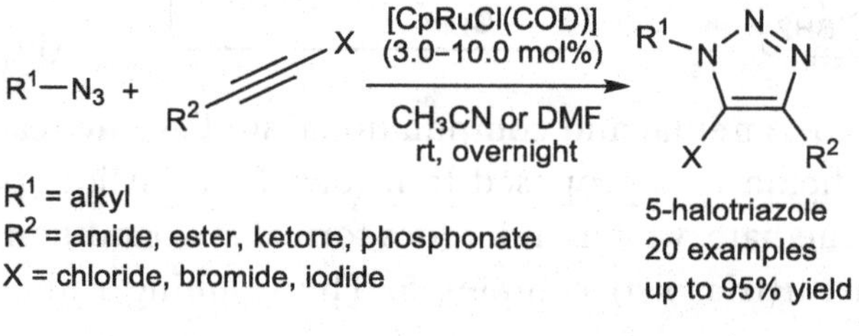

R^1 = alkyl
R^2 = amide, ester, ketone, phosphonate
X = chloride, bromide, iodide

5-halotriazole
20 examples
up to 95% yield

(Eq. 46)

2.4. Synthesis of lactams and their derivatives

In 1984, Itoh and co-workers reported an efficient Ruthenium-catalyzed cyclization of N-allyl trichloroacetamides. By heating benzene solution of N-allyl trichloroacetamides at 140°C with the aid of $RuCl_2(PPh_3)_3$ catalysis, α,α,γ-trichloro-γ-butyrolactams were obtained in high yields (Eq. 47).[51(a)] Of note, a copper salt (CuCl, 30 mol%) could also be used as a catalyst for this transformation but led to lower yield.

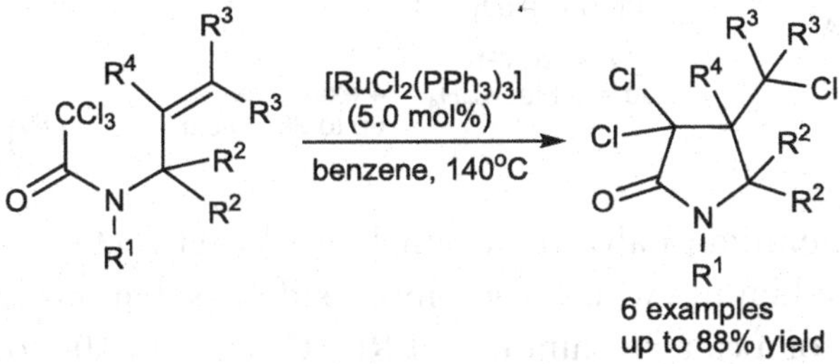

6 examples
up to 88% yield   (Eq. 47)

Using the same catalytic system, the corresponding trichlorinated γ-lactams having spiro or bicyclic systems could be easily cyclized in a highly stereoselective manner to form hexahydro-oxindole derivatives including the mesembrine alkaloid skeleton (Eq. 48).[52]

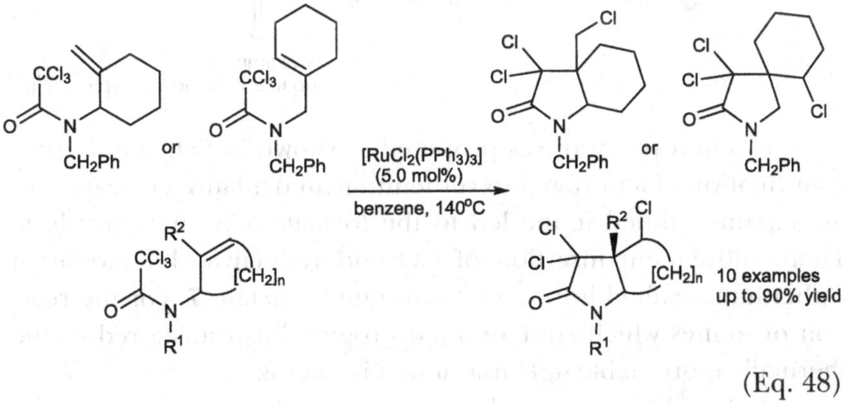

10 examples
up to 90% yield

(Eq. 48)

A $Ru_3(CO)_{12}$-catalyzed $[2+2+1]$ cycloaddition of 1,6-yne-imines with carbon monoxide was achieved by Murai and co-workers, which led to the bicyclic α,β-unsaturated lactams by incorporating the acetylene π-bond, the imine π-bond, and the carbon atom of CO (Eq. 49).[53] It should be noted that substituents on the acetylenic terminal carbon (for example: alkyl, aryl, and silyl) is essential for the reaction to proceed.

$$E = CO_2Et$$
$$Ar = p\text{-MeOC}_6H_4$$

4 examples
up to 66% yield   (Eq. 49)

Then a ruthenium-catalyzed carbonylative [4+1] cycloaddition of α,β-unsaturated imines with carbon monoxide was demonstrated. In the presence of a catalytic amount of $Ru_3(CO)_{12}$ with 10 atom of CO in toluene at 180°C, the reaction underwent smoothly to form unsaturated γ-lactams (Eq. 50).[54]

9 examples
up to 96% yield   (Eq. 50)

A reaction pathway was proposed as shown in Scheme 8. After coordination of a nitrogen to ruthenium, an oxidative cyclization of the α,β-unsaturated imine led to the formation of metallacycle **6**. Then, subsequent insertion of CO and reductive elimination of ruthenium resulted in the β,γ-unsaturated γ-lactam **7**. For the reaction of imines which contain a β-hydrogen, **7** is transferred to the thermally more stable α,β-unsaturated isomer **8**.

Imhof and co-workers demonstrated a three-component coupling reaction of α,β-unsaturated imines with CO and alkenes catalyzed by $Ru_3(CO)_{12}$, which produced α,α-disubstituted β,γ-unsaturated

γ-butyrolactams.[55] Similar reactions were also independently explored by Chatani and co-workers (Eq. 51).[56] Ethylene, vinylsilane, as well as norbornene could be used as the alkene partners, but the terminal alkenes were not compatible with the reaction system. The imine substrates bearing EWG showed low reactivity.

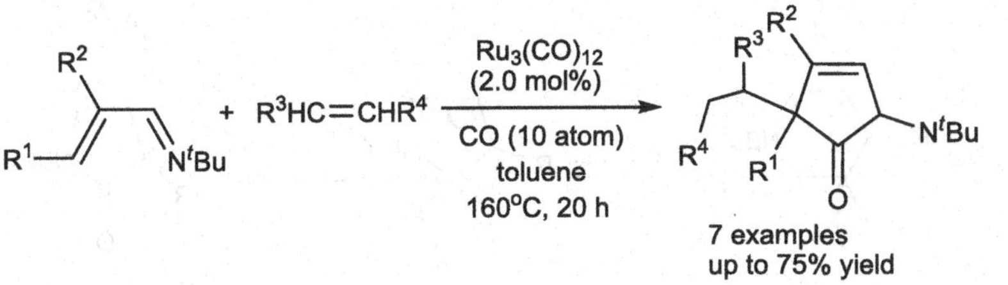

$$\text{(Eq. 51)}$$

The reaction is proposed to proceed *via* a two-step sequence: (i) catalytic carbonylation of the β-olefinic C–H bonds of α,β-unsaturated imines after insertion of ethylene, (ii) intramolecular nucleophilic attack of the imine nitrogen on the ketonic carbon followed by a 1,2-ethyl migration (Scheme 9).

In 2013, Carreira and co-workers presented an attractive strategy for the synthesis of chiral pyrrolidones *via* ruthenium-catalyzed intramolecular hydrocarbamoylation of allylic formamides under CO atmosphere (Eq. 52).[57] A formal ruthenium-catalyzed insertion into the formamide C–H bond and concomitant C–C bond formation by

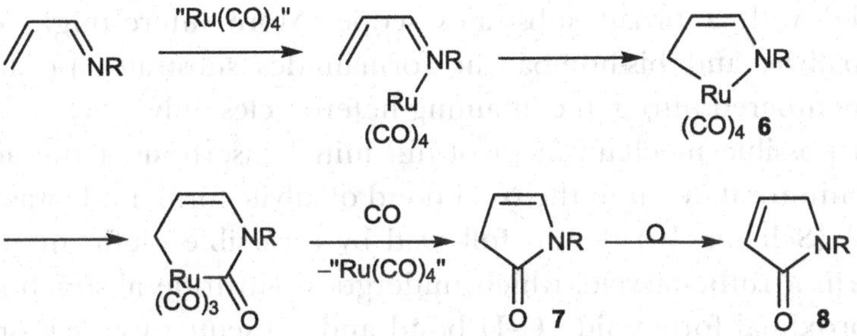

Scheme 8. Proposed mechanism for ruthenium-catalyzed unsaturated γ-lactams synthesis from α,β-unsaturated imines with carbon monoxide

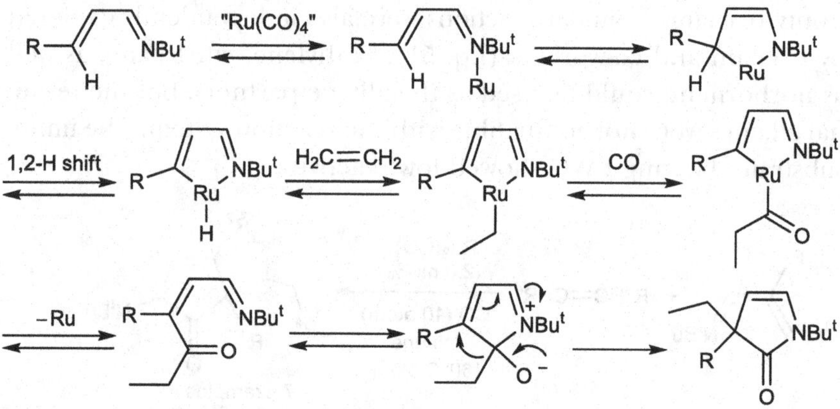

Scheme 9. Proposed mechanism for ruthenium-catalyzed α,α-disubstituted β,γ-unsaturated γ-butyrolactams synthesis from α,β-unsaturated imines with CO and alkenes

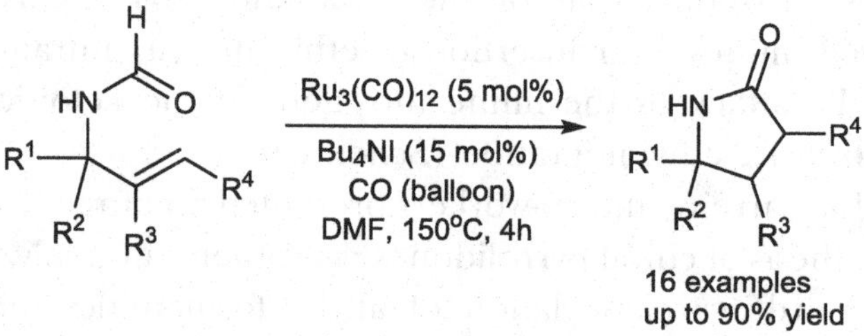

16 examples
up to 90% yield

$$(Eq.\ 52)$$

olefin hydrocarbamoylation was involved in the reaction, which made the reaction complete atom economy. The cyclization performed with a broad substrates scope. More interestingly, even homoallylic and bis-homoallylic formamides substrates produced five-membered nitrogen containing heterocycles only.

A possible mechanism involving initial insertion of the active ruthenium catalyst into the N–H bond of allylic formamide was proposed (Scheme 10). It was followed by reversible olefin insertion to form a ruthenacycle, which undergoes β-hydride abstraction of the proximal formamide C–H bond and subsequent attack of the nucleophilic alkyl moiety onto the electrophilic carbonyl carbon to give intermediate **9**. The active catalyst species was regenerated after release of the product with proton transfer.

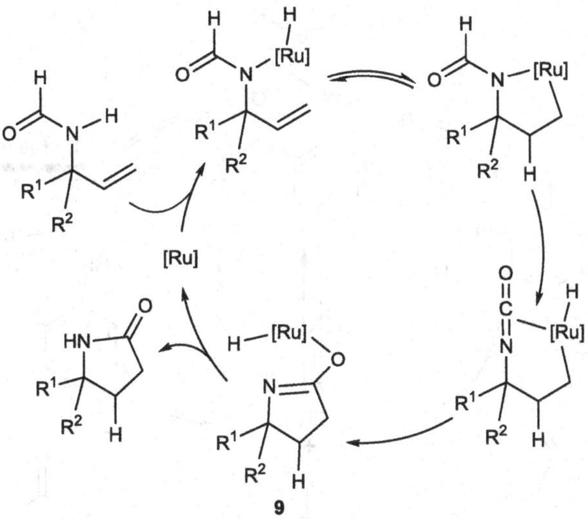

Scheme 10. Proposed mechanism for ruthenium-catalyzed pyrrolidones synthesis from allylic formamides

A Ru-catalyzed intramolecular olefin hydrocarbamoylation for the regiodivergent synthesis of indolin-2-ones and 3,4-dihydroquinolin-2-ones without requiring external CO atmosphere was disclosed by Chang and co-workers (Eq. 53).[58] In the presence of combined catalyst of $Ru_3(CO)_{12}$/Bu_4NI, a 5-*exo*-type cyclization proceeds favorably to form indolin-2-ones as a major product in good to excellent yield in DMSO/toluene cosolvent. An excellent level of regioselectivity was observed with a variety of substrates to deliver 5-*exo*-cyclized lactams. Interestingly, 3,4-dihydroquinolin-2-ones are obtained in major in moderate to high yield *via* a 6-*endo* cyclization process when the reactions are conducted in the absence of halide additives in DMA/PhCl cosolvent.

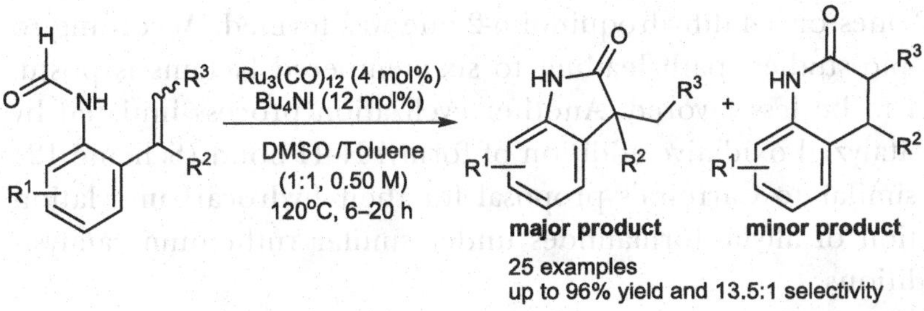

major product   minor product

25 examples
up to 96% yield and 13.5:1 selectivity

(Eq. 53)

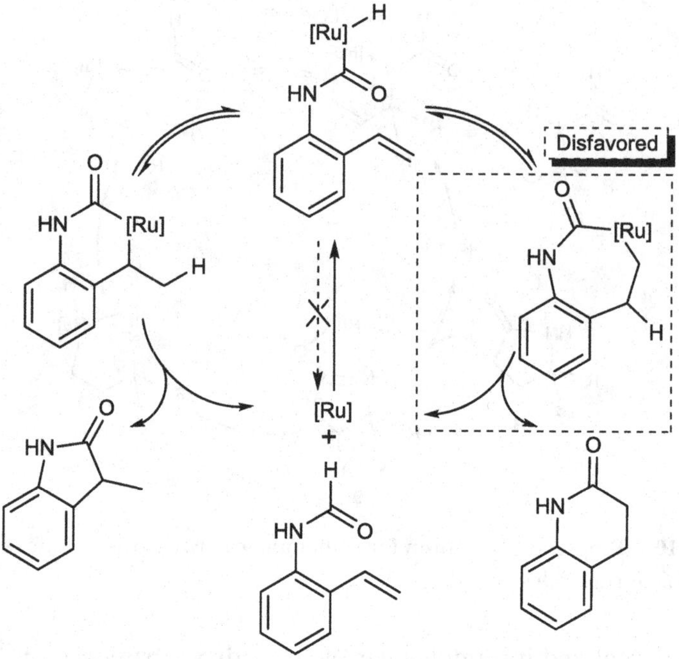

Scheme 11. Proposed mechanism for ruthenium-catalyzed intramolecular olefin hydrocarbamoylation through direct activation of the formyl C–H bond

Two mechanistic pathways, which differed in the way of ruthenium-mediated initial cleavage of formyl C–H or amido N–H bond, were proposed for the catalytic cycle. As shown in Scheme 11, an irreversibly cleavage of formyl C–H bond by the active ruthenium complex was followed by reversible insertion of the olefin into the Ru–H bond, which afforded either six-membered or seven-membered ruthenacycle. After reductive elimination, indolin-2-ones or 3,4-dihydroquinolin-2-one was formed. According to isotopic studies, path leading to six-membered lactams is postulated to be less favored. Another cyclization process initiated by Ru-catalyzed oxidative addition of formyl N–H bond (Scheme 12) was similar to Carreira's proposal for their hydrocarbamoylation reaction of allylic formamides under similar ruthenium catalysis conditions.[57]

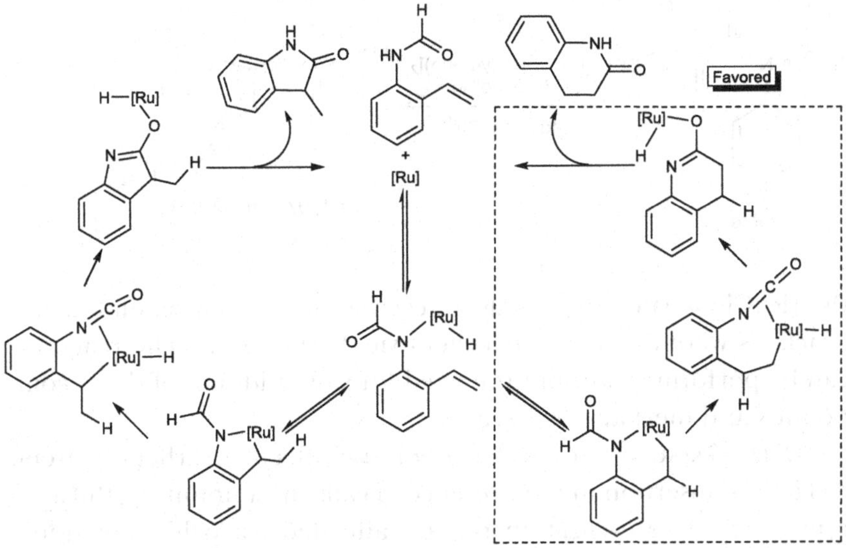

Scheme 12. Proposed mechanism for ruthenium-catalyzed intramolecular olefin hydrocarbamoylation through initial activation of the N–H bond

During the investigation of the Ru(II)-catalyzed ortho-alkenylation of arylpyrazoles with acrylates, Satoh and Miura found that benzylanilide reacted with *n*-butyl acrylate in o-xylene to deliver lactam *via* oxidative alkenylation and subsequent intramolecular aza-Michael addition (Eq. 54).[59] Similar cyclization product was also observed in the reaction of *N*-pentafluorophenyl benzamide with ethyl acrylate on activation with Ru(II) catalyst by Ackermann.[60]

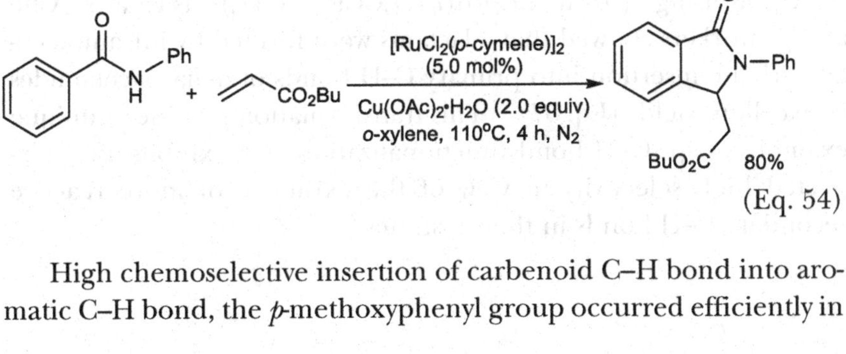

(Eq. 54)

High chemoselective insertion of carbenoid C–H bond into aromatic C–H bond, the *p*-methoxyphenyl group occurred efficiently in

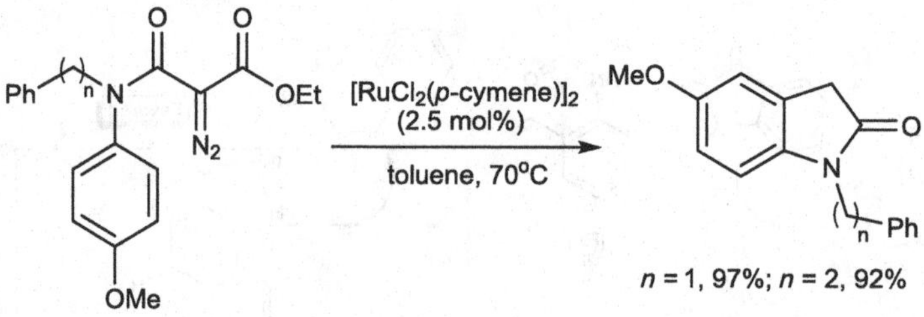

$$n = 1, 97\%; n = 2, 92\% \qquad \text{(Eq. 55)}$$

the $[RuCl_2(p\text{-cymene})]_2$-catalyzed cyclization of α-diazoanilides, and γ-lactams were isolated in excellent yields (Eq. 55).[61] The reactions can be performed without the need for slow addition of diazo compounds and inert atmosphere.

Diazo-β-ketoanilides underwent intramolecular carbenoid arene C–H bond insertion in the presence of catalytic amount of $[RuCl_2(p\text{-cymene})]_2$. This cyclization process afforded 3-alkylideneoxindole derivatives in high yields under mild conditions with excellent chemoselectivity (Eq. 56).[62] Mechanism involving cyclopropanation of the arene is proposed based on primary kinetic isotope effects (KIE) experiment.

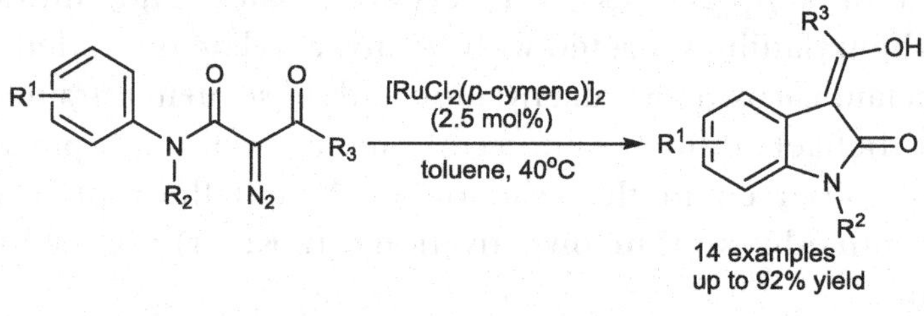

14 examples
up to 92% yield   (Eq. 56)

Upon using $[(p\text{-cymene})Ru(\eta^1\text{-}O_2CCF_3)_2(OH_2)]$ as catalyst, Che and co-workers showed that γ-lactams were formed by intramolecular carbene insertion into primary C–H bonds of α-diazoacetamides in excellent yields (Eq. 57).[63] This transformation provides a unique example of sp^3 C–H bond functionalization and exhibits an unexpected high selectivity in view of the existence of more reactive secondary C–H bonds in the substrates.

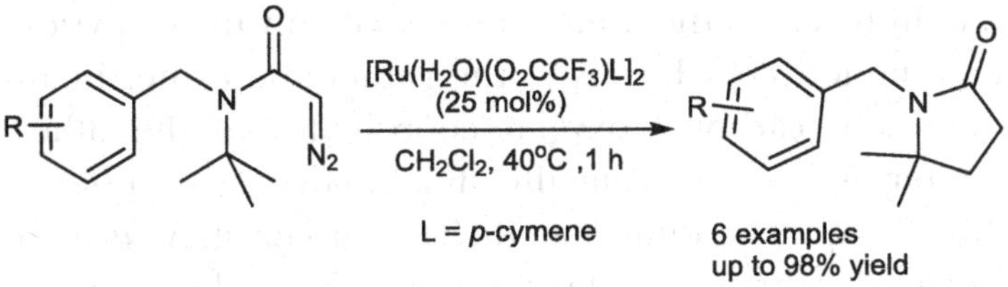

$$\text{(Eq. 57)}$$

2.5. Synthesis of succinimides and their derivatives

A catalytic intermolecular [2 + 2 + 1] co-cyclization of alkynes, isocyanates, and CO by utilization of $Ru_3(CO)_{12}$ as catalyst was developed by Kondo and co-workers (Eq. 58).[64] This process provides a rapid and atom-economical method for the preparation of a variety of unsymmetrically polysubstituted maleimides in excellent yields with high selectivity.

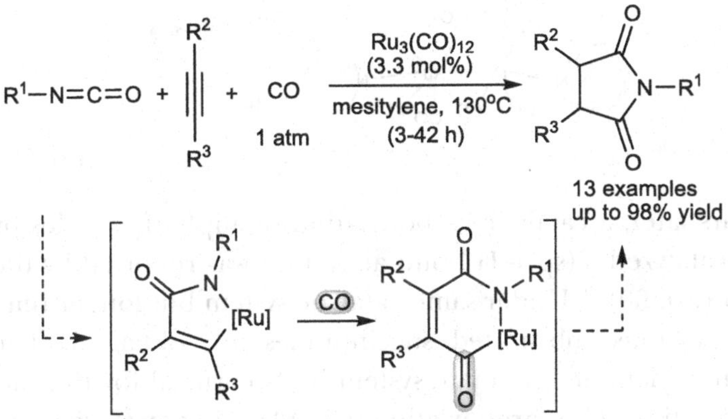

$$\text{(Eq. 58)}$$

In 2009, Chatani and co-workers developed a ruthenium-catalyzed regioselective carbonylation of aromatic amides *via* $C(sp^2)–H$ bond activation (Eq. 59).[65] It should be mentioned that the 2-pyridinylmethylamine moiety of the aromatic amides is required for this transformation to proceed. Under CO (10 atm) and ethylene (7 atm) together with small amount of water at 160°C in toluene for 24 h, diverse phthalimides are formed in high yield.

Interestingly, a dinuclear ruthenium complex **10**, in which 2-pyridi-nylmethylamine moiety coordinates to the ruthenium center in an N,N-fashion and the carbonyl oxygen coordinates to the other ruthenium center, is isolated from the stoichiometric reaction of amide with $Ru_3(CO)_{12}$. The dinuclear Ru(I) complex showed high catalytic activity under the standard reaction conditions but no reactivity without H_2O, thus indicating complex **10** does not participate in the main catalytic cycle but rather exists in a resting state.

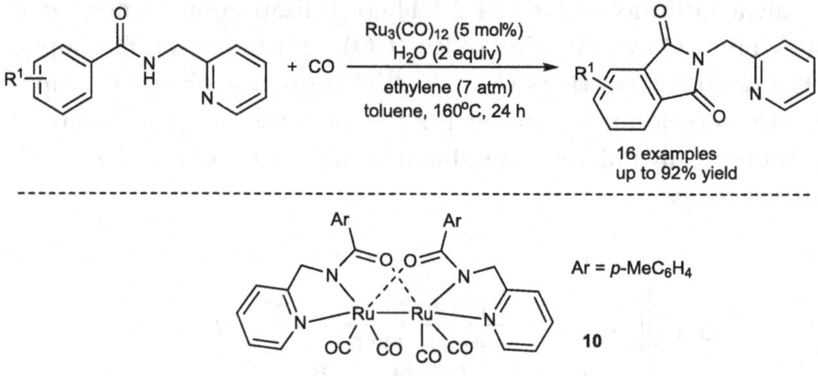

(Eq. 59)

Two years later, a catalytic carbonylation of aliphatic amides by ruthenium-catalyzed $C(sp^3)$–H bond activation was reported by the same group (Eq. 60).[66] Under same catalytic system but longer time (five days), various substituted succinimides are obtained. The 2-pyridinylmethylamine bidentate system is also crucial for the success of the reaction. The carbonylation took place in a preference as the following sequence: $C(sp^2)$–H bond > cyclopropyl $C(sp^3)$–H bond > methyl $C(sp^3)$–H > methylene $C(sp^3)$–H bond.

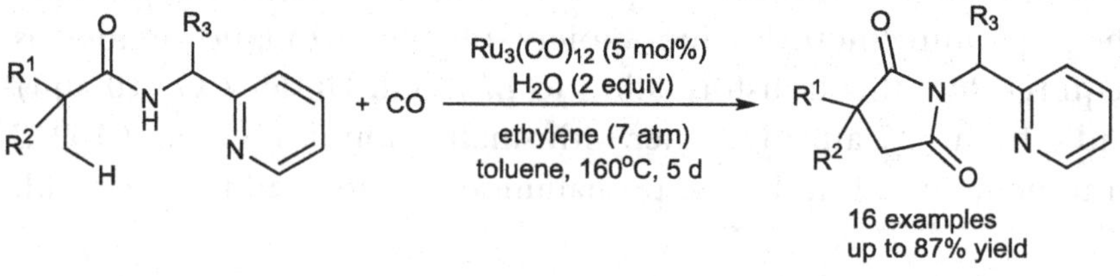

(Eq. 60)

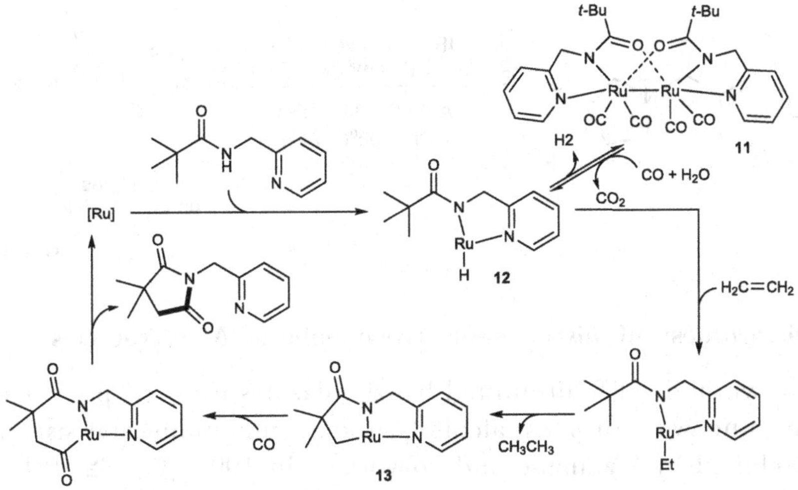

Scheme 13. Proposed mechanism for ruthenium-catalyzed carbonylation and cyclization of aliphatic amides

A dinuclear Ru(I) complex **11**, which is analogue to **10**, was also isolated from the stoichiometric reaction of aliphatic amide with $Ru_3(CO)_{12}$. As shown in Scheme 13, the complex **11** is assumed to be reduced into a mononuclear species **12** with the aid of H_2O. Then followed by insertion of ethylene and release of ethane *via* irreversible C–H bond activation of one methyl group, the cyclo-metalated intermediate **13** is formed. Finally, CO insertion into Ru–C bond and subsequent reductive elimination lead to the final product with regeneration of the ruthenium catalyst.

A novel route to synthesize phthalimide derivatives through ruthenium-catalyzed C–H bond functionalization of aromatic amides was developed by Ackermann and co-workers (Eq. 61).[67] This method is applicable to generate a potent COX-2 enzyme inhibitor in step-economical way. The reaction features by the insertion of a cycloruthenated species into a C–Het multiple bond of isocyanate and cleavage of pyrrolidinyl group. Notably, electron-rich amides which favors the reaction suggests a base-assisted IES-type C–H activation mode. Also, an initial reversible C–H bond metalation step was observed.

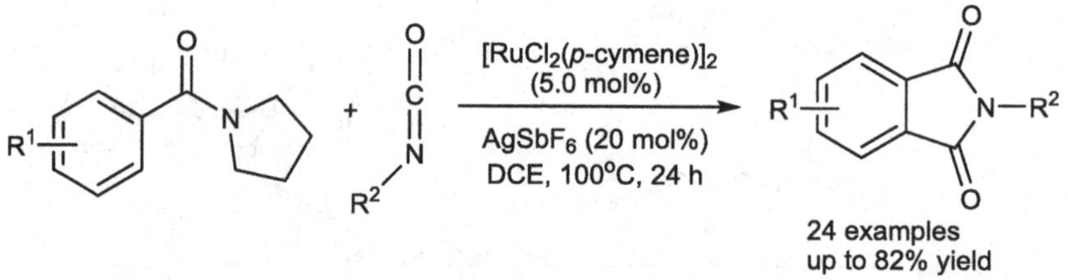

24 examples
up to 82% yield

(Eq. 61)

2.6. Synthesis of miscellaneous five-membered *N*-heterocycles

The synthesis of 2-substituted benzimidazoles form 1,2-phenylenediamine with primary alcohols under ruthenium-catalysis was developed by Watanabe and co-workers in 1991 (Eq. 62).[68] The ruthenium complex acts as a dual catalyst for the oxidation of primary alcohols to aldehydes, as well as for the subsequent cyclization step.

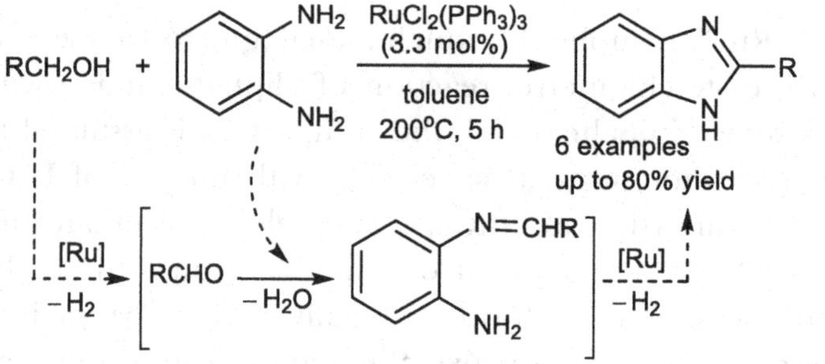

6 examples
up to 80% yield

(Eq. 62)

The reaction of *N*-sulfonylimines with methyl isocyanoacetate using $RuH_2(PPh_3)_4$ as catalyst proceeded efficiently which afforded trans-2-imidazolines stereoselectively under neutral, mild conditions (Eq. 63).[69] Hydrolysis of the resulting imidazolines give 2,3-Diamino acids in excellent yields.

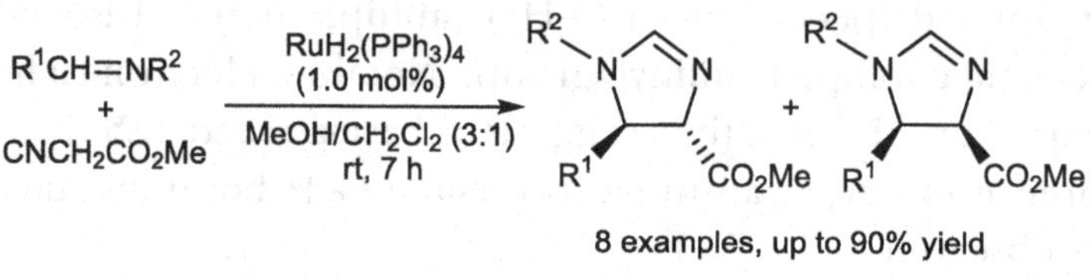

8 examples, up to 90% yield   (Eq. 63)

Cho and Kim have shown a straightforward methodology for the synthesis of 1,2-disubstituted benzimidazoles from readily available *N*-alkyl-1,2-diaminobenzenes (Eq. 64).[70] The reaction proceeds well in the presence of a ruthenium catalyst along with a hydrogen acceptor (acetophone) *via* an alkyl group transfer followed by cyclization.

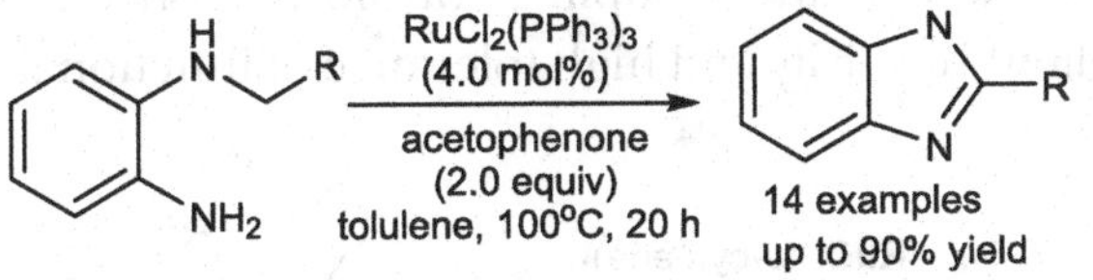

$$\text{(Eq. 64)}$$

The combination of $Ru(PPh_3)_3(CO)H_2$ with the bidentate ligand Xantphos has been successfully used in the hydrogen-transfer reactions for the reactions of alcohols with *o*-aminoaniline (Eq. 65).[71] A range of benzimidazoles were formed in good yields with the aid of piperidinium acetate as an additive and crotononitrile as a hydride acceptor.

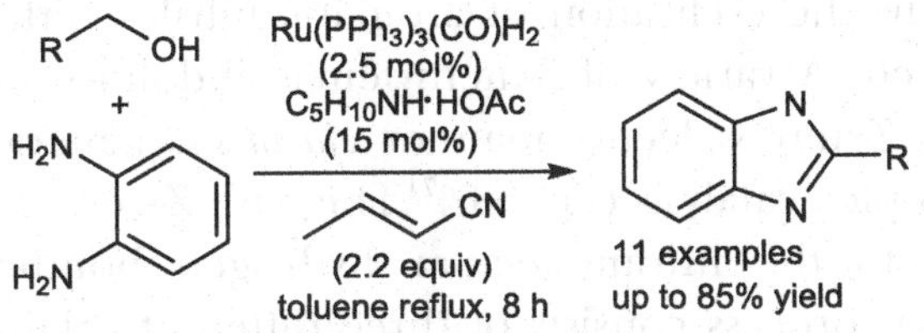

$$\text{(Eq. 65)}$$

In the presence of catalytic amount of $RuCl_3 \cdot nH_2O$, γ-carbolinium ions were produced as single isomers *via* formation of C4–N bond from the intramolecular reactions of 3-pyridyl substituted aryl azides (Eq. 66).[72] The resulting ions can be reduced with $NaBH_4$ to form γ-carboline. Furthermore, dimebolin was synthesized in a concise and stereoselective manner by employing this method.

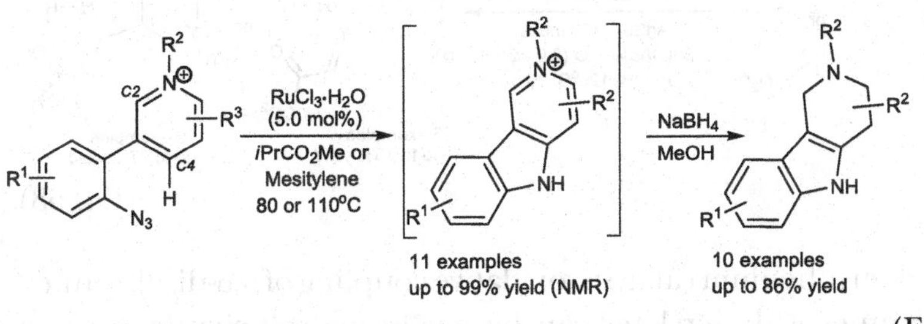

$$\text{(Eq. 66)}$$

Multisubstituted pyrazoles are commonly difficult to prepare with conventional methods. The synthesis of a variety of tri- and tetrasubstituted pyrazoles was achieved by ruthenium(II)-catalyzed intramolecular oxidative C–N coupling of easily accessible starting materials (Eq. 67).[73] Notably, dioxygen gas is employed as the oxidant for this catalytic process. The oxidative coupling transformation was also featured by its excellent reactivity and high tolerance of functional groups.

R[1] = Aryl,
R[2], R[3], R[4] = H, Alkyl, Alkenyl or Aryl

35 examples
up to 94% yield

$$\text{[RuCl}_2(p\text{-cymene)]}_2 \text{ (5.0 mol%)}$$
$$\text{NaHCO}_3 \text{ (2 equiv)}$$
$$\text{DMSO, 1 atm O}_2$$
$$60\text{–}100°C, 5\text{–}6\text{ h}$$

(Eq. 67)

Very recently, the cyclization of aromatic nitriles with alkenes was also explored. A variety of 3-methyleneisoindolin-1-ones were formed in high Z-stereoselective manner *via in situ* generated cationic Ru(II) complex catalysis (Eq. 68).[74] The high Z-stereoselectivity was attributed to the intramolecular hydrogen bonding. The whole cyclization process consists of three different catalytic reactions: (i) copper-catalyzed formation of a benzamide intermediate, (ii) [RuCl$_2$(p-cymene)]$_2$/AgSbF$_6$-catalyzed oxidative alkenylation of benzamide, (iii) Ru-catalyzed intramolecular aza-Michael addition and subsequent β-hydride elimination.

$$\text{[RuCl}_2(p\text{-cymene)]}_2 \text{ (5.0 mol%)}$$
$$\text{AgSbF}_6 \text{ (20 mol%)}$$
$$\text{Cu(OAc)}_2\cdot\text{H}_2\text{O (2.0 equiv)}$$
$$\text{AcOH, 120°C, 36 h}$$

6 examples
up to 75% yield

19 examples
up to 77% yield

(Eq. 68)

Under ruthenium catalysis, oxidative coupling of α,α-disubstituted benzylamines with acrylates can be performed efficiently at room temperature to produce (isoindol-1-yl)acetic acid derivatives (Eq. 69).[75] The reaction takes place accompanied by free amino group directed ortho-alkenylation and successive intramolecular cyclization.

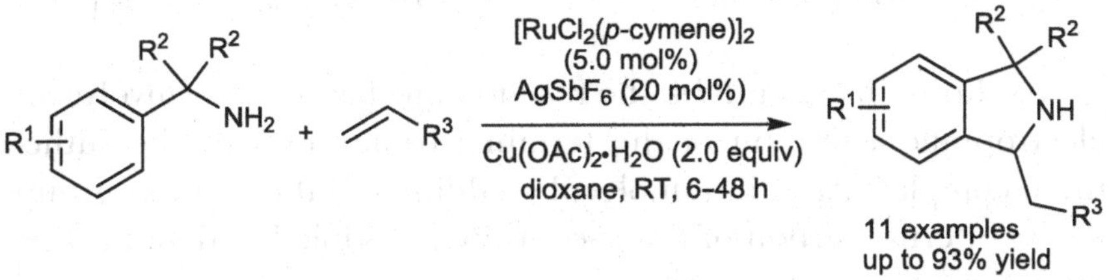

$$\text{11 examples} \atop \text{up to 93\% yield}$$

(Eq. 69)

Aryl and heteroaryl amidines underwent oxidative C–H bond functionalization with alkynes and alkenes with the aid of ruthenium catalysis. By using cationic ruthenium complexes derived from AgOAc, oxidative alkenylations with substituted acrylates provided diversely substituted 1-imino isoindolines (Eq. 70).[76] High site-, regio- and, chemoselectivity were accomplished in this transformation. Mechanistic studies revealed a reversible C–H bond activation step.

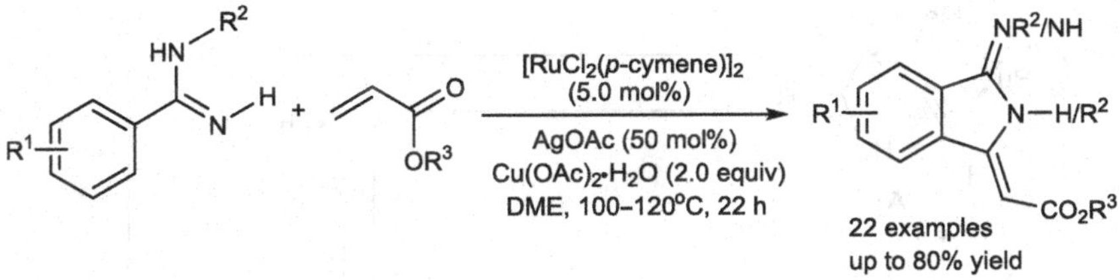

$$\text{22 examples} \atop \text{up to 80\% yield}$$

(Eq. 70)

3. Ru-Catalyzed Synthesis of Five-Membered O-Heterocycles

3.1. Synthesis of furans and their derivatives

A ruthenium-catalyzed synthesis of furan based on the intramolecular addition of an alcohol functionality to a terminal triple bond was reported by Bruneau and co-workers (Eq. 71).[77] The reaction is specific to terminal alkynes and tolerates functional groups sensitive to bases.

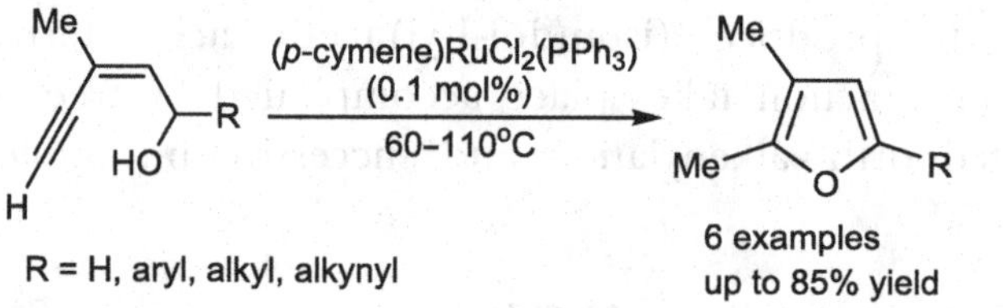

(Eq. 71)

As shown in Scheme 14, the reaction mechanism may involve an electrophilic activation of the terminal triple bond by the ruthenium complex. After intramolecular addition of the hydroxy group to the internal carbon of the η^2-coordinated triple bond and subsequent proton transfer, the furan ring was formed.

The combination of $RuCl_3 \cdot nH_2O$ (10 mol%) and AgOTf (30 mol%) acted as a catalyst for cyclization of 2-allylphenols to

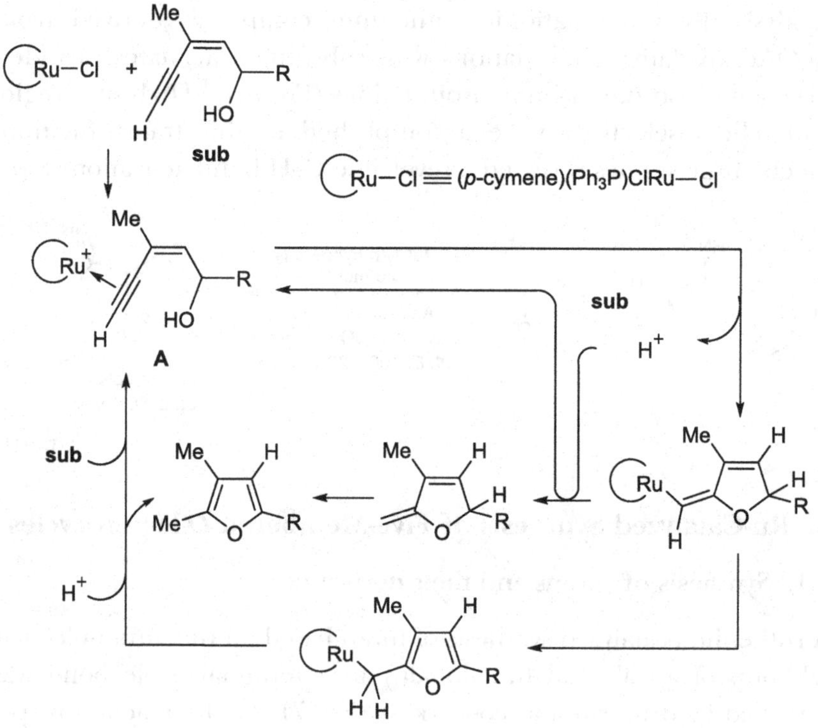

Scheme 14. Proposed mechanism for ruthenium-catalyzed cyclization of (Z)-pent-2-en-4-yn-1-ols

2,3-dihydro-2-methyl benzofurans in good yield in the presence of $Cu(OTf)_2$ as a co-catalyst and PPh_3 as a ligand (Eq. 72).[78] Later, a new catalyst system $(Cp^*RuCl_2)_2/4AgOTf/4PPh_3$, was found to be more active even in the absence of $Cu(OTf)_2$.[79]

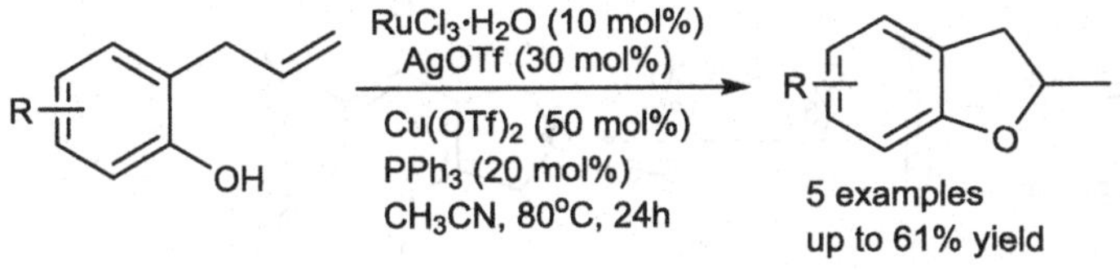

RuCl₃·H₂O (10 mol%)
AgOTf (30 mol%)

Cu(OTf)₂ (50 mol%)
PPh₃ (20 mol%)
CH₃CN, 80°C, 24h

5 examples
up to 61% yield

(Eq. 72)

Trost and Pinkerton have realized a ruthenium-catalyzed alkylative cycloetherification of vinyl ketones with allene bearing hydroxy group. In the presence of $[CpRu(CH_3CN)_3]PF_6$ as catalyst and $CeCl_3 \cdot 7H_2O$ as co-catalyst, a wide range of cyclic ethers including tetrahydrofurans (THFs) can be formed in an atom economical manner (Eq. 73).[80] This transformation establishes a mechanistic support for the involvement of allyl ruthenium intermediates which generate from unactivated allenes and enones *via* ruthenacycle formation.

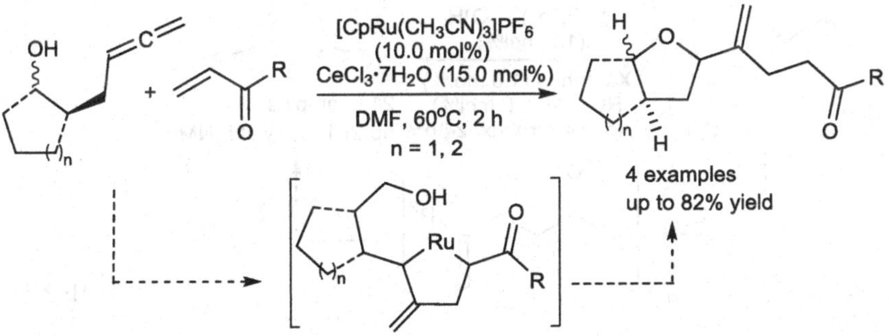

[CpRu(CH₃CN)₃]PF₆
(10.0 mol%)
CeCl₃·7H₂O (15.0 mol%)

DMF, 60°C, 2 h
n = 1, 2

4 examples
up to 82% yield

(Eq. 73)

Under $Cp^*RuCl(COD)$ catalysis, an ether tethered iodo-1,6-diynes underwent [2 + 2 + 2] cycloaddition with alkynes to furnish iodo-1,3-dihydroisobenzofurans in an efficient and regioselective manner (Eq. 74).[81] The corresponding iododiynes was accessed by silver-catalyzed Csp–H iodination of readily available 1,6-diynes. Importantly, the obtained iodobenzofurans could be utilized as a halide component for further cross-coupling reactions to give highly conjugated molecules.

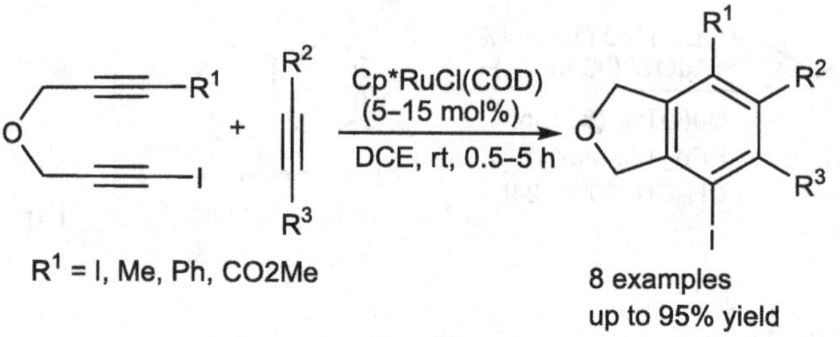

R^1 = I, Me, Ph, CO2Me

Cp*RuCl(COD)
(5–15 mol%)
DCE, rt, 0.5–5 h

8 examples
up to 95% yield

(Eq. 74)

Various 2,5-disubstituted furans were synthesized from readily available 1,4-alkynediols by using $[Ru(PPh_3)_3(CO)H_2]$ and Xantphos catalytic system with an organic acid co-catalyst in one tandem reaction (Eq. 75).[82] The present strategy can overcome some of the problems often associated with the classical Paal–Knorr synthesis such as availability and stability of the 1,4-diketone substrates.

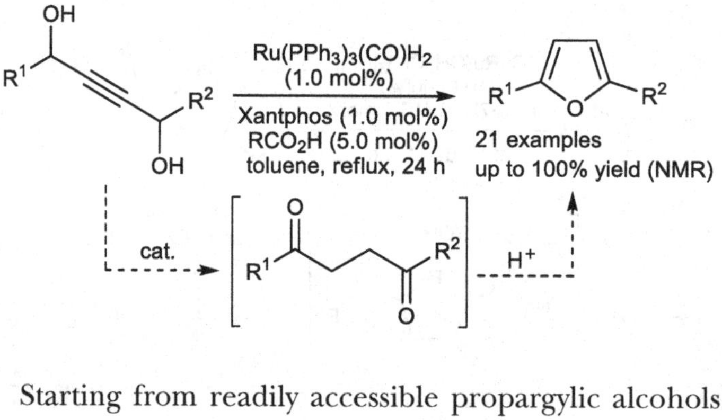

Ru(PPh3)3(CO)H2
(1.0 mol%)
Xantphos (1.0 mol%)
RCO2H (5.0 mol%)
toluene, reflux, 24 h

21 examples
up to 100% yield (NMR)

(Eq. 75)

Starting from readily accessible propargylic alcohols and commercially available 1,3-dicarbonyl compounds, a novel one-pot method for fully substituted furans catalyzed by a ruthenium(II)/trifluoroacetic acid system was developed by Cadierno and co-workers (Eq. 76).[83] The process which undergoes under solvent-free conditions involves the following two steps: (i) an initial CF_3CO_2H-promoted propargylic substitution of the secondary alkynol by the

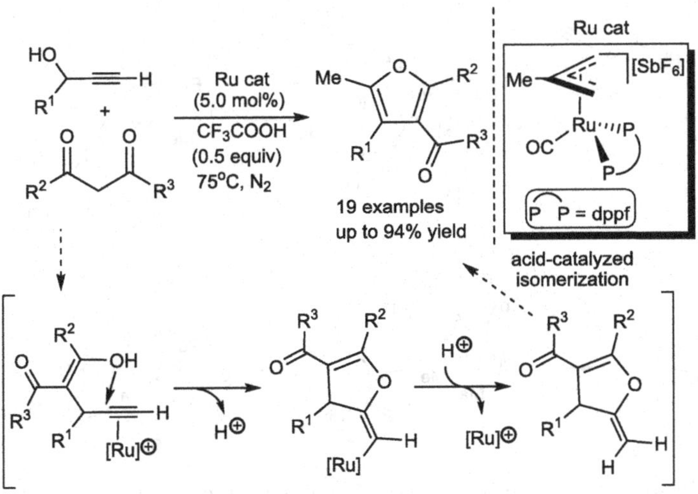

(Eq. 76)

1,3-dicarbonyl compound, and (ii) subsequent cycloisomerization of the resulting γ-ketoalkyne intermediate catalyzed by ruthenium (II) complex $[Ru(\eta^3\text{-}2\text{-}C_3H_4Me)(CO)(dppf)][SbF_6]$. This cyclization reaction is quite general being applicable both to terminal and internal monosubstituted alkynols, as well as to a variety of β-dicarbonyl compounds.

Intramolecular cyclization of 3-butyne-1,2-diols catalyzed by the methanethiolate-bridged diruthenium complex $[Cp*RuCl(\mu_2\text{-}SMe)]_2$ was reported by Nishibayashi and co-workers, which produced the corresponding substituted furans in good to high yields (Eq. 77).[84]

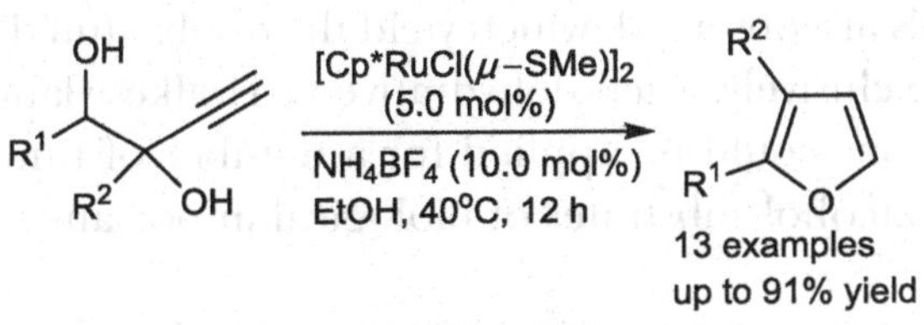

(Eq. 77)

The catalytic reaction mechanism is proposed to proceed *via* ruthenium-allenylidene complexes as key intermediates **14**, followed by tautomerization to give a vinylic vinylidene complex **15**. Finally, intramolecular nucleophilic attack of the hydroxy oxygen

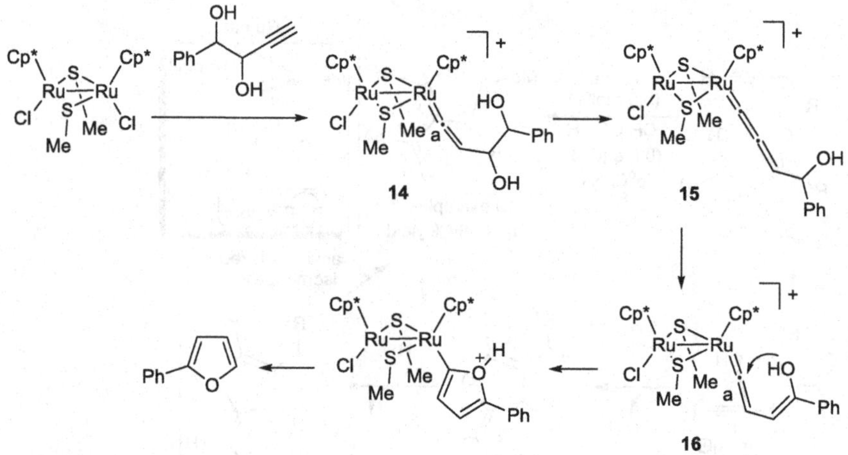

Scheme 15. Proposed mechanism for ruthenium-catalyzed intramolecular cyclization of 3-butyne-1,2-diols

atom results in the formation of furan product through an alkenyl complex **16** (Scheme 15).

Treatment of phenol with 1,2-diols and excess cyclopentene (3 equiv) in the presence of a well-defined cationic ruthenium hydride complex $[(C_6H_6)(PCy_3)(CO)RuH]^+BF_4^-$ (1 mol%) lead to the formation of benzofuran derivatives (Eq. 78).[85] The catalytic C–H coupling method exhibits a broad substrate scope, tolerated carbonyl and amine functional groups, obviated the use of any expensive and often toxic metal oxidants, and liberated water as the only by-product. Furthermore, excellent regioselective addition of the linear 1,2-diols are observed, which yield the α-substituted benzofuran products exclusively. Such dehydrative C–H alkenylation and annulation reactions could be applied for a number of functionalized phenol and alcohol substrates of biological importance.

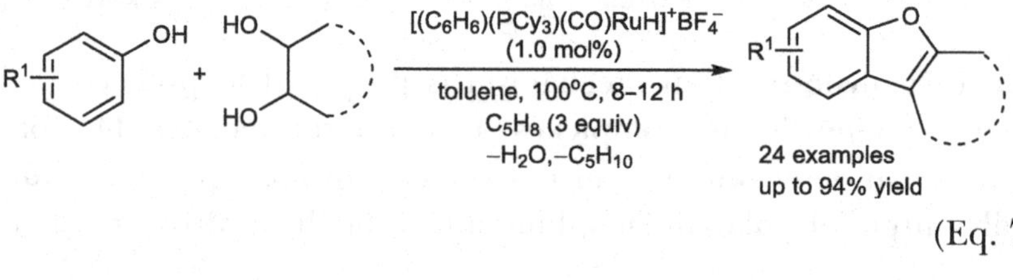

(Eq. 78)

When unsymmetrical 2-aryl cyclic 1,3-dicarbonyl compounds, containing two distinct, non-adjacent sites for C–H bond functionalization, was applied for the oxidative annulation reactions with alkenes, catalyst-controlled divergent C–H functionalization was achieved (Eq. 79).[86] With the aid of ruthenium catalyst, functionalization of a hydrogen atom four bonds away from the oxygen of the directing group mainly occurs, leading to benzofurans from electron-deficient terminal alkenes. In contrast, a palladium-based catalyst results in functionalization of a hydrogen atom five bonds away from the oxygen of the enol/enolate directing group exclusively, producing benzopyrans from electron-deficient alkenes.

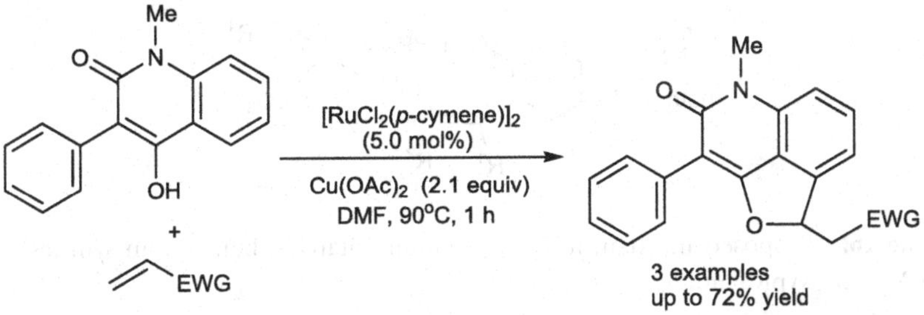

(Eq. 79)

In 2014, Liu and Lu disclosed a Ru-catalyzed C–H functionalization with alkynes for the synthesis of benzofuran derivatives using –ONHPiv as an oxidizing directing group (Eq. 80).[87] The reaction showed good functional group tolerance and high regioselectivity. Low yields were obtained with the N-phenoxypivalamide substrates bearing EWGs. Competition experiments revealed that the electron-deficient diarylalkynes were more reactive.

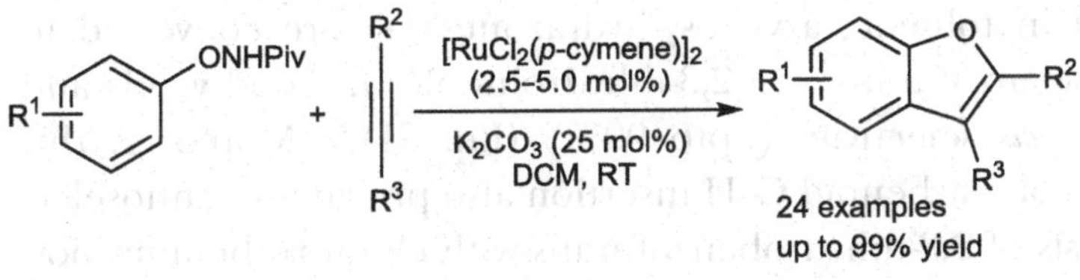

(Eq. 80)

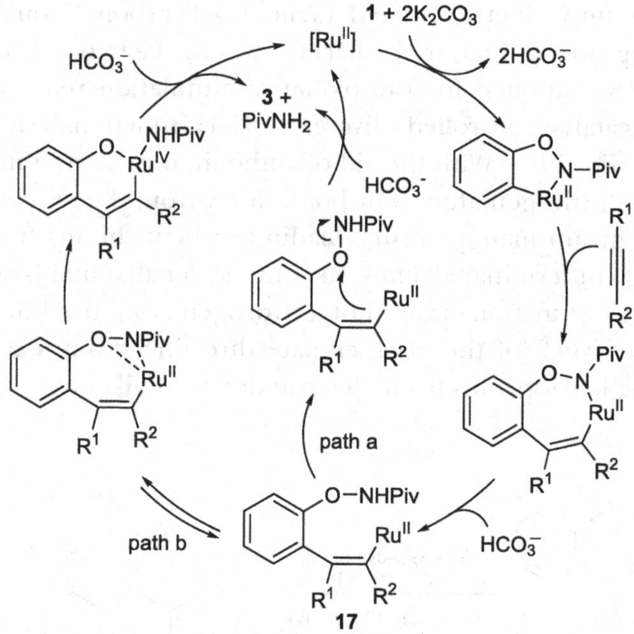

Scheme 16. Proposed mechanism for ruthenium-catalyzed benzofuran synthesis from N-phenoxypivalamide

As shown in Scheme 16, the catalytic cycle initiates with an irreversible C–H cleavage by Ru(II) to yield a five-membered ruthenacycle intermediate. Subsequent alkyne insertion and protolysis led to intermediate **17**. Then the following two pathways might occur: (i) the C–O bond formation and simultaneously the O–N bond cleavage *via* intramolecular substitution (path a); (ii) intramolecular oxidative addition and reductive elimination involving Ru(IV) intermediate (path b).

Using [Ru^{II}(TTP)(CO)] as catalyst and *n*-Bu$_4$NBr as phase transfer catalyst in toluene, aryl tosylhydrazones salt are converted to 2,3-dihydrobenzofurans and 2,3-dihydroindoles in good yields and remarkable *cis* selectivity (upto 99%) (Eq. 81).[88] Moreover, this intramolecular carbenoid C–H insertion also provides enantioselective synthesis of 2,3-dihydrobenzofurans with chiral ruthenium porphyrin complex as catalyst.

$$1) \text{ LiHMDS, THF}$$
$$-78°C, 30 \text{ min}$$
$$2) [Ru(TTP)(CO)] (1.0 \text{ mol \%})$$
$$n\text{-Bu}_4\text{NBr} (10.0 \text{ mol \%})$$
$$\text{toluene, } 60-70°C, 48 \text{ h}$$

9 examples
up to 89% yield
up to 99% *cis* selectivity

X = O, N

(Eq. 81)

With the same ruthenium-porphyrin catalyst, N-tosylhydrazones (*in situ* generating the corresponding alkyl diazomethanes) underwent intramolecular $C(sp^3)–H$ insertion of an alkyl carbene (Eq. 82).[89] A variety of substituted THFs were formed in high efficiency, with excellent selectivity and good function-group tolerance. Moreover, the operational procedure is simple without the need for slow addition with a syringe pump. Meanwhile, pyrrolidine derivatives can also be produced from the corresponding N-tosylhydrazones by applying this method.

$$[Ru(TTP)(CO)]$$
$$(1.0 \text{ mol\%})$$
$$K_2CO_3 (3 \text{ equiv})$$
$$1,4\text{-dioxane}$$
$$105°C, 10 \text{ h}$$

12 examples
up to 94% yield
up to >99:1 *cis/trans*

(Eq. 82)

3.2. Synthesis of lactones and their derivatives

The synthesis of furanones were achieved by ruthenium-catalyzed oxidative cyclocarbonylation of 1,1-disubstituted allyl alcohols (Eq. 83).[90] The present oxidative cyclocarbonylation can also be applied to the synthesis of phthalides from 1,1-disubstituted benzyl alcohols.

$$+ \text{ CO}$$
$$(10 \text{ kg cm}^{-2})$$

R^1, R^2, R^3 = aryl, alkyl

$$\text{RuCl}_2(\text{PPh}_3)_3$$
$$(5.0 \text{ mol\%})$$
$$K_2CO_3 (2.5 \text{ equiv})$$
$$\text{OAc}$$
$$(7.5 \text{ equiv})$$
$$\text{THF, } 200°C, 15 \text{ h}$$

6 examples
up to 77% yield

(Eq. 83)

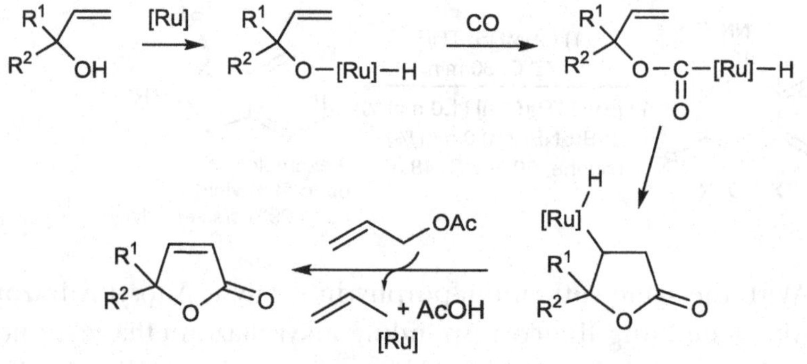

Scheme 17. Proposed mechanism for ruthenium-catalyzed oxidative cyclocarbonylation of 1,1-disubstituted allyl alcohols

As shown in Scheme 17, the reaction is proposed to proceed as follows: oxidative addition of the hydroxy group to an active ruthenium center firstly occurs, which followed by insertion of carbon monoxide into O–Ru bond. After insertion of an olefin and subsequent beta-hydride elimination affords 2(5H)-furanone, together with regeneration of an active ruthenium species via removal of hydrogen by hydrogen transfer to allyl acetate.

The addition of alkenes to 4-hydroxy-2-alkynoates in an Alder-ene-type mode produces butenolides in the presence of CpRu (COD)Cl as catalyst (Eq. 84).[91] The reaction proceeds with excellent chemoselectivity. The regioselectivity with respect to the alkene is with clean allyl inversion, and the regioselectivity with respect to the alkyne places the allyl group preferentially at the α-carbon. The sequence retains the stereochemical integrity of the propargylic position of the starting alkyne which becomes the 5-position of the product 2(5H)-furanones. Notably, the ready availability of 4-hydroxy-2-alkynoates by carbonyl addition of lithiated ethylpropiolate makes this approach very practical.

Murai and co-workers showed a ruthenium-catalyzed cyclocarbonylation of yne-aldehydes in 1998 (Eq. 85).[92] In the presence of a catalytic amount of $Ru_3(CO)_{12}$, the reactions of yne-aldehyde with CO (10 atm) in toluene at 160°C give a series of α,β-unsaturated

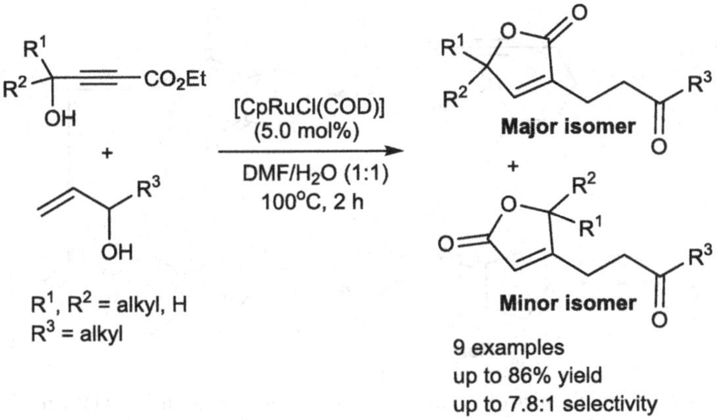

(Eq. 84)

bicyclic γ-butenolides in high yields. It is noteworthy that polyfunctional compounds could be formed in a single step by employing this method.

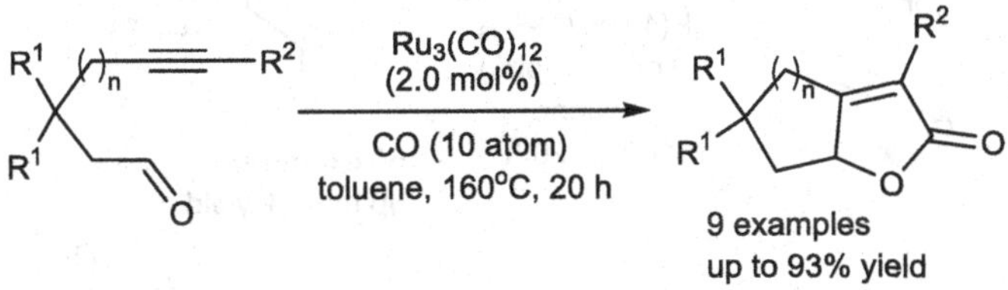

(Eq. 85)

Two possible mechanisms were proposed. One is transition metal-catalyzed hetero-Pauson–Khand process. Another pathway is shown in Scheme 18, which was initiated by the oxidative addition of an aldehyde C–H bond to ruthenium.

The ruthenium-catalyzed intermolecular cyclocoupling of ketones (or aldehydes), alkenes (or alkynes), and CO, which leads to γ-butyrolactones, was described by Murai and co-workers (Eq. 86).[93] This intermolecular carbonylative [2 + 2 + 1] cycloaddition allows the usage of a wide variety of ketones, such as α-dicarbonyl compounds and *N*-heterocyclic ketones. In the case of the cycloaddition of α-dicarbonyl compounds, the addition of $P(4\text{-}CF_3C_6H_4)_3$ is necessary to promote the reaction efficiency. Moreover, a variety of cyclic olefins, unpolarized terminal olefins, and internal alkynes can be successfully applied in the synthesis of highly functionalized lactones.

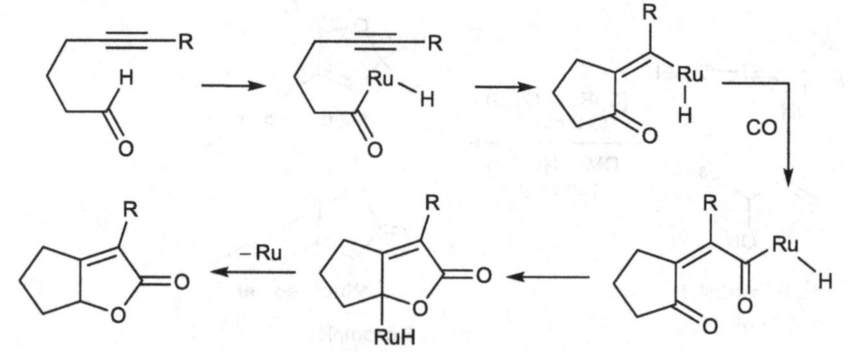

Scheme 18. Proposed mechanism for ruthenium-catalyzed cyclocarbonylation of yne-aldehydes

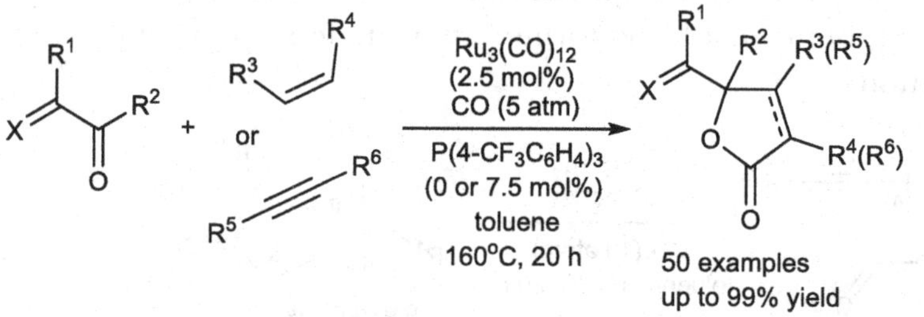

50 examples
up to 99% yield

(Eq. 86)

Interestingly, remarkable differences in additive effects, substituent effects, and a dependence on reaction parameters, such as the pressure of ethylene and CO, were observed between the reactions of α-dicarbonyl compounds and those of N-heterocyclic ketones with ethylene. Such differences can be rationalized by assuming that the rate-limiting step in the catalytic cycle is different for these two types of reactions.

The proposed reaction mechanism is shown in Scheme 19. The first step is the coordination of the substrate to a coordinatively unsaturated ruthenium species, forming a σ, σ-chelate ruthenium complex **18**. Then it reacts with an alkene (or an alkyne) to give the oxametallacycle **19** (route A). After CO insertion into a Ru–O bond and subsequent reductive elimination, the final product was formed. It should be noted that an alternative mechanism (route B), which

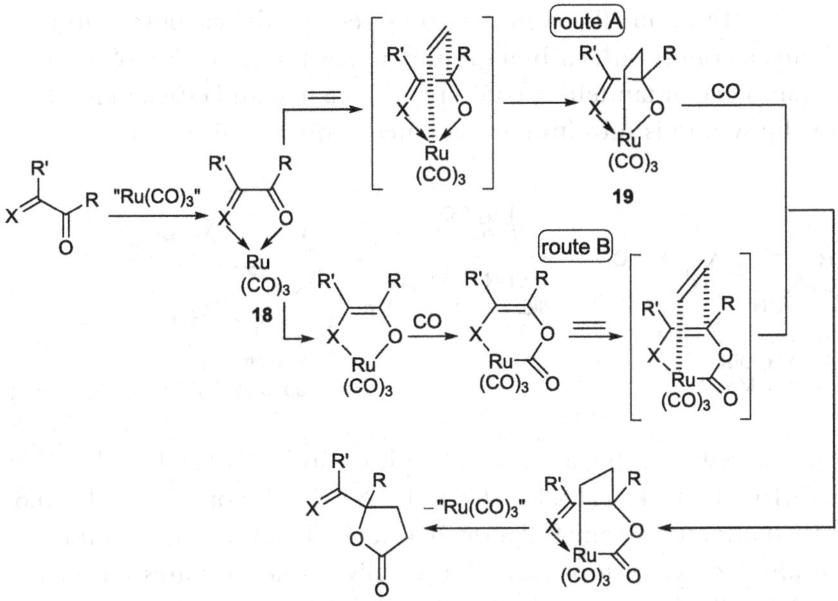
route A
route B
"Ru(CO)3"
CO
-"Ru(CO)3"
18
19

Scheme 19. Proposed mechanism for ruthenium-catalyzed cyclocoupling of ketones (or aldehydes), alkenes (or alkynes), and CO to form γ-butyrolactones

consists of the initial CO insertion and the subsequent addition of an alkene, cannot be excluded.

An efficient cycloisomerization which employs CpRu(COD)Cl and trifuryl phosphine as the precatalyst in the presence of *n*-Bu_4NBr or *n*-Bu_4NPF_6 with *N*-hydroxysuccinimide as the oxidant in DMF-water cosolvent was developed (Eq. 87).[94] In this way, a wide diversity of homopropargyl alcohols were converted to γ-butyrolactones with excellent chemoselectivity. A vinylidene metal species was proposed as reactive intermediates in this catalytic cycle.

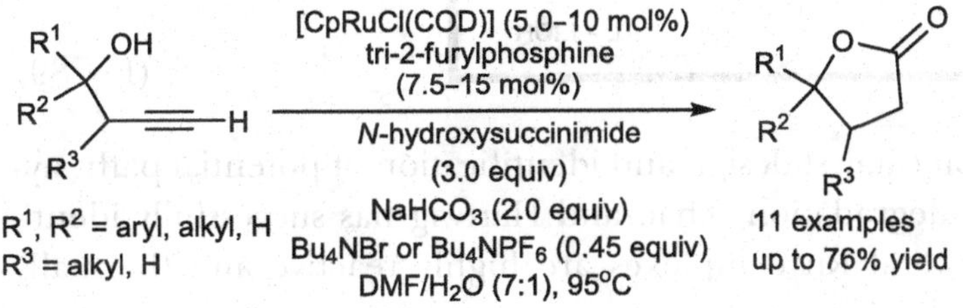
[CpRuCl(COD)] (5.0–10 mol%)
tri-2-furylphosphine
(7.5–15 mol%)
N-hydroxysuccinimide
(3.0 equiv)
NaHCO3 (2.0 equiv)
Bu4NBr or Bu4NPF6 (0.45 equiv)
DMF/H2O (7:1), 95°C
R1, R2 = aryl, alkyl, H
R3 = alkyl, H
11 examples
up to 76% yield

(Eq. 87)

A ruthenium(0) complex-catalyzed cyclic carbonylation of allenyl alcohols with carbon monoxide was reported (Eq. 88).[95] This intramolecular tandem reaction provides a new and efficient method for the synthesis substituted γ-lactones in quantitative yield.

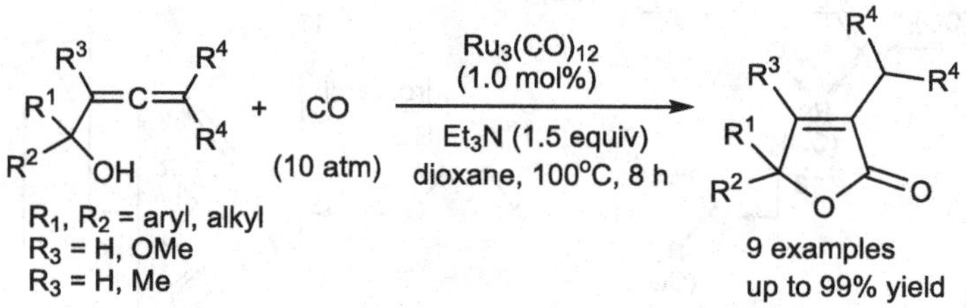

R_1, R_2 = aryl, alkyl
R_3 = H, OMe
R_3 = H, Me

9 examples
up to 99% yield   (Eq. 88)

The catalytic behavior of complex $\{Ru[P^*][CO][EtOH]\}\{P^* = 5,10,15,20$-tetrakis$[(1S,4R,5R,8S)$-1,2,3,4,5,6,7,8-octahydro-1,4:5,8-dimethanoanthracene-9-yl]porphyrinato dianion$\}$ on asymmetric intramolecular cyclopropanation of allylic diazoacetates was examined by Che and co-workers (Eq. 89).[96] In this context, the corresponding cyclopropyl lactones were obtained in moderate yields with up to 85% ee.

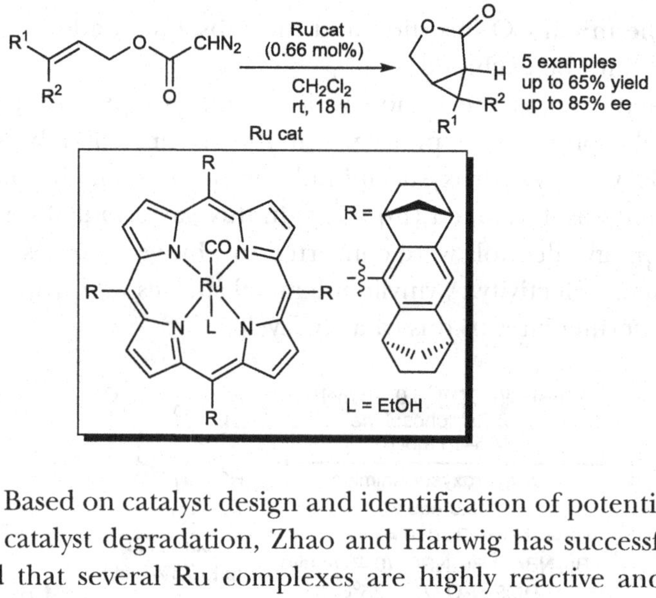

5 examples
up to 65% yield
up to 85% ee

(Eq. 89)

Based on catalyst design and identification of potential pathways for catalyst degradation, Zhao and Hartwig has successfully identified that several Ru complexes are highly reactive and thermally stable catalysts for the dehydrogenative cyclization of 1,4-butanediol to γ-butyrolactone without hydrogen acceptor or solvent (Eq. 90).[97] In particular, a ruthenium complex **20** containing an aliphatic phosphine and a diamine generates the product with high conversion and selectivity and high (17,500) turnovers. This transformation is simple to conduct, environmentally friendly, and highly efficient.

catalyst	time	yield	TON
RuH$_2$(PMe$_3$)$_4$	40 h	87%	3780
RuH$_2$(CO)(PMe$_3$)$_4$	40 h	73%	3170
complex **20**	40 h	100%	4360
complex **20** (0.0058 mol%)	48 h	100%	17000

$$HO\diagdown\diagdown\diagdown OH \xrightarrow[205°C]{\text{catalyst (0.023 mol\%)}} \text{(γ-butyrolactone)}$$

complex **20**

(Eq. 90)

Alkylative lactonization and carbocyclization of monosubstituted allene carboxylic acids and α,β-unsaturated olefins was promoted by cationic ruthenium complex [CpRu(NCCH$_3$)$_3$]PF$_6$, yielding five- and six-membered lactones (Eq. 91).[98] The reaction is featured by mild reaction conditions, lower catalyst loadings, and the tolerance of several functional groups.

$$HO_2C\text{...} + \text{...}R^2 \xrightarrow[\substack{CeCl_3\ H_2O \\ (5.0-10.0\ mol\%) \\ DMF,\ 25-60°C,\ 0.5-2\ h}]{\substack{[CpRu(NCCH_3)_3]PF_6 \\ (5.0-10.0\ mol\%)}} \text{...}R^2$$

n = 1, 2

n = 1, 14 examples
up to 82% yield

(Eq. 91)

As shown in Scheme 20, the reactions proceeded in a reaction sequence consisting of coordination of the allene and the α,β-unsaturated olefin to the ruthenium catalyst, oxidative coupling, nucleophilic trapping by the tethered carboxylic acid, and protonation of the metal and reductive elimination.

A new electron-rich PNN-type ruthenium(II) hydrido borohydride pincer complex [RuH(BH$_4$)(tBu-PNN)] (tBu-PNN = 2-ditert-butylphosphinomethyl-6-diethylaminomethylpyridine) were

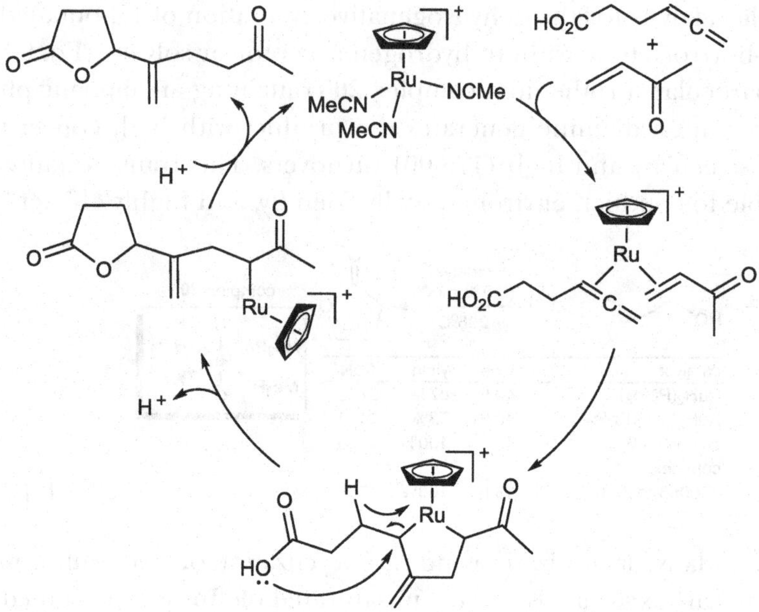

Scheme 20. Proposed mechanism for ruthenium-catalyzed alkylative lactonization and carbocyclization of monosubstituted allene carboxylic acids with α,β-unsaturated olefins

prepared. It can catalyze dehydrogenation reactions of alcohols efficiently including dehydrogenative cyclization of diols to lactones (Eq. 92).[99]

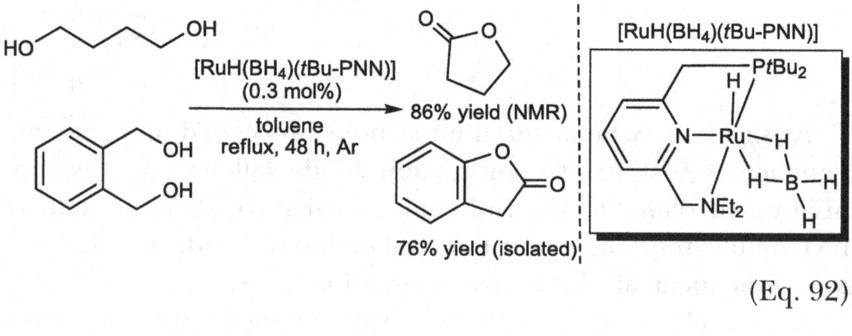

(Eq. 92)

After the seminal work reported by Satoh and Miura on ruthenium-catalyzed oxidative vinylation of heteroarene carboxylic acids with alkenesin early 2011,[100] Ackermann demonstrated a ruthenium(II)-catalyzed cross-dehydrogenative C–H bond alkenylations of benzoic acid derivatives with acrylonitrile or alkyl acrylates. Following the oxidative C–H bond alkenylation reaction, subsequent intramolecular *oxa*-Michael reaction occurred leading to phthalides in good yields (Eq. 93).[101] The reactions took place with water as an environmentally benign medium under mild conditions.

$$[RuCl_2(\textit{p}\text{-cymene})]_2 \ (2.0 \ mol\%)$$
$$Cu(OAc)_2 \cdot H_2O \ (2.0 \ equiv)$$
$$H_2O, \ 80°C, \ 16\text{--}48 \ h$$

20 examples
up to 97% yield

(Eq. 93)

A ruthenium-catalyzed phthalides synthesis from the reaction of mandelic acids with acrylates was established (Eq. 94).[102] A sequence of dialkenylation, decarboxylation, and subsequent intramolecular cyclization led to the final product. Of note, the aromatic rings of the mandelic acids with strong EWG or conjugated aryl group led to good yields of the corresponding products.

$$[RuCl_2(\textit{p}\text{-cymene})]_2 \ (5.0 \ mol\%)$$
$$Cu(OAc)_2 \ (1.1 \ equiv)$$
$$DMF, \ 110°C, \ 11 \ h, \ air$$

21 examples
up to 82% yield

(Eq. 94)

Under $Ru_3(CO)_{12}$ and 1,3-Bis(diphenylphosphino)propane (dppp) catalysis, γ-butyrolactones, as well as spiro- and α-methylene-γ-butyrolactones are produced in high yields *via* C–C coupling of vicinal diols and acrylic esters (Eq. 95).[103] A catalytically competent ruthenium(II) complex, $Ru(CO)(dppp)(\eta^1\text{-}O_2CC_{10}H_{15})$

$(\eta^3\text{-}O_2CC_{10}H_{15})$, was isolated and characterized by single-crystal X-ray diffraction. A catalytic cycle involving 1,2-dicarbonyl-acrylate oxidative coupling to form oxaruthenacyclic intermediates is postulated.

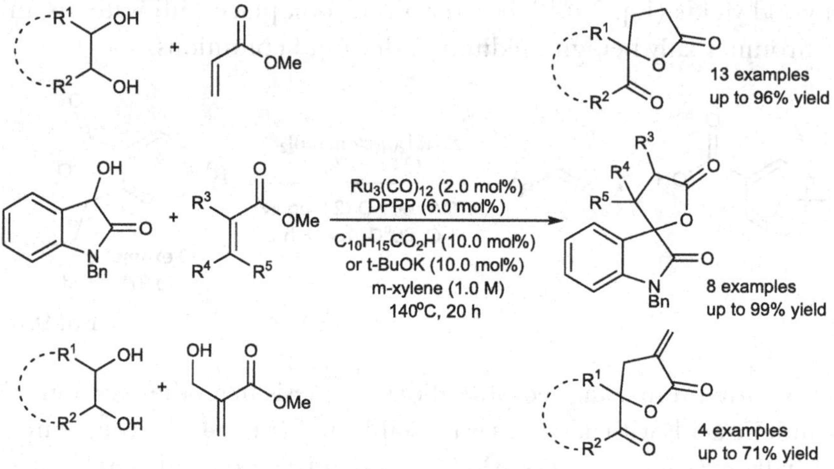

(Eq. 95)

4. Ru-Catalyzed Synthesis of Five-Membered N,O-Heterocycles

Watanabe and co-workers disclosed the first ruthenium-catalyzed reaction of 2-aminophenol with primary alcohols (Eq. 96).[68] Using $RuCl_2(PPh_3)_3$ as catalyst, the corresponding 2-substituted benzoxazoles were formed in toluene at 215°C. Then, Huh and Shim used this catalyst system for the synthesis of benzoxazoles under milder conditions (dioxane, 180°C).[104]

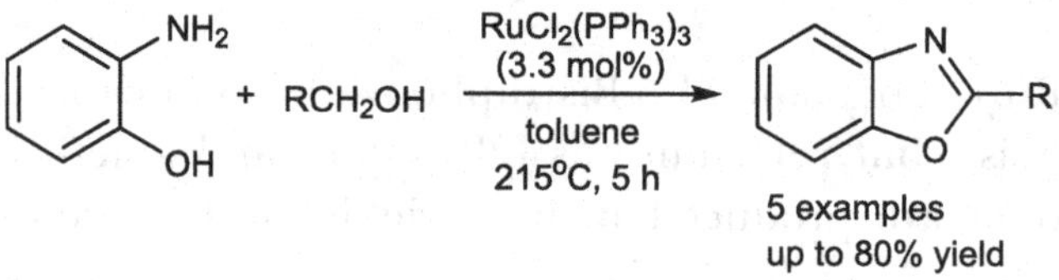

(Eq. 96)

Based on the discovery of the ruthenium(II)-catalyzed azide–alkyne cycloaddition reaction, Fokin and co-workers reported a ruthenium(II)-catalyzed cycloaddition reaction of nitrile oxides (*in situ* generated from hydroximoyl chlorides by treatment with Et_3N) and terminal or internal alkynes, producing 3,5-di- and 3,4,5-trisubstituted isoxazoles with excellent regioselectivity at room temperature (Eq. 97).[105] This method together with the copper(I)-catalyzed process allows regioselective and efficient preparation of all isomers of isoxazoles.

R^1 = alkyl or aryl
R^2 = aryl, CH_2NRR', CR_2OR'
R^3, R^4 = aryl, ester, amide, CR_2OH

9 examples
up to 93% yield

7 examples
up to 99% yield

(Eq. 97)

The proposed mechanism initiates from replacement of the spectator cyclooctadiene ligand from the [Cp*RuCl(COD)] catalyst by an alkyne and nitrile oxide produces the activated complex **21**. Then oxidative coupling of a nitrileoxide and alkyne results in ruthenacycle **22**, and this step controls the regioselectivity of the overall process. Ruthenacycle **22** undergoes reductive elimination giving **23**, and release of the isoxazole product (Scheme 21).

Under CpRuCl(COD) catalysis, reaction between nitrile oxides and electronically deficient 1-halo (choro, bromo, and iodo) alkynes was described by the same group, leading to the formation of 4-haloisoxazoles in high yield and with excellent regioselectivity (Eq. 98).[50]

Benzoxazoles can also be synthesized from direct oxidative condensation of primary amines with *o*-aminophenols under hydrogen transfer catalysis, in which the amines act as the source of the

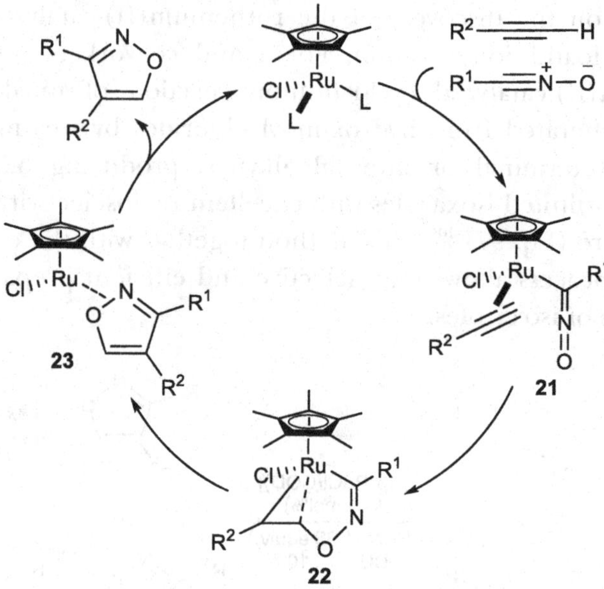

Scheme 21. Proposed mechanism for ruthenium-catalyzed 3,5-di- and 3,4,5-trisubstituted isoxazoles synthesis from nitrile oxides with alkynes

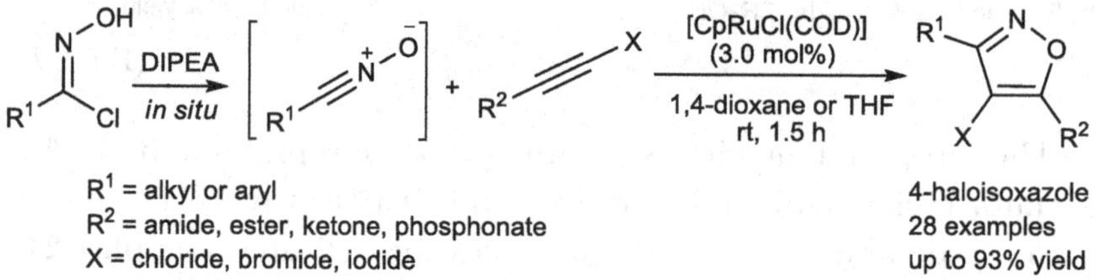

R[1] = alkyl or aryl
R[2] = amide, ester, ketone, phosphonate
X = chloride, bromide, iodide

4-haloisoxazole
28 examples
up to 93% yield

(Eq. 98)

C2–carbon (Eq. 99).[106] The Shvo catalyst $\{[(\eta^5\text{-Ph}_4\text{C}_4\text{CO})]_2\text{H}\}$ $\{\text{Ru}_2(\text{CO})_4(\mu\text{-H})\}$ (1 mol%) shows high reactivity, with dimethoxybenzoquinone (DMBQ) as the hydrogen-accepting terminal oxidant.

Khalafi-Nezhad and Panahi reported an efficient ruthenium-catalyzed acceptorless dehydrogenative strategy for the synthesis of benzoxazoles under heterogeneous conditions (Eq. 100).[107] In this process, $\text{Ru}_2\text{Cl}_4(\text{CO})_6$ as Ru precursor in the presence of phosphine-functionalized magnetic nanoparticles (PFMNP, $\text{Fe}_3\text{O}_4@\text{SiO}_2@$

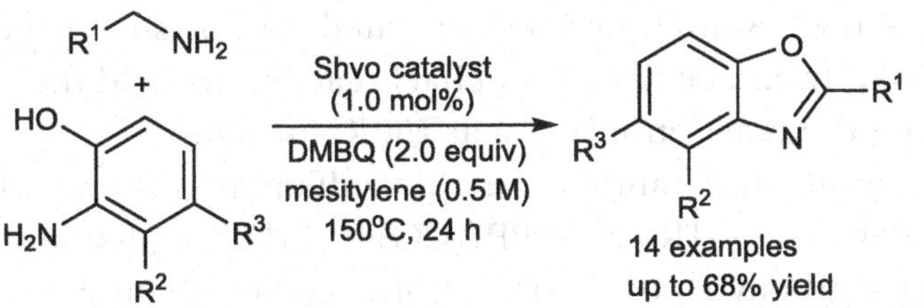

14 examples
up to 68% yield

(Eq. 99)

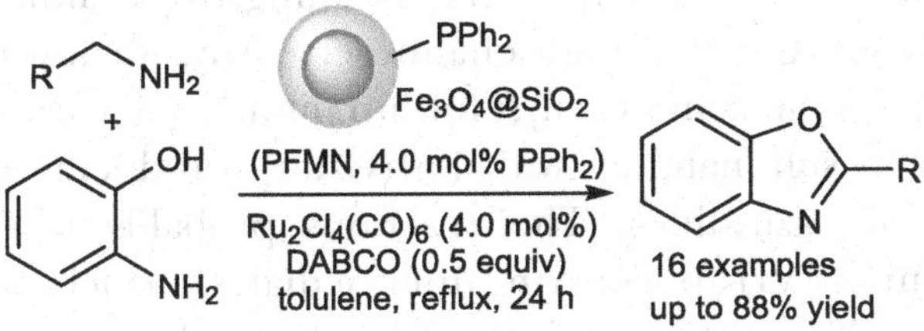

16 examples
up to 88% yield

(Eq. 100)

PPh_2) as a magnetic recyclable phosphorus ligand was found to be an efficient heterogeneous catalytic system for the coupling of primary alcohols with 2-aminophenol. A variety of substrates undergo the designed protocol smoothly to give the corresponding products in moderate to good yields.

5. Ru-Catalyzed Synthesis of Five-Membered *N,S* and *N,O,S*-Heterocycles

In 2014, Ackermann and co-workers reported the oxidative C–H alkenylation of sulfonamides (Eq. 101).[108] After ruthenium(II) catalyst enables oxidative alkenylation of sulfonamides with acrylates at

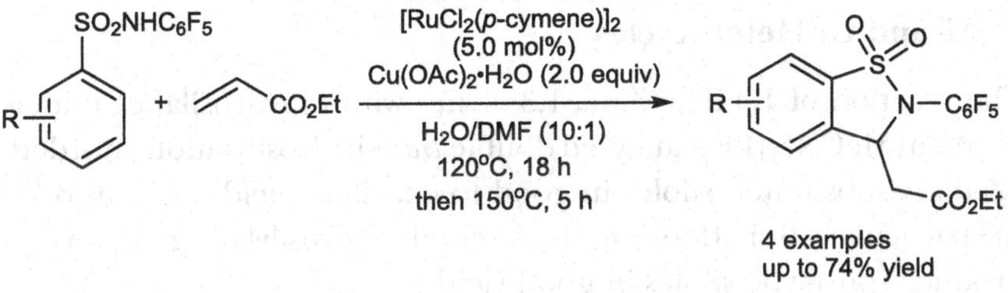

4 examples
up to 74% yield

(Eq. 101)

120°C for 18 h, the resulting ortho-alkenylated products undergo a chemoselective intramolecular aza-Michael reaction to yield the sultams by heating the reaction mixture at 150°C for 5 h.

The intramolecular amidation of sulfamate esters with PhI(OAc)$_2$ catalyzed by [Ru(F$_{20}$-TPP)(CO)][F$_{20}$-TPP = dianion of tetra(pentafluorophenyl) porphyrin] affords cyclic sulfamidates in high yields (Eq. 102).[109] The use of chiral ruthenium porphyrins catalyst allows the formation of the corresponding five-membered cyclic sulfamidates in good to high enantioselectivity. Meanwhile, [Ru(F$_{20}$-TPP)(CO)] is an active catalyst for intramolecular aziridination of unsaturated sulfonamides with PhI(OAc)$_2$, producing corresponding bicyclic aziridines. The reactions probably involve bis(imido)ruthenium(VI) porphyrin intermediates, which converted to cyclic sulfamidates via intramolecular hydrogen atom abstraction.

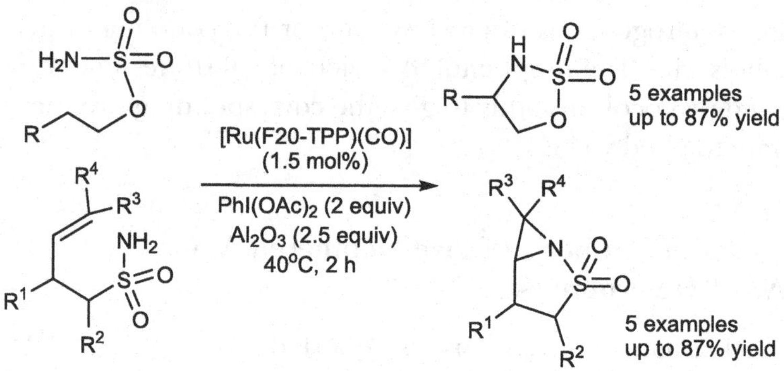

(Eq. 102)

6. Ru-Catalyzed Synthesis of Five-Membered Si- and Ge-Heterocycles

The reaction of 1,4-diarylbuta-1,3-diynes with dihydrosilanes under [Cp*Ru(MeCN)$_3$]PF$_6$-catalyzed double trans-hydrosilylation afforded 2,5-diarylsubstituted siloles in good to excellent yields (Eq. 103).[110] In particular, 9-silafluorene is a good hydrosilylating agent to produce spiro-type siloles in good yield.

$$R^1 \!-\!\!\equiv\!\!-\!\!\equiv\!\!-R^2 \quad \xrightarrow[\substack{ClCH_2CH_2Cl \\ rt, 10\ h}]{\substack{[Cp^*Ru(MeCN)_3]PF_6 \\ (20.0\ mol\%)}}$$

$+$

$H_2Si \overset{Ar^1}{\underset{Ar^2}{<}}$

19 examples
up to 79% yield

(Eq. 103)

Later, this catalytic system was applied for the [4 + 1]-type annulations reactions of 1,3-diynes with dihydrogermanes. 2,5-Disubstituted germoles are formed in good to excellent yields under $[Cp^*Ru(MeCN)_3]PF_6$-catalyzed *trans*-hydrogermylation through a two fold addition process (Eq. 104).[111] Interestingly, the highly efficient double hydrogermylation allows quadruple hydrogermylation of tetraynes to give 2,2'-bigermole (Eq. 105).

$$R^1 \!-\!\!\equiv\!\!-\!\!\equiv\!\!-R^2 \quad \xrightarrow[\substack{ClCH_2CH_2Cl \\ rt, 10\ h}]{\substack{[Cp^*Ru(MeCN)_3]PF_6 \\ (10.0\ mol\%)}}$$

$+$

R_2GeH_2

16 examples
up to 94% yield

(Eq. 104)

Ph_2GeH_2

$+$

$Ph\!-\!(\!\!\equiv\!\!)_4\!-\!Ph \quad \xrightarrow[\substack{ClCH_2CH_2Cl \\ rt, 10\ h}]{\substack{[Cp^*Ru(MeCN)_3]PF_6 \\ (20.0\ mol\%)}}$

56%

(Eq. 105)

Summary

Ruthenium-catalyzed heterocycles synthesis has received an increasing interest in the past few years. In this chapter, the synthesis of five-membered heterocycles under ruthenium catalysis have been summarized. On the basis of the solid and exciting progress, no doubt that more reactive ruthenium catalysts and new innovative transformations will be discovered in this field in the near future.

References

1. (a) J. A. Joule and K. Mills, 2000. *Heterocyclic Chemistry*, 4th Edition. Blackwell, Oxford. (b) T. Eicher and S. Hauptmann, 2003. *The*

Chemistry of Heterocycles, Wiley-VCH, Weinheim. (c) A. R. Katrizky and A. F. Pozharskii, 2000. *Handbook of Heterocyclic Chemistry*, 2nd Edition. Pergamon, Amsterdam.

2. (a) T. Naota, H. Takaya and S.-I. Murahashi, 1998. *Chem. Rev.*, 98, 2599–2660.(b) B. M. Trost, F. D. Toste and A. B. Pinkerton, 2001. *Chem. Rev.*, 101, 2067–2096. (c) G. Maas, 2004. *Chem. Soc. Rev.*, 33, 183–190. (d) V. Cadierno and J. Gimeno, 2009.*Chem. Rev.*, 109, 3512–3560. (e) P. B. Arockiam, C. Bruneau and P. H. Dixneuf, 2012. *Chem. Rev.*, 112, 5879–5918. (f) C. Bruneau and P. H. Dixneuf, 2014. Ruthenium in catalysis. *Top. Organomet. Chem.*, 48, Springer, Heidelberg. (g) L. Ackermann, 2014. *Acc. Chem. Res.*, 47, 281–295.

3. The related examples concerning heterocycles synthesis *via* Ruthenium-catalyzed ring-opening metathesis and ring-closing metathesis are not discussed here, for reviews, see: (a) G. C. Vougioukalakis and R. H. Grubbs, 2010. *Chem. Rev.*, 110, 1746–1787. (b) A. Furstner, 2000. *Angew. Chem. Int. Ed.*, 39, 3012–3043. (c) R. H. Grubbs and T. M. Trnka, 2001. *Acc. Chem. Res.*, 34, 18–29. For selective examples containing five-membered heterocycles, see: (d) D. F. Finnegan and M. L. Snapper, 2011. *J. Org. Chem.*, 76, 3644–3653. (e) E. Ascic, J. F. Jensen and T. E. Nielsen, 2011. *Angew. Chem. Int. Ed.*, 50, 5188–5191. (f) Z.-B. Zhu and M. Shi, 2010. *Org. Lett.*, 12, 4462–4465. (g) H. Wakamatsu, Y. Sato, R. Fujita and M. Mori, 2007. *Adv. Synth. Catal.*, 349, 1231–1246. (h) B. G. Kim and M. L. Snapper, 2006. *J. Am. Chem. Soc.*, 128, 52–53. (i) B. A. Seigal, C. Fajardo and M. L. Snapper, 2005. *J. Am. Chem. Soc.*, 127, 16329–16332.

4. (a) W. D. Jones and W. P. Kosar, 1986. *J. Am. Chem. Soc.*, 108, 5640–5641. (b) G. C. Hsu, W. P. Kosar and W. D. Jones, 1994. *Organomet.*, 13, 385–396.

5. (a) Y. Tsuji, K.-T. Huh, Y. Yokoyama and Y. Watanabe, 1986. *J. Chem. Soc., Chem. Commun.*, 1575–1576. (b) Y. Tsuji, S. Kotachi, K.-T. Huh and Y. Watanabe, 1990. *J. Org. Chem.*, 55, 580–584.

6. (a) Y. Tsuji, K.-T. Huh and Y. Watanabe, 1986. *Tetrahedron Lett.*, 27, 377–380. (b) Y. Tsuji, K.-T. Huh and Y. Watanabe, 1987. *J. Org. Chem.*, 52, 1673–1680. (c) Using $RuCl_3 \cdot xH_2O$/phosphine (phosphine = PPh_3 or Xantphos) catalytic system for this transformation, see: M. Tursky, L. L. R. Lorentz-Petersen, L. B. Olsen and R. Madsen, 2010. *Org. Biomol. Chem.*, 8, 5576–5582. (d) Using dinuclear complex $[Ru(CO)_2(Xantphos)]_2$ as a catalyst for this transformation,

see: M. Zhang, F. Xie, X. Wang, F. Yan, T. Wang, M. Chen and Y. Ding, 2013. *RSC Adv.*, 3, 6022–6029.

7. S. C. Shim, Y. Z. Youn, D. Y. Lee, T. J. Kim, C. S. Cho, S. Uemura and Y. Watanabe, 1996. *Synth. Commun.*, 26, 1349–1353.

8. (a) C. S. Cho, H. K. Lim, S. C. Shim, T.-J. Kim and H.-J. Choi, 1998. *Chem. Commun.*, 995–996. (b) C. S. Cho, J. H. Kim and S. C. Shim, 2000. *Tetrahedron Lett.*, 41, 1811–1814. (c) C. S. Cho, J. H. Kim, T.-J. Kim and S. C. Shim, 2001. *Tetrahedron*, 57, 3321–3329.

9. C. S. Cho, J. H. Kim, H.-J. Choi, T.-J. Kim and S. C. Shim, 2003. *Tetrahedron Lett.*, 44, 2975–2977.

10. M. Peña-López, H. Neumann and M. Beller, 2014. *Chem. Eur. J.*, 20, 1818–1824.

11. A. Penoni and K. M. Nicholas, 2002. *Chem. Commun.*, 484–485.

12. A. Penoni, J. Volkmann and K. M. Nicholas, 2002. *Org. Lett.*, 4, 699–701.

13. R. R. Burton and W. Tam, 2007. *Org. Lett.*, 9, 3287–3290.

14. A. Tenaglia and S. Marc, 2008. *J. Org. Chem.*, 73, 1397–1402.

15. W. G. Shou, J. Li, T. Guo, Z. Lin and G. Jia, 2009. *Organomet.*, 28, 6847–6854.

16. S. Maity and N. Zheng, 2012. *Angew. Chem. Int. Ed.*, 51, 9562–9566.

17. X.-D. Xia, J. Xuan, Q. Wang, L.-Q. Lu, J.-R. Chen and W.-J. Xiao, 2014. *Adv. Synth. Catal.*, 356, 2807–2812.

18. P. Zardi, A. Savoldelli, D. M. Carminati, A. Caselli, F. Ragaini and E. Gallo, 2014. *ACS Catal.*, 4, 3820–3823.

19. A. Porcheddu, M. G. Mura, L. De Luca, M. Pizzetti and M. Taddei, 2012. *Org. Lett.*, 14, 6112–6115.

20. C.-Y. Wu, M. Hu, Y. Liu, R.-J. Song, Y. Lei, B.-X. Tang, R.-J. Li and J.-H. Li, 2012. *Chem. Commun.*, 48, 3197–3199.

21. L. Ackermann and A. V. Lygin, 2012. *Org. Lett.*, 14, 764–767.

22. Z. Zhang, H. Jiang and Y. Huang, 2014. *Org. Lett.*, 16, 5976–5979.

23. Y. Tsuji, Y. Yokoyama, K.-T. Huh and Y. Watanabe, 1987. *Bull. Chem. Soc. Jpn.*, 60, 3456–3458.

24. Y. Watanabe, J. Yamamoto, M. Akazome, T. Kondo and T.-A. Mitsudo, 1995. *J. Org. Chem.*, 60, 8328–8329.

25. B. M. Trost, A. B. Pinkerton and D. Kremzow, 2000. *J. Am. Chem. Soc.*, 122, 12007–12008.

26. (a) F. Monnier, D. Castillo, S. Dérien, L. Toupet and P. H. Dixneuf, 2003. *Angew. Chem. Int. Ed.*, 42, 5474–5477. (b) M. Eckert,

F. Monnier, G. T. Shchetnikov, I. D. Titanyuk, S. N. Osipov, L. Toupet, S. Dérien and P. H. Dixneuf, 2005. *Org. Lett.*, 7, 3471–3473.

27. (a) B. M. Trost and M. Rudd, 2003. *J. Am. Chem. Soc.*, 125, 11516–11517. For Ru-catalyzed diyne cyclization leading to pyrroles and their derivatives, also see: (b) B. M. Trost and M. Rudd, 2002. *J. Am. Chem. Soc.*, 124, 4178–4179. (c) B. M. Trost and M. Rudd, 2005. *J. Am. Chem. Soc.*, 127, 4763–4776. (d) J. A. Varela, González-C. Rodríguez, S. G. Rubín, L. Castedo and C. Saá, 2006. *J. Am. Chem. Soc.*, 128, 9576–9577. (e) C. González-Rodríguez, J. A. Varela, L. Castedo and C. Saá, 2007. *J. Am. Chem. Soc.*, 129, 12916–12917.

28. G.-Y. Li, J. Chen, W.-Y. Yu, W. Hong and C.-M. Che, 2003. *Org. Lett.*, 5, 2153–2156.

29. H.-C. Shen, C.-W. Li and R.-S. Liu, 2004. *Tetrahedron Lett.*, 45, 9245–9247.

30. (a) S. J. Pridmore, P. A. Slatford, A. Daniel, M. K. Whittlesey and J. M. J. Williams, 2007. *Tetrahedron Lett.*, 48, 5115–5120. (b) S. J. Pridmore, P. A. Slatford, J. E. Taylor, M. K. Whittlesey and J. M. J. Williams, 2009. *Tetrahedron*, 65, 8981–8986. (c) Using ruthenium diamine diphosphine complexes as catalyst for dehydrogenative Paal–Knorr pyrrole synthesis, also see: N. D. Schley, G. E. Dobereiner and R. H. Crabtree, 2011. *Organomet.*, 30, 4174–4179.

31. S. K. De, 2008. *Catal. Lett.*, 124, 174–177.

32. B. M. Trost, N. Maulide and R. C. Livingston, 2008. *J. Am. Chem. Soc.*, 130, 16502–16503.

33. Q.-H. Deng, H.-W. Xu, A. W.-H. Yuen, Z.-J. Xu and C.-M. Che, 2008. *Org. Lett.*, 10, 1529–1532.

34. V. Cadierno, J. Gimeno and N. Nebra, 2007. *Chem. Eur. J.*, 13, 9973–9981.

35. N. Thies, M. Gerlach and E. Haak, 2013. *Eur. J. Org. Chem.*, 7354–7365.

36. B. Li, N. Wang, Y. Liang, S. Xu and B. Wang, 2013. *Org. Lett.*, 15, 136–139.

37. L. Wang and L. Ackermann, 2013. *Org. Lett.*, 15, 176–179.

38. M. Zhang, H. Neumann and M. Beller, 2013. *Angew. Chem. Int. Ed.*, 52, 597–601.

39. (a) M. Zhang, X. Fang, H. Neumann and M. Beller, 2013. *J. Am. Chem. Soc.*, 135, 11384–11388. (b) S. Chandrasekhar, V. Patro, L. N. Chavan, R. Chegondi and R. Grée, 2014. *Tetrahedron Lett.*, 55, 5932–5935.

40. D. Srimani, Y. Ben-David and D. Milstein, 2013. *Angew. Chem. Int. Ed.*, 52, 4012–4015.
41. K. Iida, T. Miura, J. Ando and S. Saito, 2013. *Org. Lett.*, 15, 1436–1439.
42 (a) L. Zhang, X. Chen, P. Xue, H. H. Y. Sun, I. D. Williams, K. B. Sharpless, V. V. Fokin and G. Jia, 2005. *J. Am. Chem. Soc.*, 127, 15998–15999. For cycloaddition of azides with internal alkynes, see: (b) M. M. Majireck and S. M. Weinreb, 2006. *J. Org. Chem.*, 71, 8680–8683. For selective applications of RuAAC, see: (c) D. Imperio, T. Pirali, U. Galli, F. Pagliai, L. Cafici, P. L. Canonico, G. Sorba, A. A. Genazzani and G. C. Tron, 2007. *Bioorg. Med. Chem.*, 15, 6748–6757. (d) A. W. Kelly, J. Wei, K. Kesavan, J.-C. Marie, N. Windmon, D. W. Young and L. A. Maucaurelle, 2009. *Org. Lett.*, 11, 2257–2260. (e) M. Chemama, M. Fonvielle, M. Arthur, J. M. Valery and M. Etheve-Quelquejeu, 2009. *Chem. Eur. J.*, 15, 1929–1938. (f) W. S. Horne, C. A. Olsen, J. M. Beierle, A. Montero and M. R Ghadiri, 2009. *Angew. Chem., Int. Ed.*, 48, 4718–4724. (g) H. Nulwala, K. Takizawa, A. Odukale, A. Khan, R. J. Thibault, B. R. Taft, B. H. Lipshutz and C. J. Hawker, 2009. *Macromolecules*, 42, 6068–6074. (h) M. R. Krause, R. Goddard and S. Kubik, 2010. *Chem. Commun.*, 46, 5307–5309. (i) K. Takasu, T. Azuma and Y. Takemoto, 2010. *TetrahedronLett.*, 51, 2737–2740. (j) D. S. Pederson and A. Abell, 2011. *Eur. J. Org. Chem.*, 2399–2411. (k) E. Chardon, G. L. Puleo, G. Dahm, G. Guihard and S. Bellemin-Laponnaz, 2011. *Chem. Commun.*, 47, 5864–5866. (l) Empting, M., O. Avrutina, R. Meusinger, S. Fabritz, M. Reinwarth, M. Biesalski, S. Voigt, G. Buntkowsky and H. Kolmar, 2011. *Angew. Chem., Int. Ed.*, 50, 5207–5211. (m) J. Zhang, J. Kemmink, D. T. S. Rijkers and R. M. J. Liskamp, 2011. *Org. Lett.*, 12, 3438–3441.
43. B. C. Boren, S. Narayan, L. K. Rasmussen, L. Zhang, H. Zhao, Z. Lin, G. Jia and V. V. Fokin, 2008. *J. Am. Chem. Soc.*, 130, 8923–8930.
44. S. Oppilliart, G. Mousseau, L. Zhang, G. Jia, P. Thuéry, B. Rousseaua and J.-C. Cintrat, 2007. *Tetrahedron*, 63, 8094–9098.
45. C.-T. Zhang, X. Zhang and F.-L. Qing, 2008. *Tetrahedron Lett.*, 49, 3927–3930.
46. L. K. Rasmussen, B. C. Boren and V. V. Fokin, 2007. *Org. Lett.*, 9, 5337–5339.
47. J. R. Johansson, P. Lincoln, B. Nordén and N. Kann, 2011. *J. Org. Chem.*, 76, 2355–2359.

48. P. N. Liu, H. X. Siyang, L. Zhang, S. K. S. Tse and G. Jia, 2012. *J. Org. Chem.*, 77, 5844–5849.

49. P. N. Liu, J. Li, F. H. Su, K. D. Ju, L. Zhang, C. Shi, H. H. Y. Sung, I. D. Williams, V. V. Fokin, Z. Lin and G. Jia, 2012. *Organomet.*, 31, 4904–4915.

50. J. S. Oakdale, R. K. Sit and V. V. Fokin, 2014. *Chem. Eur. J.*, 20, 11101–11110.

51. (a) H. Nagashima, H. Wakamatsu and K. Itoh, 1984. *J. Chem. Soc., Chem. Commun.*, 652–653. For ruthenium-catalyzed chlorine atom transfer cyclization, also see: (b) H. Ishibashi, N. Uemura, H. Nakatani, M. Okazaki, T. Sato, N. Nakamura and M. Ikeda, 1993. *J. Org. Chem.*, 58, 2360–2368. For a stereoselective transformation, see: (c) H. Nagashima, N. Ozaki, K. Seki, M. Ishii and K. Itoh, 1989. *J. Org. Chem.*, 54, 4497–4499.

52. H. Nagashima, K.-I. Ara, H. Wakamatsu and K. Itoh, 1985. *J. Chem. Soc., Chem. Commun.*, 518–519.

53. N. Chatani, T. Morimoto, A. Kamitani, Y. Fukumoto and S. Murai, 1999. *J. Organomet. Chem.*, 579, 177–181.

54. T. Morimoto, N. Chatani and S. Murai, 1999. *J. Am. Chem. Soc.*, 121, 1758–1759.

55. (a) D. Berger and W. Imhof, 1999. *Chem. Commun.*, 1457–1458. (b) D. Berger and W. Imhof, 2000. *Tetrahedron*, 56, 2015–2023. (c) A. Göbel and W. Imhof, 2001. *Chem. Commun.*, 593–594. (d) W. Imhof, D. Berger, M. Kötteritzsch, M. Rost and B. Schönecker, 2001. *Adv. Synth. Catal.*, 343, 795–801 (e) D. Dönnecke and W. Imhof, 2003. *Tetrahedron*, 59, 8499–8507.

56. N. Chatani, A. Kamitani and S. Murai, 2002. *J. Org. Chem.*, 67, 7014–7018.

57. N. Armanino and E. M. Carreira, 2013. *J. Am. Chem. Soc.*, 135, 6814–6817.

58. B. Li, Y. Park and S. Chang, 2014. *J. Am. Chem. Soc.*, 136, 1125–1131.

59. Y. Hashimoto, T. Ueyama, T. Fukutani, K. Hirano, T. Satoh and M. Miura, 2011. *Chem. Lett.*, 40, 1165–1166.

60. L. Ackermann, L. Wang, R. Wolfram and A. V. Lygin, 2012. *Org. Lett.*, 14, 728–731.

61. M. K.-W. Choi, W.-Y. Yu and C.-M. Che, 2005. *Org. Lett.*, 7, 1081–1084.

62. W.-W. Chan, T.-L. Kwong and W.-Y. Yu, 2012. *Org. Biomol. Chem.*, 10, 3749–3755.

63. (a) W.-W. Chan, T.-L. Kwong and W.-Y. Yu, 2012. *J. Am. Chem. Soc.*, 134, 7588–7591. Also see: (b) M. Grohmann and G. Maas, 2007. *Tetrahedron*, 63, 12172–12178. (c) W.-W. Chan, M.-H. So, C.-Y. Zhou, Q.-H. Deng and W.-Y Yu, 2008. *Chem. Asian J.*, 3, 1256–1265.

64. T. Kondo, M. Nomura, Y. Ura, K. Wada and T.-A. Mitsudo, 2006. *J. Am. Chem. Soc.*, 128, 14816–14817.

65. S. Inoue, H. Shiota, Y. Fukumoto and N. Chatani, 2009. *J. Am. Chem. Soc.*, 131, 6898–6899.

66. (a) N. Hasegawa, V. Charra, S. Inoue, Y. Fukumoto and N. Chatani, 2011. *J. Am. Chem. Soc.*, 133, 8070–8073. (b) K. Shibata, N. Hasegawa, Y. Fukumoto and N. Chatani, 2012. *ChemCatChem*, 4, 1733–1736. (c) N. Hasegawa, K. Shibata, V. Charra, S. Inoue, Y. Fukumoto and N. Chatani, 2013. *Tetrahedron*, 69, 4466–4472.

67. S. De Sarkar and L. Ackermann, 2014. *Chem. Eur. J.*, 20, 13932–13936.

68. (a) T. Kondo, S. Yang, K.-T. Huh, M. Kobayashi, S. Kotachi and Y. Watanabe, 1991. *Chem. Lett.*, 1275–1278. The method has been applied to polycondensation of 3,3′-diaminobenzidine and 1, 12-dodecanediol to give poly(alkylenebenzimidazole), see: (b) I. Yamaguchi, K. Osakada and T. Yamamoto, 1996. *J. Am. Chem. Soc.*, 118, 1811–1812. (c) I. Yamaguchi, K. Osakada and T. Yamamoto, 1997. *Macromolecules*, 30, 4288–4294. For Ru(0)-catalyzed polyaddition to reaction of α,ω-diynes with 3,3′-diamino-4,4′-dihydroxybiphenyl and 3,3′-diaminobenzidine to provide poly(benzimidazole) and poly(benzoxazole), see: (d) I. Yamaguchi, K. Osakada and T. Yamamoto, 1999. *Polym. Bull.*, 42, 141–147.

69. Y.-R. Lin, X.-T. Zhou, L.-X. Dai and J. Sun, 1997. *J. Org. Chem.*, 62, 1799–1803.

70. C. S. Cho and J. U. Kim, 2008. *Bull. Korean Chem. Soc.*, 29, 1097–1098.

71. A. J. Blacker, M. M. Farah, Hall, M. I. S. P. Marsden, O. Saidi and J. M. J. Williams, 2009. *Org. Lett.*, 11, 2039–2042.

72. H. Dong, R. T. Latka and T. G. Driver, 2011. *Org. Lett.*, 13, 2726–2729.

73. J. Hu, S. Chen, Y. Sun, J. Yang and Y. Rao, 2012. *Org. Lett.*, 14, 5030–5033.

74. M. C. Reddy and M. Jeganmohan, 2014. *Org. Lett.*, 16, 4866–4869.

75. C. Suzuki, K. Morimoto, K. Hirano, T. Satoh and M. Miura, 2014. *Adv. Synth. Catal.*, 356, 1521–1526.

76. J. Li, M. John and L. Ackermann, 2014. *Chem. Eur. J.*, 20, 5403–5408.

77. B. Seiller, C. Bruneau and P. H. Dixneuf, 1994. *J. Chem. Soc., Chem. Commun.*, 493–494.

78. K. Hori, H. Kitagawa, A. Miyoshi, T. Ohta and I. Furukawa, 1998. *Chem. Lett.*, 1083–1084.

79. T. Ohta, Y. Kataoka, A. Miyoshi, Y. Oe, I. Furukawa and Y. Ito, 2007. *J. Organomet. Chem.*, 692, 671–677.

80. B. M. Trost and A. B. Pinkerton, 1999. *J. Am. Chem. Soc.*, 121, 10842–10843.

81. (a) Y. Yamamoto, K. Hattori and H. Nishiyama, 2006. *J. Am. Chem. Soc.*, 128, 8336–8340. For related cycloaddition reaction, also see: (b) Y. Yamamoto, K. Kinpara, R. Ogawa, H. Nishiyama and K. Itoh, 2006. *Chem. Eur. J.*, 12, 5618–5631.

82. S. J. Pridmore, P. A. Slatford and J. M. J. Williams, 2007. *Tetrahedron Lett.*, 48, 5111–5114.

83. V. Cadierno, J. Gimeno and N. Nebra, 2007. *Adv. Synth. Catal.*, 349, 382–394.

84. Y. Yada, Y. Miyake and Y. Nishibayashi, 2008. *Organomet.*, 27, 3614–3617.

85. D.-H. Lee, K.-H. Kwon and C. S. Yi, 2012. *J. Am. Chem. Soc.*, 134, 7325–7328.

86. J. D. Dooley, S. R. Chidipudi and H. W. Lam, 2013. *J. Am. Chem. Soc.*, 135, 10829–10836.

87. Z. Zhou, G. Liu, Y. Shen and X. Lu, 2014. *Org. Chem. Front.*, 1, 1161–1165.

88. (a) W.-H. Cheung, S.-L. Zheng, W.-Y. Yu, G.-C. Zhou and C.-M. Che, 2003. *Org. Lett.*, 5, 2535–2538. (b) S.-L. Zheng, W.-Y. Yu, M.-X. Xu and C.-M. Che, 2003. *Tetrahedron Lett.*, 44, 1445–1447.

89. A. R. Reddy, C.-Y. Zhou, Z. Guo, J. Wei and C.-M. Che, 2014. *Angew. Chem. Int. Ed.*, 53, 14175–14180.

90. T. Kondo, K. Kodoi, T.-a. Mitsudo and Y. Watanabe, 1994. *J. Chem. Soc., Chem. Commun.*, 755–756.

91. B. M. Trost, T. J. J. Muller and J. Martinez, 1995. *J. Am. Chem. Soc.*, 117, 1888–1899.

92. N. Chatani, T. Morimoto, Y. Fukumoto and S. Murai, 1998. *J. Am. Chem. Soc.*, 120, 5335–5336.

93. N. Chatani, M. Tobisu, T. Asaumi, Y. Fukumoto and S. Murai, 1999. *J. Am. Chem. Soc.*, 121, 7160–7161.

94. B. M. Trost and Y. H. Rhee, 1999. *J. Am. Chem. Soc.*, 121, 11680–11683.

95. E. Yoneda, T. Kaneko, S.-W. Zhang, K. Onitsuka and S. Takahashi, 2000. *Org. Lett.*, 2, 441–443.

96. C.-M. Che, J.-S. Huang, F.-W. Lee, Y. Li, T.-S. Lai, H.-L. Kwong, P.-F. Teng, W.-S. Lee, W.-C. Lo, S.-M. Peng and Z.-Y Zhou, 2001. *J. Am. Chem. Soc.*, 123, 4119–4129.

97. J. Zhao and J. F. Hartwig, 2005. *Organomet.*, 24, 2441–2446.

98. B. M. Trost and A. McClory, 2006. *Org. Lett.*, 8, 3627–3629.

99. J. Zhang, E. Balaraman, G. Leitus and D. Milstein, 2011. *Organomet.*, 30, 5716–5724.

100. T. Ueyama, S. Mochida, T. Fukutani, K. Hirano, T. Satoh and M. Miura, 2011. *Org. Lett.*, 13, 706–708.

101. L. Ackermann and J. Pospech, 2011. *Org. Lett.*, 13, 4153–4155.

102. L. Chen, H. Li, F. Yu and L. Wang, 2014. *Chem. Comm.*, 50, 14866–14869.

103. E. L. McInturff, J. Mowat, A. R. Waldeck and M. J. Krische, 2013. *J. Am. Chem. Soc.*, 135, 17230–17235.

104. K.-T. Huh and S. C. Shim, 1993. *Bull. Korean Chem. Soc.*, 14, 449–452.

105. S. Grecian and V. V. Fokin, 2008. *Angew. Chem. Int. Ed.*, 47, 8285–8287.

106. A. J. Blacker, M. M. Farah, S. P. Marsden, O. Saidi and J. M. J. Williams, 2009. *Tetrahedron Lett.*, 50, 6106–6109.

107. A. Khalafi-Nezhad and F. Panahi, 2014. *ACS Catal.*, 4, 1686–1692.

108. W. Ma, R. Mei, G. Tenti and L. Ackermann, 2014. *Chem. Eur. J.*, 20, 15248–15251.

109. (a) J.-L. Liang, S.-X. Yuan, J.-S. Huang, W.-Y. Yu and C.-M. Che, 2002. *Angew. Chem. Int. Ed.*, 41, 3465–3468. (b) J.-L. Liang, S.-X. Yuan, J.-S. Huang and C.-M. Che, 2014. *J. Org. Chem.*, 69, 3610–3619. Using ruthenium(II)-pybox(pybox: pyridine bisoxaline) complexes for asymmetric C–H amination to the construction of five-membered cyclic sulfamidates, see: (c) E. Milczek, N. Boudet and S. Blakey, 2008. *Angew. Chem. Int. Ed.*, 47, 6825–6828.

110. T. Matsuda, S. Kadowaki and M. Murakami, 2007. *Chem. Commun.*, 2627–2629.

111. T. Matsuda, S. Kadowaki, Y. Yamaguchi and M. Murakami, 2010. *Org. Lett.*, 12, 1056–1058.

CHAPTER 2

Ruthenium Catalysis in the Synthesis of Six-Membered Heterocycles

Min Zhang

School of Chemistry & Chemical Engineering,
South China University of Technology, Wushan Rd-381,
Guangzhou 510641, People's Republic of China

Owing to the cost-effectiveness and versatility of ruthenium in catalysis, the utilization of ruthenium for developing atom and step-economic methodologies, allowing for the synthesis of six-membered heterocycles, is of particular importance in synthetic chemistry because of the important value of such compounds employed in discovering biologically and pharmacologically active compositions, the preparation of novel materials with specific functions, etc. This chapter highlights the recent 15 years' advances on ruthenium-catalyzed synthesis of six-membered heterocycles, with particular focus on the related approaches and mechanistic basis, which includes the synthesis of *N*-heterocycles, *O*-heterocycles, *N*,*O*-heterocycles, and other type of heteroatom-heterocycles.

1. Introduction

Among various transition metals, the utilization of ruthenium has recently attracted much attention owing to its relative cost-effectiveness and the unique properties in catalysis. Moreover, the six-membered heterocycles are frequently found in numerous natural and synthetic compounds that exhibit a broad spectrum of biological and pharmacological activities. Moreover, six-membered heterocycles have been extensively employed as target-specific intermediates for various synthetic purposes including the preparation of functional materials. Hence, the development of ruthenium-catalyzed atom and step-economic methodologies for the synthesis of six-membered heterocycles constitutes an interesting topic and is of significant importance in the organic chemistry. This chapter highlights the past 15 years' advances on such a topic with particular focus on the skills of constructing heterocycles and the mechanistic basis as well.

2. Ruthenium-Catalyzed Synthesis of Six-Membered aza-Heterocycles

2.1 Synthesis of pyridine derivatives

2.1.1 Synthesis of pyridine derivatives via [2 + 2 + 2] cycloaddition

Among various methods for the synthesis of pyridine derivatives, the transition metal-catalyzed [2 + 2 + 2] cycloaddition of two alkyne units with one nitrile provides an interesting approach to access such a class of compounds in atom-economical fashion. The key issue of such a transformation is to suppress the competing side reactions, such as dimerization and trimerization of the diynes. To date, the catalysis by Co, Ru, Rh, Ni, Ti, and Fe species have been successfully employed for the synthesis of substituted pyridines. Herein, two representative examples are demonstrated here *via* ruthenium catalysis.

In 2006, Yamamoto and the co-workers have reported a chemo- and regio-selective synthesis of bicyclic pyridines.[1] By using catalytic amount of [Cp*RuCl(cod)] (Cp* = pentamethylcyclopentadienyl, cod = 1,5-cyclooctadiene), a variety of 1,6-diynes could be efficiently converted in combination with nitriles bearing a coordinating group, such as dicyanides or a-halonitriles, into various desired products at ambient temperature. Upon a careful screening of the nitrile components, it was revealed that a carbon–carbon (C–C) triple bond or heteroatom substituents, such as methoxy and methylthio groups, acting as the coordinating groups, whereas C=C or C=O double bonds and amino groups were ineffective for the transformation. The result suggests that coordinating groups with multiple π-bonds or lone electron-pairs are essential for the nitrile coupling components (Scheme 1).

Later, a similar contribution was reported by the Wan group.[2] By combining a water-soluble phosphine ligand tppts (for structure see Eq. 1) with [Cp*RuCl(cod)], the cycloaddition reaction of diynes and nitriles could be performed in pure water. Interestingly, both hydrophobic and hydrophilic diynes were suitable for the transformation, furnishing the bicyclic pyridine products in moderate to high yields (Eq. 1).

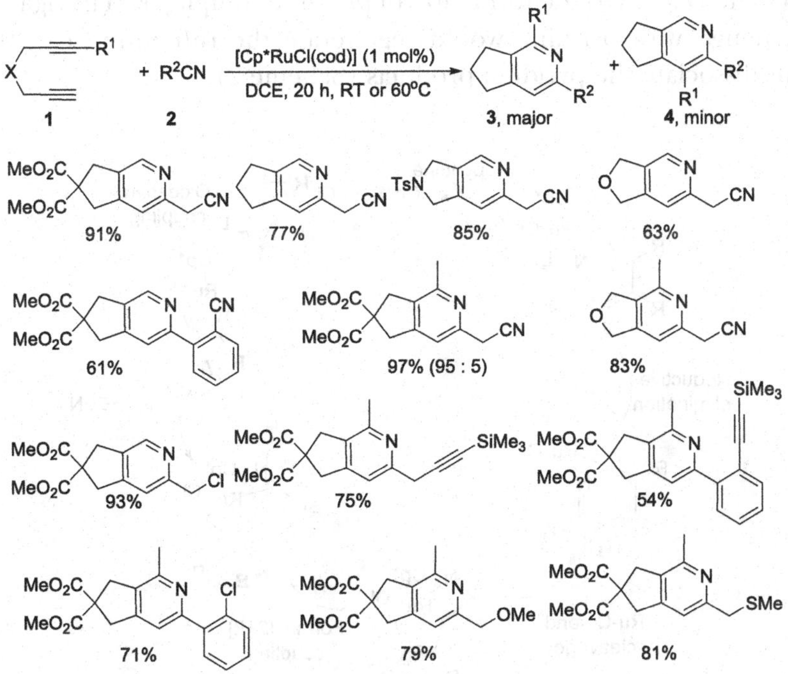

Scheme 1. Synthesis of bicyclic pyridines

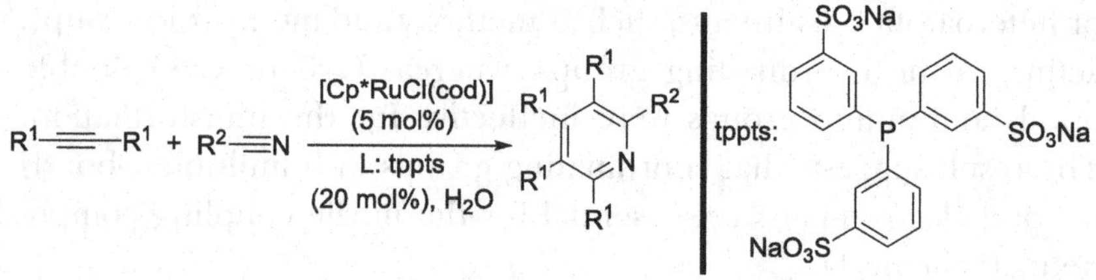

$$R^1 \!\!=\!\! -R^1 + R^2 \!\!-\!\!\equiv\!\! N \xrightarrow[\substack{\text{L: tppts} \\ \text{(20 mol\%), H}_2\text{O}}]{\substack{\text{[Cp*RuCl(cod)]} \\ \text{(5 mol\%)}}}$$

tppts:

(Eq. 1)

The possible mechanism for the formation of pyridines is as follows: the ligand exchange with alkynes and subsequent oxidative coupling of diyne leads to ruthenacyclopentatriene complex **A** with mixed Fischer- and Schrock-type behavior. Then, the coordination of the nitrile to the ruthenium center and further formal [2 + 2] cycloaddition would lead to azaruthenatricycle **C**. Finally, the Ru–C bond cleavage in **C** followed by the reductive elimination of intermediate **D** leads to the formation of pyridine complex **E**. The ligand exchange with pyridine would regenerate the ruthenium catalyst and dissociate the pyridine products (Scheme 2).

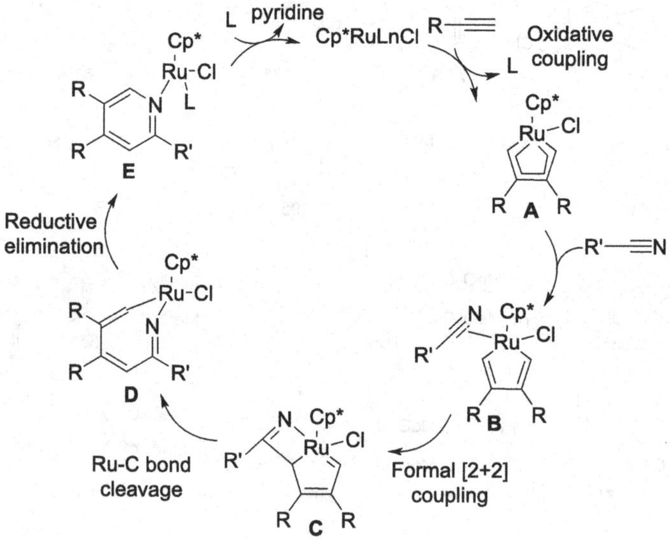

Scheme 2. Proposed mechanism for the [2 + 2 + 2] cycloaddition

2.1.2 Synthesis of pyridine derivatives via [4 + 2] cycloaddition

Very recently, Yu and co-workers reported a ruthenium-catalyzed dehydrative [4 + 2] cycloaddition of enamides and alkynes, offering economic method to access highly substituted pyridines under mild conditions.[3] Moreover, the synthesis features of broad substrate scope, high efficiency, good functional group tolerance, and excellent regioselectivities. Through density functional theory (DFT) calculations, the reaction initiates with a concerted metalation of the enamide *via* proton abstraction by the acetate ligand of the Ru catalyst, generating a six-membered ruthenacycle intermediate **A**. Then alkyne inserts into the ruthenium–carbon bond which gives an eight-membered ruthenacycle **B**. The carbonyl group then inserts into the ruthenium–carbon bond followed by protonation and further dehydrative aromatization to produce the final pyridine product. The high regioselectivity of the dehydrative [4 + 2] cycloaddition is determined by the alkyne insertion step to avoid the steric repulsion of aryl group with the enamide moiety in the six-membered ruthenacycle and to have a good conjugation between the aryl group and the C–C triple bond of the used alkyne. Hence, the aryl group of the alkynes is in the β-position of the formed pyridines (Scheme 3).

2.1.3 Synthesis of pyridines via ruthenium-catalyzed cycloisomerization

Mohammad Movassaghi and co-workers reported a two-step procedure for the synthesis of substituted pyridines from N-vinyl and N-aryl amides in 2006.[4] Through a single-step conversion of amides into the C-silyl alkynyl imines, which were directly transformed into the pyridine-based heterocycles *via* Ru-catalyzed protodesilylation and cycloisomerization. And the deuterium-labelling experiments revealed that the reaction proceeds *via* the straightforward conversion of the silyl alkynylimine to the C-silyl metal vinylidene followed by protodesilylation and cycloisomerization produce the pyridine products (Scheme 4).

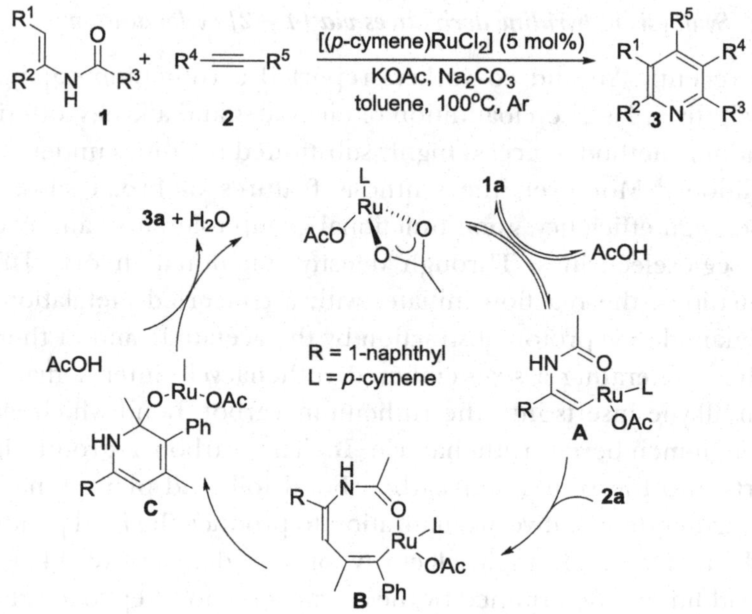

Scheme 3. Mechanism for synthesis of pyridines *via* [4 + 2] cycloaddition

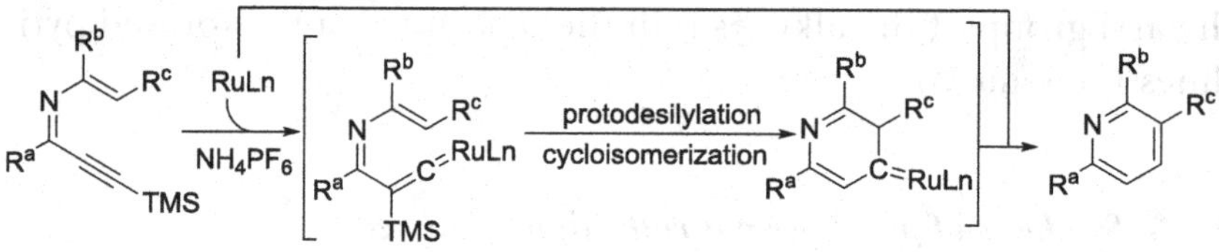

Scheme 4. Ruthenium-catalyzed synthesis pyridines from C-silyl alkynyl imines

In the presence of catalytic amount of $[CpRu(CH_3CN)_3]PF_6$, the Trost group utilized the conventional cycloisomerization of primary and secondary propargyl diynols to afford unsaturated ketones and aldehydes, the introduction of hydroxyl amines would *in situ* give 1-azatriene intermediates, which subsequently undergoes an intramolecular 6π-electrocyclization and dehydration to provide substituted pyridines with excellent regio-selectivity (Scheme 5).[5]

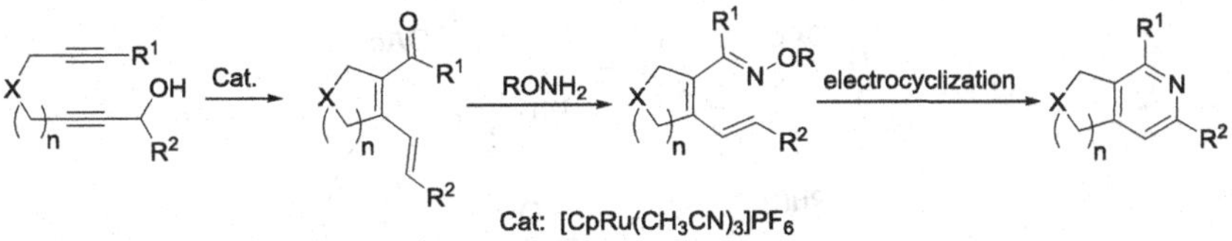

Scheme 5. Cycloisomerization and 6π-electrocyclization to access pyridines

2.1.4 Other representative method for the synthesis of pyridines

A novel ruthenium-catalyzed cyclization of ketoxime carboxylates with N,N-dimethylformamide (DMF) for the synthesis of tetrasubstituted symmetrical pyridines has been developed. A methyl carbon on DMF performed as a source of a one carbon synthon. And $NaHSO_3$ plays a crucial role in the transformation (Eq. 2).[6]

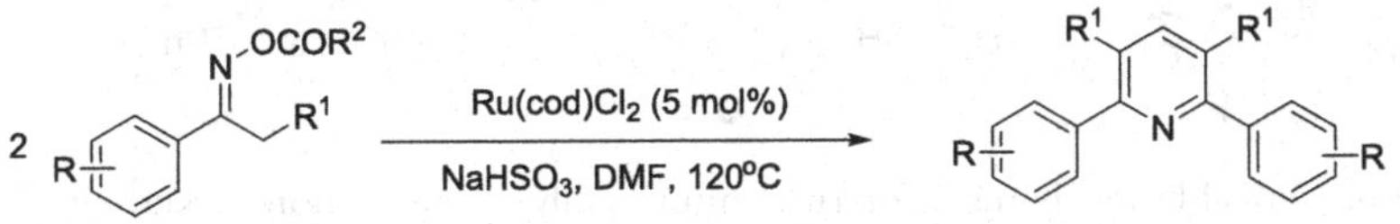

$$(Eq.\ 2)$$

The pathway for the formation of product is depicted as follows: The oxidation of DMF by Ru(II) gives an iminium species **A** and Ru(0). Then, oxidative addition of ketoxime acetate to Ru(0) generates an imino-Ru(II)complex **B**, which leads to form anenamino-Ru(II) complex **C** *via* tautomerization. Then, nucleophilic addition of **C** to iminium **A** produces an imine intermediate **D**. The condensation of **D** with a second ketoxime acetate results in intermediate **E**. Nucleophilic substitution of **E** by $NaHSO_3$ followed by intramolecular cyclization **F** gives a dihydropyridine intermediate **G**. Finally, the Ru-catalyzed oxidative aromatization of **G** in the presence O_2 produces the pyridine product. Alternatively, nucleophilic substitution by the $NaHSO_3$ might occur prior to the condensation step (Scheme 6).

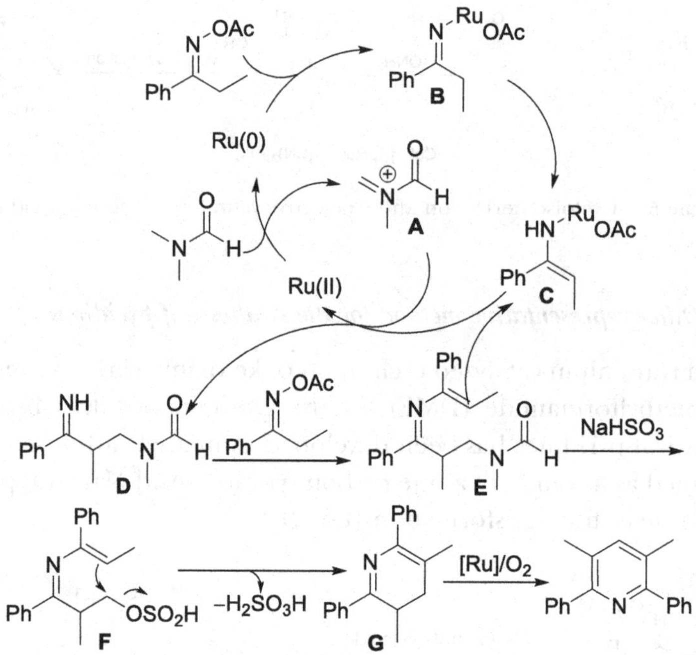

Scheme 6. Possible mechanism for ruthenium-catalyzed cyclization of ketoxime carboxylates with DMF

2.2 Ruthenium-catalyzed synthesis of piperidine derivatives

Through a key ruthenium-catalyzed oxidative cyclization of the easily available allylic alcohols and propargylic amines to form a key ruthenacyclic intermediate, followed by β-hydride elimination to form the unsaturated ketone, which leads to the formation of highly functionalized piperidine derivatives *via* an intramolecular dehydrative cyclization step with the generation of water or alcohol as the by-product.[7] This new and straightforward synthetic method was reported by Sylvie Derien and co-workers in 2012. Noteworthy, owing to the obtained products possess multiple reactive positions, which has the potential for further elaboration of complex molecules including various alkaloids through C–C coupling, hydrogenation and the nucleophilic displacement of the alkoxy group on the hemiaminal ethers (Scheme 7).

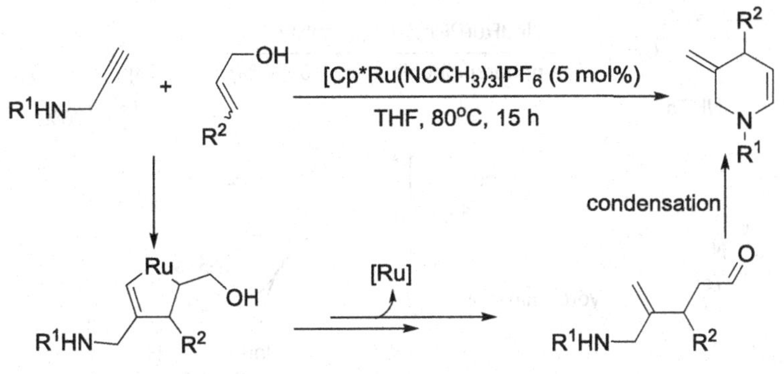

Scheme 7. Direct access to piperidine derivatives

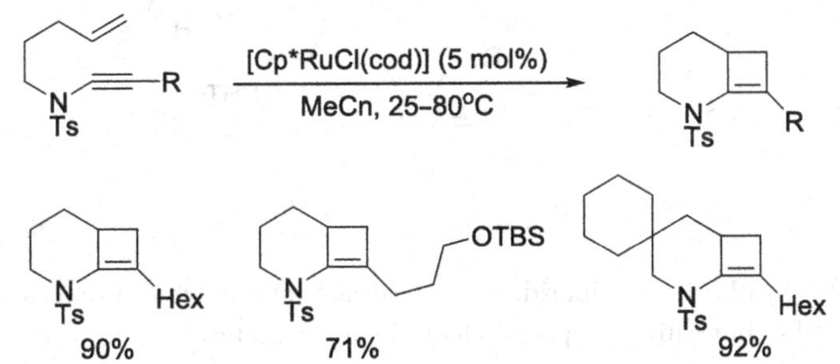

Scheme 8. Synthesis of 6,4-fused cyclobutenes from 1,7-enynamides

By using [Cp*RuCl(cod)] as a catalyst under mild conditions, the Anderson group reported the cycloaddition of 1,7-enynamides to afford 6,4-fused cyclobutene products with excellent chemoselectivity (Scheme 8). Noteworthy, it is well established that the ynamides bearing either electron-deficient or strained-alkene groups are essential for the formal [2 + 2] cycloadditions, whereas in this intramolecular transformation high efficiency is also observed with unstrained, electron-neutral alkenes. The synthesis under mild reaction conditions is believed due to a beneficial Thorpe–Ingold effect.[8]

In 2008, Trost and co-workers employed substituted nitrogenized tether with a propargyl alcohol for the direct synthesis of piperidines through ruthenium-catalyzed redox isomerization of reactant to enone intermediate followed by intramolecular hydroamination to the alkene unit. For such transformation, the acid co-catalyst

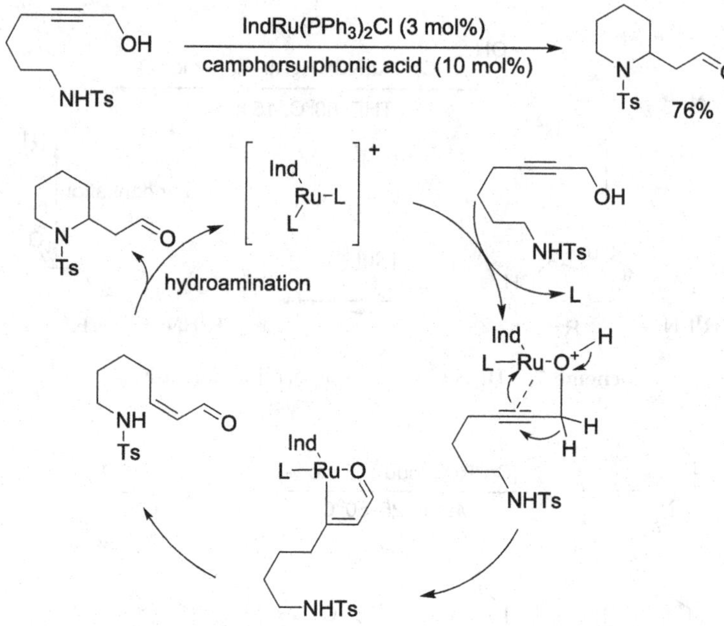

Scheme 9. Synthesis of piperidines *via* ruthenium-catalyzed isomerization of nitrogenated tether with a propargyl alcohol and cyclization

(camphor sulphonic acid) was crucial to obtain a full conversion for substrates to the piperidine products (Scheme 9).[9]

Interestingly, shifting the propargyl alcohol position through the carbon backbone of the substrates results in interesting variations. Upon a simple one-pot, two-stage operation (addition of methanolic potassium carbonate to the mixture), the internal propargyl alcohol furnishes the tetrahydropiperidone in good yield (Eq. 3). In contrast, if the cyclization step is carried out under acidic conditions, the analogous N-Boc-piperidine ketal was obtained in 65% overall yield (Eq. 4).[9]

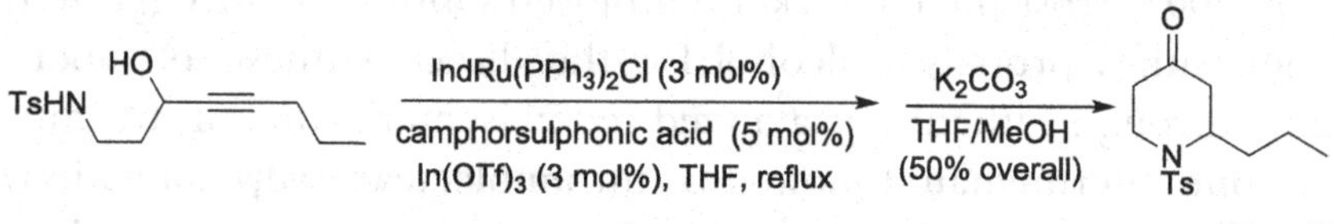

$$(\text{Eq. 3})$$

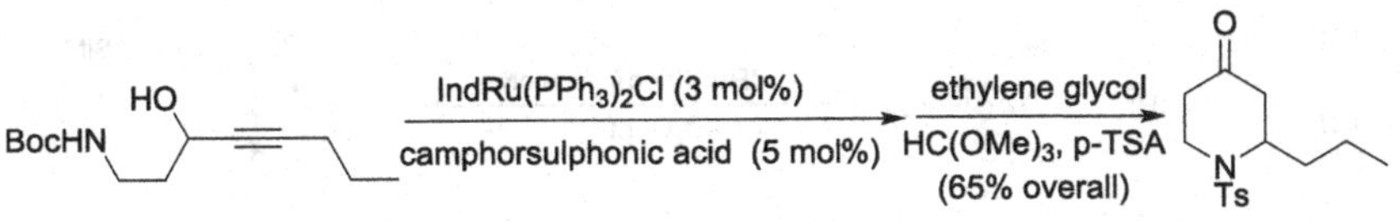

(Eq. 4)

Through a ruthenium-catalyzed intramolecular allylic dearomatization reaction of indole derivatives, an efficient synthesis of spiroindolenine derivatives was developed by the You group in 2013 under mild condition. In addition, the method has the advantages of using cheap and accessible catalyst, wide substrate scope, operational simplicity, and insensitivity to water (Scheme 10).[10,11]

With the use of [Cp*RuCl(cod)] as an efficient catalyst, Dixneuf and co-workers reported the reaction of 1,7-enynes with diazo compounds for the synthesis of piperidine core containing bicyclicamino acid derivatives. The catalytic transformation proceeds *via* tandem carbene addition/cyclopropanation sequences. The cyclization tolerates with various *N*-protecting groups, while the presence of hydrogen on the nitrogen atom gives no cyclization. High Z-selectivity for the newly formed alkenyl chain is observed. The synthesis has the advantages of easily accessible substrates, mild reaction conditions and good product yields upon isolation (Scheme 11).[12,13]

Through mechanistic investigations, the catalytic transformation of 1,7-enynes with diazoalkane in the presence of catalyst precursor Cp*RuCl (cod) initially generates the catalytic species Cp*RuCl(=CHY) **A**, which first reacts with the terminal C–C triple bond to give the intermediate **B** *via* [2 + 2] cycloaddition, the subsequent sigma-metathesis would lead to the ruthenium vinyl carbene **C**.

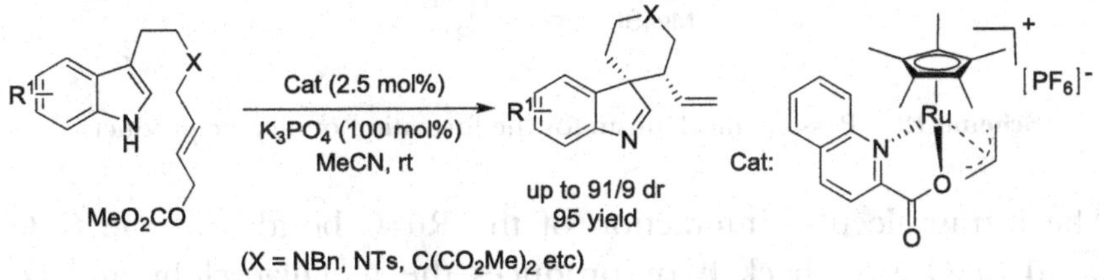

Scheme 10. Intramolecular allylic dearomatization reaction of indole derivatives

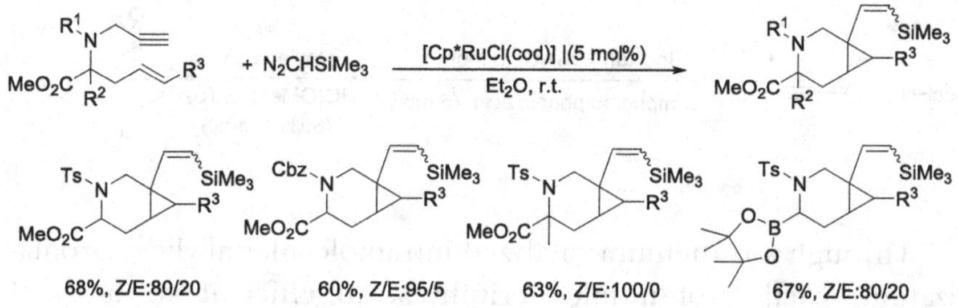

Scheme 11. Synthesis of piperidine containing bicyclicamino acids from 1,7-enynes and trimethylsilydiazomethane

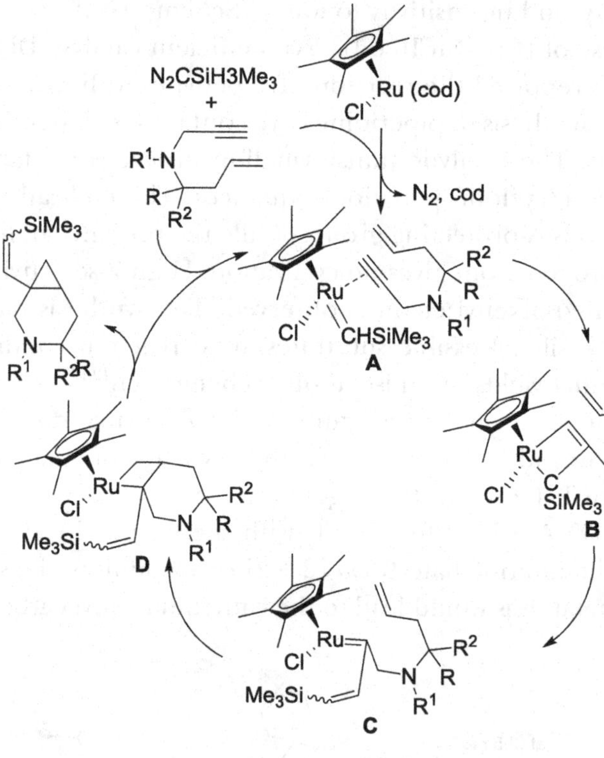

Scheme 12. Possible mechanism for the formation of bicyclic products

The intramolecular interaction of the Ru=C bond with the C=C bond of **C** gives back **B** or produces the metallacyclobutane **D**, which is subjected to reductive elimination to give the bicyclic product and regenerate the catalytic species (Scheme 12).[12,13]

2.3 Ruthenium-catalyzed synthesis of quinolines

2.3.1 *Synthesis of quinolines via dehydrogenative coupling reactions*

In the presence of acatalytic amount of $RuCl_2(PPh_3)_3$ by using 1-dodecene as a sacrificial hydrogen acceptor, a synthetic approach for the synthesis of quinolines was established in 2003 by the Cho group. 2-Aminobenzyl alcohol is oxidatively condensed with secondary alcohols to afford the desired products in moderate to good isolated yields and the method is applicable to a wide range of alcohols (Eq. 5).[14]

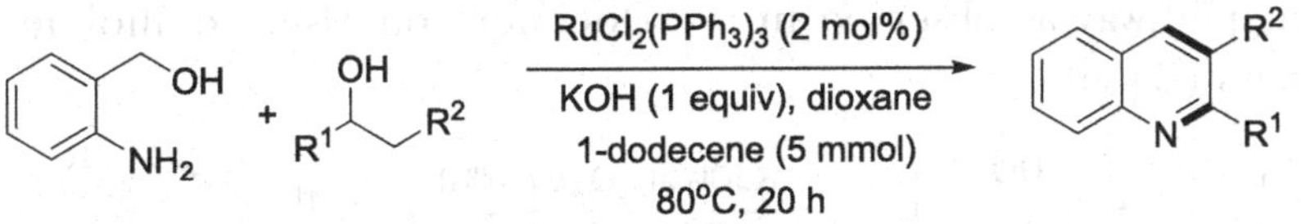

$$\text{(Eq. 5)}$$

The possible pathway for the formation quinolines is depicted in Scheme 13. The initial oxidations of both substrates leads to the carbonyl intermediates **1′** and **2′**, then the cross aldol condensation under KOH and transfer hydrogenation of the C–C double bond afford α,β-unsaturated ketone **A** and ketone **B**, respectively. Finally, the cyclodehydration of **B** gives dihydroquinoline **C** the subsequent dehydrogenation of **C** in the presence of sacrificial reagent 1-dodecene would generate the quinoline product. Alternatively, the direct intramolecular condensation of **A** to the quinolines also could not be ruled out.[14]

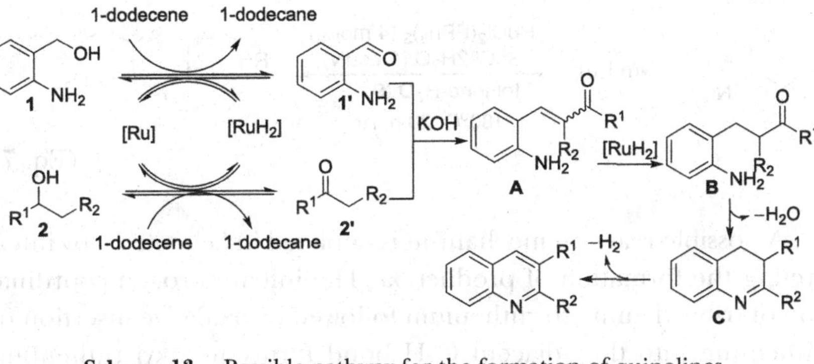

Scheme 13. Possible pathway for the formation of quinolines

By introducing with catalytic amounts of $RuCl_3 \cdot xH_2O$, PBu_3 and $MgBr_2 \cdot OEt_2$, Madsen and co-workers developed a straightforward synthesis of substituted quinolines from anilines and 1,3-diols. The reaction proceeds without need for any stoichiometric additives and generation of water and dihydrogen as by-products. Anilines containing methyl, methoxy, and chloro substituents as well as naphthylamines were compatible for the transformation. The reaction mechanism is believed to involve dehydrogenation of the 1,3-diol to the 3-hydroxyaldehyde which eliminates water to the corresponding a,β-unsaturated aldehyde, which then reacts with anilines through a similar pathway as observed in the Doebner–von Miller quinoline synthesis (Eq. 6).[15]

(Eq. 6)

2.3.2 Synthesis of quinolines via transfer hydrogenative coupling reactions

In the presence of a catalytic amount of a ruthenium catalyst together with $SnCl_2 \cdot 2H_2O$, the early example on synthesis of substituted quinolines from nitroarenes and trialkylamines was reported by the Cho group in 2002. The synthesis proceeded at high temperature (180°C) in an aqueous medium (toluene–H_2O) and afforded the products in moderate to good yields (Eq. 7).[16]

(Eq. 7)

A possible reaction mechanism is depicted in Scheme 14 by interpreting the formation of product 3a. The initial nitrogen coordination of tributylamine to ruthenium followed by oxidative insertion of ruthenium into the adjacent C–H bond forms an alkyl ruthenium

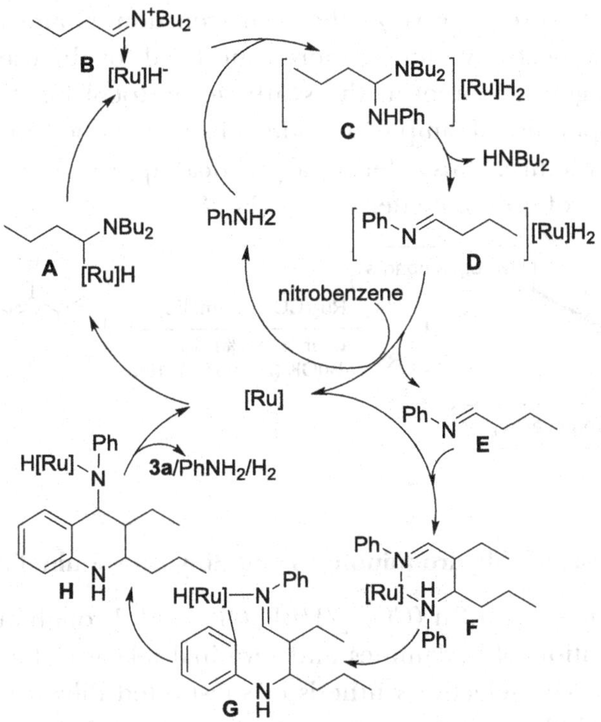

Scheme 14. Proposed reaction mechanism

intermediate **A**, which equilibrates with an iminium ion complex **B**. The nucleophilic addition of aniline to **B** followed by thermodynamically favorable dissociation of the aminal dihydridoruthenium complex **C** generates imine **D** and *N*-butylaniline. Noteworthy, the transfer hydrogenation of nitrobenzeneto aniline might be induced by cooperative actions of $SnCl_2 \cdot 2H_2O$ in an aqueous medium and dihydridoruthenium. Subsequently, the known Schiff-based imerization and cyclization would form intermediates **F** and **H**, respectively. Finally, quinoline **3a** is afforded by reductive elimination, deamination, and dehydrogenative aromatization, along with regeneration of ruthenium catalytic species.[16]

By employing the transfered hydrogenative coupling strategy, the Zhang group has provided a ruthenium-catalyzed straight forward synthesis of quinolines from a variety of α-2-nitroaryl alcohols and alcohols. In such as a synthetic protocol, two alcohol units and the nitro-group serve as the hydrogen donors and hydrogen acceptor, respectively. Hence, there is no need for the use external reducing agents. Moreover, the synthetic protocol has the advantages of operational simplicity, wide substrate scope, tunability of product structure, thus offering a practical approach for versatile preparation of quinoline derivatives (Eq. 8).[17]

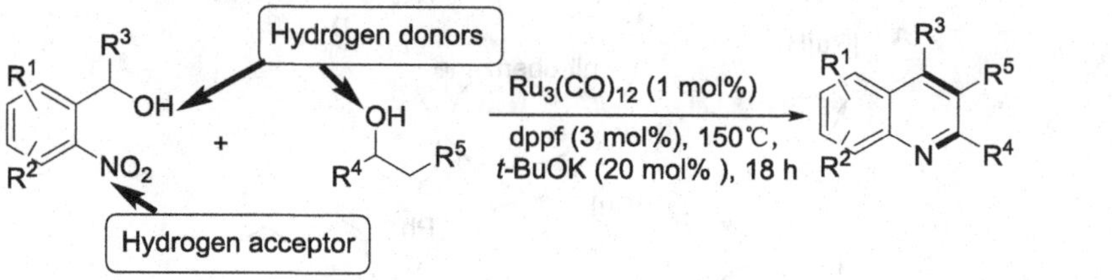

$$ \text{(Eq. 8)} $$

2.4 Synthesis of dihydroquinolines and dihydroisoquinolines

By using the catalyst $Ru_3(CO)_{12}/HBF_4.OEt_2$ and through ortho-C–H bond activation of arylamines and terminal alkynes, the Yi group achieved a regioselective synthesis of substituted dihydroquinoline and related derivatives in 2005. The isotope effect experiment revealed that the ortho-C–H bond activation is a rate-determining step, whereas the deuterium labeling showed that the alkyne C–H bond activation step is reversible (Eq. 9).[18]

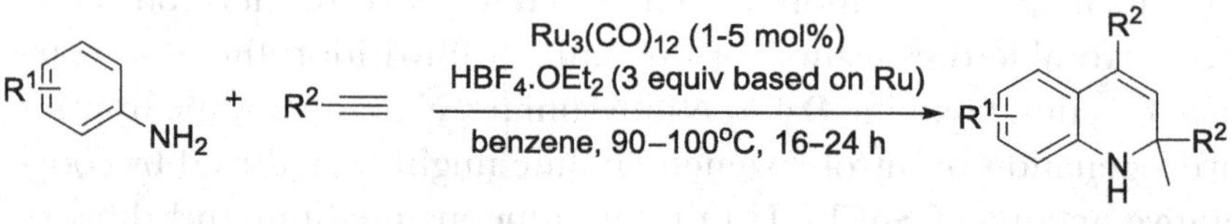

$$ \text{(Eq. 9)} $$

The mechanistic investigations support that cationic ruthenium acetylide 10 is a key reaction intermediate, which would lead to the cationic enaminyl species B in the presence of aniline. The subsequent ortho-arene C–H bond activation and the reductive elimination of the vinyl group of C would form the ortho-metalated species D. The second alkyne insertion to the Ru–C bond and the regioselective migratory insertion to the C–C double bond lead to the cationic

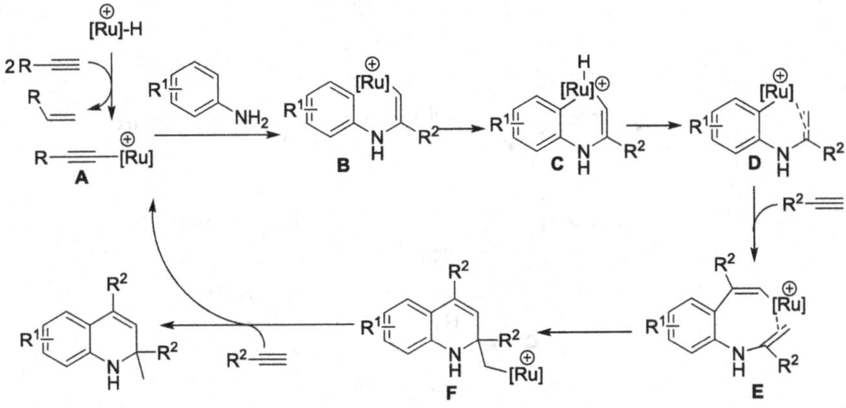

Scheme 15. Proposed pathway for the formation dihydroquinoline derivatives

alkyl species **F**. Due to lack of any α-hydrogens in complex **E**, either the oxidative addition/reductive elimination or the σ-bond metathesis of the terminal alkyne must be involved for the formation of the terminal product and the regeneration of the active acetylide species **A** (Scheme 15).[18]

The Saa group developed a ruthenium-catalyzed cycloisomerizations of aromatic bis-homopropargylic amines/amides to 1,2-dihydroisoquinolines in 2011. The synthesis underwent *via* regioselective 6-endo cyclizations from key Ru vinylidene intermediates. The presence of an amine/ammonium base–acid pair accelerates the cyclization and facilitates the catalytic turnover (Eq. 10). Notably, the synthetic protocol could also be employed for the synthesis of 1,2-dihydroisoquinolines while using aromatic bis-homopropargylic amide substrate (Eq. 11).[19]

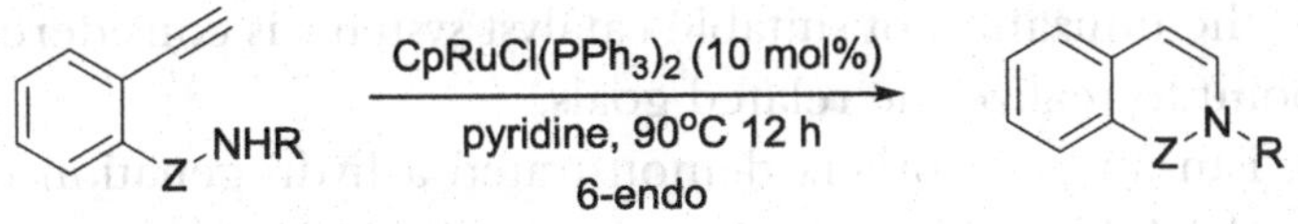

CpRuCl(PPh$_3$)$_2$ (10 mol%)

pyridine, 90°C 12 h

6-endo

Z = C and heteroatom substituents

(Eq. 10)

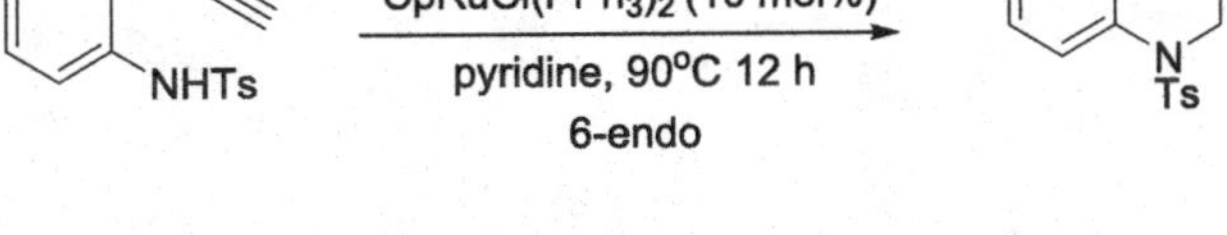

CpRuCl(PPh$_3$)$_2$ (10 mol%)

pyridine, 90°C 12 h

6-endo

(Eq. 11)

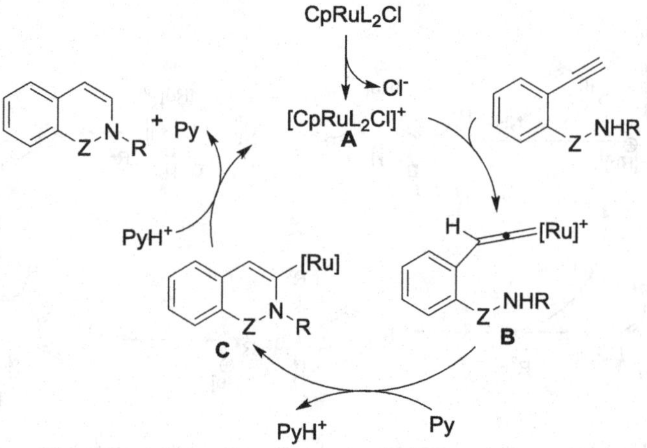

Scheme 16. Proposed mechanism for the Ru-catalyzed cycloisomerization

The regioselective 6-endo cyclization leading to the dihydroiso-quinoline products is interpreted in Scheme 16. Dissociation of anion ligand Cl from the Ru pre-catalyst followed by coordination of the alkyne and subsequent rearrangement leads to Ru vinylidene complex **B**. Such a key intermediate would undergo a nucleophilic attack by the amine or amide group with concurrent removal of a proton by pyridine to give the alkenyl Ru species **C**. Finally, protonation of the Ru–C bond by the pyridinium salt affords the final dihydroisoquinolines, with regeneration of the active catalytic species.[19]

2.5 Synthesis of tetrahydroquinolines and tetrahydroisoquinolines

The synthesis of tetrahydroquinolines and tetrahydroisoquinolines were usually prepared by the hydrogenation of quinolines and isoqinolines, the utilization of suitable catalyst systems is considered as the key point to realize the related goals.

In 2013, Fan and co-workers demonstrated a hydrogenation of quinolines and 3,4-dihydroisoquinolines in neat ionic liquids by using cationic Ru(diamine) complexes, affording the hydrogenated products with excellent enantioselectivity (up to >99% ee). They observed

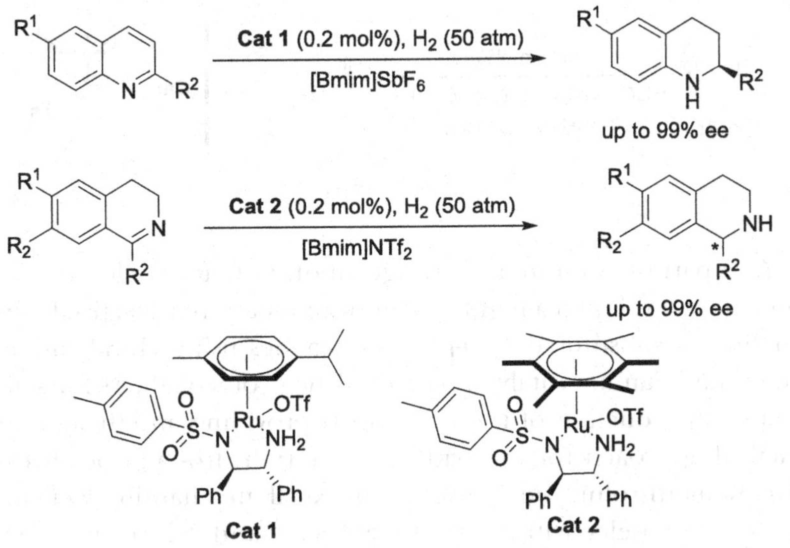

Scheme 17. Enantioselective synthesis of chiral tetrahydroquinolines and tetrahydroisoquinolines

that the catalytic performance was influenced by the anion of the ionic liquids in both cases. Interestingly, the hydrogenation of quinoline derivatives bearing a carbonyl group was selective for C=N (quinoline) over C=O (ketone) bonds. Moreover, the use of ionic liquid could stabilize the ruthenium catalyst and thus facilitate its recycling (Scheme 17).[20]

Similar to Fan's contribution, Ratovelomanana-Vidal and co-workers reported a highly enantioselective Ru-catalyzed transfer hydrogenation of 3,4-dihydroisoquinoline derivatives in which, the azeotropic 5:2 formic acid and trimethylamine mixture was used as the hydrogen source. The synthesis offers several advantages of mild reaction conditions, low catalyst loading, and operational simplicity, offering an attractive method for the synthesis of the valuable 1-aryl-tetrahydroisoquinolines with excellent enantioselectivities (up to 99% ee) and in high yields upon isolation (Eq. 12).[21]

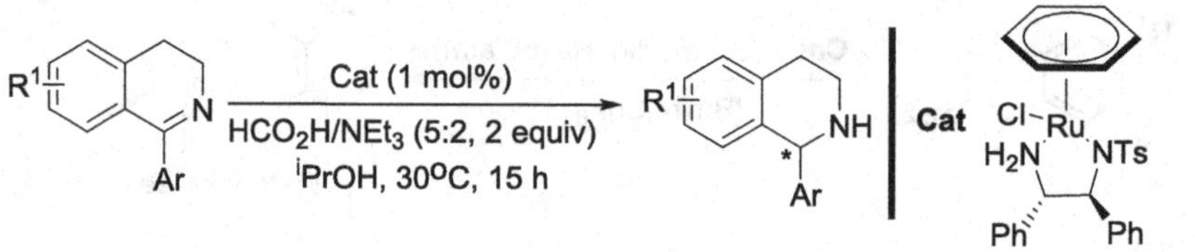

(Eq. 12)

As a part of continuous hydrogenation of heterocycles, the Fan group also developed a highly enantio- and diastereoselective hydrogenation of substituted 1,10-phenanthrolines using chiral cationic ruthenium diamine catalysts. The catalytic protocol allows for selective or full reduction of the two pyridyl providing an efficient and practical approach for the synthesis of tetrahydro- and octahydro-1,10-phenanthroline derivatives with excellent enantioselectivities and diastereoselectivities (up to >99%ee and >20:1 d.r.). The obtained compounds are of important significance in the developing of new chiral ligands (Scheme 18).[22]

2.6 Synthesis of 1,2,3,4-tetrahydronaphthyridines (THNADs)

1,2,3,4- THNADs constitute an important class of N-containing heterocycles, exhibiting diversely interesting biological and therapeutic activities. In addition, THNADs serve as useful building blocks for various synthetic purposes including the preparation of functionalized materials. However, the development shortcut for versatile

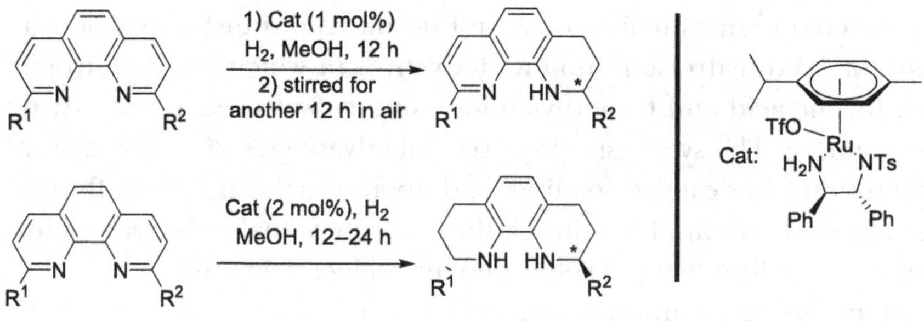

Scheme 18. Enantioselective transfer hydrogenation of 1,10-phenanthroline derivatives

synthesis of this type of compound still remains a challenging goal so far. By using chiral cationic ruthenium diamine complexes, the Fan group demonstrated an asymmetric hydrogenation of disubstituted 1,5-naphthyridines to 1,2,3,4-tetrahydro-1,5-naphthyridines with up to 99% ee and full conversions (Eq. 13).[23]

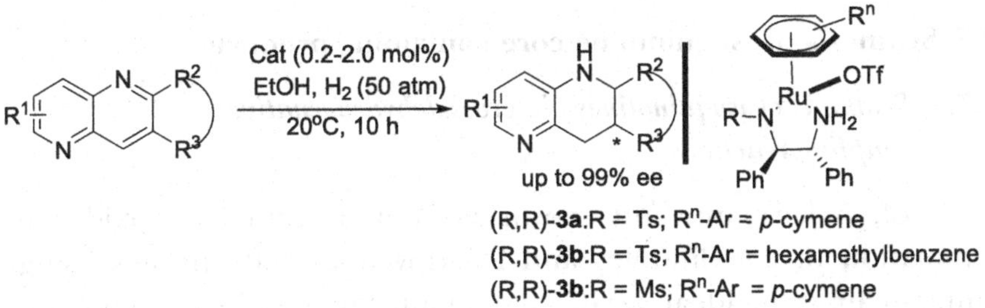
Cat (0.2-2.0 mol%)
EtOH, H2 (50 atm)
20°C, 10 h
up to 99% ee
(R,R)-3a:R = Ts; Rn-Ar = p-cymene
(R,R)-3b:R = Ts; Rn-Ar = hexamethylbenzene
(R,R)-3b:R = Ms; Rn-Ar = p-cymene

(Eq. 13)

Through a ruthenium-catalyzed selective transfer hydrogenative coupling reaction, the Zhang group has developed a novel straightforward synthesis of 1,2,3,4-tetrahydronaphthyridines from ortho-aminopyridyl methanols and alcohols. The synthetic protocol proceeds in an atom- and step-economic fashion together with the advantages of operational simplicity, broad substrate scope, production of water as the only by-product. In the redox pair, the pyridyl ring and two alcohol units serve as the hydrogen acceptor (oxidant) and hydrogen donors (reductants), respectively. Hence, there is no need for using any external reductants (Scheme 19). Noteworthy,

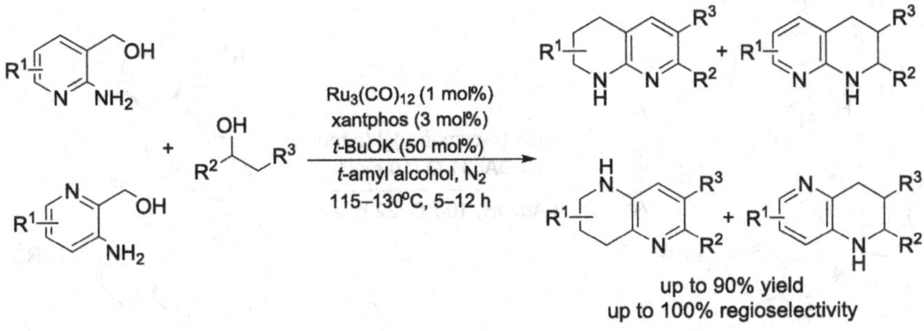
Ru3(CO)12 (1 mol%)
xantphos (3 mol%)
t-BuOK (50 mol%)
t-amyl alcohol, N2
115–130°C, 5–12 h
up to 90% yield
up to 100% regioselectivity

Scheme 19. Synthesis of 1,8- and 1,5-THNADs *via* selective transfer hydrogenation of the pyridyl ring

the transfer hydrogenation mainly occurred at the sterically less-hindered pyridyl ring, affording the desired products exclusively or with excellent regioselectivity and the theoretical calculations for contrastive energy of the regioisomers reveal that the product selectivity correlates with their thermodynamic stability.[24]

2.7 Synthesis of isoquinoline core containing products

2.7.1 Synthesis of isoquinolines via cross-dehydrogenative coupling reactions

In 2012, Ackermann and co-workers demonstrated an oxidative annulation of 2-arylindoles and 2-arylpyrroles with alkynes using ambient air as the ideal sacrificial oxidant. Notably, the introduction of co-catalytic amounts of $Cu(OAc)_2.H_2O$ is essential for the aerobic annulation reactions, delivering a wide range of isoquinoline core containing products, frequently existing in many structural analogs of bioactive marine alkaloids. The experimental mechanistic studies support a concerted deprotonative metalation through acetate assistance (Scheme 20).[25]

Similarly, Ackermann and co-workers also reported the ruthenium-catalyzed oxidative annulations of 1H-pyrazoles with alkynes, affording the isoquinoline core containing products in moderate to high yields with excellent chemo and regioselectivities. The direct C–H/N–H functionalizations of aryl-, heteroaryl-, and alkenyl-substituted

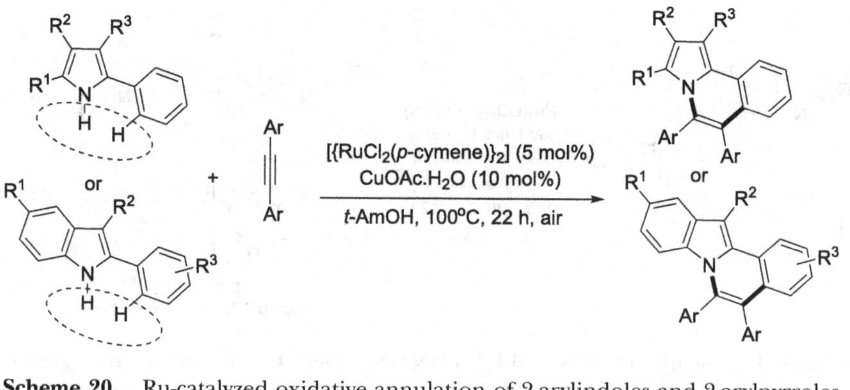

Scheme 20. Ru-catalyzed oxidative annulation of 2-arylindoles and 2-arylpyrroles with alkynes

1H-pyrazoles with aryl- and alkyl-alkynes has demonstrated ample substrate scope of the developed method. Detailed mechanistic studies suggest a reversible C–H bond metalation step with the cationic ruthenium(II)catalyst (Eq. 14).[26]

$$\begin{array}{c} \text{[\{RuCl}_2(p\text{-cymene})\}_2] \text{ (5 mol\%)} \\ \text{AgSbF}_6 \text{ (20 mol\%)} \\ \hline \text{CuOAc.H}_2\text{O} \\ \text{DCE, 100°C, air} \end{array}$$

(Eq. 14)

Similar to the concept of the cross dehydrogenative coupling methods for the synthesis of six-membered heterocycles from the Ackermann group, Chandrasekhar and co-workers have developed a ruthenium-catalyzed oxidative annulation of 2-aryl benzimidazoles with alkynes by using copper salt as the oxidant, affording the bis-heterocyclic framework, benzimidazoisoquinoline, with high regioselectivity (Eq. 15).[27]

$$\begin{array}{c} \text{[\{RuCl}_2(p\text{-cymene})\}_2] \text{ (5 mol\%)} \\ \text{Cu(OAc)}_2.\text{H}_2\text{O} \text{ (0.5 euqiv)} \\ \hline \text{toluene, reflux, 12 h} \end{array}$$

(Z = O, S)

(Eq. 15)

In addition to the above-described examples, 2-phenyl imidazoles were also employed as effective coupling partners to react with alkynes in the presence of {[RuCl$_2$(p-cymene)]$_2$} catalyst and benzoquinone as an oxidant, providing efficient access to various isoquinolines. Such a valuable process is highlighted by the use of cost-effective catalyst system, remarkably high chemo and regioselectivity and the mechanistic investigations showed that the C–H bond metalation was the rate-limiting step. Noteworthy, the conversion between electron-deficient alkynes gives increased efficacy and leads to improved isolated yields (Eq. 16).[28]

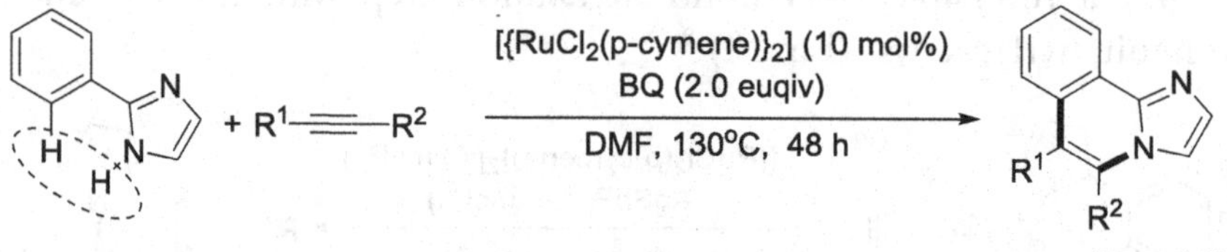

(Eq. 16)

Through imino group directed C–H bond functionalization, the Ackermann developed a ruthenium-catalyzed oxidative annulation of alkynes with ketimines, furnishing1-methylene-1,2-dihydroisoquinolines in high yields with excellent chemo-, site-, and regioselectivities under an ambient atmosphere of air. Particularly, the carboxylate-assisted ruthenium(II) catalysis was proved to be key to success for the reaction, and the C–H bond metalation step is a reversible step *via* mechanistic investigations (Eq. 17).[29]

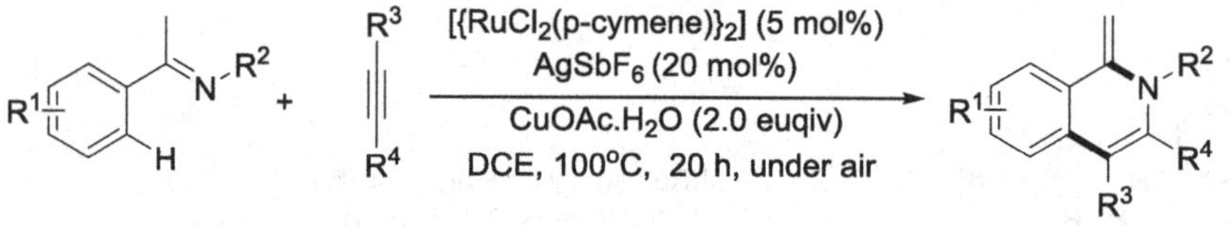

(Eq. 17)

Through the free amino group directed regioselective C–H bond activation, Urriolabeitia and co-workers established a ruthenium-catalyzed oxidative coupling of primary amines with internal alkynes in 2015, offering an interestingly atom-economic approach for the construction of various isoquinoline core containing products in 15 minutes under microwave irradiation or in 24 h with conventional heating (Eqs. 18 and 19).[30]

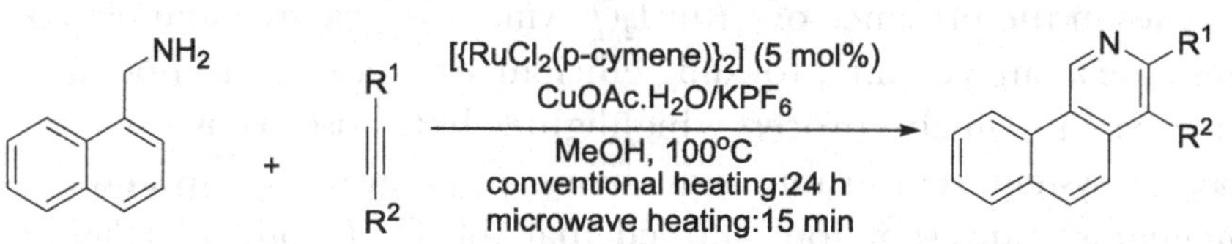

(Eq. 18)

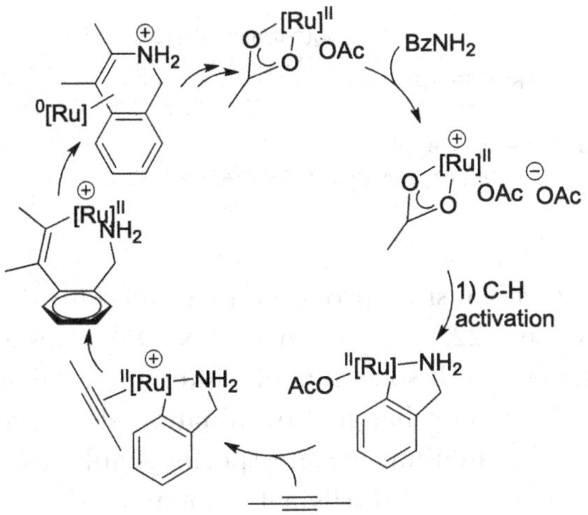

Scheme 21. Proposed reaction mechanism

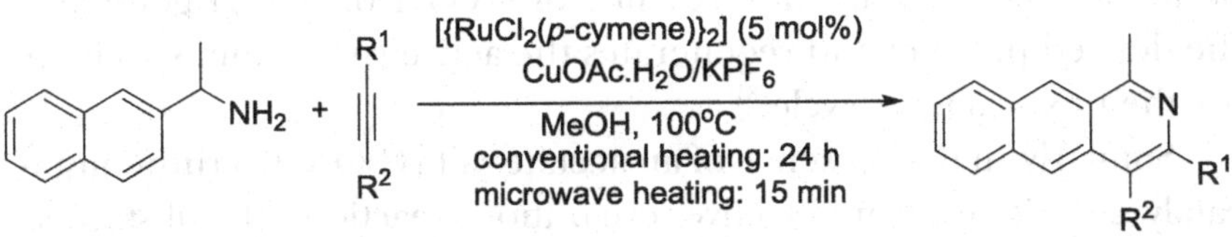

$$(\text{Eq. } 19)$$

The DFT calculations on the reaction of a benzylamine with but-2-yne support a mechanism that consists of acetate-assisted C–H bond activation, migratory insertion, and reductive elimination to C–N bond formation (Scheme 21).[30]

2.7.2 *Synthesis of isoquinolines via cross-coupling reactions*

The ruthenium-catalyzed cross-coupling reaction of substituted aromatic and heteroaromatic ketoximes with internal as well as terminal alkynes was reported by Jeganmohan and co-workers in 2012, such a cyclization reaction regioselectively afforded the isoquinoline derivatives in good to excellent yields in the presence of catalytic amount of NaOAc (Eq. 20).[31]

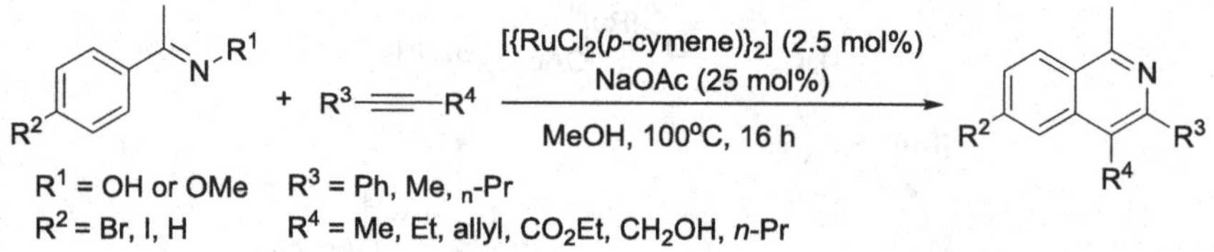

$$\text{(Eq. 20)}$$

A possible mechanism is proposed to account for the cyclization reaction in Scheme 22. The presence of NaOAc gives a ruthenium acetate species **A** *via* dissociation of dimer pre-catalyst {[RuCl$_2$(p-cymene)]$_2$} and anion exchange. Coordination of the nitrogen atom of oxime to the ruthenium acetate species **A** followed by acetate-assisted ortho-metalation affords a five-membered metallacycle **B**. The coordination of alkyne to Ru center followed by regioselective C–C triple bond insertion into the Ru–C bond provides a seven-membered intermediate **C**. Then, C–N bond formation and N–O bond cleavage of **C** in the presence of MeOH or AcOH generates the desired product and regenerates the active ruthenium species **A** for the next catalytic cycle.[31]

Similarly, in the presence of an acetate salt (CsOAc), a ruthenium-catalyzed highly regioselective cyclization reaction of substituted

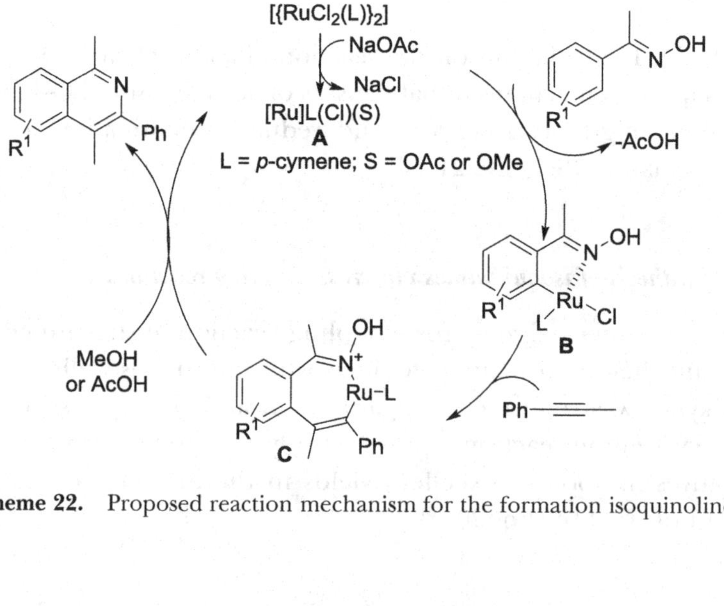

Scheme 22. Proposed reaction mechanism for the formation isoquinolines

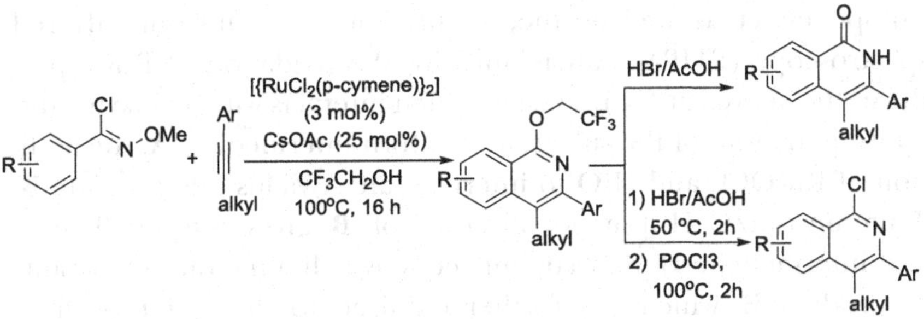

Scheme 23. Synthesis of isoquinolines and the further transformations

N-methoxy benzimidoyl halides with alkynes was reported by Jegan-mohan and co-workers. The method allows to regioselectively synthe-size a wide range of isoquinoline products. Interestingly, the obtained products could be conveniently further transformed into isoquino-line derivatives and substituted 1-halo-isoquinolines in good to excel-lent yields with suitable acid conditions (Scheme 23).[32]

2.8 Synthesis of quinolinones, isoquinolines and pyridines

2.8.1 *Synthesis of quinolinones via oxidative coupling reactions*

Using simple $RuCl_3$ as a catalyst and $CuCl_2$ as a regenerated oxidant in the presence of oxygen, Li and the co-workers developed a novel intramolecular oxidative hydration–deprotonation–cyclization method for the synthesis of substituted quinolinones. This tandem method has the potential for the preparation of bioactive amino-substituted quinolinones with high functional tolerance (Eq. 21).[33]

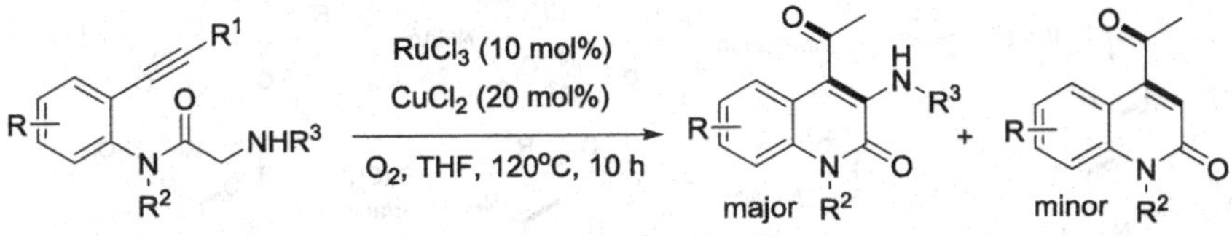

$$\text{(Eq. 21)}$$

A possible mechanism was proposed on the basis of the control experiments, determination of the value of intermolecular kinetic isotope effect as well as the, *in situ* fourier transform infrared spectroscopy (FTIR) analysis: initially, the oxidation of $RuCl_3$ produces the active Ru^VOCl_3 species, which interacts with an alkyne and a nitrogen atom of the substrate to afford intermediate **A**, the addition of Ru^VOCl_3 and H_2O to intermediate **A** yields intermediate **B**. Tautomerism/C–H functionalization of **B** gives intermediate **C** and intermediate **D**. Subsequent reductive elimination of **C** results in product **E**, which was further oxidized to the major product 3-aminoquinolinone in the presence of $RuCl_3$ or/and $CuCl_2$. The reductive elimination of **D** might have two pathways: (1) α-H elimination leading to the major product and (2) α-NHAc elimination to the minor product (Scheme 24).[33]

In 2014, Jeganmohan and co-workers reported a Ru-catalyzed cyclization of anilides with acrylates *via* oxidation process and with propiolates *via* non-oxidation process, affording polyfunctional 2-quinolinones in good to excellent yields. Such a synthesis is different with the cross-dehydrogenative coupling protocols for the

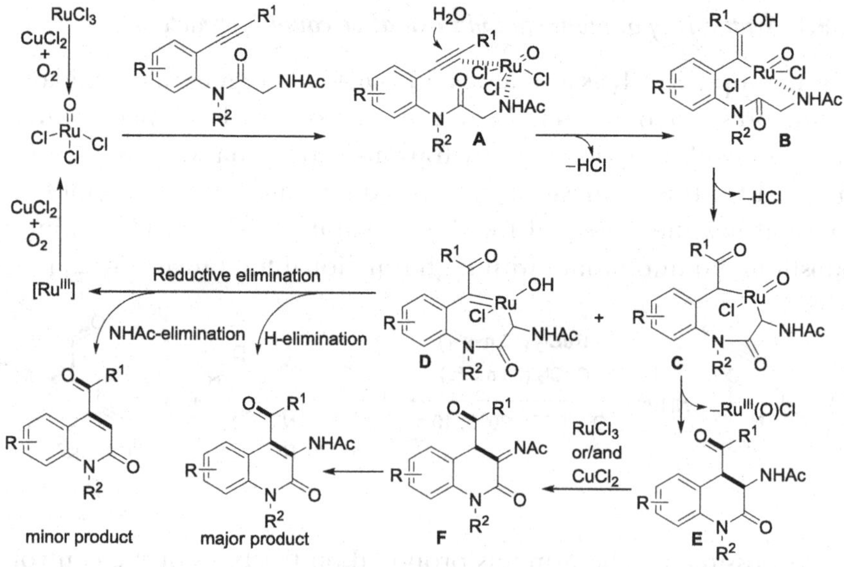

Scheme 24. Proposed reaction mechanism

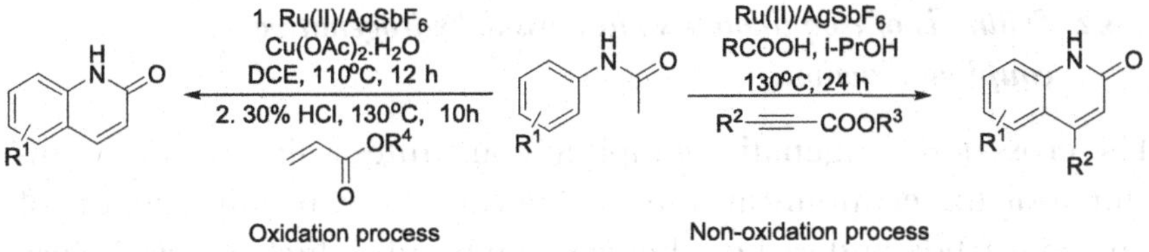

Scheme 25.   Oxidative coupling of anilides with propiolates or acrylates for the synthesis of 2-quinolinones

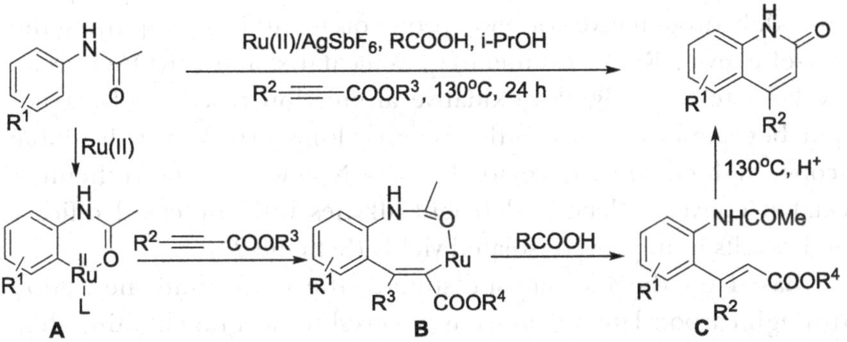

Scheme 26.   Proposed mechanism for the formation of 2-quinolinones

construction of isoquinolones, the addition of acid is essential for the final cyclization (Scheme 25).[34]

A possible product forming process is depicted in Scheme 26 based on the experimental evidence and deuterium labeling studies. The presence of AgSbF$_6$ leads to anion exchange and forms an active cationic species. Coordination of the carbonyl group to the Ru followed by ortho-metalation provides ruthenacycle **A**. Coordinative insertion of alkyne into the Ru–C bond of intermediate gives intermediate **B**. Protonation of the Ru–C bond of **B** by acid additive affords ortho-alkenylated anilide **C** and regenerates the active ruthenium species. Finally, the acid or solvent *i*-PrOH accelerates *trans* to *cis* isomerization the subsequent intramolecular nucleophilic addition of the N–H unit to the ester moiety leads to generate the product and liberates the acetyl group. In which, the organic acid plays multiple roles including serving as a proton source, accelerating the alkenyl isomerization and deacylation of anilide to aniline.[34]

2.8.2 Synthesis of isoquinolones via cross-dehydrogenative coupling reactions

The cross-dehydrogenative coupling constitutes a significant important tool for economical and ecologically benign construction of carbon–carbon and carbon–heteroatom bonds, which avoids the use of pre-functionalized starting materials and thereby allows for a streamlining of organic synthesis. The first ruthenium-catalyzed synthesis isoquinolones was reported by the Ackermann group *via* cross-dehydrogenative coupling protocols in 2011. By employing the cost-effective {[RuCl$_2$(p-cymene)]$_2$} as a catalyst and Cu(OAc)$_2$ as an oxidant, respectively, the oxidative annulation reaction of alkynes and benzamides afforded the isoquinolone products with ample scope, high chemo and regioselectivity. Noteworthy, the ruthenium catalyst converts electron-deficient alkynes with increased efficacy and results in improved isolated yields (Scheme 27).[35]

Based on the mechanistic studies, the C–H bond metalation through carboxylate assistance is believed to be a rate-limiting step, which forms intermediate **A**. The subsequent insertion of C–C triple bond of the alkynes to the Ru–C bond would lead to a seven-membered ruthenacycle **B**. The isoquinolone products is finally afforded *via* reductive elimination and the catalytic species is regenerated in the presence of oxidant Cu(OAc)$_2$ (Scheme 28).[35]

In the same year, Ackermann and co-workers proposed another annulation of alkynes with benzamides *via* C–H bond cleavages with water as a green reaction medium. The carboxylate assistance allows

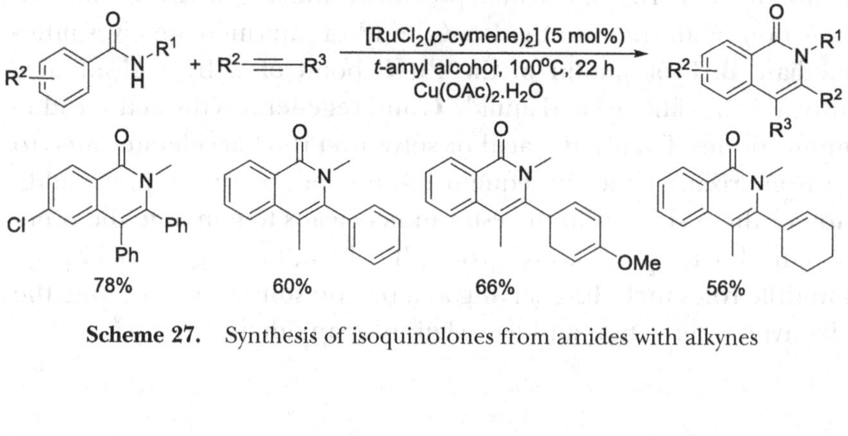

Scheme 27. Synthesis of isoquinolones from amides with alkynes

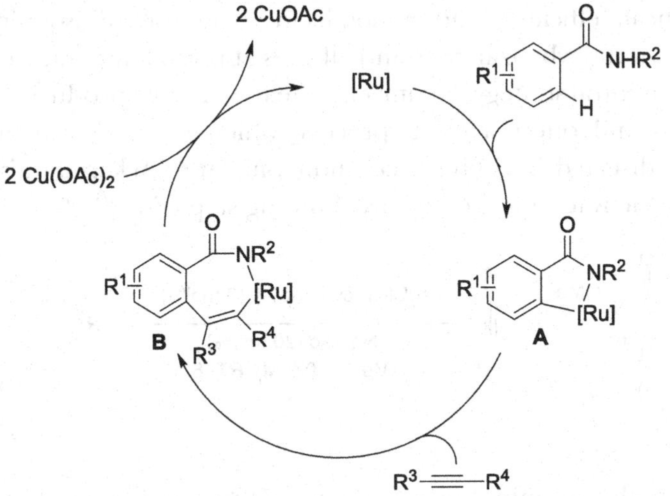

Scheme 28. Proposed mechanism for the oxidative annulation of amides with alkynes.

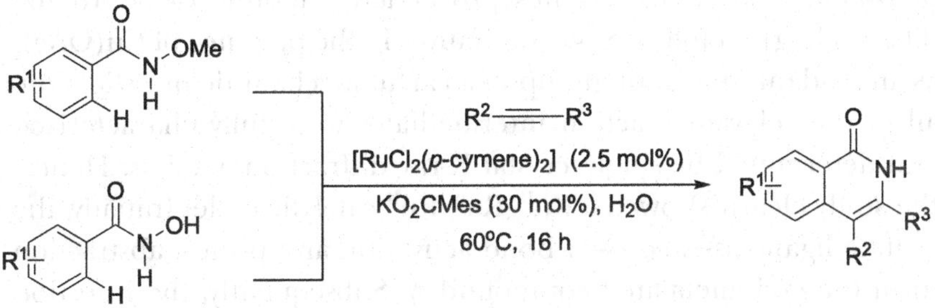

Scheme 29. Ruthenium-catalyzed C–H/N–O bond functionalization in water

for a broadly applicable ruthenium-catalyzed isoquinolone synthesis from *N*-methoxybenzamides. Moreover, the direct use of free hydroxamic acids is also applicable using the same catalytic system (Scheme 29). In comparison with the previous work, this contribution offers a convenient synthesis of nitrogen non-substituted isoquinolones in a chemoselective manner.[36,37]

Interestingly, by employing the reaction system {[RuCl$_2$(p-cymene)]$_2$}/NaOAc in 0.2 m methanol, Wang and co-workers reported a practical, efficient, and regioselective synthesis of isoquinolone from *N*-methoxybenzamides and alkynes at room temperature. Such a redox neutral strategy circumvents the use of waste-producing metal oxidants and offers a clean process, which proceeds *via* internal oxidant directed C–H bond activation. Similar to Ackermann's work, the C–H activation is the turnover-limiting step (Eq. 22).[38]

$$\text{(Eq. 22)}$$

With the isoquinolones in hands, Wang and co-workers then employed this type of compound for further elaboration of complex molecules *via* direct cross-dehydrogenative coupling protocol. Still using {[RuCl$_2$(p-cymene)]$_2$} as an efficient catalyst, the annulations of isoquinolones with alkynes proceeded smoothly to afford the dibenzo-[a,g]quinolizin-8-one derivatives in the presence of Cu(OAc)$_2$ as an oxidant. Interestingly, upon a carful mechanistic investigation, all of the relevant reaction intermediates were fully characterized and determined by single crystal X-ray diffraction analysis. Hence, the study clearly showed a Ru^{II}–Ru^0–Ru^{II} catalytic cycle. Initially, the acetate ligand-assisted C–H bond activation and proton abstraction form the cyclometalated compound **A**. Subsequently, the insertion of alkyne to the Ru–C bonds of **A** gives a ring-expanded intermediate **B**. Finally, oxidative coupling of the C–N bond of **B** affords Ru^0 sandwich complex **C**, which undergoes oxidation to regenerate the active Ru^{II} complex in the presence of copper oxidant and releases the desired product (Scheme 30).[39]

As a continuing interest in developing cross-dehydrogenative coupling methods for the synthesis of isoquinolones, the Jegan-mohan group has exhibited an alternative approach to achieve the related end in 2013 from aryl nitriles and alkynes. Such a annulation reaction is realized *via* cooperative catalysis of copper and ruthenium. Cu(OAc)$_2$ serves as a Lewis acid to activate the C–N group of benzonitrile, which benefits the *in situ* hydrolysis of

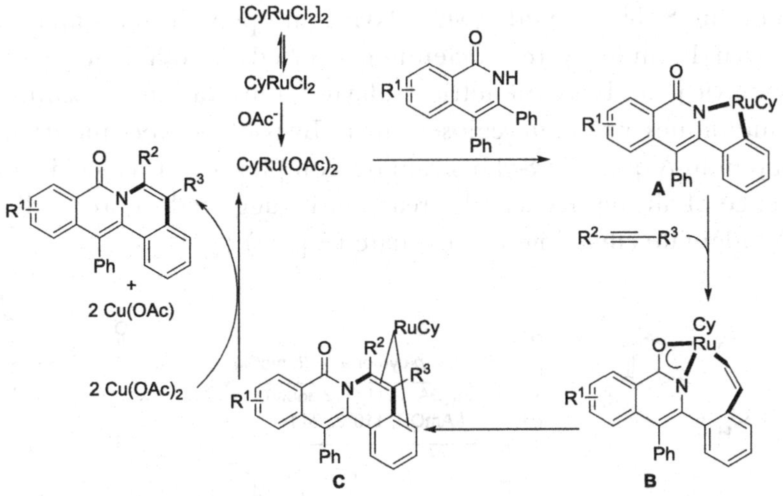

Scheme 30. Possible mechanism for the formation of isoquinolones

benzonitriles to benzamides. KPF_6 likely replaces the chloride ligand from the $\{[RuCl_2(p\text{-cymene})]_2\}$ followed by the ligand exchange with $Cu(OAc)_2$, giving an cationic ruthenium species. Similar to the mechanisms of Ackermann and Wang, the ortho-metalation of benzamides and insertion of alkyne into the Ru–C bond followed by reductive elimination would afford the desired isoquinolones. The presence of oxidant $Cu(OAc)_2$ regenerates the active ruthenium species. Interestingly, the reaction requires only catalytic amount of $Cu(OAc)_2$ (30 mol%) because the $Cu(OAc)_2$ is regenerated under oxygen or air from the reduced copper source under acid conditions (Eq. 23).[40]

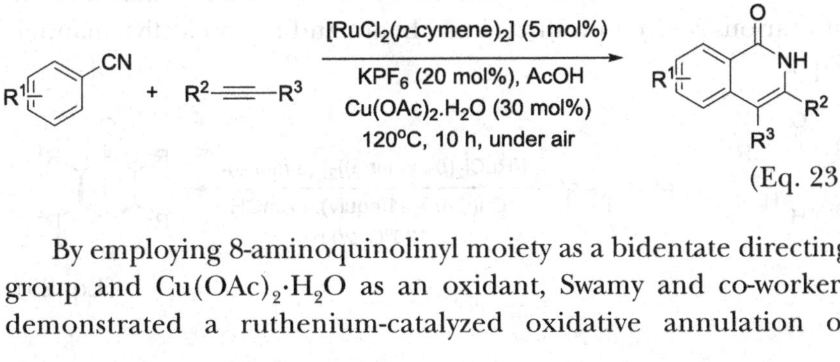

$$\text{(Eq. 23)}$$

By employing 8-aminoquinolinyl moiety as a bidentate directing group and $Cu(OAc)_2 \cdot H_2O$ as an oxidant, Swamy and co-workers demonstrated a ruthenium-catalyzed oxidative annulation of

N-quinolin-8-yl-benzamides with alkynes in open air. Both aryl and heteroaryl amides were efficiently converted in combination with symmetrical and unsymmetrical alkynes into various substituted isoquinolones with high regioselectivity. By control experiments, the ruthenium-*N*-quinolin-8-yl-benzamide complex was isolated in the absence of alkyne. Hence, the reaction is suggested to proceed *via* *N*,*N*-bidentate chelating intermediate (Eq. 24).[41]

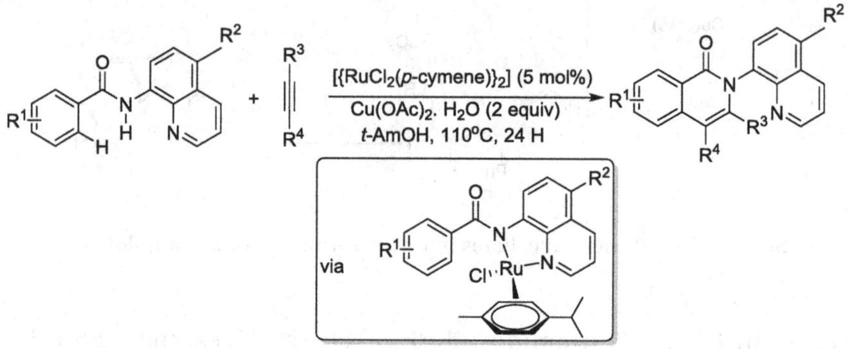

(Eq. 24)

2.8.3 Synthesis of pyridones via cross-dehydrogenative coupling reactions

Similar to the concept on the utilization of cross-dehyrogenative coupling protocol for the synthesis of isoquinolones, Ackermann and co-workers developed an economical synthesis of pyridines through oxidative annulation of alkynes with acrylamides using {[RuCl$_2$(*p*-cymene)]$_2$} and Cu(OAc)$_2$ as the catalyst and oxidant, respectively. Both electron-rich and electron-deficient acrylamides as well as (di) aryl- and (di)alkyl-substituted alkynes could be efficiently transformed into various desired products in a chemo and regioselective manner (Eq. 25).[42]

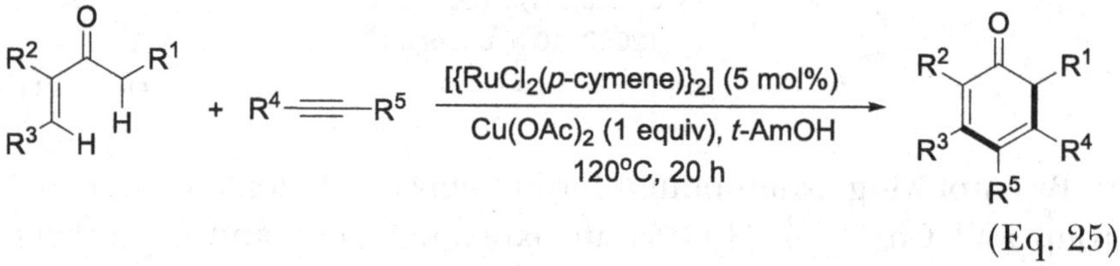

(Eq. 25)

2.9 Ruthenium-catalyzed synthesis of quinoxalines

The oxidative cyclization reaction of *o*-phenylenediamines with vicinal diols was reported in 2006 by the Cho group. Using catalytic amount of ruthenium catalyst $RuCl_2(PPh_3)_3$ along with KOH gave the quinoxalines in high yields. Noteworthy, the introduction of benzalacetone as a sacrificial hydrogen acceptor was essential to obtain a desired product yield (Eq. 26).[43]

$$
\begin{array}{c}
R^1 \quad NH_2 \\
R^1 \quad NH_2
\end{array}
+
\begin{array}{c}
HO \\
HO \quad R^2
\end{array}
\xrightarrow[\substack{\text{benzalacetone (2 equiv)} \\ \text{diglyme, reflux, 20 h}}]{\substack{RuCl_2(PPh_3)_3 \text{ (4 mol\%)} \\ \text{KOH (4 equiv)}}}
\begin{array}{c}
R^1 \quad N \\
R^1 \quad N \quad R^2
\end{array}
$$

(Eq. 26)

Through a ruthenium-catalyzed hydrogen transfer strategy, Zhang and co-workers have demonstrated a one-pot method for efficient synthesis of quinoxalines from 2-nitroanilines and biomass-derived vicinal diols. The synthetic protocol employs the diols and the nitro group as the hydrogen suppliers and acceptors, respectively. Hence, there is no need for the use of external reducing agents. Advantageously, such a method has the advantages of operational simplicity, broad substrate scope and the use of renewable reactants, which complements the existed methods to access quinoxaline derivatives (Eq. 27).[44]

$$
\underbrace{
\begin{array}{c}
R^1 \quad NO_2 \\
R^2 \quad NH_2
\end{array}
}_{\text{Hydrogen acceptor}}
+
\underbrace{
\begin{array}{c}
HO \quad OH \\
R^3 \quad R^4
\end{array}
}_{\text{Hydrogen donors}}
\xrightarrow[\substack{CsOH \text{ (50 mol\%)} \\ 150^\circ C, 8\text{ h}}]{\substack{Ru_3(CO)_{12} \text{ (1 mol\%)} \\ dppp \text{ (3 mol\%)}}}
\begin{array}{c}
R^1 \quad N \quad R^4 \\
R^2 \quad N \quad R^3 \\
\text{or} \\
R^1 \quad N \quad R^3 \\
R^2 \quad N \quad R^4
\end{array}
$$

(Eq. 27)

2.10 Ruthenium-catalyzed synthesis of quinazoline derivatives

By employing $Ru(PPh_3)_3(CO)H_2/Xantphos$ as the catalyst and crotononitrile as a sacrificial hydrogen acceptor, Williams and co-workers developed a ruthenium-catalyzed oxidative cyclization reaction of 2-aminobenzamide with a variety of alcohols to dihydroquinazolines (Eq. 28), and it was found that the addition of NH_4Cl play a crucial role in affording the products. However, the cyclization reactions of 2-aminobenzamide and 2-aminosulfonamide with alcohols in the absence of NH_4Cl gave the quinazoline products, selectively (Eqs. 29 and 30).[45]

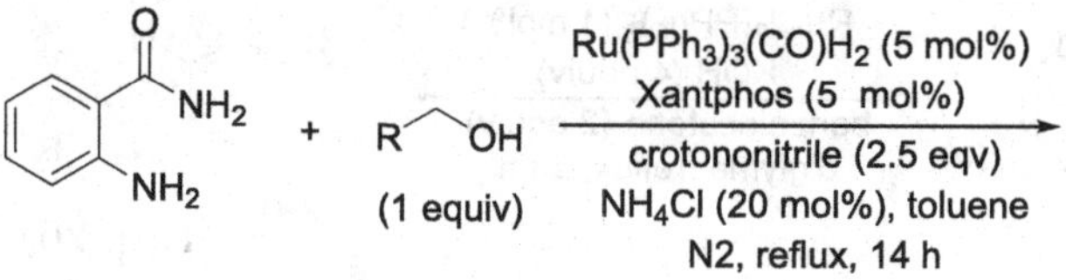

$$(Eq. 28)$$

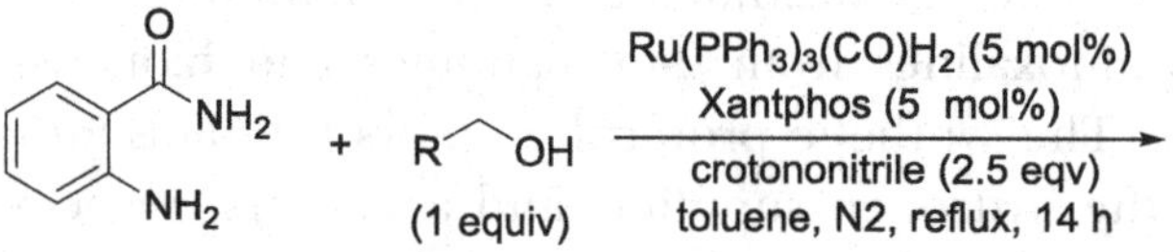

$$(Eq. 29)$$

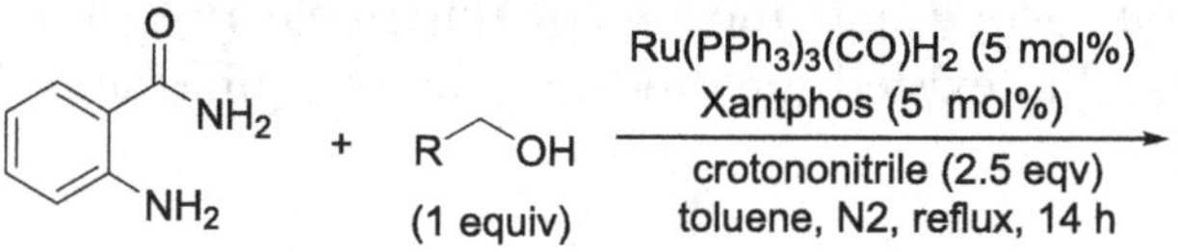

$$(Eq. 30)$$

Through an acceptorless dehydrogenative coupling (ADC) process, Zhang and co-workers demonstrated straightforward ruthenium-catalyzed dehydrogenative synthesis of 2-arylquinazolines. A series of 2-aminoaryl methanols were efficiently converted in combination with different types of benzonitriles into various desired products in moderate to good yields. The advantages of the synthetic protocol involve operational simplicity, high atom efficiency, broad substrate scope, and no need for the use of less environmentally benign halogenated reagents (Eq. 31).[46]

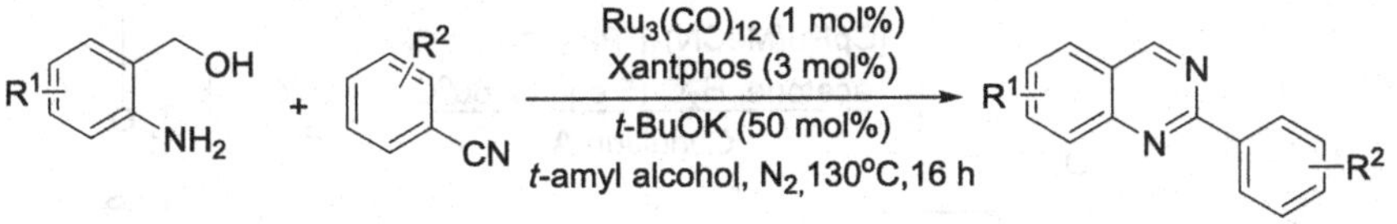

(Eq. 31)

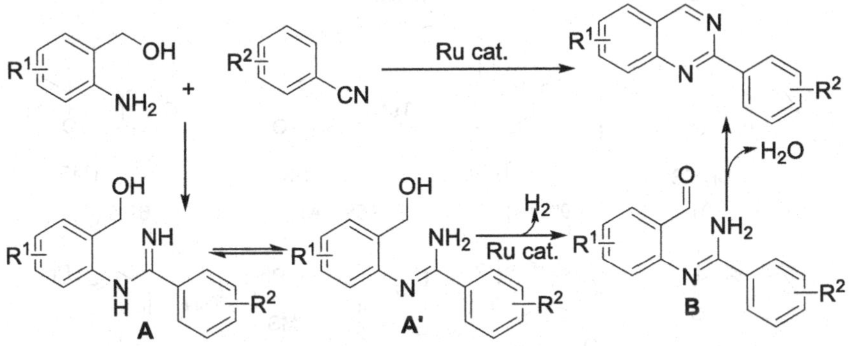

Scheme 31. Possible pathway to access quinazolines

The mechanistic investigations suggests that the reaction might undergo the following tandem sequences: (1) The nucleophilic addition of the amino group of **1** to the benzonitrile **2** forms an amidine intermediate **A**; (2) Then, the ruthenium-catalyzed dehydrogenation of the alcohol unit of **A** gives an *o*-carbonyl amidine **B**; (3) Finally, the thermodynamically favorable intramolecular condensation of **B** afforded the desired quinazoline product **3** (Scheme 31).[46]

3. Ruthenium-Catalyzed Synthesis of Six-Membered Oxa-Heterocycles

3.1 Ruthenium-catalyzed synthesis of pyran derivatives

The cycloisomerization of diyne-ols containing an alkynylsilane catalyzed by cationic ruthenium complex $[CpRu(MeCN)_3]^+PF_6^-$ was reported by Trost and co-workers in 2004, which affords the 2-silyl-[6*H*]-pyrans or isomeric dihydropyrans in high yield by the ratio of water. The regioselectivity of the reaction depends on the substrate

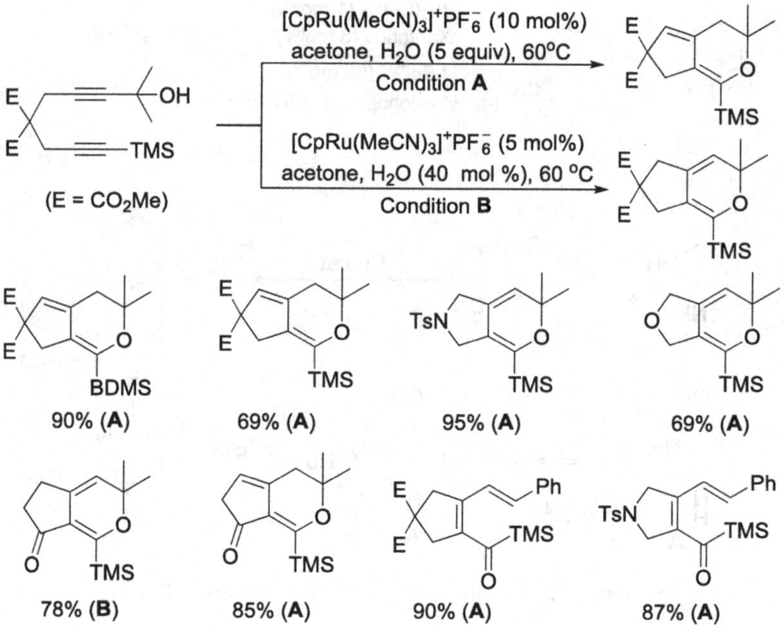

Scheme 32. Ruthenium-catalyzed synthesis of pyran derivatives

structure (presence/absence of a heteroatom in the tether) or by solvent selection. Noteworthy, the cyclization to a five-membered ring is much more facile. However, in the demonstrated synthesis, the reactants bearing tertiary propargylic and secondary benzylic alcohols tends to form six-membered ring. If additional conjugation is present in the molecule, the acylsilane itself is the stable isomeric form (Scheme 32).[47]

By utilization *in situ* formed cationic ruthenium species, the Jeganmohan group developed a ruthenium-catalyzed regioselective intermolecular multistep homo and heterodimerization of substituted propiolates, affording α-pyrone-5-carboxylates and α-pyrone-6-carboxylates in an atom-economic fashion (Scheme 33).[48]

A possible reaction mechanism is depicted in Scheme 34. Initially, the active cationic ruthenium species **A** forms in the presence of {[$RuCl_2$(p-cymene)]$_2$} and silver salt $AgSbF_6$ *via* anion exchange. Then, the regioselective coordination of propiolates to

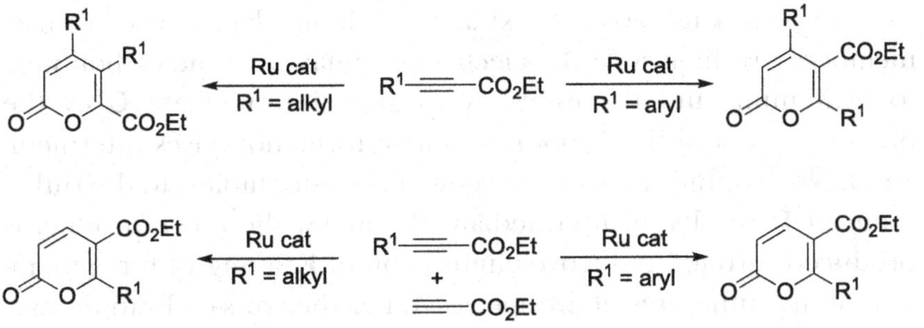

Reaction conditions: [{RuCl$_2$(*p*-cymene)}$_2$] (5 mol%), AgSbF$_6$ (20 mol%), pivalic acid (10 equiv), 1,4-dioxane, 110°C, 12 h

Scheme 33.   Regioselective ruthenium-catalyzed synthesis of pyrone derivatives

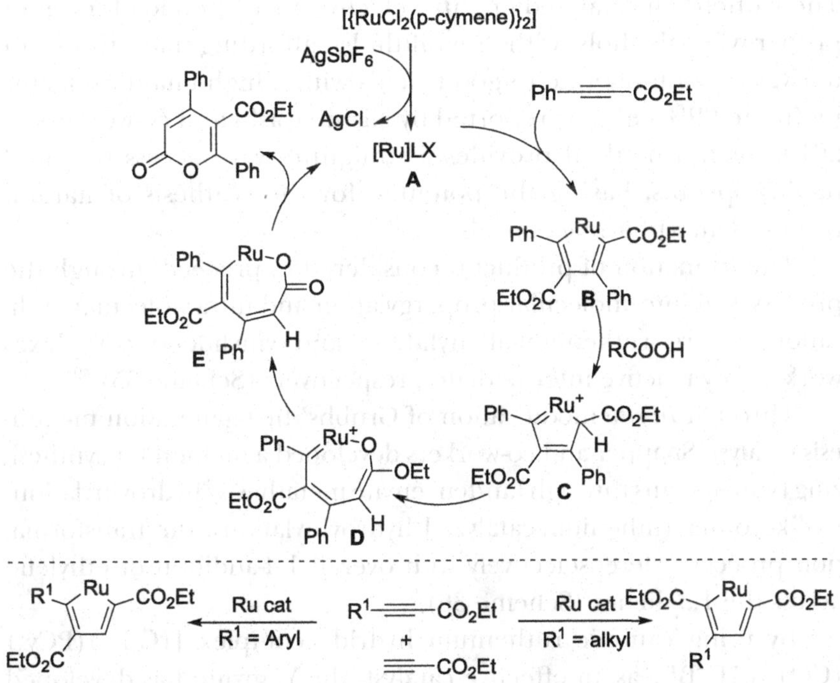

Scheme 34.   Possible reaction mechanism

the complex **A** followed by oxidative cyclometalation leads to five-membered ruthenacycle **B**. Selective protonation at the carbon next to ruthenium and the ester group gives intermediate **C** by the organic acid, and the O-bound enolate formation gives intermediate **D**. Nucleophilic attack of oxygen of the ester moiety to the ruthenium of **D** results in intermediate **E**. Finally, the cyclic product is produced through reductive elimination of **E**, along with regeneration of the ruthenium active species **A**. For the cross-cyclization reaction, only the ester moiety of terminal alkynes was involved in the C–O bond formation, which might be attributed to the steric hindrance of terminal alkyne ester group which is much less compared with the internal alkynes.[48]

3.2 Ruthenium-catalyzed synthesis of benzopyran and isobenzopyran derivatives

The ruthenium-catalyzed enantioselective [3 + 3] cycloaddition of propargylic alcohols with 2-naphthols, affording naphthopyran derivatives in moderate to good yields with a high enantioselectivity (up to 99% ee), was reported by Nishibayashi and co-workers in 2010. Such a method provides a straightforward access to chiral naphthopyrans, having the potential for the synthesis of natural products and pharmaceuticals.

The formation of product is considered to proceed through the processes of intermolecular propargylation and intramolecular cyclization, where ruthenium-allenylidene and vinylidene complexes work as key reactive intermediates, respectively (Scheme 35).[49]

Through *in situ* modification of Grubbs' first-generation metathesis catalyst, Snapper and co-workers developed a method for synthesizing benzopyrans through tandem enyne metathesis/hydrovinylation. Unlike other ruthenium-catalyzed hydrovinylations, the transformation proceeds stereoselectively with overall 1,4-addition of ethylene across the 1,3-dienes (Scheme 36).[50]

By using cationic ruthenium hydride complex $[(C_6H_6)(PCy_3)(CO)RuH]^+BF_4^-$ as an effective catalyst, the Yi group has developed an efficient method for the synthesis of flavene derivatives through oxidative C–H coupling reaction of phenols with α, β-unsaturated

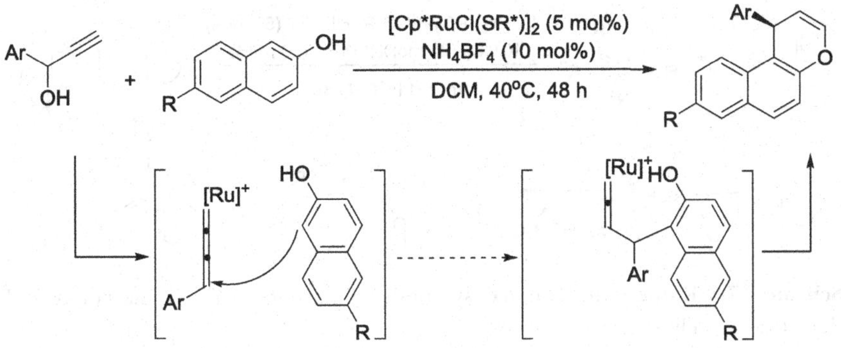

Scheme 35. Ruthenium-catalyzed [3 + 3] cycloaddition for the synthesis of benzopyrans

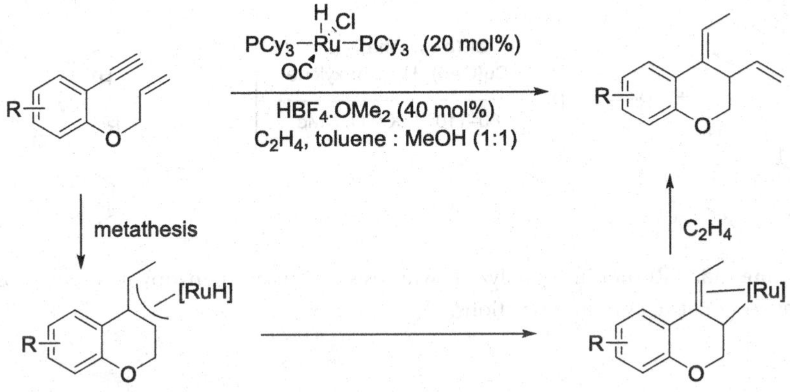

Scheme 36. Ruthenium-catalyzed benzopyrans *via* metathesis and hydrovinylation

aldehydes, intramolecular addition, and dehydration processes. The synthesis proceeds without generation of any waste by-products and does not need any metal oxidants (Scheme 37).[51]

With the utilization cross-dehydrogenative coupling strategy, the Ackermann group demonstrated a ruthenium-catalyzed synthesis of fused benzopyran derivatives hydroxyl group directed C–H bond functionalization. Thus, the method gives step- and atom-economical access to diversely decorated products with ample substrate scope. The mechanistic studies support a carboxylate-assisted reversible C–H bond ruthenation (Scheme 38).[52]

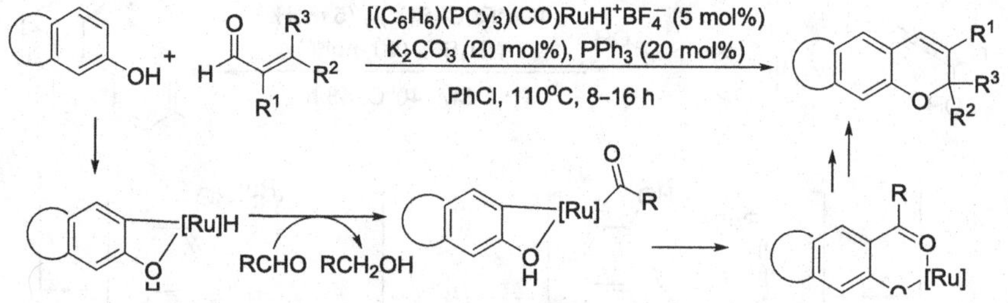

Scheme 37. Ruthenium-catalyzed synthesis of flavenes *via* C–H acylation and dehydrative cyclization

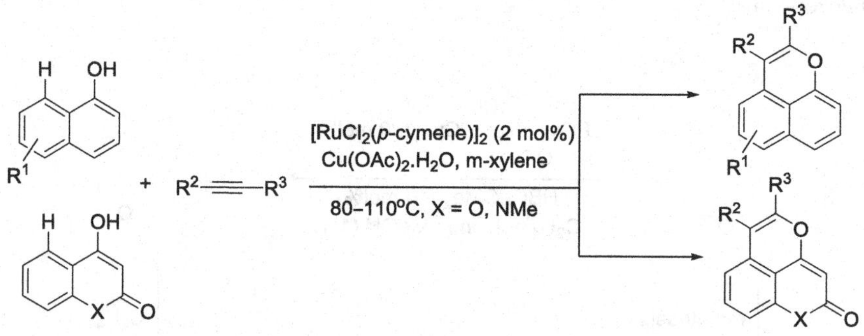

Scheme 38. Ruthenium-catalyzed synthesis of fused benzopyrans *via* cross-dehydrogenative coupling reactions.

3.3 Ruthenium-catalyzed synthesis of coumarin and isocoumarin derivatives

In the presence of catalytic amounts of {[$RuCl_2$(p-cymene)]$_2$}, $AgSbF_6$ and copper catalysts, the regioselectively oxidative cyclization of benzoic acids with alkynes was developed by the Jeganmohan group in 2012, which allows to synthesize various isocoumarin derivatives in good to excellent yields. Noteworthy, the silver salt ($AgSbF_6$) plays a crucial role in controlling the reaction regioselectivity and completely suppress the formation of decarboxylative naphthalene derivatives. Noteworthy, a similar work was also contributed by the Ackermann group in the same year (Eq. 32).[53,54]

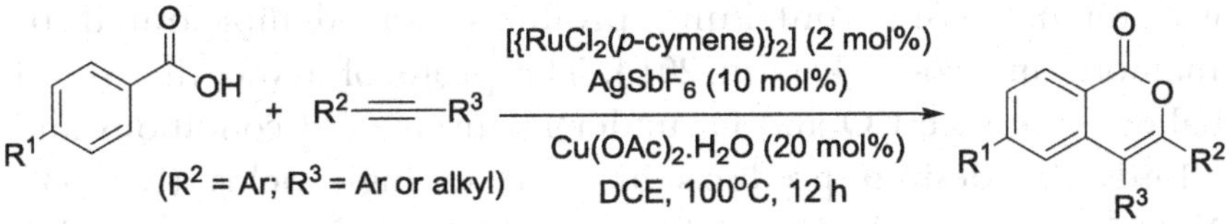

$$(Eq.\ 32)$$

The catalytic reaction is likely initiated by anion exchange between Cl⁻ and SbF$_6$⁻ and affords a cationic complex. Coordination of the carboxylate oxygen to the ruthenium species followed by o-metalation affords a five-membered ruthenacycle **A**. Then, regioselective coordinative insertion of alkyne into the Ru–C bond of **A** provides a seven-membered metal cycle **B**. Subsequent reductive elimination of intermediate **B** in the presence of Cu(OAc)$_2$ affords the final product and regenerates the active ruthenium species for the next catalytic cycle. In the reaction, the catalytic amount of Cu(OAc)$_2$ could be regenerated under oxygen, which is essential for the regeneration of the active ruthenium species. The presence of AgSbF$_6$ leading to a cationic ruthenium species favors the o-metalation over the decarboxylation (Scheme 39).[53,54]

The ruthenium-catalyzed carbonylative C–H cyclization of 2-arylphenols leading to 6H-dibenzo[b,d]pyran-6-one, a class of

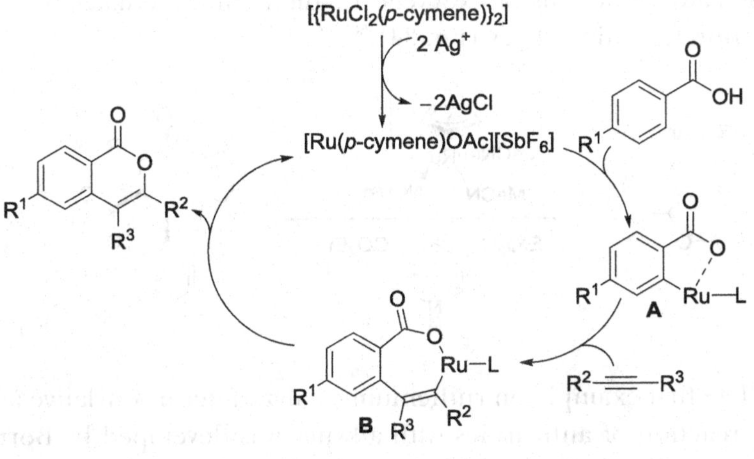

Scheme 39. Proposed mechanism for the formation of isocoumarins

isocoumarin core containing products, was demonstrated by Inamoto and co-workers in 2013. The protocol allows the use of balloon pressure CO and O_2 under relatively mild conditions and affords the desired products in good to high yields (Eq. 33). Noteworthy, the electron-rich substrates give the corresponding products in higher yields comparing to the electron-deficient ones.[55]

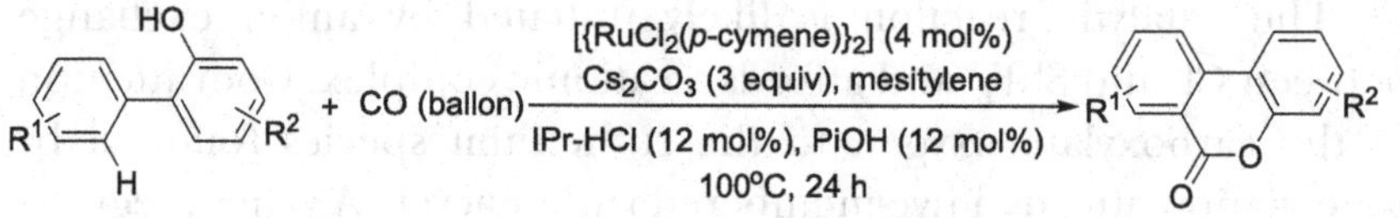

$$\text{(Eq. 33)}$$

In 2014, the Yamamoto group developed a new ruthenium-catalyzed method for the synthesis of dihydrocoumarin-fused poly-cyclic products from enediyne substrates containing 1,6-diyne, acrylate dienophile, and phenol tether moieties. Such a tandem reaction comprises the sequences of transfer-hydrogenative cyclization and subsequent intramolecular Diels–Alder annulation in which, the cationic complex $[CpRu(MeCN)_3]^+PF6^-$ was employed as an efficient catalyst, and a Hantzsch ester served as the H_2 surrogate. Based on DFT calculations, the relative stereochemistry of the tandem reaction products was rationalized that the intramolecular Diels–Alder reactions of exocyclic 1,3-diene intermediates proceed *via* endo transition states (Eq. 34).[56]

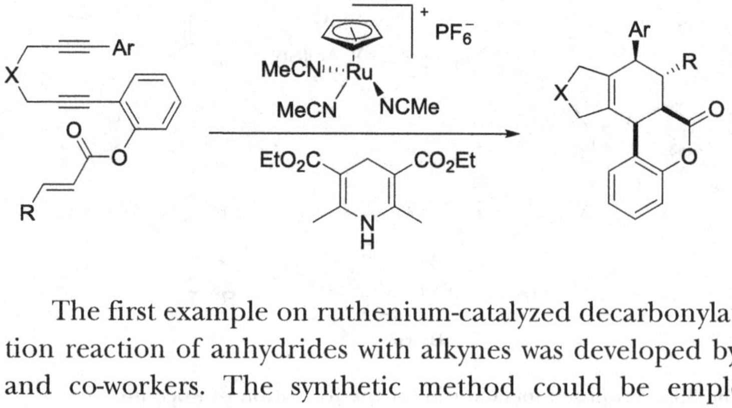

$$\text{(Eq. 34)}$$

The first example on ruthenium-catalyzed decarbonylative addition reaction of anhydrides with alkynes was developed by Boruah and co-workers. The synthetic method could be employed for regioselective preparation of a variety of isocoumarins and α-pyrones in high yields. The synthesis features broad substrate scope, operational simplicity, low catalyst loading and high yield of products (Scheme 40).[57]

A possible pathway for the formation of isocoumarins is shown in Scheme 41. The oxidative addition of the ruthenium catalyst to the anhydride oxygen–acyl bond gives a six-membered ruthenium cycle **A**. Decarbonylation of **A** and subsequent insertion of C–C triple bond to the Ru–C bond generates a seven-membered ruthenacycle

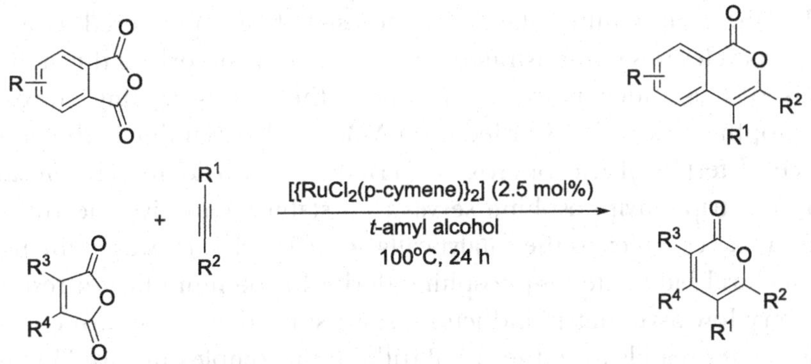

Scheme 40. Ruthenium-catalyzed decarbonylative addition reaction for the synthesis of isocoumarins and α-pyrones

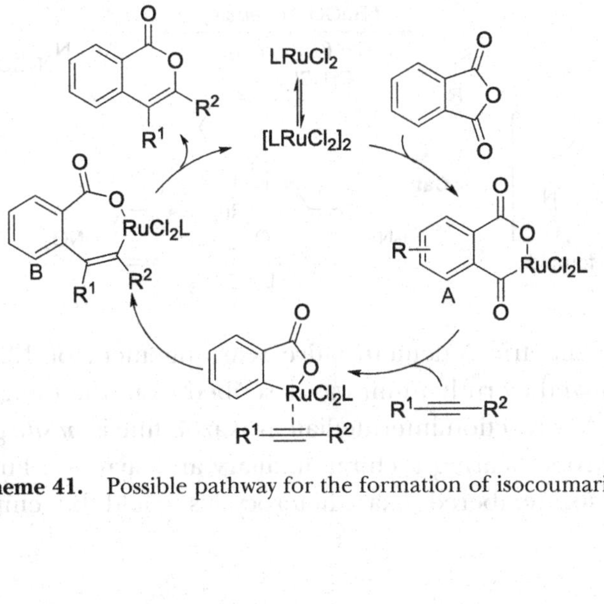

Scheme 41. Possible pathway for the formation of isocoumarins

B, the subsequent reductive elimination of would afford the desired product and regenerate the active ruthenium species. The improved product yield using tert-amyl alcohol as the solvent might be due to the electrostatic interaction between the ruthenium species (hard acid) and the alcohol (hard base), which favors to stabilize the ruthenium intermediates.[57]

4. Ruthenium-Catalyzed Synthesis of Six-Membered Oxaza-Heterocycles

The first ruthenium-catalyzed synthesis of six-membered oxaza-heterocycles was demonstrated by Whiting and co-workers through a hetero-Diels–Alder reaction in which, the corresponding nitroso dienophile was *in situ* afforded from *N*-Boc-hydroxylamine in the presence of tert-butyl hydroperoxide (TBHP) as an oxidant. The *in situ* formed triphenylphosphine serves as a stabilizer to give the ruthenium oxo-complex as the catalytically active species. However, the use of a chiral bidentate bis-phosphine derived ruthenium ligand results in very low asymmetric induction, suggesting that the intermediate dissociates readily from the chiral ruthenium complex involved in the oxidation step prior to Diels–Alder cycloaddition (Eq. 35).[58]

(Eq. 35)

An asymmetric *N*-demethylative rearrangement of 1,2-isoxazolidines catalyzed by ruthenium was described from the Kang group in 2014. The key reaction intermediate isoxazolidine is *in situ* generated from a nitrone bearing a chiral auxiliary and styrenes. Further, the obtained six-membered oxaza-heterocycles could be employed as useful building blocks for the preparation of enantio enriched *syn*-1,3-aminoalcohols as well as *cis*-1,3-oxazinanes in good yields (Eq. 36).[59]

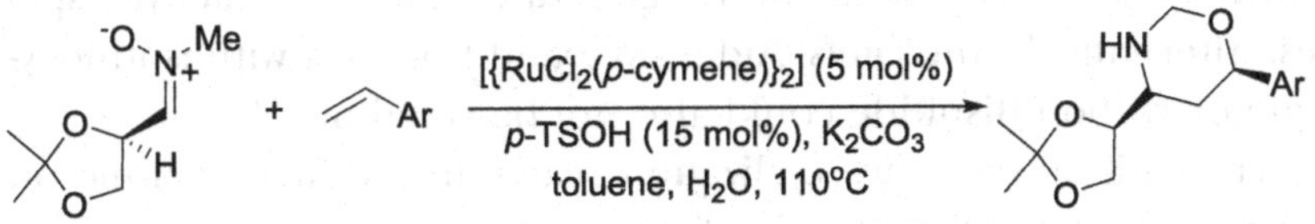

Me
H
Ar
[{RuCl2(p-cymene)}2] (5 mol%)
p-TSOH (15 mol%), K2CO3
toluene, H2O, 110°C
HN
O
Ar

(Eq. 36)

With the use of a cost-effective ruthenium catalyst system, the oxygen substituted aminoallene underwent smooth intramolecular *exo*-hydroamination reactions, yielding the corresponding six-membered 1,3-oxaza-heterocyclic compound in good yield (Eq. 37).[60]

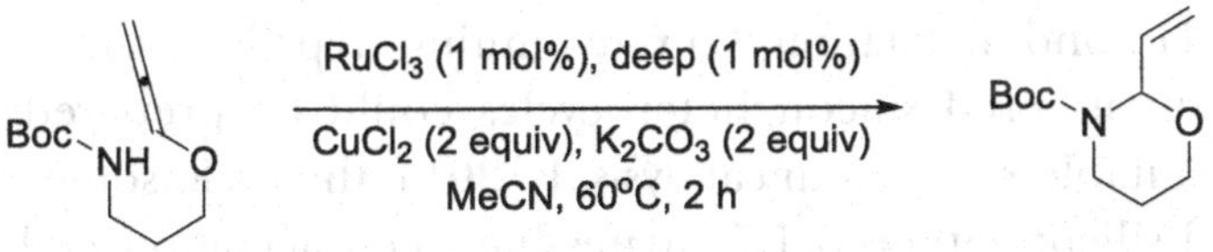

Boc
NH
O
RuCl3 (1 mol%), deep (1 mol%)
CuCl2 (2 equiv), K2CO3 (2 equiv)
MeCN, 60°C, 2 h
Boc
N
O

(Eq. 37)

5. Synthesis of Other Type of Six-Membered Heterocycles

The early ruthenium-catalyzed cycloaddition of 1,6-diynes with isothiocyanates or carbon disulfide, leading to sulfur atom containing six-membered heterocycles, was reported by Itoh and co-workers. The possible mechanism for the formation products was outlined in Scheme 42. Initially, the oxidative cyclization of a 1,6-diyne with ruthenium catalytic species forms a ruthenacyclopentadiene **A**.

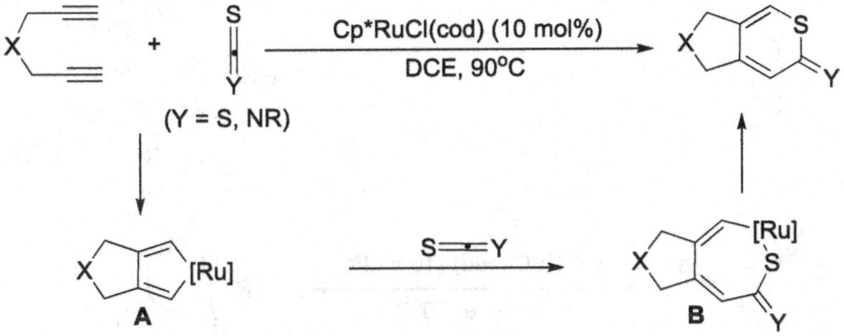

X
S
Y
(Y = S, NR)
Cp*RuCl(cod) (10 mol%)
DCE, 90°C
S
Y
[Ru]
A
S=Y
[Ru]
S
B
Y

Scheme 42. Ruthenium-catalyzed synthesis of sulfur-containing heterocycles

M. Zhang

Subsequently, the insertion of isothiocyanate or carbon disulfide generates intermediate **B**. Finally, the reductive elimination of **B** affords the desired product and regenerates the active catalytic species. Alternatively, the Diels–Alder cycloaddition of **A** with isothiocyanate or carbon disulfide could also not be ruled out.[61]

Through an oxidative cyclization reaction, the first example on ruthenium-catalyzed synthesis of phosphaisocoumarins from phosphonic acid monoesters or phosphinic acids with alkynes under aerobic conditions was developed by the Lee group in 2013. The reaction proceeded with the advantages of using cost-effective catalyst system, good product yields, broad substrate scope, etc. and the competition experiments showed that the electronic effects of alkynes do not affect the oxidative cyclizations. Further, the mechanistic studies revealed C–H bond metalation is a rate-limiting step (Eq. 38).[62]

The six-membered silicon heterocycles could be prepared by employing suitable ruthenium catalysis. In 2014, the cycloisomerization of vinyl silicon-tethered 1,7-enynes has been accomplished by the Clark group in 2014 with the use of catalytic amount of Cp*Ru (COD)Cl, affording a new silane moiety and trisubstituted alkenes as part of the products. The reaction scope includes aryl, alkyl, vinyl, and cyclopropyl alkyne functionalities (Eq. 39).[63]

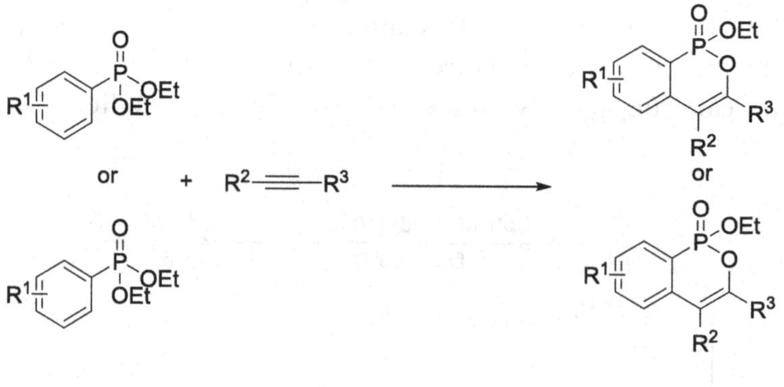

(Eq. 38)

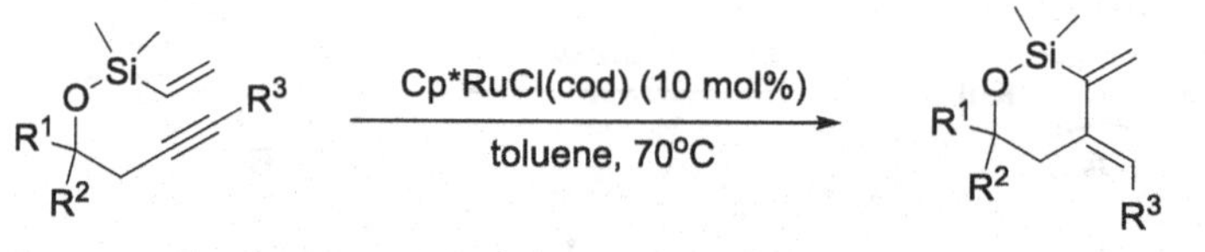

(Eq. 39)

6. Conclusions

By employing suitable ruthenium catalyst, a variety of methodologies for the synthesis of six-membered heterocycles, including nitrogen-, oxygen-, oxaza- and other heteroatom-ones, have been developed in an atom- and step-economic manner. The construction of six-membered heterocycles is based on the key cycloaddition, cycloisomerization, oxidative cyclization (or oxidative coupling), cross-dehydrogenative coupling, transfer hydrogenative coupling, hydrogenation, addition, Diels–Alder, and other related reactions. These fundamental examples have proved the versatility and great potential of ruthenium in catalysis.

The continued research efforts to understand the structure–activity relationship will definitely result in developing more active task-specific catalysts. The development of new synthetic methodologies by continued evolution of ruthenium catalysis is expected to tackle a wide range of challenging and highly valuable organic synthetic issues. Moreover, it is conceivable that getting insight into the reaction modes including the chemical bond forming processes will help to realize the related end and enrich the connotation of ruthenium catalysis.

References

1. Y. Yamamoto, K. Kinpara, R. Ogawa, H. Nishiyama and K. Itoh, 2006. *Chem. Eur. J.*, 12, 5618–5631.
2. F. Xu, C. X. Wang, X. C. Li and B. S. Wan, 2012. *ChemSusChem*, 5, 854–857.
3. J. C. Wu, W. B. Xu, Z. X. Yu and J. Wang, 2015. *J. Am. Chem. Soc.*, 137, 9489–9496.
4. M. Movassaghi and M. D. Hill, 2006. *J. Am. Chem. Soc.*, 128, 4592–4593.
5. B. M. Trost and A. C. Gutierrez, 2007. *Org. Lett.*, 9, 1473–1476.
6. M. N. Zhao, R. R. Hui, Z. H. Ren, Y. Y. Wang and Z. H. Guan, 2014. *Org. Lett.*, 16, 3082–3085.
7. S. Murugesan, F. Jiang, M. Achard, C. Bruneau and S. Derien, 2012. *Chem. Commun.*, 48, 6589–6591.

8. P. R. Walker, C. D. Campbell, A. Suleman, G. Carr and E. A. Anderson, 2013. *Angew. Chem. Int. Ed.*, 52, 9139–9143.

9. B. M. Trost, N. Maulide and R. C. Livingston, 2008. *J. Am. Chem. Soc.*, 130, 16502–16503.

10. X. Zhang, W. B. Liu, Q. F. Wu and S. L. You, 2013. *Org. Lett.*, 15, 3746–3749.

11. Z. P. Yang, C. X. Zhuoa and S. L. You, 2014. *Adv. Synth. Catal.*, 356, 1731–1734.

12. C. Vovard-Le Bray, H. Klein, P. H. Dixneuf, A. Mace, F. Berree, B. Carboni and S. Derien, 2012. *Adv. Synth. Catal.*, 354, 1919–1925.

13. F. Monnier, C. Vovard-Le Bray, D. Castillo, V. Aubert, S. Derien, P. H. Dixneuf, L. Toupet, A. Ienco and C. Mealli, 2007. *J. Am. Chem. Soc.*, 129, 6037–6049.

14. C. S. Cho, B. T. Kim, H. J. Choi, T. J. Kimb and S. C. Shim, 2003. *Tetrahedron*, 59, 7997–8002.

15. R. N. Monrad and R. Madsen, 2011. *Org. Biomol. Chem.*, 9, 610–615.

16. C. S. Cho, T. K. Kim, B. T. Kim, T. J. Kim and S. C. Shim, 2002. *J. Organomet. Chem.*, 650, 65–68.

17. F. Xie, M. Zhang, M. M. Chen, W. lv and H. F. Jiang, 2015. *ChemCatChem*, 7, 349–353.

18. C. S. Yi and S. Y. Yun, 2005. *J. Am. Chem. Soc.*, 127, 17000–17006.

19. A. Varela-Fernandez, J. A. Varela and C. Saa, 2011. *Adv. Synth. Catal.*, 353, 1933–1937.

20. Z. Y. Ding, T. L. Wang, Y. M. He, F. Chen, H. F. Zhou, Q. H. Fan, Q. X. Guo and A. S. C. Chan, 2013. *Adv. Synth. Catal.*, 355, 3727–3735.

21. Z. Wu, M. Perez, M. Scalone, T. Ayad and V. Ratovelomanana-Vidal, 2013. *Angew. Chem. Int. Ed.*, 52, 4925–4928.

22. T. L. Wang, F. Chen, J. Qin, Y. M. He and Q. H. Fan, 2013. *Angew. Chem. Int. Ed.*, 52, 7172–7176.

23. J. W. Zhang, F. Chen, Y. M. He and Q. H. Fan, 2015. *Angew. Chem. Int. Ed.*, 54, 4622–4625.

24. B. Xiong, Y. Li, W. Lv, Z. D. Tan, H. F. Jiang and M. Zhang, 2015. *Org. Lett.*, 17, 4054–4057.

25. L. Ackermann, L. H. Wang and A. V. Lygin, 2012. *Chem. Sci.*, 3, 177–180.

26. W. B. Ma, K. Graczyk and L. Ackermann, 2012. *Org. Lett.*, 14, 6318–6321.

27. N. Kavitha, G. Sukumar, V. P. Kumar, P. S. Mainkar and S. Chandrasekhar, 2013. *Tetrahedron Lett.*, 54, 4198–4201.

28. R. Wang and J. R. Falck, 2014. *J. Organomet. Chem.*, 759, 33–36.

29. J. Li and L. Ackermann, 2014. *Tetrahedron*, 70, 3342–3348.

30. S. Ruiz, P. Villuendas, M. A. Ortuno, A. Lledos and E. P. Urriolabeitia, 2015. *Chem. Eur. J.*, 21, 8626–8636.

31. R. K. Chinnagolla, S. Pimparkara and M. Jeganmohan, 2012. *Org. Lett.*, 14, 3032–3035.

32. R. K. Chinnagolla, S. Pimparkar and M. Jeganmohan, 2013. *Chem. Commun.*, 49, 3703–3705.

33. B. X. Tang, R. J. Song, C. Y. Wu, Z. Q. Wang, Y. Liu, X. C. Huang, Y. X. Xie and J. H. Li, 2011. *Chem. Sci.*, 2, 2131–2134.

34. R. Manikandan and M. Jeganmohan, 2014. *Org. Lett.*, 16, 3568–3571.

35. L. Ackermann, A. V. Lygin and N. Hofmann, 2011. *Angew. Chem. Int. Ed.*, 50, 6379–6382.

36. L. Ackermann and S. Fenner, 2011. *Org. Lett.*, 13, 6548–6551.

37. M. Deponti, S. I. Kozhushkov, D. S. Yufitb and L. Ackermann, 2013. *Org. Biomol. Chem.*, 11, 142–148.

38. B. Li, H. L. Feng, S. S. Xu and B. Q. Wang, 2011. *Chem. Eur. J.*, 17, 12573–12577.

39. B. Li,. H. L. Feng, N. C. Wang, J. F. Ma, H. B. Song, S. S. Xu and B. Q. Wang, 2012. *Chem. Eur. J.*, 18, 12873–12879.

40. M. C. Reddy, R. Manikandan and M. Jeganmohan, 2013. *Chem. Commun.*, 49, 6060–6062.

41. S. Allu and K. C. K. Swamy, 2014. *J. Org. Chem.* 79, 3963–3972.

42. L. Ackermann, A. V. Lygin and N. Hofmann, 2011. *Org. Lett.*, 13, 3278–3281.

43. C. S. Cho and S. G. Oh, 2006. *Tetrahedron Lett.*, 47, 5633–5636.

44. F. Xie, M. Zhang, H. F. Jiang, M. M. Chen, W. Lv, A. B. Zheng and X. J. Jian, 2015. *Green. Chem.*, 17, 279–284.

45. A. J. A. Watson, A. C. Maxwell and J. M. J. Williams, 2012. *Org. Biomol. Chem.*, 10, 240–243.

46. M. M. Chen, M. Zhang, B. Xiong, Z. D. Tan, W. Lv and H. F. Jiang, 2014. *Org. Lett.*, 16, 6028–6031.

47. B. M. Trost, M. T. Rudd, M. G. Costa, P. I. Lee and A. E. Pomerantz, 2004. *Org. Lett.*, 6, 4235–4238.

48. R. Manikandan and M. Jeganmohan, 2014. *Org. Lett.*, 16, 652–655.

49. K. Kanao, Y. Miyake and Y. Nishibayashi, 2010. *Organomet.*, 29, 2126–2131.

50. J. Gavenonis, R. V. Arroyo and M. L. Snapper, 2010. *Chem. Commun.*, 46, 5692–5694.

51. H. Lee and C. S. Yi, 2015. *Eur. J. Org. Chem.*, 1899–1904.

52. V. S. Thirunavukkarasu, M. Donati and L. Ackermann, 2012. *Org. Lett.*, 14, 3416–3419.

53. R. K. Chinnagolla and M. Jeganmohan, 2012. *Chem. Commun.*, 48, 2030–2032.

54. Ackermann, J. Pospech, K. Graczyk and K. Rauch, 2012. *Org. Lett.*, 14, 930–933.

55. K. Inamoto, J. Kadokawa and Y. Kondo, 2013. *Org. Lett.*, 15, 3962–3965.

56. Y. Yamamoto, K. Matsui and M. Shibuya, 2014. *Org. Lett.*, 16, 1806–1809.

57. R. Prakash, K. Shekarrao, S. Gogoi and R. Boruah, 2015. *Chem. Commun.*, 51, 9972–9974.

58. K. R. Flower, A. P. Lightfoot, H. Wan, and A. Whiting, 2002. *J. Chem. Soc., Perkin Trans.*, 1, 2058–2064.

59. Z. F. Xiao, C. Z. Yao and Y. B. Kang, 2014. *Org. Lett.*, 16, 6512–6514.

60. G. Broggini, G. Poli, E. M. Beccalli, F. Brusa, S. Gazzola and J. Oble, 2015. *Adv. Synth. Catal.*, 357, 677–682.

61. Y. Yamamoto, H. Takagishi and K. Itoh, 2002. *J. Am. Chem. Soc.*, 124, 28–29.

62. Y. Park, I. Jeon, S. Shin, J. Min and P. H. Lee, 2013. *J. Org. Chem.*, 78, 10209–10220.

63. L. Kaminsky and D. A. Clark, 2014. *Org. Lett.*, 16, 5450–5453.

CHAPTER 3

Rh-Catalyzed Five-Membered Heterocycle Synthesis

Subban Kathiravan* and Ian A. Nicholls[†]

*Bioorganic & Biophysical Chemistry Laboratory,
Linnaeus University Centre for Biomaterials Chemistry
and Department of Chemistry & Biomedical Sciences,
Linnaeus University, Kalmar, SE-391 82, Sweden
suppan.kathiravan@lnu.se
[†]Department of Chemistry — BMC, Uppsala University,
Uppsala, SE-751 23, Sweden
ian.nicholls@lnu.se

1. Introduction

Heterocycles constitute the largest and most diverse family of organic compounds. Among them, five-membered heterocycles are structural motifs found in a great number of biologically active natural and synthetic compounds, with a broad spectrum of uses, e.g. as drugs, and agrochemicals. Moreover, five-membered heterocycles are widely used for synthesis of dyes and polymeric materials of high value.[1] There are numerous reports on employment of five-membered heterocycles as intermediates in organic synthesis.[2] Although a variety of highly efficient methodologies for the synthesis of aromatic heterocycles and their derivatives has been reported in the past, the development of novel methodologies is in continuous demand. Of particular interest is the development of new synthetic approaches towards five-membered heterocycles, aiming at achieving greater levels of molecular complexity and better versatility with respect to functional group compatibilities, their use in convergent and atom-economical fashions, readily accessible starting materials and under mild reaction conditions, which are ever goals for modern synthetic organic chemistry. Transition metal-catalyzed transformations, which often help to meet the above criteria, are among the most attractive synthetic tools.

The transition metals palladium, ruthenium, and rhodium are especially important in this context, not only because they are used very frequently, but especially because of the wide variety of reactions that they can catalyze. While the importance of palladium and ruthenium is reflected in the large body of literature describing their use in chemical catalysis, the importance of rhodium remains comparatively neglected. The primary use of the element is as a catalyst in automotive exhausts, so-called three-way catalytic converters, the production of which consumes around 80% of world rhodium production.[3] Rhodiums can adopt oxidation states ranging from +0 to + 6, with + 3 being the most commonly observed. In this chapter, we present an overview of the use of rhodium in catalyzing the synthesis of five-membered heterocycles with a particular focus on recent methodological developments.

Focus of this chapter

We focus here on the rhodium-catalyzed synthesis of five-membered heterocycles. We have divided this area according to the heteroatom present, namely, N, O, S, P, and Si. Accordingly, this chapter is divided into five major sections. Section 1.2 covers rhodium-catalyzed nitrogen containing five-membered heterocycle synthesis and includes a particularly broad range of reactions and here we have presented some of the most significant reports where the major objective of the work was to develop novel five-membered heterocycle syntheses. The second (Section 1.3), deals with rhodium catalyzed oxygen containing five-membered heterocycle synthesis with the exception of oxazole and oxazolidinones since both the heterocycles also contain nitrogen in it. Section 1.4 describes the five-membered sulfur containing heterocycle synthesis. Finally Sections 1.5 and 1.6 cover the Rh catalyzed synthesis of P and Si containing five-membered heterocycles.

1.2 Rhodium-catalyzed nitrogen containing five-membered heterocycle synthesis

1.2.1 *Pyrrole*

Pyrroles are an significant class of heterocyclic compounds that are broadly used in synthetic organic chemistry and materials science.[4] Due to their distinctive properties, extensive investigations have been made to develop preparative methods for substituted pyrroles.[5] In general, 1,2,3,4-tetrasubstituted pyrroles have been prepared by Knorr reactions, Hantzsch pyrrole synthesis, or the 1,3-dipole addition of azomethine ylides with alkynes.[6] There are many efficient methods for the synthesis of this important class of heterocycle, but all of the various approaches have certain restrictions regarding the scope and placement of the substitution pattern around the heterocycle core.

In 2011, Gevorgyan and co-workers revealed the first transannulation of 1,2,3-triazoles with terminal alkynes into pyrroles under intriguingly mild reaction conditions when binary $Rh_2(oct)_4$/ $AgOCOCF_3$ catalyst system providing a straightforward approach to 1,2,4-trisubstituted pyrroles in good to excellent yields. A mechanistic rationale, involving a direct nucleophilic attack of an electron-rich alkyne at the Rh–iminocarbene intermediate, followed by cyclization step was proposed for this new transannulation reaction (Scheme 1).[7]

Davies and co-workers developed rhodium-catalyzed conversion of furans to highly functionalized pyrroles. The reaction features an initial [3 + 2] cycloaddition across the furan C2–C3 bond, to form bicyclic hemiaminals, followed by termination of the cascade by

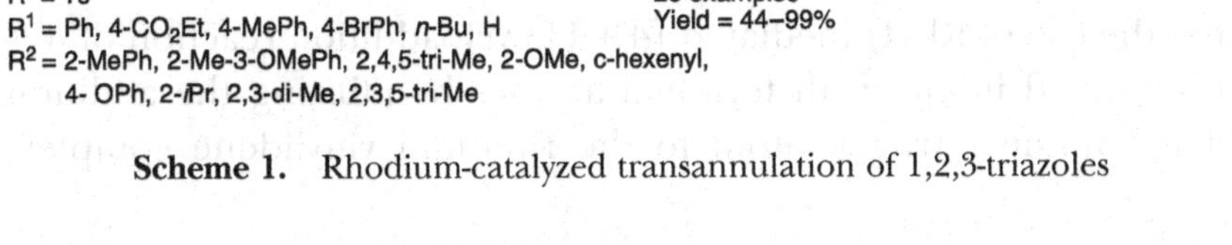
Rh2(oct)2 (2.5 mol%)
AgOCOCF3 (5 mol%)
Hexane, 70°C, 3–20 h
R = Ts
R1 = Ph, 4-CO2Et, 4-MePh, 4-BrPh, n-Bu, H
R2 = 2-MePh, 2-Me-3-OMePh, 2,4,5-tri-Me, 2-OMe, c-hexenyl, 4- OPh, 2-iPr, 2,3-di-Me, 2,3,5-tri-Me
26 examples
Yield = 44–99%
Rh2(oct)2

Scheme 1. Rhodium-catalyzed transannulation of 1,2,3-triazoles

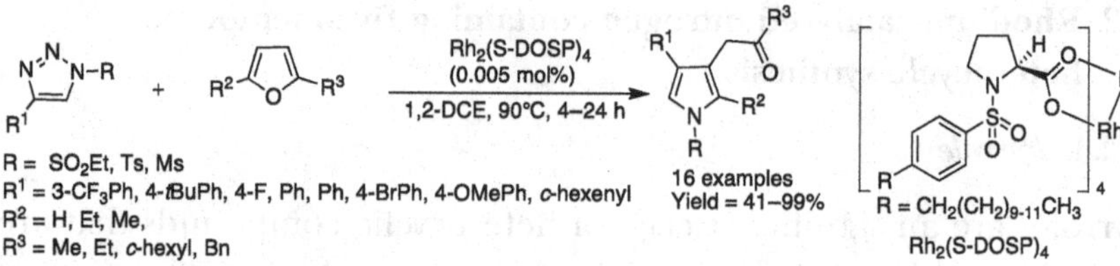

R = SO$_2$Et, Ts, Ms
R^1 = 3-CF$_3$Ph, 4-tBuPh, 4-F, Ph, Ph, 4-BrPh, 4-OMePh, c-hexenyl
R^2 = H, Et, Me,
R^3 = Me, Et, c-hexyl, Bn

Rh$_2$(S-DOSP)$_4$
(0.005 mol%)
1,2-DCE, 90°C, 4–24 h

16 examples
Yield = 41–99%

R = CH$_2$(CH$_2$)$_{9-11}$CH$_3$
Rh$_2$(S-DOSP)$_4$

Scheme 2. Rhodium-catalyzed conversion of furan into pyrroles

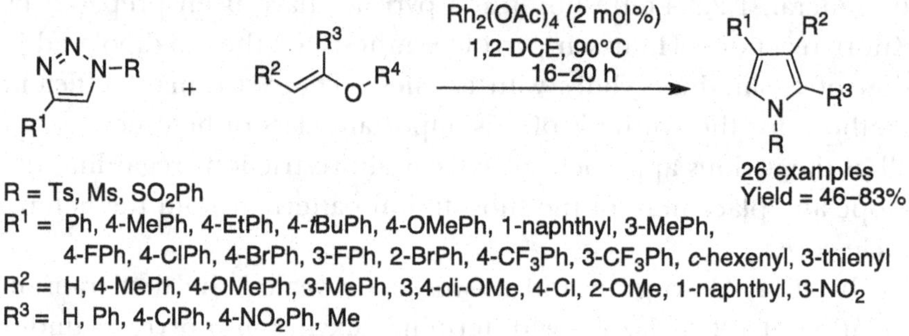

R = Ts, Ms, SO$_2$Ph
R^1 = Ph, 4-MePh, 4-EtPh, 4-tBuPh, 4-OMePh, 1-naphthyl, 3-MePh,
4-FPh, 4-ClPh, 4-BrPh, 3-FPh, 2-BrPh, 4-CF$_3$Ph, 3-CF$_3$Ph, c-hexenyl, 3-thienyl
R^2 = H, 4-MePh, 4-OMePh, 3-MePh, 3,4-di-OMe, 4-Cl, 2-OMe, 1-naphthyl, 3-NO$_2$
R^3 = H, Ph, 4-ClPh, 4-NO$_2$Ph, Me

Rh$_2$(OAc)$_4$ (2 mol%)
1,2-DCE, 90°C
16–20 h

26 examples
Yield = 46–83%

Scheme 3. Rhodium-catalyzed transannulation of 1,2,3-triazoles with vinyl ethers

concomitant elimination/aromatization to generate trisubstituted pyrrole nucleus. The reaction was found to be highly dependent on both the solvent and the dirhodium catalyst (Scheme 2).[8]

A similar mechanistic scenario was proposed by Anbarasan and co-workers who forced the transannulation of 1,2,3-triazoles to access the polysubstituted pyrroles by applying rhodium catalysts under mild condition with substituted vinyl ethers. This method allows the synthesis of mono, di, and trisubstituted pyrroles with appropriate substitutions. Furthermore, the developed methodology was applied in the formal synthesis of neolamellarin A, an antitumor agent (Scheme 3).[9]

The application of rhodium catalysts in pyrrole synthesis was further explored by Iwasawa and co-workers, who demonstrated that the use of Rh(I) mediated [4 + 1] cycloaddition reaction of α,β-unsaturated imines with terminal alkynes by utilizing the addition of the imine nitrogen atom to the rhodium vinylidene complex.

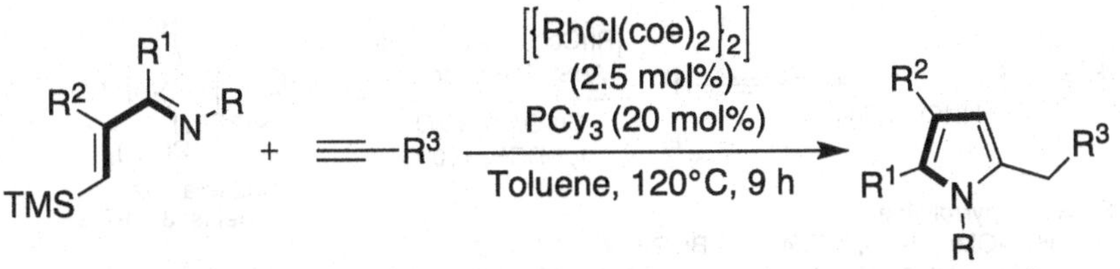

R = Me, Bn
R^1 = Ph, C=C(Me)$_2$, propynyl, 2-furyl, 2-thienyl, Ph
R^2 = H, Me
R^3 = c-hexyl, CH$_2$CH$_2$OTBS, n-hexyl

12 examples
Yield = 50–72%

Scheme 4. Rhodium-catalyzed [4 + 1] cycloaddition reaction of α,β-unsaturated imines with terminal alkynes

This reaction also demonstrates another use of the rhodium vinylidene complex, namely as a reaction substrate. Moreover, the reaction proved to be tolerant towards an array of functional groups as well as different, more sterically demanding substitution patterns in the substrate (Scheme 4).[10]

1.2.2 *Indole*

Indole is one of the most common motifs in natural bioactive products, drugs, and other functional materials.[11] There is a continued interest in the development of simple and general synthetic methods to access this structural moiety. Of the various methods developed, the Fischer indole synthesis has remained one of the most widely adopted approaches.[12] Over the past few decades, substantial research has led to the introduction of efficient procedures that allow for the selective synthesis of indoles through rhodium-catalyzed reactions.[13] Based on Larock-type Pd-catalyzed cross-coupling indole synthesis,[14] many rhodium-catalyzed oxidative couplings towards substituted indoles have been reported by our group (Scheme 5),[15] Fagnou (Scheme 6),[16] Ackermann,[17] and others.[18] More recently the redox-neutral coupling with an oxidizing directing group as an internal oxidant has emerged as an attractive strategy in indole synthesis (Scheme 7).[19,20]

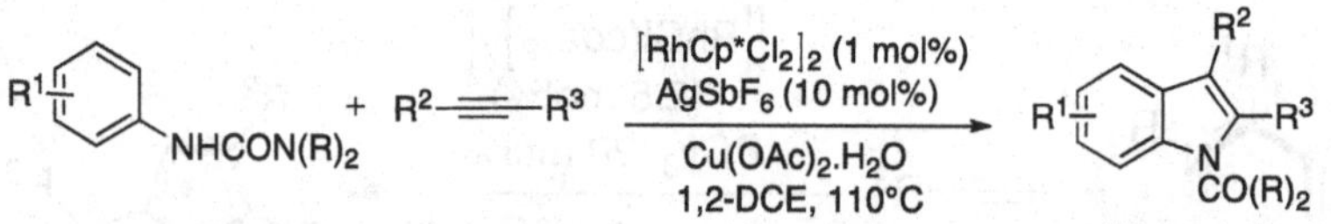

33 examples
Yields: 30–92%

R = Me, pyrrolidine
R_1 = H, 4-OMe, 2-Me, 4-F, 4-Cl, 4-Br, 2-OMe, 3,4-di-OMe, 4-ethyl, 4-COOMe, 4-NO_2, 4-CN, 4-Me, 3-Me, 3,4-di-Me, 2-F, 2.F, 3,4-di-Cl
R_2 = Ph, ethyl, *n*-propyl, *n*-butyl, 4-MePh, 4-OMePh, 4-ClPh, 2-naphthyl, 1-naphthyl, thiophene
R_3 = Ph, ethyl, *n*-propyl, *n*-butyl, 4-MePh, 4-OMePh, 4-ClPh, 2-naphthyl, 1-naphthyl, thiophene

Scheme 5. Rhodium(III)-catalyzed synthesis of Indole from N-arylurea

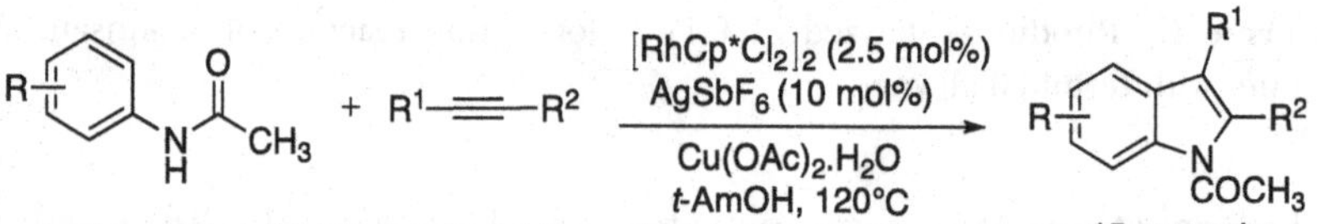

R = 4-OMe, 4-F, 4-COOMe, 4-Cl, 2-Me, naphthyl, 2,4,-di-OMe, 3-OMe
R_1 = Me, Ph, *n*-propyl, *n*-hexyl
R_2 = Ph, *n*-propyl, *n*-hexyl, thiophene, 4-N-Ts-indole

15 examples
Yields: 47–83%

Scheme 6. Rhodium(III)-catalyzed synthesis of Indole

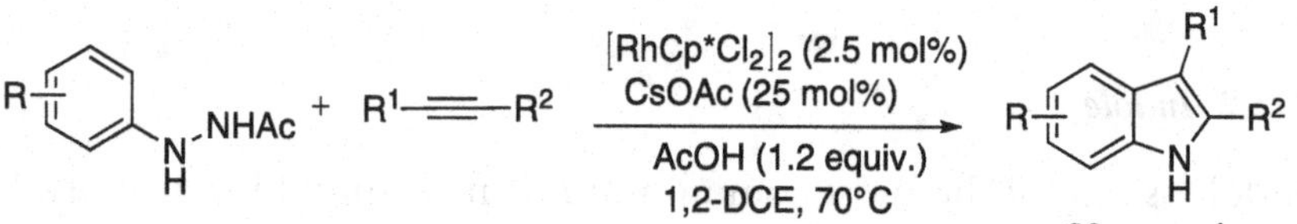

R = H, 3-Me, 4-me, 3,4-di-Me, 4-OMe, 4-Ph, 2-F, 4-F, 4-Cl, 4-Br, 3,4-di-Cl, 3-I-4-me, 4-OCF_3, 4-CO_2Me, 4-CN
R^1 = 4-MePh, 4-OMePh, 3-OMePh, 4-ClPh, 3-ClPh, 4-FPh, Et, n-Pr, n-Bu, Me
R^2 = 4-MePh, 4-OMePh, 3-OMePh, 4-ClPh, 3-ClPh, 4-FPh, Et, n-Pr, n-Bu, Ph

30 examples
Yields: 41–92%

Scheme 7. Rhodium(III)-catalyzed redox-neutral synthesis of Indole

1.2.3 *Dihydropyrrole*

With the notion that a rhodium(II) catalyst such as Rh_2(*S*-NTTL)$_4$, Rh_2(*S*-4-Br-NTTL)$_4$, or Rh_2(*S*-TFPTTL)$_4$, Fokin and co-workers explored the rhodium-catalyzed asymmetric annulation of *N*H-1,2,3-triazoles with olefins to develop a convenient one-pot synthesis of 2,3-dihydropyrroles. This procedure further expands the utility of azavinyl carbene chemistry and provides access to an important class

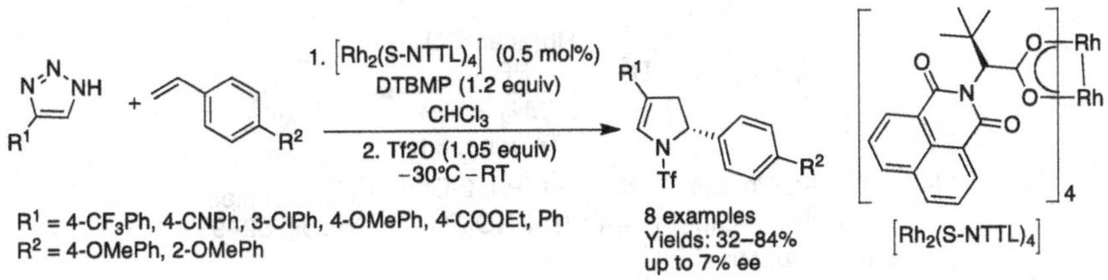

Scheme 8. Rhodium-catalyzed asymmetric transannulation of NH-1,2,3-triazoles

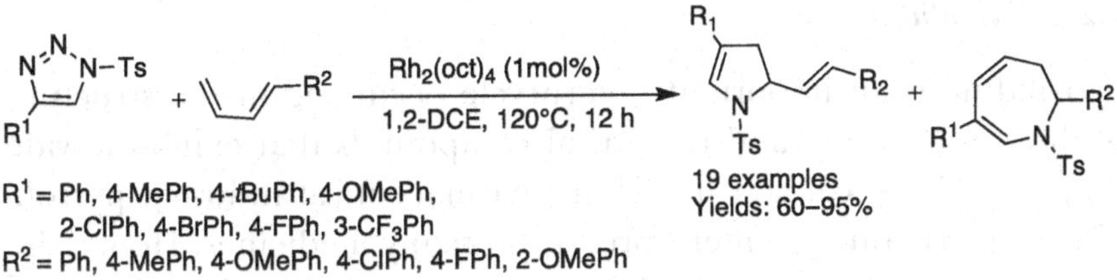

Scheme 9. Divergent synthesis of nitrogen heterocycles by Rh(II) catalyst

of cyclic enamides. In their report from 2014, some mechanistic insight could be obtained when subjecting *in situ* generated triflated triazoles to the reaction conditions (Scheme 8).[21]

In addition to NH-1,2,3-triazoles, the benefit of five-membered heterocycles in rhodium(II) catalysis the N-tosyl triazoles was also equally recognized. In 2014, Shang and co-workers noted that 1-sulfonyl 1,2,3-triazoles undergo transannulation most likely in a fast [3 + 2] cycloaddition of rhodium azavinyl intermediate when bound to the Rh(II) catalyst to yield valuable dihydropyrroles. However, this reactivity was also resulted in [4 + 3] cycloadditions and also provide 2,5-dihydroazepines *via* aza-Cope rearrangement (Scheme 9).[22]

Zhang and co-workers developed the rhodium-catalyzed intramolecular hydroaminomethylation reaction, for the generation of 4-aryl-2,3-dihydropyrroles. Triphenylphosphine has proved to be an excellent ligand for preparing dihydropyrrole derivatives in up to 99% yield. This procedure provides an efficient and simple alternative route to synthesize 4-aryl-2,3-dihydropyrroles only in one step (Scheme 10).[23]

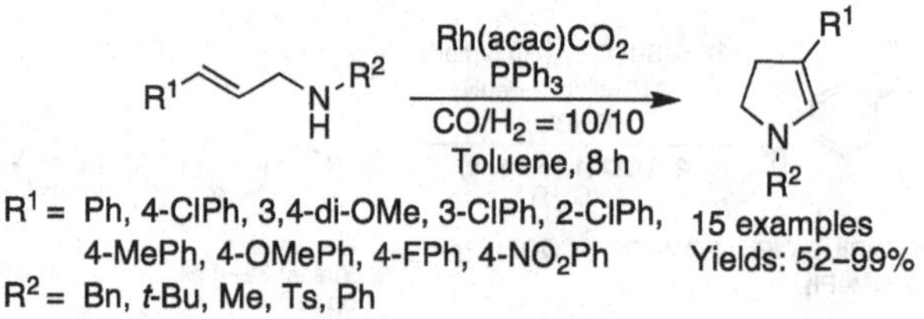

$$R^1 \diagdown \diagdown_N^{R^2} \xrightarrow[\text{CO/H}_2 = 10/10]{\substack{\text{Rh(acac)CO}_2 \\ \text{PPh}_3}} \overset{R^1}{\underset{\underset{R^2}{N}}{\bigcirc}}$$

$R^1 =$ Ph, 4-ClPh, 3,4-di-OMe, 3-ClPh, 2-ClPh,   15 examples
      4-MePh, 4-OMePh, 4-FPh, 4-NO$_2$Ph   Yields: 52–99%
$R^2 =$ Bn, t-Bu, Me, Ts, Ph

Scheme 10. Rhodium-catalyzed intramolecular hydroaminomethylation reaction

1.2.4 Pyrrolidine

Pyrrolidine is an important heterocycle occurring in the structure of different natural and unnatural compounds that exhibit a wide range of biological activities.[24] In pharmaceutical industry, pyrrolidine is a structural element present in many medicines. Hence the development of new methodologies towards the synthesis of functionalized pyrrolidines remains a highly appreciated research area for synthetic organic chemists.[25]

In 2008, Hartwig and co-workers introduced rhodium-catalyzed intramolecular hydroamination of unactivated terminal and internal alkenes with primary and secondary amines for the synthesis of pyrrolidines. In addition to spanning a broader scope of amine and alkene than other late metal catalysts, this system operates in the presence of functional groups, including esters and free hydroxyl groups, which do not tolerate early metal and lanthanide catalysts. These results demonstrate that complexes of group nine metals, which are classic catalysts for alkene hydrogenation and hydrosilylation, can also be made useful for alkene hydroamination. Moreover, the rhodium catalyst also increased the scope of amines that could be added to alkenes (Scheme 11).[26]

The synthetic utility of rhodium with unactivated alkenes was demonstrated by Buchwald and co-workers in the synthesis of a series of enantioenriched 2-methylpyrrolidines. Enantioselectivities up to 91% ee were obtained. The nature of the nitrogen atom-protecting group has a pronounced influence on the outcome of the reaction. Presumably, the substituent at the 2-position results in a more ordered transition state. Increasing the size of this

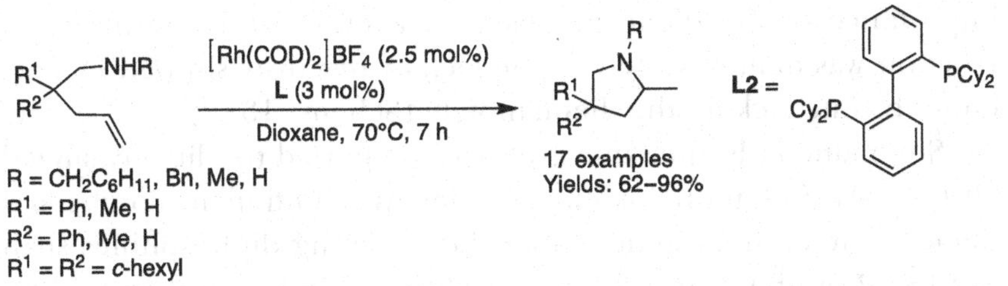
[Rh(COD)2]BF4 (2.5 mol%)
L (3 mol%)
Dioxane, 70°C, 7 h
17 examples
Yields: 62–96%
L2 =
PCy2
Cy2P
R = CH2C6H11, Bn, Me, H
R1 = Ph, Me, H
R2 = Ph, Me, H
R1 = R2 = c-hexyl

Scheme 11. Rhodium-catalyzed intramolecular hydroamination

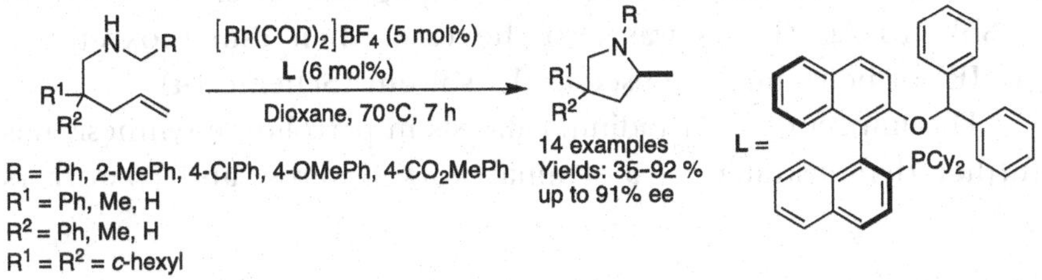
[Rh(COD)2]BF4 (5 mol%)
L (6 mol%)
Dioxane, 70°C, 7 h
14 examples
Yields: 35–92 %
up to 91% ee
L =
PCy2
R = Ph, 2-MePh, 4-ClPh, 4-OMePh, 4-CO2MePh
R1 = Ph, Me, H
R2 = Ph, Me, H
R1 = R2 = c-hexyl

Scheme 12. Rhodium-catalyzed asymmetric intramolecular hydroamination

substituent or using a 2,6-dimethylbenzyl protecting group, however elicited a profound decrease in reactivity, as a result of inhibition of metal binding. Varying the para substituent on the aryl ring of the nitrogen atom-protecting group has little effect on the enantioselectivity, but electron-donating substituents retarded the reaction and resulted in a lower yield (Scheme 12).[27]

In 2010, Li and co-workers noted the potential application of rhodium catalysts in the synthesis of pyrroles *via* conjugated diene-assisted activation of allylic C–H bonds and its addition to the alkene of the conjugated diene moiety in ene-2-diene substrates. The reaction of nitrogen-tethered ene-2-diene substrate in the presence of Rh(I) catalyst (generated from $RhCl(PPh_3)_3$ and $AgSbF_6$) underwent cyclization to form substituted pyrrole rather than the [4 + 2] cycloadduct. The proposed mechanism involves intermediate A, a Rh complex coordinated by both diene and ene, followed by allylic C–H activation to generate allyl–Rh–H complex B, followed by reductive elimination to afford desired product. Moreover, in this reaction highly chemoselective C–H activation was achieved as only the allylic C–H activation of the ene component was activated. The importance of the diene in this C–H activation/alkene insertion reaction was demonstrated by the lack of reactivity showed by ene-ene substrate lacking the diene moiety (Scheme 13).[28]

Simultaneously, the same group also reported rhodium-catalyzed allylic C–H activation for enantioselective addition to conjugated dienes. The authors noticed that the chelating diphosphines such as BINAP inhibited the reaction, whereas chiral phosphoramidites gave high yield and promising ee value. Moreover, enhanced enantioselectivity was obtained upon changing the silver source from $AgSbF_6$ to AgOTf, as was also the case when dimethoxyethane (DME) or benzene were used as the solvent (Scheme 14).[29]

The impact of the rhodium catalysts in pyrrolidine synthesis was further demonstrated in a seminal work by Frost and co-workers

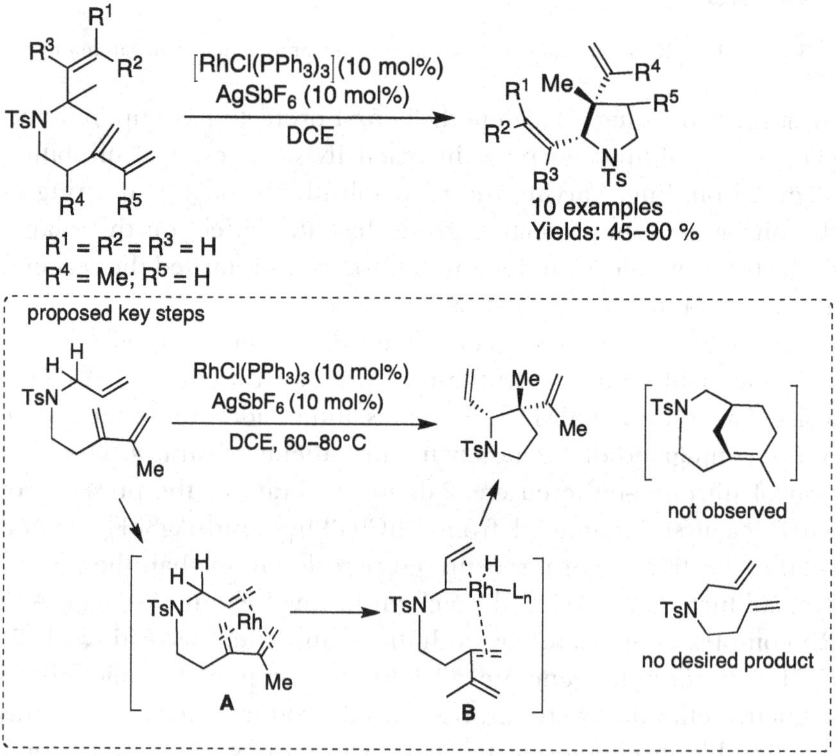

Scheme 13.   Rhodium-catalyzed allylic C–H bond activation

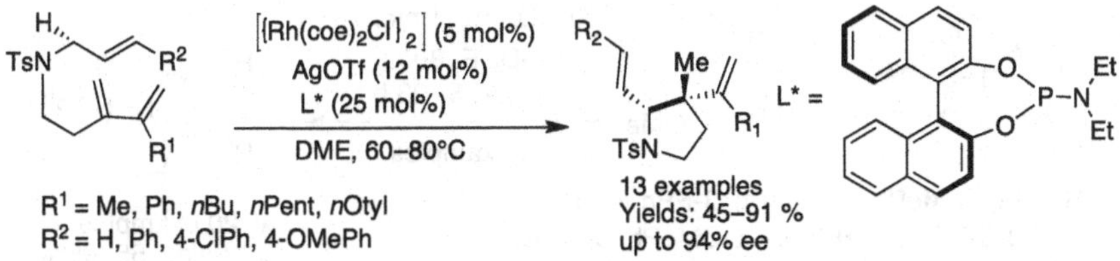

R^1 = Me, Ph, nBu, nPent, nOtyl
R^2 = H, Ph, 4-ClPh, 4-OMePh

[Rh(coe)$_2$Cl]$_2$ (5 mol%)
AgOTf (12 mol%)
L* (25 mol%)

DME, 60–80°C

13 examples
Yields: 45–91 %
up to 94% ee

Scheme 14. Rhodium-catalyzed asymmetric allylic C–H bond activation

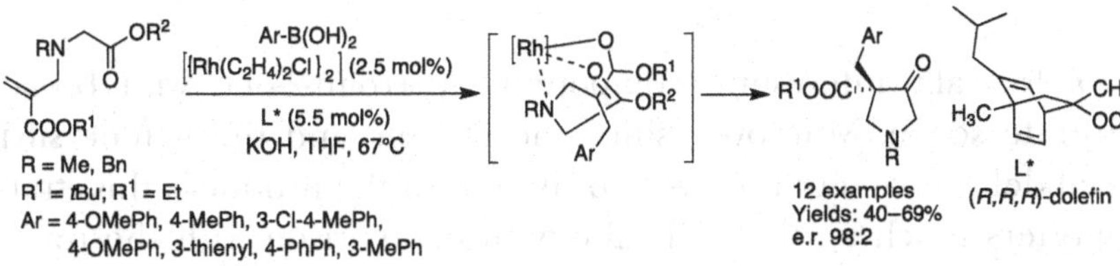

R = Me, Bn
R^1 = tBu; R^1 = Et
Ar = 4-OMePh, 4-MePh, 3-Cl-4-MePh,
4-OMePh, 3-thienyl, 4-PhPh, 3-MePh

Ar-B(OH)$_2$
[Rh(C$_2$H$_4$)$_2$Cl]$_2$ (2.5 mol%)

L* (5.5 mol%)
KOH, THF, 67°C

12 examples
Yields: 40–69%
e.r. 98:2

L*
(R,R,R)-dolefin

Scheme 15. Rhodium-catalyzed asymmetric enantioselective Dieckmann type annulation

(Scheme 15).[30] There in, intramolecular reaction of a rhodium enolate with an ester to sequentially install an aryl group and a ketone across an activated alkene with the concomitant formation of quaternary stereogenic center. The principal challenge in this asymmetric process is that the enantioselectivity is determined at the acylation step and not as common in enantioselective conjugate addition reactions, at the insertion step.

1.2.5 *Pyrrolin-2-one*

3-pyrrolin-2-ones, related to pyrroles, are useful building blocks for the construction of pyrroles or γ-lactam derivative, and they are also essential structures of bioactive natural products and pharmaceuticals.[31] Li and co-workers have synthesized 3-pyrrolin-2-ones by rhodium-catalyzed transannulation of 1-sulfonyl-1,2,3-triazole with ketene silyl acetal. The α-imino rhodium carbenoids generated from 1-sulfonyl-1,2,3-triazole were applied to the [3 + 2] cycloaddition with ketene silyl acetal, offering a novel and straightforward synthesis

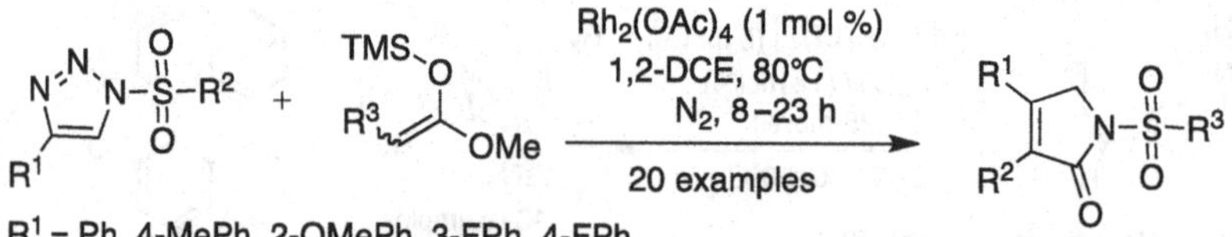

R^1 = Ph, 4-MePh, 2-OMePh, 3-FPh, 4-FPh,
    3-CF$_3$Ph, 4-BrPh, 4-COOMePh, 3-thienyl
R^2 = Me, 4-BrPh, 4-OMePh, 2-naphthyl, 2,4,6-tri-*i*prop-Ph,
    TMSCH$_2$CH$_2$, 4-MePh
R^3 = *n*-Bu, *n*-hexyl, Ph, 4-OMePh

20 examples
Yields: 55–95%

Scheme 16. Rhodium-catalyzed transannulation with vinyl ether

of biologically interesting compound 3-pyrrolin-2-one with broad substrate scope. Moreover, since the triazole and the ketene silyl acetal deliver each of the two substituents of the products, this strategy offers much synthetic flexibility in comparison with the traditional methods (Scheme 16).[32]

In 2014, Rovis and co-workers described enantioselective rhodium-catalyzed isomerization of 4-iminocrotonates. This reaction uses widely available amines to couple with 4-oxocrotonate to provide a convenient access to central chiral building block in good yield and high enantioselectivity. Although the mechanism of this new transformation remains unclear, both Rh and the phosphoramidate play a central role. A possible mechanism for the transformation could be the initial coordination of 4-iminocrotonate with rhodium(I)/phoroamidate complex, **A** isomerizes from *E* to *Z* configuration. At this point, the imine nitrogen is poised for intramolecular attack on the carbonyl group, affording *N*-acyliminium intermediate **B**. A resultant trapping of **B** by alkoxide delivers 5-alkoxy-3-pyrrolin-2-one and regenerates catalyst. Rhodium precatalysts bearing norbornadiene or cyclooctadiene ligands gave no product. Presumably, these strongly coordinating diene ligands impede coordination of the 4-iminocrotonate, thereby preventing the catalyst from entering into the catalytic cycle. Although 5-alkoxy-3-pyrrolin-2-ones generally produce *N*-acyliminiums in the presence of a Lewis acid catalyst, control experiments revealed that alkoxide attack on acyliminium ion **B** was irreversible under the reaction conditions. It is worth noting that intermediate **B** in the absence of

Rh coordination is putatively antiaromatic, which may suggest a further role for Rh in this reaction (Scheme 17).[33]

Hou and co-workers developed Rh(III) catalyzed addition of alkenyl C–H bond to isocyanates and intramolecular cyclization to synthesis pyrrolin-2-ones. This atom economical reaction affords simple and straightforward access to the biologically relevant 5-ylidene pyrrol-2(*5H*)-ones and can be carried out under mild and neutral conditions in the absence of any additives (Scheme 18).[34]

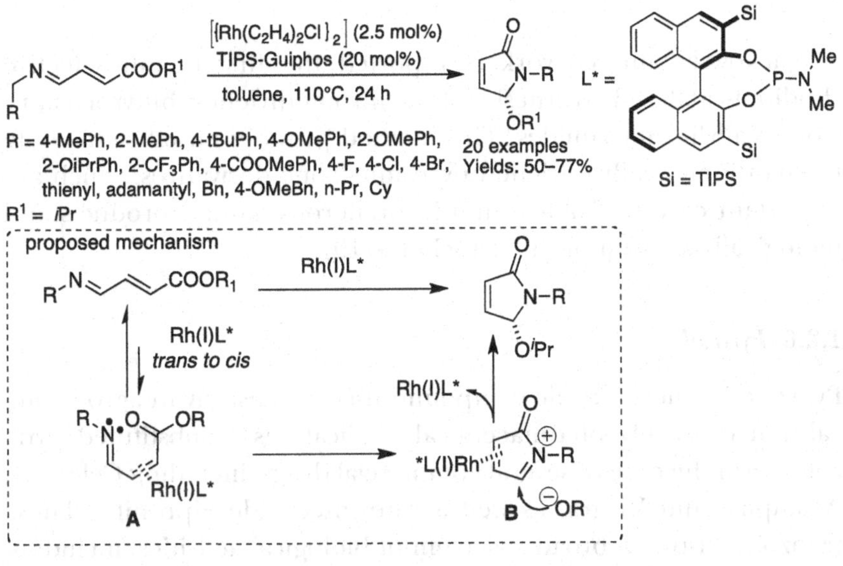
[{Rh(C2H4)2Cl}2] (2.5 mol%)
TIPS-Guiphos (20 mol%)
toluene, 110°C, 24 h
COOR1
N-R
OR1
L* =
Si
Me
P-N
Me
R = 4-MePh, 2-MePh, 4-tBuPh, 4-OMePh, 2-OMePh, 2-OiPrPh, 2-CF3Ph, 4-COOMePh, 4-F, 4-Cl, 4-Br, thienyl, adamantyl, Bn, 4-OMeBn, n-Pr, Cy
R1 = iPr
20 examples
Yields: 50–77%
Si = TIPS
proposed mechanism
Rh(I)L*
COOR1
Rh(I)L*
trans to cis
O^iPr
Rh(I)L*
*L(I)Rh
N-R
OR
A
B

Scheme 17. Rhodium-catalyzed intramolecular isomerization of iminocrotonates

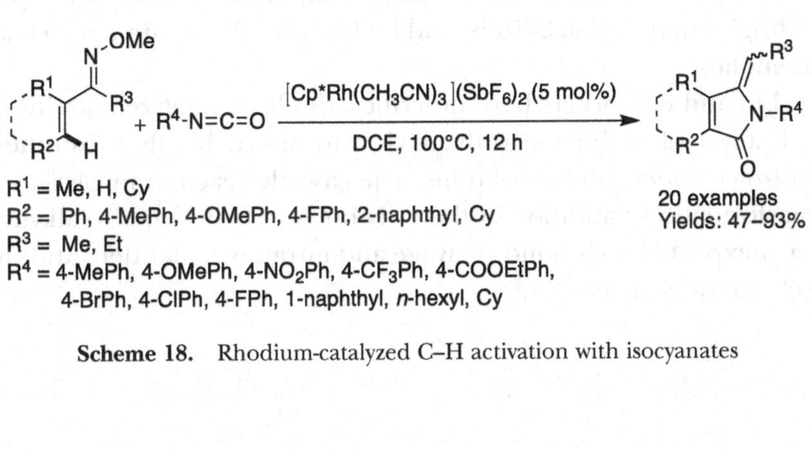
OMe
R1
R3
R2
H
+ R4-N=C=O
[Cp*Rh(CH3CN)3](SbF6)2 (5 mol%)
DCE, 100°C, 12 h
N-R4
R1 = Me, H, Cy
R2 = Ph, 4-MePh, 4-OMePh, 4-FPh, 2-naphthyl, Cy
R3 = Me, Et
R4 = 4-MePh, 4-OMePh, 4-NO2Ph, 4-CF3Ph, 4-COOEtPh, 4-BrPh, 4-ClPh, 4-FPh, 1-naphthyl, n-hexyl, Cy
20 examples
Yields: 47–93%

Scheme 18. Rhodium-catalyzed C–H activation with isocyanates

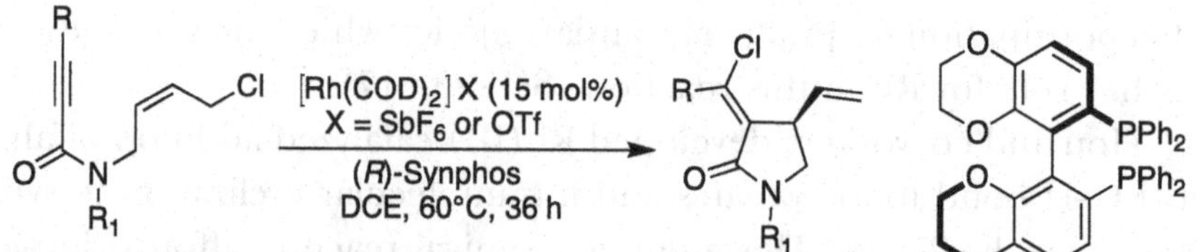

R = Me, Ph, 4-OMePh, H, c-propyl, Et, *n*-Pr, C_2H_4OBn, C_2H_4Cl
R_1 = Ts, Bn

13 examples
Yield = 35–91%

Scheme 19. Rhodium-catalyzed conversion of *N*-allylic alkynamides into pyrroline-2-one

Jackowski and co-workers reported the first enantioselective rhodium-catalyzed synthesis of α-chlromethylene-γ-butyrolactams from *N*-allylic alkynamides. This method provides an efficient route to enantiomerically enriched pyrroline-2-one derivatives, which are important core scaffolds found in numerous natural products and biologically active molecules (Scheme 19).[35]

1.2.6 *Pyrazole*

Pyrazole is a heterocyclic compound used extensively in agrochemical and especially pharmaceutical applications.[36] Substituted pyrazoles form the core of several commercial drugs, including Celebrex, Acomplia, and Viagra, as well as the insecticide Fipronil.[37] These pyrazoles display a broad spectrum of biological activities, including anti-inflammatory, analgesic, sedative, and hypnotic properties.[38] They can also act as ligand in coordination compounds, and serve as optical brighteners, UV stabilizers, and building blocks in supramolecular assemblies.

Liu and co-workers have described rhodium-catalyzed addition-cyclization of hydrazine with alkynes to afford highly substituted pyrazoles under mild conditions. The cascade reaction involves two transformations: addition of the C–N bond of hydrazines to alkynes *via* unexpected C–N bond cleavage and intramolecular dehydration cyclization (Scheme 20).[39]

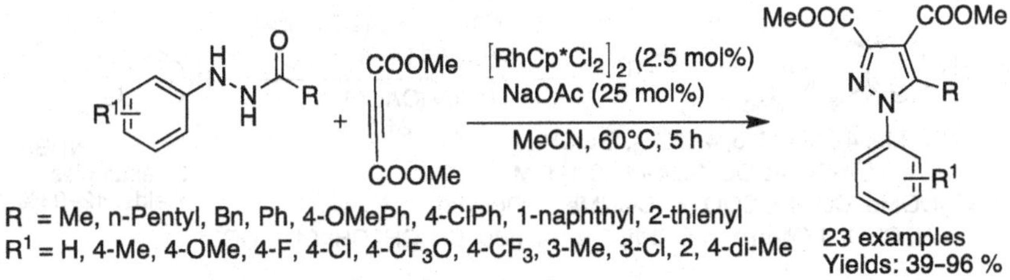
[RhCp*Cl2]2 (2.5 mol%)
NaOAc (25 mol%)
MeCN, 60°C, 5 h
R = Me, n-Pentyl, Bn, Ph, 4-OMePh, 4-ClPh, 1-naphthyl, 2-thienyl
R1 = H, 4-Me, 4-OMe, 4-F, 4-Cl, 4-CF3O, 4-CF3, 3-Me, 3-Cl, 2, 4-di-Me
23 examples
Yields: 39–96 %

Scheme 20. Rhodium-catalyzed pyrazole synthesis

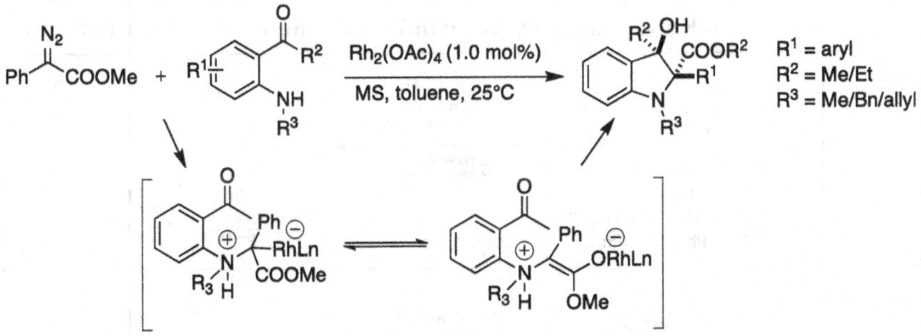
Rh2(OAc)4 (1.0 mol%)
MS, toluene, 25°C
R1 = aryl
R2 = Me/Et
R3 = Me/Bn/allyl

Scheme 21. Rhodium(II)-catalyzed indoline synthesis

1.2.7 *Indoline*

In 2014, Hu and co-workers developed a $Rh_2(OAc)_4$ catalyzed diazo decomposition reaction of diazo esters with 2-aminophenyl ketones. A series of 3-hydroxy-2,2,3-trisubstituted indolines are produced in good yields with excellent diastereoselectivities *via* an intramolecular aldol-type trapping of ammonium ylides with ketone units (Scheme 21).[40]

Recently Zhao and co-workers reported a Rh(III) catalyzed C–H activation/cyclative capture approach, involving a nucleophilic addition of $C(sp^3)$–Rh species to a polarized double bond. This constitutes the intermolecular catalytic method to directly access 1-aminoindolines with a broad substituent scope under mild conditions. However, they have utilized this method for the synthesis of

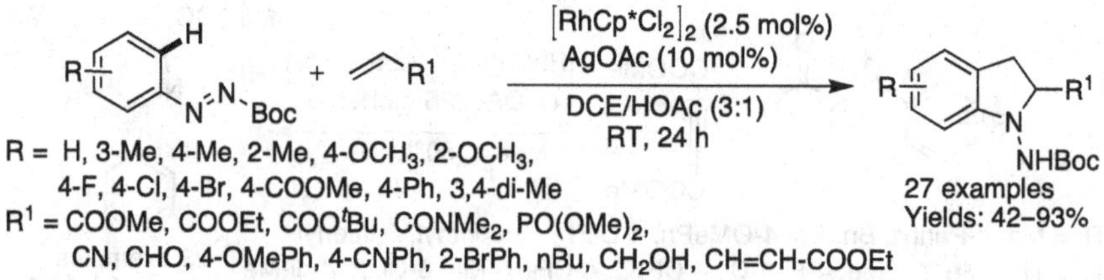

R = H, 3-Me, 4-Me, 2-Me, 4-OCH$_3$, 2-OCH$_3$,
4-F, 4-Cl, 4-Br, 4-COOMe, 4-Ph, 3,4-di-Me
R^1 = COOMe, COOEt, COO^{t}Bu, CONMe$_2$, PO(OMe)$_2$,
CN, CHO, 4-OMePh, 4-CNPh, 2-BrPh, nBu, CH$_2$OH, CH=CH-COOEt

Scheme 22. Rhodium(III)-catalyzed indoline synthesis

Table 1. Rhodium-catalyzed oxindole synthesis

S. No.	Rhodium-catalyzed oxindole synthesis	Ref. No.
23	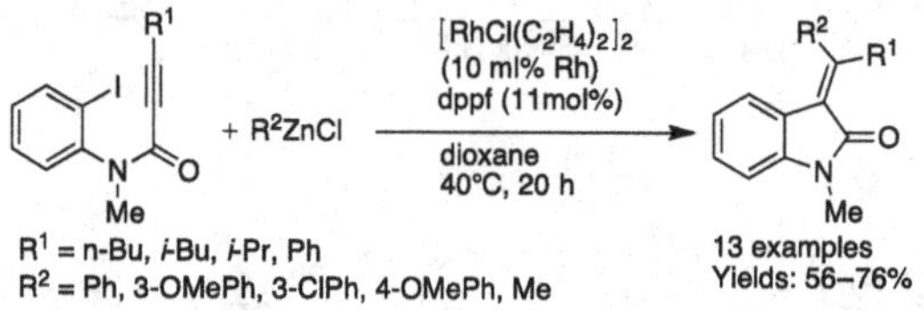 R^1 = n-Bu, i-Bu, i-Pr, Ph R^2 = Ph, 3-OMePh, 3-ClPh, 4-OMePh, Me	42
24	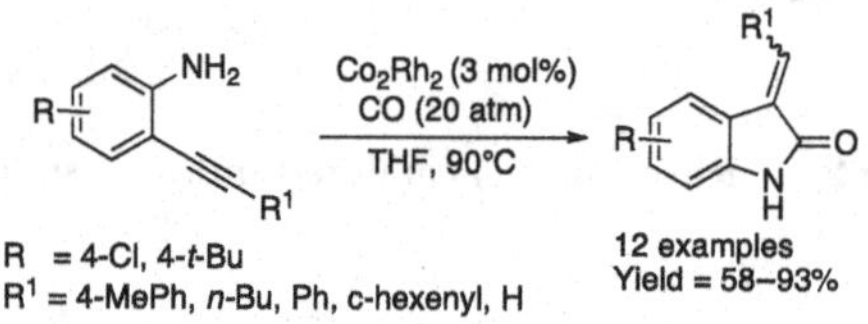 R = 4-Cl, 4-t-Bu R^1 = 4-MePh, n-Bu, Ph, c-hexenyl, H	43

1-amino indoles by introducing additional oxidant such as AgOAC (Scheme 22).[41]

1.2.8 *Oxindole*

Hayashi and co-workers reported rhodium-catalyzed multicomponent coupling reactions involving a carborhodation cross-coupling sequence. Through a series of experimental investigations of the underlying mechanism, it was demonstrated that the reaction most likely proceeds *via* a carborhodation-oxidative addition-reductive elimination pathway, which clearly contrasts to the corresponding palladium catalyzed processes (Table 1, Scheme 23).[42] In another report, Chung and co-workers reported cobalt–rhodium heterobime-tallic nanoparticle-catalyzed synthesis of oxindoles from 2-alkynylani-lines in the presence of carbon monoxide (Table 1, Scheme 24).[43]

1.2.9 *Isoindoline*

Isoindolines are a significant class of nitrogen containing heterocycle, which are abundant in natural products and biologically active com-pounds.[44] Zhu and co-workers have reported rhodium-catalyzed tan-dem annulation *via* C–H olefination of *N*-benzoyl sulfonamides with internal olefins followed by C–N bond formation. A *N*-substituted quaternary center is formed during the reaction thus providing effi-cient access to a series of 3,3-disubstituted isoindolinones (Table 2, Scheme 25).[45] Kim and co-workers have reported rhodium-catalyzed oxidative acylation between secondary benzamides and aryl alde-hydes *via* sp^2 C–H bond activation followed by an intramolecular cyclization. This method results in the direct and efficient synthesis of 3-hydroxyindolin-1-one building blocks (Table 2, Scheme 26).[46] Direct synthesis of alkylidene pyrrolo(3,4-*b*)pyridine-7-one deriva-tives *via* Rh(III) catalyzed cascade oxidative alkenylation/annulation of picolinamides was also described by Carretrero and co-workers (Table 2, Scheme 27).[47] Cramer and co-workers developed asymmet-ric synthesis of isoindolones by chiral cyclopentadienyl-Rhodium(III) catalyzed C–H functionalization. Rhodium complexes comprising a chiral Cp* ligand with an atropchiral biaryl backbone enables an asymmetric synthesis on isoindolones from arylhydroxamates and weakly alkyl donor/acceptor diazo derivatives as one-carbon compo-nent under mild conditions. The complex guides the substrates with a high double facial selectivity yielding the chiral isoindolones in good yields and excellent enantioselectivities (Table 2, Scheme 28).[48] Miura and co-workers described that the dehydrogenative direct cou-pling of α,α-disubstituted benzylamines with acrylates takes place effi-ciently at room temperature under rhodium catalysis, accompanied by free amino group directed ortho-alkenylation and successive

Table 2. Rhodium-catalyzed isoindoline synthesis

S.No.	Rhodium-catalyzed isoindoline synthesis	Ref. No
25	$[RhCp^*Cl_2]_2$ (4 mol%), $Cu(OAc)_2$ (2.1 equiv.), toluene, 130°C, 24 h. R^1 = H, naphthyl, 4-OMePh, 4-MePh, 4-FPh, 4-BrPh, 4-CF$_3$Ph, 3-OMePh, 3,5-di-OMe, 2-F; R^2 = CO$_2$Et, COEt, CN; R^3 = CO$_2$Et, COEt. 15 examples, Yields: 41–85%	45
26	$[RhCp^*Cl_2]_2$ (2.5 mol%), AgSbF$_6$ (20 mol%), THF, 150°C. R^1 = H, Ph, naphthyl, 3-OMePh, 3-OBnPh, 4-BrPh, 4-ClPh, 3-ClPh, 2-OMePh, 2-FPh; R^2 = 4-CF$_3$, CO$_2$Me, NO$_2$, COMe, CN, F, Cl, Br, OMe. 22 examples, Yields: 30–83%	46
27	MeO$_2$C, $[RhCp^*Cl_2]_2$ (2.5 mol%), AgSbF$_6$ (10 mol%), Cu(OAc)$_2$ (2.0 equiv.), p-Xylene, 120°C, 4 h, N$_2$. R^1 = 3-MePh, 3-ClPh, 4-CF3Ph, 5-MePh, 2-Me-thienyl, Ph; R^2 = CO$_2$^{n}Bu, CO$_2$^{t}Bu, CN, P(O)(OMe)$_2$, COMe, H, CO$_2$Me; R^3 = Bn, PMP. 23 examples, Yields: 10–96%	47
28	N$_2$, Rh4 (5 mol% of Rh), 5 mol% (BzO)$_2$, MeCN. R^1 = 3-CF3Ph, H, 4-MePh, 4-OMePh, 4-FPh, 4-ClPh, naphthyl, 3-MePh, 3-OMePh, 4-BrPh; R^2 = ^{i}Pr, Bn, CH$_2$Bn, allyl, CH$_2$allyl, CH$_2$CH$_2$-c-pentenyl, CH$_2$CH$_2$OMe, CF$_3$; R^3 = CH(^{i}Pr). 20 examples, Yields: 52–94% up to 96:5 e.r. Rh1 = ; Rh2 (R = OMe); Rh3 (R = O^{i}Pr); Rh4 (R = OTIPS)	48
29	$[RhCp^*Cl_2]_2$ (2 mol%), Cu(OAc)$_2$.H$_2$O (2 equiv.), o-Xylene, 80°C. R$_1$ = Me, Cl; R$_2$ = H; R$_3$ = Me, Ph, Et; R$_4$ = Me, Ph, Et; R$_5$ = CO$_2$^{n}Bu, CO$_2$^{t}Bu, CO$_2$^{t}Bu, CO$_2$Cy, CO$_2$Et. 22 examples, Yields: 5–91%	49

cyclization to produce (isoindol-1-yl)acetic acid derivatives. The reactions using styrene's in place of acrylates proceed effectively in the presence of a rhodium catalyst rather than rhodium (Table 2, Scheme 29).[49]

1.2.10 *Indazole*

1*H*-indazoles are widely used as anti-cancer, anti-inflammatory, anti-HIV, and anti-microbial drugs.[50] The efficient synthesis of 1*H*-indazoles has attracted much attention for a long time. Glorius and co-workers have reported Rh(III) and Cu(II) co-catalyzed synthesis of 1*H*-indazoles through C–H amidation and N–N bond formation. This was the first report for N–H imidates that are demonstrated to be good directing groups in C–H activation, and also capable of undergoing intramolecular N–N bond formation. The process is scalable and green, with O_2 as the terminal oxidant and N_2 and H_2O formed as by-products. Moreover, the products were transformed into diverse important derivatives (Scheme 30).[51]

Ellman and co-workers developed an efficient, one step, and highly functional group compatible synthesis of substituted *N*-aryl-2*H*-indazoles *via* the rhodium(III)-catalyzed C–H bond addition of azobenzenes to aldehydes. The regioselective coupling of unsymmetrical azobenzenes was further demonstrated and led to the development of a new removable aryl group that allows for the preparation of indazoles without *N*-substitution. The products were also found to be a new class of fluorophores (Scheme 31).[52]

Following the initial finding of Glorius and Ellaman works on indazole synthesis, Kim and co-workers, Zhang and co-workers simultaneously reported Rh(III) catalyzed oxidative coupling of 1,2-disubstituted arylhydrazines and olefins a new strategy for 2,3-hydro-1*H*-indazoles (Scheme 32).[53]

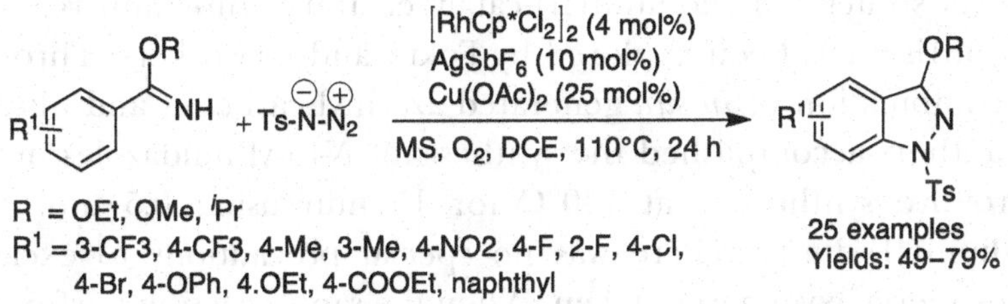
OR
NH
R1
+ Ts-N-N2
[RhCp*Cl2]2 (4 mol%)
AgSbF6 (10 mol%)
Cu(OAc)2 (25 mol%)
MS, O2, DCE, 110°C, 24 h
OR
N
N
Ts
R = OEt, OMe, iPr
R1 = 3-CF3, 4-CF3, 4-Me, 3-Me, 4-NO2, 4-F, 2-F, 4-Cl, 4-Br, 4-OPh, 4.OEt, 4-COOEt, naphthyl
25 examples
Yields: 49–79%

Scheme 30. Rhodium-catalyzed indazole synthesis

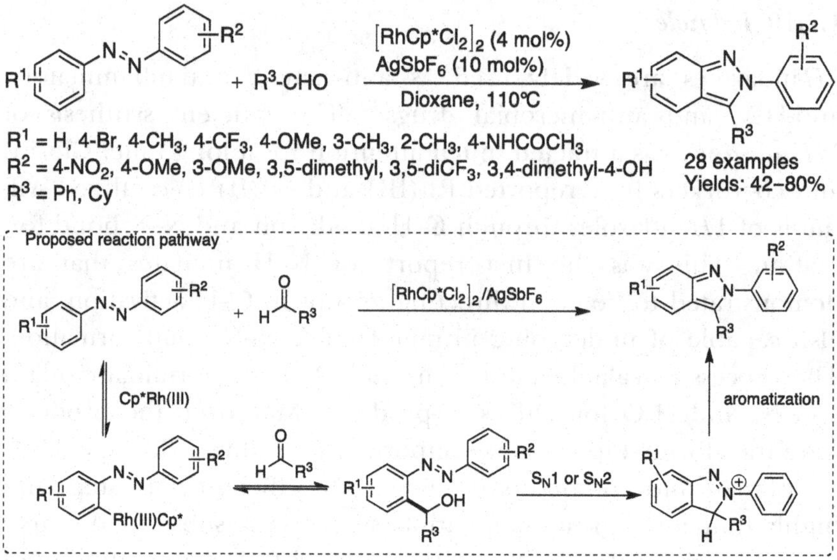

[RhCp*Cl2]2 (4 mol%)
AgSbF6 (10 mol%)
Dioxane, 110°C
+ R3-CHO
R1 = H, 4-Br, 4-CH3, 4-CF3, 4-OMe, 3-CH3, 2-CH3, 4-NHCOCH3
R2 = 4-NO2, 4-OMe, 3-OMe, 3,5-dimethyl, 3,5-diCF3, 3,4-dimethyl-4-OH
R3 = Ph, Cy
28 examples
Yields: 42–80%
Proposed reaction pathway
[RhCp*Cl2]2/AgSbF6
Cp*Rh(III)
aromatization
Rh(III)Cp*
SN1 or SN2

Scheme 31. Rhodium-catalyzed indazole synthesis

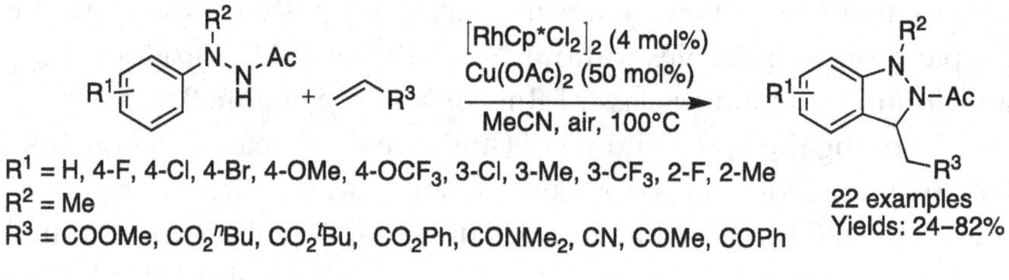

[RhCp*Cl2]2 (4 mol%)
Cu(OAc)2 (50 mol%)
MeCN, air, 100°C
R1 = H, 4-F, 4-Cl, 4-Br, 4-OMe, 4-OCF3, 3-Cl, 3-Me, 3-CF3, 2-F, 2-Me
R2 = Me
R3 = COOMe, CO2nBu, CO2tBu, CO2Ph, CONMe2, CN, COMe, COPh
22 examples
Yields: 24–82%

Scheme 32. Rhodium-catalyzed indazole synthesis

1.2.11 *Imidazole*

A novel strategy in rhodium(II)-catalyzed transannulation reaction with nitriles has been explored by Fokin and co-workers. Through the combination of *in situ* generated azavinyl carbenes and nitriles, the authors accomplished the synthesis of *N*-tosyl imidazoles in the microwave synthesizer at 140°C for 15 min using 0.5 mol% of rhodium(II)-dimer. There was no special precautions to exclude atmospheric oxygen were taken. When the same reaction performed with conventional heating at 80°C under inert atmosphere, the

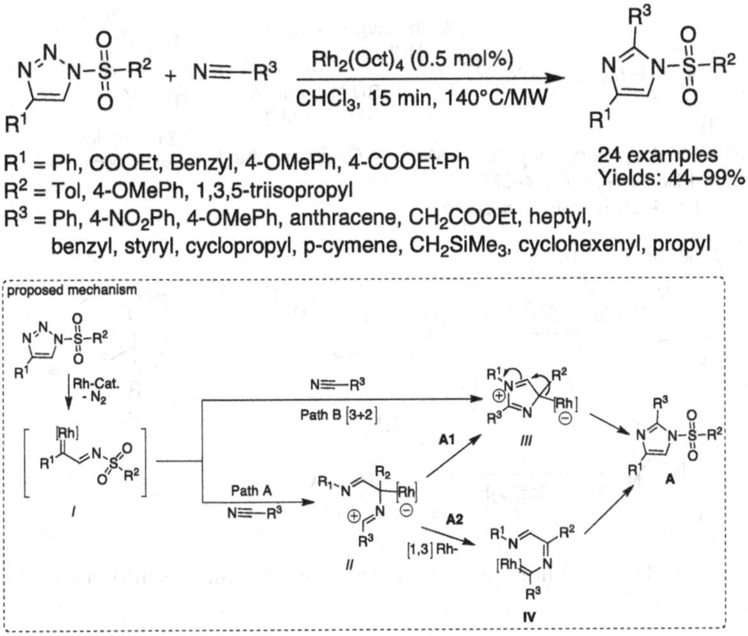
Rh2(Oct)4 (0.5 mol%)
CHCl3, 15 min, 140°C/MW
24 examples
Yields: 44–99%
R1 = Ph, COOEt, Benzyl, 4-OMePh, 4-COOEt-Ph
R2 = Tol, 4-OMePh, 1,3,5-triisopropyl
R3 = Ph, 4-NO2Ph, 4-OMePh, anthracene, CH2COOEt, heptyl,
benzyl, styryl, cyclopropyl, p-cymene, CH2SiMe3, cyclohexenyl, propyl
proposed mechanism
Rh-Cat.
- N2
Path B [3+2]
Path A
A1
A2
[1,3] Rh-

Scheme 33. Rhodium-catalyzed conversion of nitriles into imidazole

reaction proceeded to completion in 15 h and furnished imidazole in good yield. Mechanistic studies suggested that the in pathway A, a nucleophilic attack of a nitrile at the Rh-carbenoid *I* leads to ylide *II*, which upon cyclization (path A1) into a zwitterion *III*, and subsequent metal loss, produces imidazole *A*. Alternatively, ylide II, may give rise to the Rh-carbenoid *IV via* a 1,3-Rh-shift (path 2). Subsequent cyclization of *IV*, followed by the reductive elimination, furnishes **A**. A possible direct formation of *III via* a cycloaddition of I with a nitrile (path B) cannot be ruled out at this time (Scheme 33).[54]

In 2013, Fokin and co-workers expanded the synthetic utility of rhodium azavinyl carbene for a novel synthetic route to imidazoles with aldimines (Scheme 34).[55]

1.2.12 *Phthalimide*

Phthalimides are an important class of compounds with a range of biological activities, e.g. antimicrobial, anticonvulsant, antihistaminic,

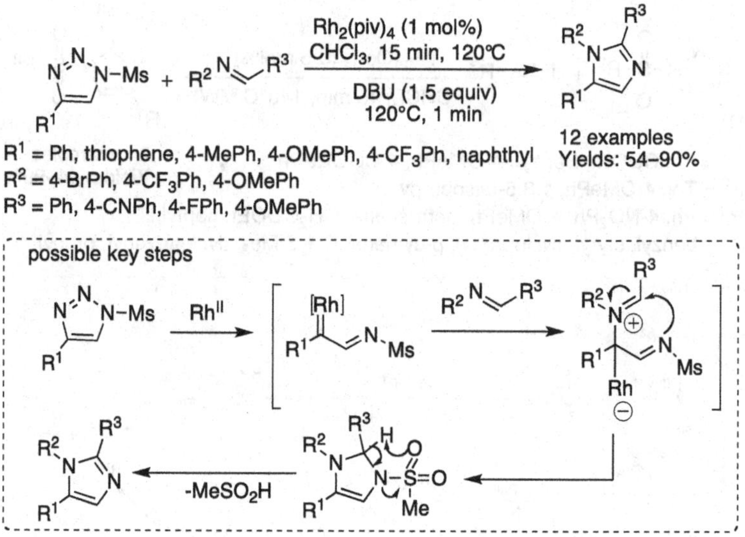
Rh2(piv)4 (1 mol%)
CHCl3, 15 min, 120°C
DBU (1.5 equiv)
120°C, 1 min
12 examples
Yields: 54–90%
R1 = Ph, thiophene, 4-MePh, 4-OMePh, 4-CF3Ph, naphthyl
R2 = 4-BrPh, 4-CF3Ph, 4-OMePh
R3 = Ph, 4-CNPh, 4-FPh, 4-OMePh
possible key steps
RhII
-MeSO2H

Scheme 34. Rhodium-catalyzed conversion of aldimines into imidazole

anti-inflammatory, and hypolipidemic.[56] They are also used as synthetic intermediates in the preparation of dyes, pesticides, liquid crystals, functional materials, and polymers. Li and co-workers reported a rhodium-catalyzed cascade cyclization for the direct synthesis of *N*-substituted phthalimides from isocyanates and benzoic acids. Aromatic isocyanates in this reaction were more reactive than aliphatic isocyanates. This observation is in contrast with the reactivity trend of isocyanates in most nucleophilic addition reactions, in which aromatic isocyanates are less reactive than aromatic ones due to electron donation from the ring, which makes the carbonyl carbon less electrophilic (Scheme 35).[57]

1.2.13 *Imidazolidinone*

The rhodium-catalyzed synthesis of biologically important imidazolidinones *via* transannulation of sulfonyl 1,2,3-triazoles was explored by Fokin and co-workers in 2013. Readily available 1-mesyl-1,2,3-triazoles are efficiently converted into a variety of imidazolones and thiazoles by Rh(II) catalyzed denitrogenative reactions with isocyanates, respectively. They proposed triazole–diazoimine equilibrium

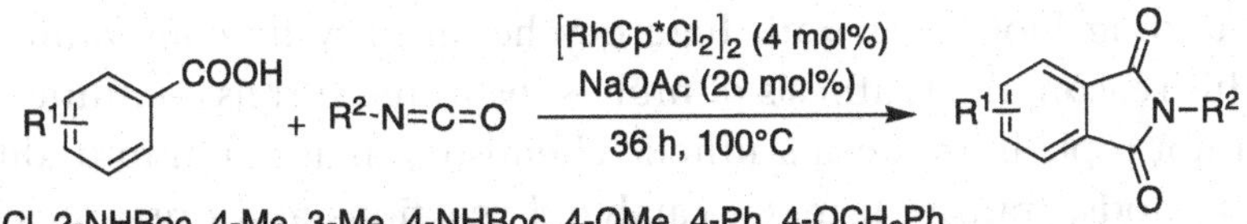

$$[RhCp^*Cl_2]_2 \text{ (4 mol\%)}$$

R_1 = 4-Cl, 2-NHBoc, 4-Me, 3-Me, 4-NHBoc, 4-OMe, 4-Ph, 4-OCH₂Ph,
3,5-di-Me, 3-Me-4-NHBoc, 2,4-di-Me, napthyl, 3-OMe, 2-OMe,
3,5-di-OCH₂Ph, 2,3,4-tri-OMe, 3,4,5-tri-OMe

27 examples
Yields: 19–91%

Scheme 35. Rhodium-catalyzed conversion of isocyanates into phthalimides

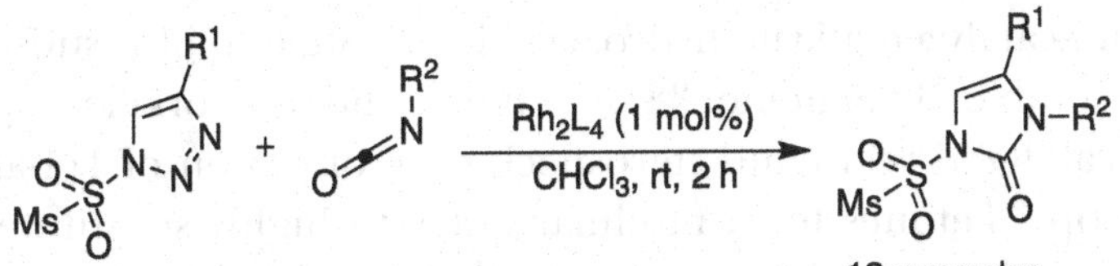

R^1 = Ph, 3FPh, 4-CF₃Ph, 4-OMePh, naphthyl, 4-Me
R^2 = Ph, 4-ClPh, 4-OMePh, 4-BrPh, 4-ClPh, 4-CF₃Ph,
4-NO₂Ph, 3-FPh, benzyl, c-hexyl, allyl, butyl, Br-ethyl, Cl-propyl

12 examples
Yields: 69–95%

Scheme 36. Rhodium-catalyzed conversion of heterocumulenes into imidazolidinone

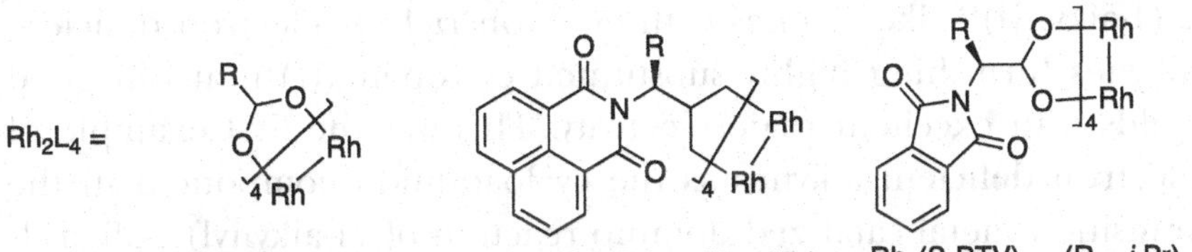

Rh_2L_4 =

Rh₂(Oct)₄ (R = *n*-Pr)
Rh₂(Piv)₄ (R = *t*-Pr)

Rh₂(S-NTV)₄ (R = *i*-Pr)
Rh₂(S-NTTL)₄ (R = *t*-Bu)

Rh₂(S-PTV)₄ (R = *i*-Pr)
Rh₂(S-PTTL)₄ (R = *t*-Bu)

Figure 1.1 Structure of catalysts

results in the formation of highly reactive azavinyl metal-carbenes, which react with heterocumulenes causing an apparent swap of the 1,2,3-triazole core for another heterocycle (Scheme 36).[58]

1.3 Rhodium-catalyzed oxygen containing five-membered heterocycle synthesis

1.3.1 *Furan*

Furans occur in a variety of biologically active natural products, pharmaceuticals and functional materials. Moreover, they are versatile building blocks for both heterocyclic and acyclic compounds. For this reason, the synthesis of multi-substituted furans continues to be a hot topic in modern synthetic chemistry. In addition to traditional methods, transition metal catalyzed reactions and organo catalytic approaches have been reported during the last decades.[59] Li and co-workers presented rhodium catalyzed synthesis of multi-substituted furans from N-sulfonyl-1,2,3-triazoles bearing a tethered carbonyl group (Table 3, Scheme 37).[60] Wills and co-workers described a rhodium-catalyzed alkyne hydroacylation route to highly substituted furans (Table 3, Scheme 38).[61] Zhang and co-workers reported Rh(I) catalyzed regio and sterospecific cabonylation of 1(1-alkyny) cyclopropyl ketones for a modular entry to highly substituted 5,6-dihydro cyclopenta(c)furan-4-ones initially by the activation of the carbon–carbon s-bond (Table 3, Scheme 39).[62] Ellman and co-wokers reported rhodium(III)-catalyzed alkenyl C–H bond functionalization for the convergent synthesis of furans (Table 3, Scheme 40).[63] Zhang and co-workers reported cationic rhodium(I)-catalyzed regioselective tandem heterocyclization/(3 + 2) cycloaddition of 2-(1-alkynyl)2-alken-1-ones with electron rich or electron deficient alkynes furnishing highly substituted cyclopenta(c)furans in good yields with excellent regioselectivity. This was the first example of electron deficient alkynes as the cycloaddition component in the transition metal catalyzed domino reaction of (1-alkynyl)-2-alken-1-ones, which represents a novel non-cyclopropane 3C component in rhodium chemistry (Table 3, Scheme 41).[64]

1.3.2 *Furanone*

Compounds encompassing the furanone moiety are substantial synthetic targets due to their presence as a subunit in many natural products and more precisely, they may play a significant role in the discovery of new therapeutic agents.[65] Arcadi and co-workers reported that sequential rhodium-catalyzed addition/lactonization reaction of organoboron derivatives to alkyl 4-hydroxy-2-alkynoates would constitute a novel methodology for the synthesis of 4-aryl/heteroaryl/vinyl-2(5H)-furanones with an excellent control of the

Table 3.   Rhodium-catalyzed furan synthesis

S. No.	Rhodium-catalyzed furan synthesis	Ref. No
37	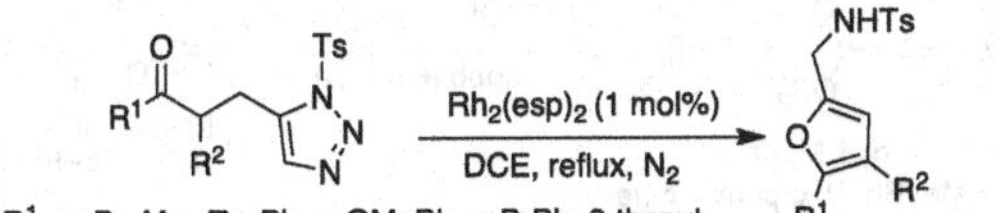 R^1 = nPr, Me, iPr, Ph, p-OMePh, p-BrPh, 2-thenyl R^2 = H, Me, Bn Rh$_2$(esp)$_2$ (1 mol%), DCE, reflux, N$_2$ 14 examples Yields: 50–92%	60
38	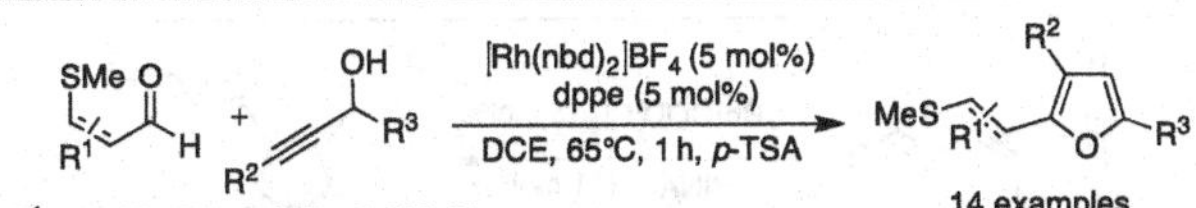 R^1 = H, Ph, Me, 3-ClPh, 2-SMePh R^2 = Me, Ph, Et, COOEt R^3 = Me, Ph, thiophene, 4-CF$_3$Ph, [Rh(nbd)$_2$]BF$_4$ (5 mol%), dppe (5 mol%), DCE, 65°C, 1 h, p-TSA 14 examples Yields: 42–93%	61
39	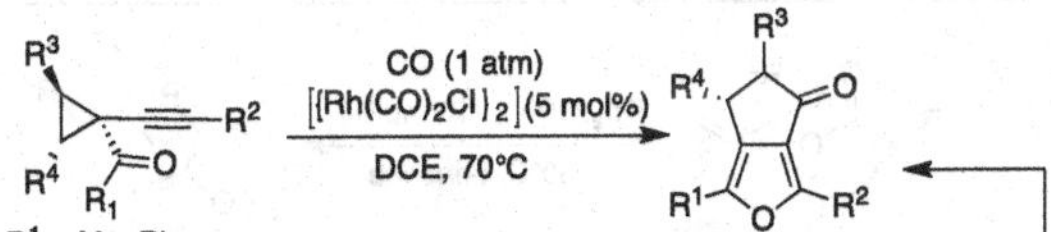 R^1 = Me, Ph R^2 = 1-naphthyl, n-C$_4$H$_9$, c-hexyl, cyclopropyl, Ph, 4-OMePh, AcOC$_2$H$_4$ R^3 = Ph, 4-OMePh; R^4 = PhCO, EtCO CO (1 atm), [{Rh(CO)$_2$Cl}$_2$](5 mol%), DCE, 70°C 21 examples Yields: 38–95% (racemic) Proposed mechanism 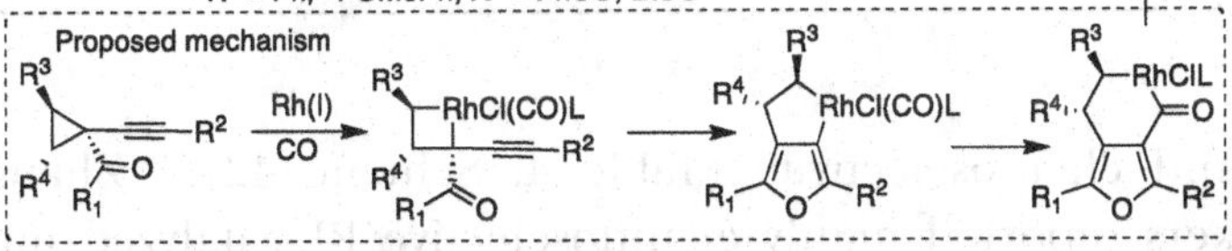	62
40	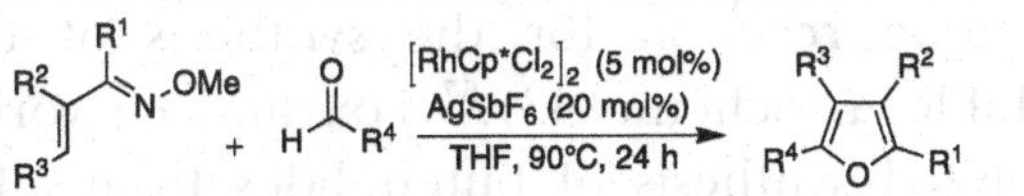 R^1 = Me, Bu, Et; R^2 = Ph, Me, Bn ; R^3 = Me, Bu R^4 = CO$_2$Et, 4-CF$_3$Ph, 4-ClPh, 4-CO$_2$MePh, 4-NO$_2$Ph, 3-FPh, 2-FPh, 3-OMePh, iPr, C-hexyl, [RhCp*Cl$_2$]$_2$ (5 mol%), AgSbF$_6$ (20 mol%), THF, 90°C, 24 h 22 examples Yields: 41–89%	63
41	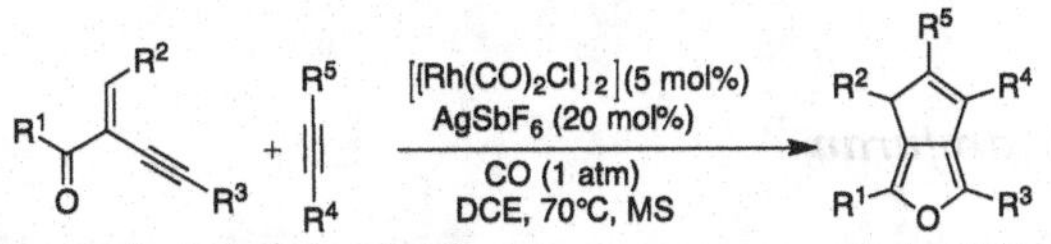 R^1 = Me, Ph; R^2 = Ph, 4-ClPh, 4-OMePh, R^3 = Ph, 4-MePh, 4-OMePh, 4-NO$_2$Ph, 1-naphthyl R^4 = Ph, 4-MePh, 4-OMePh, 1-cyclohexenyl R^4 = COOMe, COOEt, Ph, 4-MePh [{Rh(CO)$_2$Cl}$_2$](5 mol%), AgSbF$_6$ (20 mol%), CO (1 atm), DCE, 70°C, MS 16 examples Yields: 59–89%	64

Table 4. Rhodium-catalyzed furanone synthesis

S. No.	Rhodium-catalyzed furanone synthesis	Ref. No
42	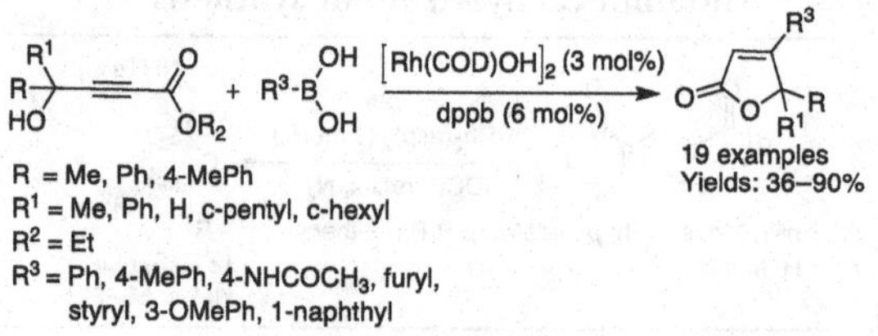 R = Me, Ph, 4-MePh R^1 = Me, Ph, H, c-pentyl, c-hexyl R^2 = Et R^3 = Ph, 4-MePh, 4-NHCOCH$_3$, furyl, styryl, 3-OMePh, 1-naphthyl 19 examples Yields: 36–90%	66
43	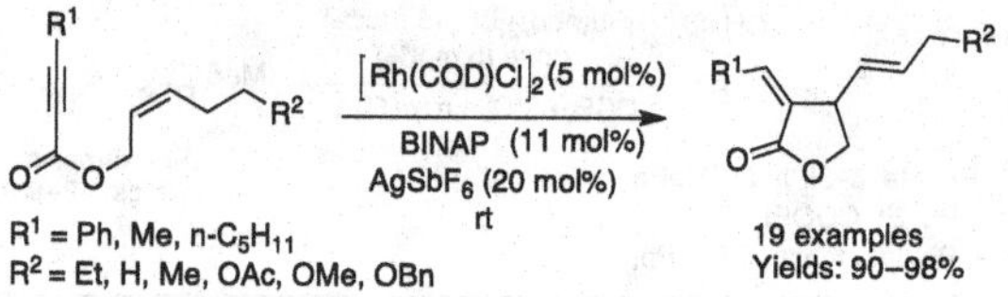 R^1 = Ph, Me, n-C$_5$H$_{11}$ R^2 = Et, H, Me, OAc, OMe, OBn 19 examples Yields: 90–98%	67
44	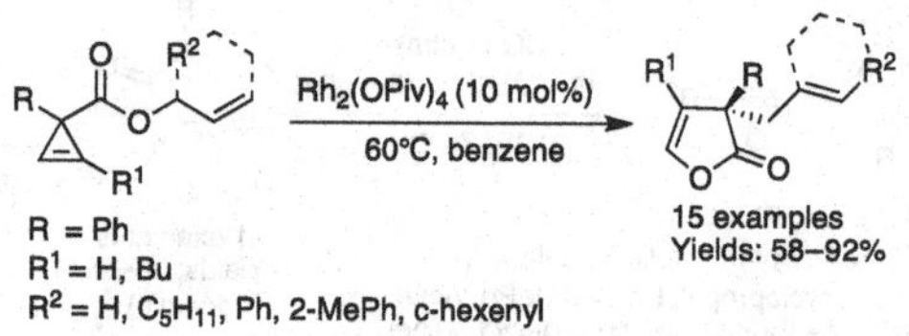 R = Ph R^1 = H, Bu R^2 = H, C$_5$H$_{11}$, Ph, 2-MePh, c-hexenyl 15 examples Yields: 58–92%	68

regio and chemoselectivity (Table 4, Scheme 42).[66] Zhang and co-workers reported highly enantioselective Rh-catalyzed intramolecular Alder-ene reaction for the synthesis of functionalized γ-lactones (Table 4, Scheme 43).[67] Fox and co-workers reported rhodium-catalyzed synthesis of butenolides from allylic cyclopropane carboxylates *via* tandem ring expansion/[3,3]-sigmatropic rearrangements (Table 4, Scheme 44).[68]

1.3.3 Tetrahydrofuran

Tetrahydrofurans are recurring motifs across several natural product families and synthetic analogues of medicinal importance that exhibit remarkable bioactivity.[69] Boyer reported a rhodium(II)-catalyzed stereo controlled synthesis of 2-tetrasubstituted saturated heterocycles from 1-sulfonyl-1,2,3-triazoles with pendent allyl and

Table 5. Rhodium-catalyzed synthesis of tetrahydrofuran

S. No.	Rhodium-catalyzed tetrahydrofuran synthesis	Ref. No
45	R^1 = n-pentyl, CH$_2$OTBDPS, CH(Me)$_2$OBn, Bn, Me, H, iPr — Rh$_2$(OAc)$_4$ (5 mol%), PhMe, 70°C — 10 examples, Yields: 75–98%	70
46	R = Ph, 2-ClPh, 3-ClPh, 4-ClPh, 4-MePh, 4-CF$_3$Ph, Me, n-Bu, COMe, COPh, COOEt, CH$_2$OH, CH$_2$OMOM, R^1 = Et, H, Me, OAc, OMe, OEt, OH — [(Rh(COD)Cl$_2$)](10 mol%), (R)-BINAP (12 mol%), AgSbF$_6$ — 16 examples, Yields: 81–96%, ee = 99.9%	71
47	R = H, 2-OMe, 4-OMe, 2,6-di-OMe, 4-Me, 2-Br, 4-Br, 2-Cl, 4-Cl, 4-CF$_3$, 4-CO$_2$Me, 4-CN, 4-OH, 4-CHO, 1-naphthyl, 2-naphthyl, furan, thiophene, benzofuran, benzothiophene — [Rh(coe)$_2$Cl]$_2$ (5 mol%), L* (5.5 mol%), MeOH, Et$_3$N, rt — 24 examples, Yields: 49–74%, ee = 90%	72

propargyl ethers to onium ylides that undergo (2,3)-sigmatropic rearrangement to give 2-tetrasubstituted heterocycles with high yield and diastereoselectivity (Table 5, Scheme 45).[70] Zhang and coworkers reported highly enantioselective rhodium-catalyzed intramolecular Alder-ene reactions for the synthesis of chiral tetrahydrofurans (Table 5, Scheme 46).[71] Lautens and co-workers described a new enantioselective rhodium-catalyzed domino reaction that gives access to fused furans by desymmetrization of alkyne-tethered cyclohexadienones. Two new C–C bonds and two stereocenters are formed in one step with good enantioselectivity. In contrast to prior reports, it was found that a vinylidene is not involved in the product formation but that syn-addition of the rhodium–aryl species onto the alkyne takes place (Table 5, Scheme 47).[72]

1.3.4 Furocoumarin

Tollari and Palmisano explored the efficient construction method of fused furocoumarins by the use of a Rh(II)-induced [3 + 2] cycloaddition of and oxygenated alkenes (Scheme 48).[73]

1.3.5. Furo(2,3,-b)furan

Fused polycyclic acetals are exemplified in various natural products. Among bicyclic acetals, the tetrahydrofuro[2,3-b]furans are of special interest since both aliphatic and benzoannelated compounds of biological activity are known.[74] Therefore, these oxabicycles are attractive targets for total synthesis. Eilbracht and co-workers reported a rhodium-catalyzed synthesis of furo[2,3-b]furans through tandem hydroformation/acetalization (Scheme 49).[75]

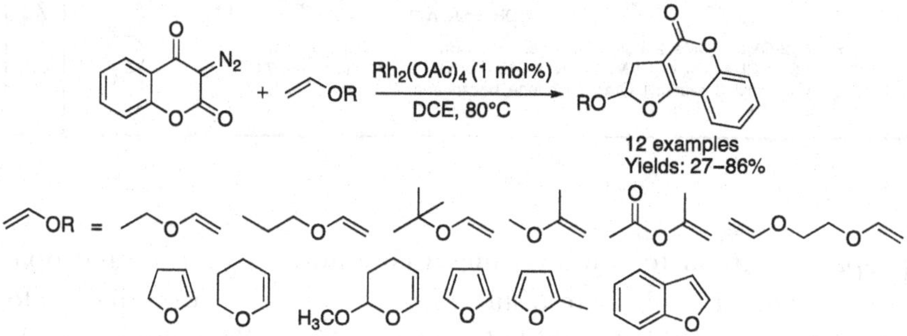

Scheme 48.　Rhodium-catalyzed synthesis of furocoumarin

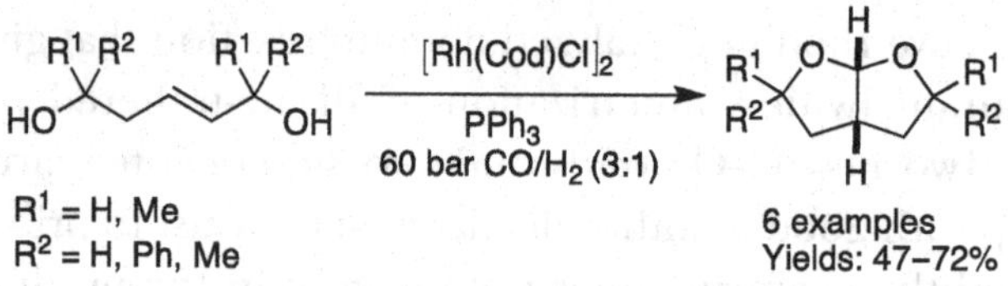

Scheme 49.　Rhodium-catalyzed synthesis of furo[2,3,-b]furan

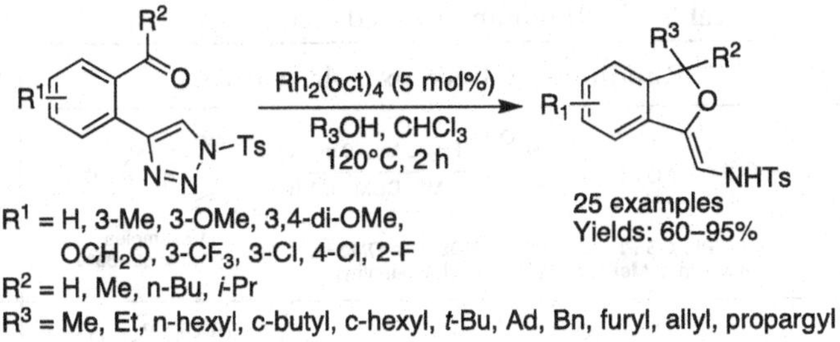
Rh2(oct)4 (5 mol%)
R3OH, CHCl3
120°C, 2 h
N-Ts
NHTs
25 examples
Yields: 60–95%
R1 = H, 3-Me, 3-OMe, 3,4-di-OMe, OCH2O, 3-CF3, 3-Cl, 4-Cl, 2-F
R2 = H, Me, n-Bu, i-Pr
R3 = Me, Et, n-hexyl, c-butyl, c-hexyl, t-Bu, Ad, Bn, furyl, allyl, propargyl

Scheme 50. Rhodium-catalyzed synthesis of isobenzofuran

1.3.6 *Isobenzofuran*

Yang and co-workers reported rhodium-catalyzed tandem synthesis of dihydroisobenzofurans from 2-triazole benzaldehydes (Scheme 50).[76]

1.3.7 *Oxazole*

Doyle and co-workers reported efficient synthesis of oxazoles by dirhodium(II)-catalyzed reactions of styryl diazoacetate with aryl oximes (Table 6, Scheme 51).[77] Moody and co-workers described that dirhodium tetraacetate catalyzed reaction of α-diazo-β-keto-carboxylates and phosphonates with arenecarboxamides gives 2-aryloxazole-4-carboxylates and 4-phosphonates by carbene N–H insertion and cyclodehydration. In stark contrast, dirhodium tetrakis(heptafluorobutyramide) catalysis results in a dramatic change of regioselectivity to give oxazole-5-carboxylates and 5-phosphonates (Table 6, Scheme 52).[78] Yuan and co-workers presented that a facile synthesis of 1,3-oxazole by rhodium-catalyzed heterocycloaddition reaction and a series of 4-phosphoryl substituted 1,3-oxazole are prepared conveniently by reaction of diethyl 1-diazo-2-oxo-alkylphosphonates and aromatic nitriles in the presence of a catalytic amount of rhodium(II)acetate (Table 6, Scheme 53).[79] Zhang and co-workers reported Rh(III) catalyzed synthesis of naphthoxazole derivatives from electron-deficient naphthoquinones and

Table 6. Rhodium-catalyzed oxazole synthesis

S. No.	Rhodium-catalyzed oxazole synthesis	Ref. No
51	Ph⌒⌒(N₂)CO₂Me + N⌒OH / R → Rh₂(OAc)₄ (2 mol%), MS, DCM, rt, 1 h → MeO oxazole product R = 4-ClPh, 4-BrPh, 4-FPh, Ph, 4-NO2Ph, 4-OMePh, 4-MePh, 3-MePh, 2-MePh, 2-furyl, 2-naphthyl 11 examples, Yields: 62–89%	77
52	R⌒C(O)NH₂ + R¹⌒C(O)R²(N₂) → Rh₂(NHCOC₃H₇)₄ (2 mol%), toluene, reflux or toluene, 135°C, MW → (R²O)₂OP oxazole product R = Ph, 2-BrPh, 2-BnOPh, 4-BrPh, 4-OMePh, 4-NO₂Ph, 4-CbzNHPh, 4-PhC₆H₄, 3,5-F₂Ph, 2-thienyl, 2-benzothiphenyl, allyl R¹ = Me, Ph; R² = PO(OMe)₂, PO(OPh)₂, Ts 25 examples, Yields: 28–70%	78
53	R¹⌒C(O)P(O)(OEt)(OEt)(N₂) + R²≡N → Rh₂(OAc)₄ (1 mol%), 85°C → (Et₂O)₂OP oxazole product R¹ = Me, OMe, OEt R² = Ph, 2-MePh, 3-MePh 7 examples, Yields: 31–83%	79
54	naphthoquinone-NH-CH₂-R³ + R¹≡R² → [RhCp*Cl₂]₂ (5 mol%), Cu(OAc)₂·H₂O (200 mol%), AgSbF₆ (20 mol%), DCE, 120°C, 16 h → fused oxazole product R¹ = Ph, 4-MePh, 4-ClPh, 4-FPh, 4-EtPh, 4-OMePh, 4-CF₃Ph, 4-thienyl, n-Pr, n-Bu, Me R² = Ph, 4-MePh, 4-ClPh, 4-FPh, 4-EtPh, 4-OMePh, 4-CF₃Ph, 4-thienyl, n-Pr R³ = Me, Et, n-Pr, i-Pr, n-Bu, Bn, C-Pr, Ad, t-Bu, NMeBoc, 27 examples, Yields: 45–95%	80

alkynes through C–H annulation and C(sp³) bond cleavage. This approach proceeds through a tandem cascade process involving substrate tautomerization, C–H activation, oxidative addition, cyclization, and aromatization. In addition, broad substrate scope, simple starting materials, and steric tolerance make this strategy of great practical importance (Table 6, Scheme 54).[80]

1.3.8 *Oxazolidinones*

Du Bois and co-worker reported a rhodium-catalyzed C–H insertion reaction for the oxidative conversion of carbamates to oxazolidinones. They explored the use of simple substituted carbamate materials as precursors for the synthesis of substituted oxazolidinones.

Aided by a series of control experiments, they concluded that AcOH, generated as a by-product from $PhI(OAc)_2$, reduced the catalytic activity of $[Rh_2(OAc)_4]$ that prompted the screening of various base additives, and of those reagents MgO proved uniquely effective (Table 7, Scheme 55).[81] Lebel and co-workers reported N-tosyloxycarbamates as a source of metal nitrenes for the rhodium-catalyzed C–H insertion and aziridination reactions for the synthesis

Table 7. Rhodium-catalyzed oxazolidinone synthesis

S. No.	Rhodium-catalyzed oxazolidinone synthesis	Ref. No
55	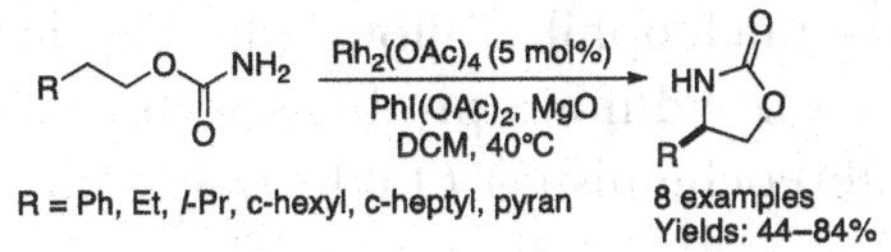 R = Ph, Et, i-Pr, c-hexyl, c-heptyl, pyran 8 examples Yields: 44–84%	81
56	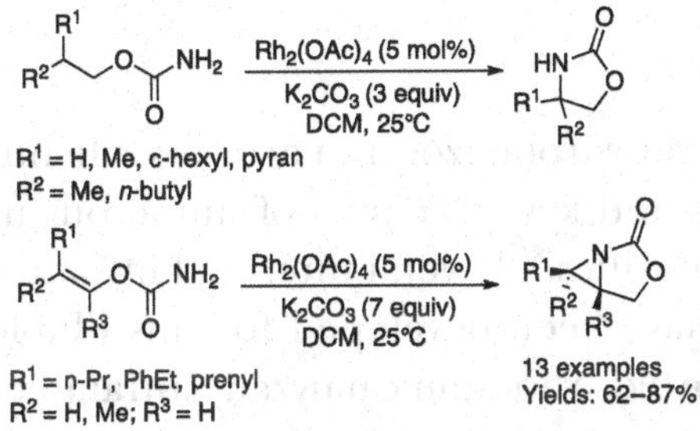 R^1 = H, Me, c-hexyl, pyran R^2 = Me, n-butyl R^1 = n-Pr, PhEt, prenyl R^2 = H, Me; R^3 = H 13 examples Yields: 62–87%	82
57	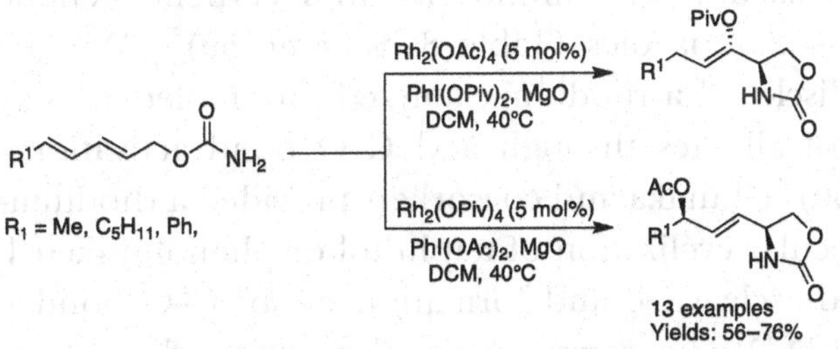 R_1 = Me, C_5H_{11}, Ph, 13 examples Yields: 56–76%	83
58	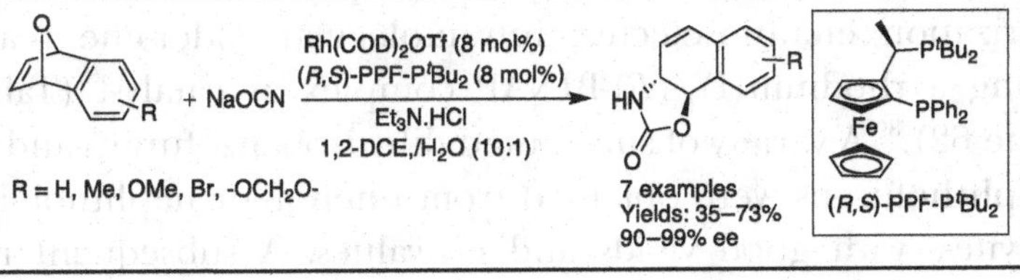 R = H, Me, OMe, Br, -OCH$_2$O- 7 examples Yields: 35–73% 90–99% ee (R,S)-PPF-P^{t}Bu$_2$	84

of oxazolidinones (Table 7, Scheme 56).[82] Castillon and co-workers reported rhodium-catalyzed tandem regio and stereoselective oxyamination of dienes *via* tandem aziridination/ring-opening of dienol carbamates to afford oxazolidinones. The carboxylate present in the iodine(III) reagent released during the reaction behaves as a nucleophile opening the aziridine intermediate. The procedure developed was further utilized to develop a straightforward access to sphingosine and derivatives (Table 7, Scheme 57).[83]

Lautens and co-workers developed a method for synthesizing chiral oxazolidinone scaffolds from readily available oxabicyclic alkenes with sodium cyanate. The reaction utilizes a domino sequence of Rh(I) catalyzed asymmetric ring-opening (ARO) with sodium cyanate as a novel nucleophile followed by intramolecular cyclization to generate oxazolidinone products with excellent enantioselectivities (*trans* stereochemistry) (Table 7, Scheme 58).[84]

1.3.9 Benzofuran

Benzofuran and dihydrobenzofuran are versatile building blocks in organic synthesis and key structures of numerous natural products with biological activities.[85] Several inter and intramolecular Rh catalyzed reactions have been developed for this (Table 8). Chen and co-workers reported rhodium-catalyzed intramolecular sp^3 C–H insertion reaction of α-imino rhodium carbene generated from *N*-sulfonyl-1,2,3-triazoles (Table 8, Scheme 59)[86]; Douglas and co-workers disclosed a rhodium-catalyzed intramolecular oxyacylation reaction of alkenes through acyl C–O bond activation (Table 8, Scheme 60)[87]; Tanaka and co-workers provided a rhodium-catalyzed intramolecular cyclization of naphthol or phenol linked 1,6-enynes through the cleavage and formation of sp^2 C–O bonds (Table 8, Scheme 61).[88] The same group also reported a cascade olefin isomerization/enantioselective intramolecular Alder-ene reaction by using a rhodium(I)/(*R*)-BINAP complex as catalyst (Table 8, Scheme 62).[89] A variety of substituted dihydrobenzofurans and dihydronaphthofurans were obtained from phenol- or naphthol-linked 1,7-enynes, with good yields and ee values. A subsequent major

Table 8. Rhodium-catalyzed Benzofuran synthesis

S. No.	Rhodium-catalyzed synthesis of benzofuran	Ref. No
59	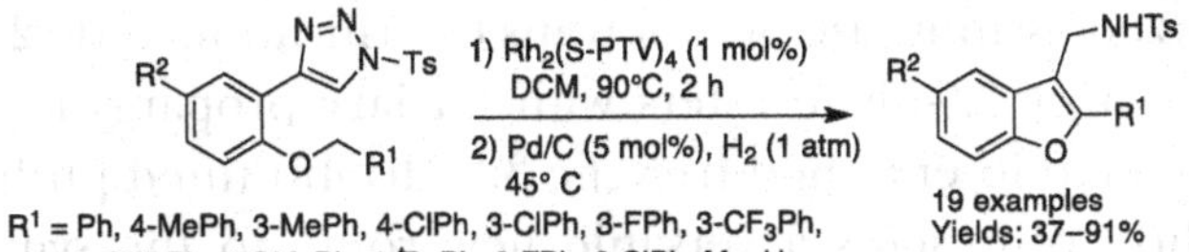 1) Rh$_2$(S-PTV)$_4$ (1 mol%) DCM, 90°C, 2 h; 2) Pd/C (5 mol%), H$_2$ (1 atm) 45°C. R^1 = Ph, 4-MePh, 3-MePh, 4-ClPh, 3-ClPh, 3-FPh, 3-CF$_3$Ph, 3-OMePh, 4-OMePh, 4-^{t}BuPh, 2-FPh, 2-ClPh, Me, H; R^2 = H, Me, Et, Cl, t-Bu. 19 examples Yields: 37–91%	86
60	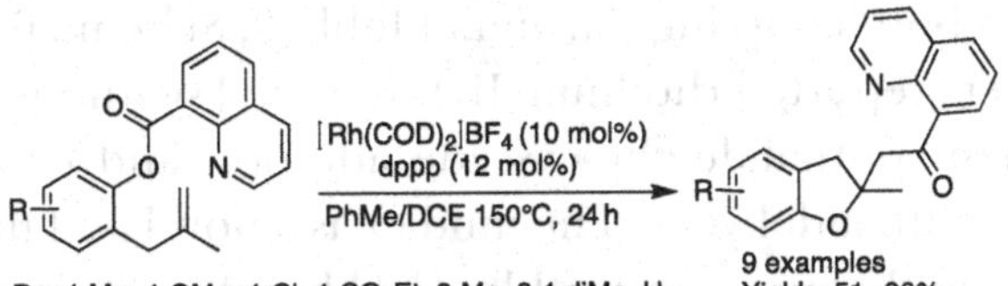 [Rh(COD)$_2$]BF$_4$ (10 mol%) dppp (12 mol%) PhMe/DCE 150°C, 24 h. R = 4-Me, 4-OMe, 4-Cl, 4-CO$_2$Et, 2-Me, 2,4-diMe, H. 9 examples Yields: 51–90%	87
61	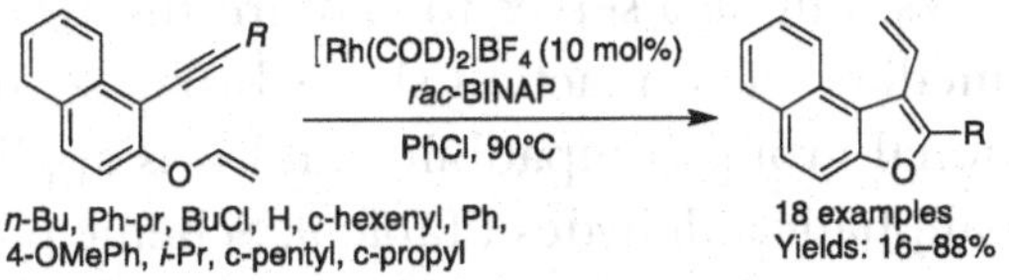 [Rh(COD)$_2$]BF$_4$ (10 mol%) rac-BINAP, PhCl, 90°C. R = n-Bu, Ph-pr, BuCl, H, c-hexenyl, Ph, 4-OMePh, i-Pr, c-pentyl, c-propyl. 18 examples Yields: 16–88%	88
62	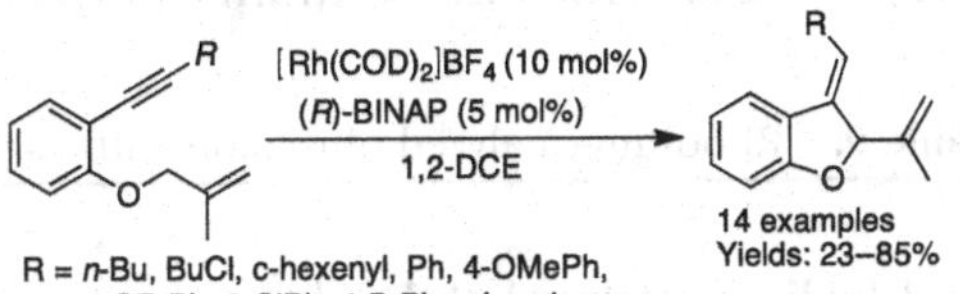 [Rh(COD)$_2$]BF$_4$ (10 mol%) (R)-BINAP (5 mol%) 1,2-DCE. R = n-Bu, BuCl, c-hexenyl, Ph, 4-OMePh, 4-CF$_3$Ph, 3-ClPh, 4-BrPh, c-hexyl. 14 examples Yields: 23–85%	89
63	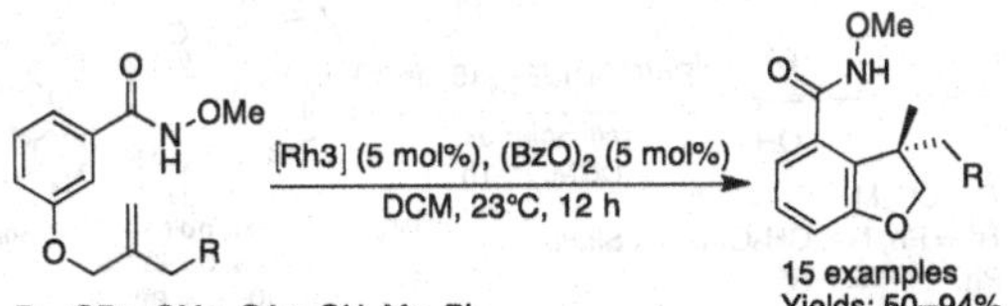 [Rh3] (5 mol%), (BzO)$_2$ (5 mol%) DCM, 23°C, 12 h. R = OBn, OMe, OAc, OH, Me, Ph. 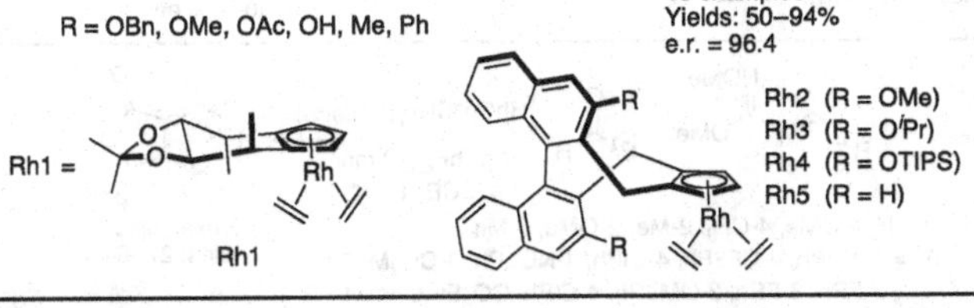 15 examples Yields: 50–94% e.r. = 96.4. Rh2 (R = OMe), Rh3 (R = O^{i}Pr), Rh4 (R = OTIPS), Rh5 (R = H)	90

achievement by Cramer and co-workers, was the chiral Cp*Rh(III) fragment (Cp* = pentamethyl cyclopentadienyl) catalyzed asymmetric hydroarylations of 1,1-disubstituted alkenes to generate the desired benzofuran (Table 8, Scheme 63).[90]

1.3.10 Phthalides

Tanaka and co-workers developed a cationic rhodium/Solphos complex catalyzed asymmetric on-pot trans esterification and $[2+2+2]$ cycloaddition of 1,6-diyne esters with tertiary propargylic alcohols leading to enantio enriched tricyclic 3,3,-disubstituted phthalides.[91] This method represents a versatile new route to the synthesis of enantioenriched tricyclic 3,3-disubstituted phthalides in view of the easy access to both coupling partners (Table 9, Scheme 64). Ellman and co-workers reported rhodium(III)-catalyzed synthesis of biologically important phthalides by cascade addition and cyclization of benzimidates with aldehydes. The imidate is a novel and unexplored directing group that not only enables C–H bond activation and addition to aldehydes, but also serves to capture the reversibly formed alcohol intermediate. The reaction shows broad scope with a high level of functional group compatibility and it is applicable to both aromatic and aliphatic aldehydes (Table 9, Scheme 65).[92] Later Luo and co-workers reported an efficient rhodium-catalyzed lactonization

Table 9. Rhodium-catalyzed phthalide synthesis

S. No.	Rhodium-catalyzed phthalide synthesis	Ref. No
64	$Z = O, NTs, CH_2$; $E = CO_2Me$ $R^1 = Me, CO_2Me$; $R^2 = Ph, Me, CH_2OMe, H, SiMe_3$ $R^3 = Me, Et$; $R^4 = Ph, \equiv\!\!-R^2$ [Rh(COD)$_2$]BF$_4$ (5 mol%), (R)-Solphos, DCM, rt, 1h 13 examples Yields: 53–87% up to 94% ee	91
65	R = H, 4-OMe, 4-CF$_3$, 2-Me, 2-OMe, 3-Me R^1 = 4-MePh, 4-CF3Ph, 4-ClPh, 4-NO2Ph, 4-CO$_2$MePh, 2-FPh, 3-FPh, 3-OMePh, 4-ClPh, CO$_2$Et, c-hexyl [RhCp*Cl$_2$]$_2$ (5 mol%), AgSbF$_6$ (20 mol%), DCE, 110°C 17 examples Yields: 27–84 %	92
66	R = Ph, 4-MePh, 3-MePh, 2-MePh, 3,5-di-MePh, 4-OMePh, 2-OMePh, 4-FPh, 4-ClPh, 1-naphthyl [Rh(COD)Cl]$_2$ (2.5 mol%), dppe (5 mol%), Li$_2$CO$_3$, DCE/H$_2$O 10 examples Yields: 67–93%	93

of phthalaldehydes with potassium organotrifluoroborates to access 3-arylphthalides (Table 9, Scheme 66).[93]

1.4 Rhodium-catalyzed sulfur containing five-membered heterocycle synthesis

The deployment of rhodium-based catalysis in the synthesis of more exotic heterocycles has recently begun to be explored, as seen in the case of sulfur containing heterocycles. While there are at present a limited number of examples in the literature, the results obtained suggest that this should prove an area for further development.

1.4.1 *Sultam*

Liu and co-workers reported benzofused five ring sultams *via* rhodium-catalyzed C–H olefination directed by an *N*-Ac-substituted sulfonamide group. The *N*-acetyl group is a key for this transformation implying that N–H acidity is the major influence. The acetyl group is removed under mild conditions in excellent yield to provide NH-free sultam that can be transformed into various benzofused five-ring sultam analogues *via* acylation nucleophilic substitution, and Mitsunobu alkylation (Scheme 67).[94]

1.4.2 *Thiazole*

In contrast to the reaction of diazocarbonyl compounds with carboxamides, the corresponding reactions with thiocarboxamides are

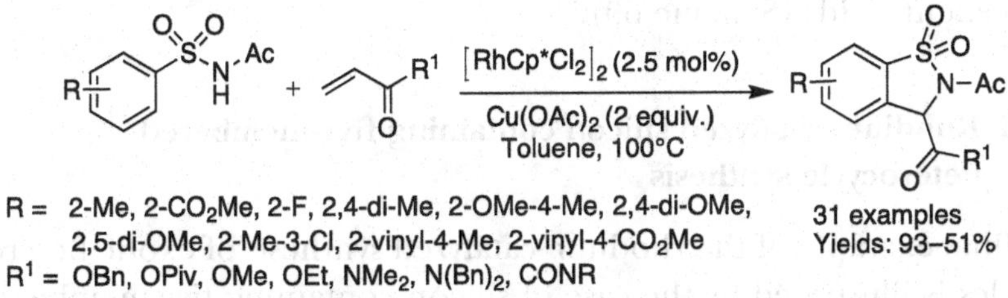
[RhCp*Cl2]2 (2.5 mol%)
Cu(OAc)2 (2 equiv.)
Toluene, 100°C
R = 2-Me, 2-CO2Me, 2-F, 2,4-di-Me, 2-OMe-4-Me, 2,4-di-OMe, 2,5-di-OMe, 2-Me-3-Cl, 2-vinyl-4-Me, 2-vinyl-4-CO2Me
R1 = OBn, OPiv, OMe, OEt, NMe2, N(Bn)2, CONR
31 examples
Yields: 93–51%

Scheme 67. Rhodium-catalyzed sultam synthesis

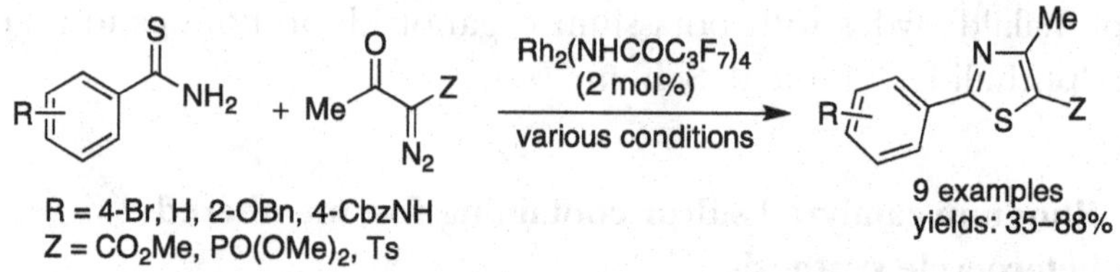
Rh2(NHCOC3F7)4
(2 mol%)
various conditions
R = 4-Br, H, 2-OBn, 4-CbzNH
Z = CO2Me, PO(OMe)2, Ts
9 examples
yields: 35–88%

Scheme 68. Rhodium-catalyzed thiazole synthesis

less well known. The reaction of thiobenzamides with methyl 2-diazo-3-oxobutanoate gave the thiazole-5-carboxylate in reasonable yield, complementing the Hantzsch synthesis of thiazole-5-carboxylate in reasonable yield. The diazophosphonate and diazosulfone reacted similarly, leading to a range of thiazoles in modest to excellent yield (Scheme 68).[78]

1.5 Rhodium-catalyzed phosphorous containing five-membered heterocycle synthesis

To date, only one example of a phosphorus containing five-membered heterocyclic structure prepared by rhodium-catalyzed synthesis has been reported.

1.5.1 *Benzophsophole 1-oxide*

Lee and co-workers reported rhodium-catalyzed tandem oxidative alkenylation and an intramolecular oxy-Michael reaction using arylphosphonic acid monomethyl esters and alkenes under aerobic conditions, which produced benzoxaphosphole-1-oxides in good to excellent yields (Scheme 69).[95]

1.6 Rhodium-catalyzed silicon containing five-membered heterocycle synthesis

A final example of the rhodium-catalyzed synthesis of exotic heterocycles is illustrated by the case of silicon containing five-membered heterocyclic structures.

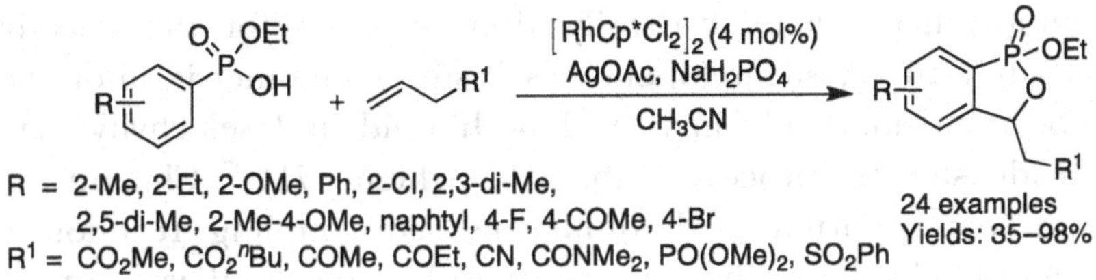

R = 2-Me, 2-Et, 2-OMe, Ph, 2-Cl, 2,3-di-Me,
2,5-di-Me, 2-Me-4-OMe, naphtyl, 4-F, 4-COMe, 4-Br
R^1 = CO$_2$Me, CO$_2^n$Bu, COMe, COEt, CN, CONMe$_2$, PO(OMe)$_2$, SO$_2$Ph

24 examples
Yields: 35–98%

Scheme 69. Rhodium-catalyzed benzophosphole-1-oxide synthesis

Table 10. Rhodium-catalyzed five membered silyl heterocycles synthesis.

S. No.	Rhodium-catalyzed five membered silyl heterocycles synthesis	Ref. No
70	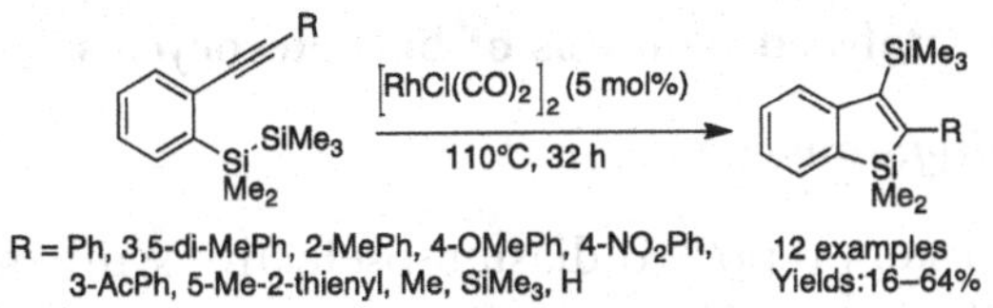 R = Ph, 3,5-di-MePh, 2-MePh, 4-OMePh, 4-NO$_2$Ph, 3-AcPh, 5-Me-2-thienyl, Me, SiMe$_3$, H 12 examples Yields:16–64%	97
71	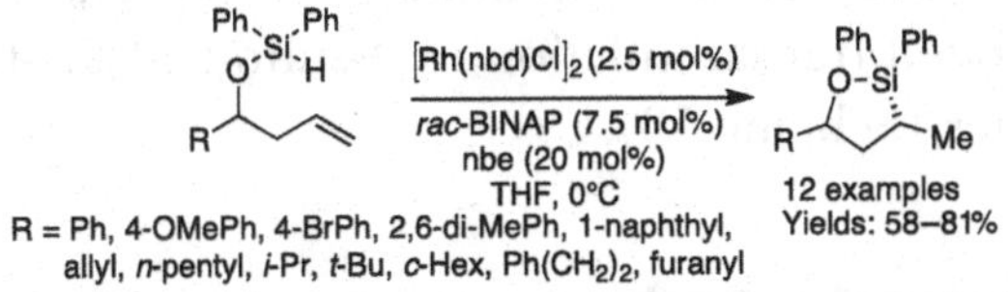 R = Ph, 4-OMePh, 4-BrPh, 2,6-di-MePh, 1-naphthyl, allyl, n-pentyl, i-Pr, t-Bu, c-Hex, Ph(CH$_2$)$_2$, furanyl 12 examples Yields: 58–81%	98
72	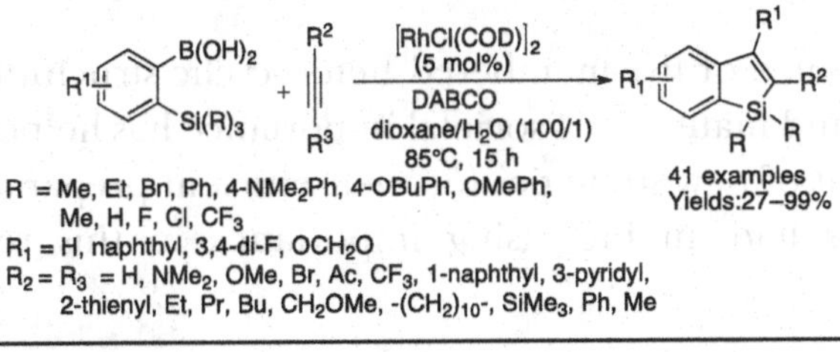 R = Me, Et, Bn, Ph, 4-NMe$_2$Ph, 4-OBuPh, OMePh, Me, H, F, Cl, CF$_3$ R$_1$ = H, naphthyl, 3,4-di-F, OCH$_2$O R$_2$ = R$_3$ = H, NMe$_2$, OMe, Br, Ac, CF$_3$, 1-naphthyl, 3-pyridyl, 2-thienyl, Et, Pr, Bu, CH$_2$OMe, -(CH$_2$)$_{10}$-, SiMe$_3$, Ph, Me 41 examples Yields:27–99%	99

1.6.1 Siloles

Matsuda and co-worker reported that intramolecular addition of a Si–Si bond across a C–C triple bond occurs in a trans fashion in the presence of rhodium (I) catalysts. The trans-bis-silylation reaction of (2-alkynylphenyl) disilanes affords 3-silyl-1-benzosiloles (Table 10, Scheme 70).[96] Jeon and co-workers presented ligand controlled norbornene-mediated regio and diastereoselective rhodium-catalyzed intramolecular alkene hydrosilylation reactions for the selective access to trans-oxasilacyclopentanes. A substoichiometric amount of norbornene markedly increased both yield and selectivity via a 'hydride shuttle' process (Table 10, Scheme 71).[97] Chatani and co-workers reported a rhodium-catalyzed coupling reaction of 2-trimethylsilylphenyl boronic acids with internal alkyne for the synthesis of 2,3-disubstituted benzosilole derivatives. A range of functional groups, encompassing ketones, esters, amines, aryl bromides, and heteroarenes, are compatible, which provides rapid access to diverse benzosiloles (Table 10, Scheme 72).[98]

1.7 Rhodium-catalyzed synthesis of bis-heterocycles

1.7.1 Tetrahydrofuranopyrrole

Lee and co-workers reported diastereoselective synthesis of tetrahydrofurano dihydropyrroles containing N, O-acetal moieties via rhodium-catalyzed transannulation of N-sulfonyl-1,2,3-triazoles with oxacycloalkenes (Scheme 73).[99]

1.8 Conclusions and outlook

The importance of five-membered heterocyclic structural motifs in substances and materials of societal importance has helped drive the development of new strategies for their efficient preparation. Metal catalysis has had an increasing importance in this regard. The

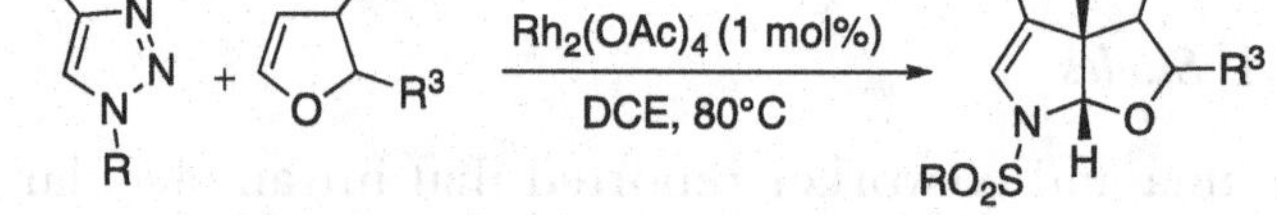

R   = Ms, Ts, SO_2iPr, SO_2-4-OMePh, SO_2-4-ClPh, SO_2-4-CF_3Ph   23 examples
R^1 = Ph, 3-MePh, 4-MePh, 2-OMePh, 4-OMePh, 3-ClPh, 4-ClPh,   Yields: 52–91%
       3-BrPh, 4-BrPh, 3-MePh, 2-OMePh, thiophene
R^3 = Ph, 4-MePh, 4-ClPh, H
R^4 = Ph, 4-OMePh, 4-ClPh

Scheme 73.   Rhodium-catalyzed conversion of 1,2,3-triazole into furanopyrrole

relatively recent growth in the exploration of rhodium's use in the synthesis of five-membered heterocyclic structures highlights the importance of these synthetic targets, and of the potential of rhodium. Looking to the future, the very recent reports of successful examples of rhodium-catalyzed syntheses of sulfur, phosphorus, and silicon containing five membered heterocyclic structures points to an area where significant activity can be anticipated.

References

1. A. R. Katritzky and C. W. Rees, 1984. *Comprehensive Heterocyclic Chemistry*, Vols. 1–8. Pergamon Press, New York. (b) A. R. Katritzky, C. W. Rees and E. F. V. Scriven, 1996. *Comprehensive Heterocyclic Chemistry II*, Vols. 1–8 Pergamon, New York. (c) A. T. Balaban, D. C. Oniciu and A. R. Katritzky, 2004. *Chem. Rev.*, 104, 2777. (d) M. A. P. Martins, Q. Xunico, C. M. P. Pereira, A. F. C. Flores, H. G. Bonacorso and N. Zanatta, 2004. *Curr. Org. Synth.*, 1, 391. (e) M. Baumann, I. R. Baxendale, S. V. Ley and N. Nikbin, 2011. *Beilstein J. Org. Chem.*, 71, 442. (f) J. A. Joule, K. Mills, and G. F. Smith, 1995. *Heterocyclic Chemistry*. Chapman & Hall, U.K.

2. (a) B. H. Lipshutz, 1986. *Chem. Rev.*, 86, 795. (b) N. A. McGrath, M. Brichacek and J. T. Njardarson, 2010. *J. Chem. Educ.*, 87, 1348–1349.

3. (a) J. P. Loferski, 2013. Commodity Report: Platinum-Group Metals, United States Geological Survey. Retrieved July 16, 2012. (b) D. A. Colby, R. G. Bergman and J. A. Ellman, 2010. *Chem. Rev.*, 110, 624. (c) N. Kuhl, N. Schröder and F. Glorius, 2014. *Adv. Synth. Catal.*, 356, 1443. (d) L. Ackermann, R. Vicente and A. R. Kapadi, 2009. *Angew. Chem. Int. Ed.*, 121, 9976. (e) X. Chen, K. M. Engle, D.-H. Wang and J.-Q. Yu, 2009. *Angew. Chem. Int. Ed.*, 48, 5094. (f) T. W. Lyons and M. S. Sanford, 2010. *Chem. Rev.*, 110, 1147. (g) P. Thansandote and M. Lautens, 2009. *Chem. Eur. J.*, 15, 5874. (h) C. S. Yeund and V. M. Dong, 2011. *Chem. Rev.*, 111, 1215.

4. (a) J. Gupton, 2006. In *Heterocyclic Antitumor Antibiotics*, M. Lee (ed.), Vol. 2, p. 53. Springer, Berlin. (b) D. Mal, B. Shome and B. K. Dinda, 2011. In *Heterocycles in Natural Product Synthesis*, K. C. Majumdar and S. K. Chattopadhyay (eds.), p. 187. Wiley-VCH, Verlag GmbH & Co. KGaA, Weinheim, Germany. (c) M. E. Mason,

B. Johnson and M. J. Hamming, 1966. *Agric. Food Chem.*, 14, 454. (d) Jr. S. Lunak, M. Vala, J. Vynuchal, I. Ouzzane, P. horakova, P. Moziskova, Z. Elias and M. Weiter, 2011. *Dyes Pigm.*, 91, 269. (e) Jr. S. Lunak, L. Havel, J. Vynuchal, P. Horakova, J. Kucerik, M. Weiter and R. Hrdina, 2010. *Dyes Pigm.*, 85, 27.

5. (a) C. V. Galliford and K. A. Scheidt, 2007. *J. Org. Chem.*, 72, 1811. (b) Y. Wang, X. Lei and Y. Tang, 2015. *Chem. Commun.*, 51, 4507. (c) H. Takaya, S. Kojima and S. I. Murahashi, 2001. *Org. Lett.*, 3, 421. (d) Y. Shi and V. Gevorgyan, 2013. *Org. Lett.*, 15, 5394. (e) J. S. Alford, J. E. Spangler and H. M. L. Davies, 2013. *J. Am. Chem. Soc.*, 135, 11712. (f) J. Feng, Y. Wang, Q. Li, R. Jiang and Y. Tang, 2014. *Tet. Lett.*, 55, 6455. (g) C. E. Kim, S. Park, D. Eom, B. Seo and P. H. Lee, 2014. *Org. Lett.*, 16, 1900. (h) W. J. Humenny, P. Kyriacou, K. Sapeta, A. Karadeolian and M. A. Kerr, 2012. *Angew. Chem. Int. Ed.*, 51, 11088. (i) R. Narayan, R. Froehlich and E.-U. Wuerthwein, 2012. *J. Org. Chem.*, 77, 1868. (j) F. Chen, T. Shen, Y. Cui and N. Jiao, 2012. *Org. Lett.*, 14, 4926.

6. (a) C. Paal, 1885. *Ber. Dtsch. Chem. TGes.*, 18, 367. (b) L. Knorr, 1884. *Ber. Dtsch. Chem. Ges.*, 17, 1635. (c) R. Huisgen, H. Gotthardt, H. O. Bayer, F. C. Schaefer, 1964. *Angew. Chem. Int. Ed.*, 3, 136. (d) A. Hantzsch, 1890. *Ber. Dtsch. Chem. Ges.*, 23, 1474.

7. B. Chattopadhyay and V. Gevorgyan, 2011. *Org. Lett.*, 13, 3746.

8. B. T. Parr, S. A. Green and H. M. L. Davies, 2013. *J. Am. Chem. Soc.*, 135, 4716.

9. S. Rajasekar and P. Anbarasan, 2014. *J. Org. Chem.*, 79, 8428.

10. A. Mizuno, H. Kusama and N. Iwasawa, 2009. *Angew. Chem. Int. Ed.*, 48, 8318.

11. (a) G. R. Humphrey and J. T. Kuthe, 2006. *Chem. Rev.*, 106, 2875. (b) R. J. Sundberg, 1996. *Indoles*. Academic Press, San Diego. (c) T. Eicher and S. Hauptmann, 2003. *The Chemistry of Heterocycles: Structure, Reactions, Syntheses, and Applications*, 2nd Edition. Wiley-VCH, Weinheim. (d) T. Kawasaki and K. Higuchi, 2005. *Nat. Prod. Rep.*, 22, 761.

12. (a) B. Robison, 1963. *Chem. Rev.*, 63, 373. (b) B. Robinson, 1969. *Chem. Rev.*, 69, 227. (c) D. L. Hughes, 1993. *Org. Prep. Proced. Int.*, 25, 607. (d) B. Robinson, 1982. *The Fischer Indole Synthesis*. Wiley-Interscience, New York.

13. (a) Z: Fan, S. Song, W. Li, K. Geng, Y. Xu, Z. H. Miao and A. Zhang, 2015. *Org. Lett.*, 17, 310. (b) B. Rajagopal, C. H. Chou, C. C. Chung

and P. C. Lin, 2014. *Org. Lett.*, 16, 3752. (c) K. Sun, S. Liu, P. M. bec and T. G. Driver, 2011. *Angew. Chem. Int. Ed.*, 50, 1702. (d) D. Shu, W. Song, X. Li and W. Tang, 2013. *Angew. Chem. Int. Ed.*, 52, 3237. (e) K. Okuro, J. Gurnham and H. Alper, 2011. *J. Org. Chem.*, 76, 4715. (f) B. M. Trost and A. McClory, 2007. *Angew. Chem. Int. Ed.*, 46, 2074.

14. G. Zeni and R. C. Larock, 2006. *Chem. Rev.*, 106, 4644.

15. S. Kathiravan and I. A. Nicholls, 2014. *Chem. Commun.*, 50, 14964.

16. D. R. Stuart, M. Bertrand-Laperle, K. M. N. Burgess and K. Fagnou, 2008. *J. Am. Chem. Soc.*, 130, 16474.

17. L. Ackermann, 2005. *Org. Lett.*, 7, 439.

18. (a) P. Tao and Y. Jia, 2014. *Chem. Commun.*, 50, 7367. (b) Y. Liang, K. Yu, B. Li, S. Xu, H. Song and B. Wang, 2014. *Chem. Commun.*, 50, 6130. (c) T. Miura, Y. Funakoshi and M. Murakami, 2014. *J. Am. Chem. Soc.*, 136, 2272. (d) B. Liu, C. Song, C. Sun, S. Zhou and J. Zhu, 2013. *J. Am. Chem. Soc.*, 135, 16625. (e) C. Wang, H. Sun, Y. Fang and Y. Huang, 2013. *Angew. Chem. Int. Ed.*, 52, 5795. (f) T. Matsuda and Y. Tomaru, 2014. *Tet. Lett.*, 55, 3302. (g) Y. Hoshino, Y. Shibata and K. Tanaka, 2014. *Adv. Synth. Catal.*, 356, 1577. (h) L. Zheng and R. Hua, 2014. *Chem. Eur. J.*, 20, 2352. (i) D. Y. Li, H. J. Chen and P. N. Liu, 2014. *Org. Lett.*, 16, 6176.

19. D. Zhao, Z. Shi and F. Glorius, 2013. *Angew. Chem. Int. Ed.*, 52, 12426.

20. K. Muralirajan and C. H. Cheng, 2014. *Adv. Synth. Catal.*, 356, 1571.

21. S. W. Kwok, L. Zhang, N. P. Grimster and V. V. Fokin, 2014. *Angew. Chem. Int. Ed.*, 53, 3452.

22. H. Shang, Y. Wang, Y. Tian, Y. Feng and Y. Tang, 2014. *Angew. Chem. Int. Ed.*, 53, 5662.

23. X. Zheng, B. Cao and X. Zhang, 2014. *Tet. Lett.*, 55, 4489.

24. (a) T. Matviiuk, F. Rodriguez. N. Saffon, S. Mallet-Ladeira, M. Gorichko, A.-L. Ribeiro, M. R. Pasca, C. Lherbet, Z. Voitenko and M. Baltas, 2013. *Eur. J. Med. Chem.*, 70, 37. (b) T. Fukuda, Y. Sudoh, Y. Tsuchiya, T. Okuda and Y. Igarashi, 2014. *J. Nat. Prod.*, 77, 813.

25. (a) J. J. Medvedev, O. S. Galkina, A. A. Klinkova, D. S. Giera, L. Hennig, C. Schneider and V. A. Nikolaev, 2015. *Org. Biomol. Chem.*, 13, 2640. (b) S. Kim and Y. K. Chung, 2014. *Org. Lett.*, 16, 4352. (c) T. Cochet, V. Bellosta, D. Roche, J. Y. Ortholand, A. Greiner, and J. Cossy, 2012. *Chem. Commun.*, 48, 10745. (d) M. H. Shaw, E. Y. Melikhova, D. P. Kloer, W. G. Whittingham and J. F. Bower, 2013.

J. Am. Chem. Soc., 135, 4992. (e) F. Serpier, B. Flamme, J. L. Brayer, B. Folleas and S. Darses, 2015. Org. Lett., 17, 1720. (f) J. Q. Zhou and H. Alper, 1992. J. Org. Chem., 57, 3328.

26. Z. Liu and J. F. Hartwig, 2008. J. Am. Chem. Soc., 130, 1570.

27. X. Shen and S. L. Buchwald, 2010. Angew. Chem. Int. Ed., 49, 564.

28. Q. Li and Z. X. Yu, 2011. Angew. Chem. Int. Ed., 50, 2144.

29. Q. Li and Z. X. Yu, 2010. J. Am. Chem. Soc., 132, 4542.

30. J. D. Hargrave, J. C. Allen, G. K. Köhn, G. Bish and C. G. Frost, 2010. Angew. Chem. Int. Ed., 49, 1825.

31. (a) G. Verniest and N. De Kimpe, 2003, 2013. Synlett. (b) K. Tonogaki, K. Itami and J. -I. Yoshida, 2006. J. Am. Chem. Soc., 128, 1464. (c) G. Casiraghi and G. Rassu, 1995. Synthesis, 607. (d) Z. Feng, F. Chu, Z. Guo, P. Sun, 2009. Bioorg. Med. Chem. Lett., 19, 2270. (e) A. G. H. Wee and S. C. Duncan, 2005. J. Org. Chem., 70, 8372. (f) T. Iwata, F. Inagaki and C. Mukai, 2013. Angew. Chem. Int. Ed., 52, 11138.

32. R. Q. Ran, J. He, S. D. Xiu, K. B. Wang and C. Y. Li, 2014. Org. Lett., 16, 3704.

33. W. Z. Zhang, J. C. K. Chu, K. M. Oberg and T. Rovis, 2015. J. Am. Chem. Soc., 137, 553.

34. W. Hou, B. Zhou, Y. Yang, H. Feng and Y. Li, 2013. Org. Lett., 15, 1814.

35. O. Jackowski, J. Wang, X. Xie, T. Ayad, Z. Zhang and V. Ratovelomanana-Vidal, 2012. Org. Lett., 14, 4006.

36. J. Elguero, 1996. In Comprehensive Heterocyclic Chemistry II, I. Shinkai, (ed.), Vol. 3, pp. 1–75. Pergamon Press, Oxford, U.K.

37. N. K. Terrett, A. S. Bell, D. Brown and P. Ellis, 1996. Bioorg. Med. Chem. Lett., 6, 1819. (b) S. R. Donohue, C. Halldin and V. W. Pike, 2006. Bioorg. Med. Chem., 14, 3712. (c) S. H. Hwang, K. M. Wagner, C. Morisseau, J. -Y. Liu, H. Dong, A. T. Wecksler and B. D. Hammock, 2011. J. Med. Chem., 54, 3037.

38. H. Maeda, Y. Ito, Y. Kusunose and T. Nakanishi, 2007. Chem. Commun., 1136. (b) J. Catalan, F. Fabero, R. M. Claramunt, M. D. Santa Maria, M. C. Foces-Foces, F. Hernandez Cano, M. Martinez-Ripoll, J. Elguero and R. Sastre, 1992. J. Am. Chem. Soc., 114, 5039. (c) A. Dorlars, C.-W. Schellhammer and J. Schroeder, 1975. Angew. Chem. Int. Ed., 14, 665.

39. D. Y. Li, X. F. Mao, H. J. Chen, G. R. Chen and P. N. Liu, 2014. Org. Lett., 16, 3476.

40. C. Jing, D. Xing and W. Hu, 2014. *Chem. Commun.*, 50, 951.

41. D. Zhao, S. Vasquez-Cespedez and F. Glorius, 2015. *Angew. Chem. Int. Ed.*, 54, 1657.

42. R. Shintani, T. Yamagami and T. Hayashi, 2006. *Org. Lett.*, 8, 4799.

43. J. H. Park, E. Kim and Y. K. Chung, 2008. *Org. Lett.*, 10, 4719.

44. (a) D. F. Ewing, C. Len, G. Mackenizie, J. P. Petit, G. Ronco and P. Villa, 2001. *J. Pharm. Pharmacol.*, 53, 945. (b) D. Berger, R. Citarella, M. Dutia, L. Greenberger, W. Hallett, R. Paul and D. Powell, 1999. *J. Med. Chem.*, 42, 2145. (c) J. G. Topliss, L. M. Konzelman, N. Sperber and F. E. Roth, 1964. *J. Med. Chem.*, 7, 453. (d) X. Wei, F. Wang, G. Song, Z. Du and X. Li, 2012. *Org. Biomol. Chem.*, 10, 5521. (e) N. K. Mishra, J. Park, S. Sharma, S. Han, M. Kim, Y. Shin, J. Jang, J. H. Kwak, Y. H. Jung and I. S. Kim, 2014. *Chem. Commun.*, 50, 2350.

45. C. Zhu and J. R. Falck, 2012. *Chem. Commun.*, 48, 1674.

46. S. Sharma, E. Park and I. S. Kim, 2012. *Org. Lett.*, 14, 906.

47. A. M. Martinez, N. Rodriguez, R. G. Arrayas and J. C. Carretero, 2014. *Chem. Commun.*, 50, 6105.

48. B. Ye and N. Cramer, 2014. *Angew. Chem. Int. Ed.*, 53, 7896.

49. C. Suzuki, K. Morimoto, K. Hirano, T. Sathoh and M. Miura, 2014. *Adv. Synth. Catal.*, 356, 1521.

50. (a) J. Magano, M. Waldo, D. Greene and E. Nord, 2008. *Org. Process Res. Dev.*, 12, 877. (b) H. Cerecetto, A. Gerpe, M. Gonzalez, V. J. Aran and C. Ochoa de Ocariz, 2005. *Mini-Rev. Med. Chem.*, 5, 869. (c) S. Caron and E. Vazquez, 1999. *Synthesis*, 588. (d) S. Han, Y. Shin, S. Sharma, N. K. Mishra, J. Park, M. Kim, M. Kim, J. Jang and I. S. Kim, 2014. *Org. Lett.*, 16, 2494.

51. D. G. Yu, M. Suri and F. Glorius, 2013. *J. Am. Chem. Soc.*, 135, 8802.

52. Y. Lian, R. G. Bergman, L. D. Lavis and J. A. Ellman, 2013. *J. Am. Chem. Soc.*, 135, 7122.

53. J. Yao, R. Feng, C. Lin, Z. Liu and Y. Zhang, 2014. *Org. Biomol. Chem.*, 12, 5469.

54. T. Horneff, S. Chuprakov, N. Chernyak, V. Gevorgyan and V. V. Fokin, 2008. *J. Am. Chem. Soc.*, 130, 14972.

55. M. Zibinsky and V. V. Fokin, 2013. *Angew. Chem. Int. Ed.*, 52, 1507.

56. (a) R. Jayakumar, R. Balaji and S. Nanjundan, 2000. *Eur. Polym. J.*, 36, 1659. (b) X. Collin, J. Robert, G. Wielgosz, G. Le Baut, C. Bobin-Dubigeon, N. Grimaud and J. Petit, 2001. *Eur. J. Med. Chem.*, 36, 639. (c) H. Teisseire and G. Vernet, Pestic, 2001. *Biochem. Physiol.*, 69, 112.

(d) U. Sharma, P. Kumar, N. Kumar and B. Singh, 2010. *Mini-Rev. Med. Chem.*, 10, 678. (e) H. Nakayama, J. Nishida, N. Takada, H. Sato and Y. Yamashita, 2012. *Chem. Mater.*, 24, 671.

57. X. Y. Shi, A. Renzetti, S. Kundu and C. J. Li, 2014. *Adv. Synth. Catal.*, 356, 723.

58. S. Chuprakov, S. W. Kwok and V. V. Fokin, 2013. *J. Am. Chem. Soc.*, 135, 4652.

59. (a) B. A: Keay and P. W. Dibble, 1996. *Furans and their Benzo Derivatives: Applications, in Comprehensive Heterocyclic Chemistry II*, A. R. Katritzky, C. W. Rees and E. F. V. Scriven, (ed.), Vol. 2, p. 395. Pergamon Press, Oxford, U.K. (b) A. S. K. Hashmi, J. P. Weyrauch, E. Kurpejovic, T. M. Frost, B. Miehlich, W. Frey and J. W. Bats, 2006. *Chem. Eur. J.*, 12, 5806. (c) L. Peng, X. Zhang, M. Ma and J. Wang, 2007. *Angew. Chem. Int. Ed.*, 46, 1905. (d) Y. Shibata, K. Noguchi, M. Hirano and K. Tanaka, 2008. *Org. Lett.*, 10, 2825. (e) M. J. Gonzalez, E. Lopez and R. Vicente, 2014. *Chem. Commun.*, 50, 5379.

60. W. B. Zhang, S. D. Xiu and C. Y. Li, 2015. *Org. Chem. Front.*, 2, 47.

61. P. Lenden, D. A. Entwistle and M. C. Willis, 2011. *Angew. Chem. Int. Ed.*, 50, 10657.

62. Y. Zhang, Z. Chen, Y. Xiao and J. Zhang, 2009. *Chem. Eur. J.*, 15, 5208.

63. Y. Lian, T. Huber, K. D. Hesp, R. G. Bergman and J. A. Ellman, 2013. *Angew. Chem. Int. Ed.*, 52, 629.

64. H. Gao and J. Zhang, 2012. *Chem. Eur. J.*, 18, 2777.

65. J. P. Bouillon, V. Kikelj, B. Tinant, D. Harakat and C. Portella, 2006. *Synthesis*, 1050. (b) S. Limura, L. E. Overman, R. Paulini and A. Zakarian, 2006. *J. Am. Chem. Soc.*, 128, 13095. (c) P. Langer, 2006. *Synlett*, 3369. (d) R. Raju, L. J. Allen, T. Le, C. D. Taylor and A. R. Howell, 2007. *Org. Lett.*, 9, 1699. (e) J. Adiro and J. C. Carretero, 2007. *J. Am. Chem. Soc.*, 129, 778.

66. M. Alfonsi, A. Arcadi, M. Chiarini and F. Marinelli, 2007. *J. Org. Chem.*, 72, 9510.

67. (a) A. Lei, M. He and X. Zhang, 2002. *J. Am. Chem. Soc.*, 124, 8198. (b) X. Tong, D. Li, Z. Zhang and X. Zhang, 2004. *J. Am. Chem. Soc.*, 126, 7601.

68. X. Xie, Y. Li and J. M. Fox, 2013. *Org. Lett.*, 15, 1500.

69. (a) F. Lovering, J. Bikker and C. Humblet, 2009. *J. Med. Chem.*, 52, 6752. (b) T. J. Ritchie, S. J. F. Macdonald, R. J. Young and

S. D. Pickett, 2011. *Drug Discovery Today*, 16, 164. (c) A. Lorente, J. Lamariano-Merketegi, F. Albericio and M. Alvarez, 2013. *Chem. Rev.*, 113, 4567. (d) J. Bartroli, E. Carceller, M. Merlos, J. Garcia-Rafanell and J. Forn, 1991. *J. Med. Chem.*, 34, 373. (e) C. E. Kim, Y. Park, S. Park and P. H. Lee, 2015. *Adv. Synth. Catal.*, 357, 210.

70. A. Boyer, 2014. *Org. Lett.*, 16, 5878.

71. A. Lei, M. He, S. Wu and X. Zhang, 2002. *Angew. Chem. Int. Ed.*, 41, 3457.

72. J. Keilitz, S. G. Newman and M. Lautens, 2013. *Org. Lett.*, 15, 1148.

73. S. Cenini, G. Cravotto, G. B. Giovenzana, A. Penoni, S. Tollari and G. Palmisano, 2000. *J. Mol. Cat. A.*, 164–165.

74. (a) H. Chen, R. Tan, Z. L. Liu and Y. Zhang, 1996, *J. Nat. Prod.*, 59, 668. (b) M. C. Elliott, 1998. *J. Chem. Soc., Perkin Trans. 1*, 4175. (c) H. Kizu, N. Sugita and T. Tomimori, 1998. *Chem. Pharm. Bull.*, 46, 988. (d) C. J. Burns and D. S. Middleton, 1996. *Contemp. Org. Synth.*, 3, 229.

75. R: Roggenbuck, A. Schmidt and P. Eilbracht, 2002. *Org. Lett.*, 4, 289.

76. H. Shen, J. Fu, J. Gong and Z. Yang, 2014. *Org. Lett.*, 16, 5588.

77. X. Xu, P. Y. Zavalij, W. Hu and M. P. Doyle, 2012. *Chem. Commun.*, 48, 11522.

78. B. Shi, A. J. Blake, W. Lewis, I. B. Campbell, B. D. Judkins and C. J. Moody, 2010. *J. Org. Chem.*, 75, 152.

79. D. Gong, L. Zhang and C. Yuan, 2004. *Syn. Commun.*, 34, 3259.

80. M. Wang, C. Zhang, L. P. Sun, C. Ding and A. Zhang, 2014. *J. Org. Chem.*, 79, 4553.

81. C. G. Espino and J. D. Bois, 2001. *Angew. Chem. Int. Ed.*, 40, 598.

82. H. Lebel, K. Huard and S. Lectard, 2005. *J. Am. Chem. Soc.*, 127, 14198.

83. J. Guasch, Y. Diaz, M. I. Matheu and S. Castillon, 2014. *Chem. Commun.*, 50, 7344.

84. G. C. Tsui, N. M. Ninnemann, A. Hosotani and M. Lautens, 2013. *Org. Lett.*, 15, 1064.

85. (a) M. D. Argade, A. Y. Mehta, A. Sarkar and U. R. Desai, 2014. *J. Med. Chem.*, 57, 3559. (b) Y. H. Liu, M. Kubo and Y. Fukuyama, 2012. *J. Nat. Prod.*, 75, 2152. (c) M. H. Sun, C. Zhao, G. A. Gfesser, C. Thiffault, T. R. Miller and M. Cowart, 2005. *J. Med. Chem.*, 48, 6482. (d) Z. Yang, H. B. Liu, C. M. Lee, H. M. Chang and H. N. C. Wong, 1992. *J. Org. Chem.*, 57, 7248. (e) T. A. Davis, T. K. Hyster and T. Rovis, 2013. *Angew. Chem. Int. Ed.*, 52, 14181.

86. X. Ma, F. Wu, X. Yi, H. Wang and W. Chen, 2015. *Chem. Commun.*, 51, 6862.

87. G. T. Hoang, V. J. Reddy, H. H. K. Nguyen and C. J. Douglas, 2011. *Angew. Chem. Int. Ed.*, 50, 1882.

88. N. Sakiyama, K. Noguchi and K. Tanaka, 2012. *Angew. Chem. Int. Ed.*, 51, 5976.

89. R: Okamoto, E. Okazaki, K. Noguchi and K. Tanaka, 2011. *Org. Lett.*, 13, 4894.

90. B. Ye, P. A. Donets and N. Cramer, 2014. *Angew. Chem. Int. Ed.*, 53, 507.

91. K. Tanaka, T. Osaka, K. Noguchi and M. Hirano, 2007. *Org. Lett.*, 9, 1307.

92. Y. Lian, R. G. Bergman and J. A. Ellman, 2012. *Chem. Sci.*, 3, 3088.

93. F. Luo, S. Pan, C. Pan, P. Qian and J. Cheng, 2011. *Adv. Synth. Catal.*, 353, 320.

94. X. Li, Y. Dong, F. Qu and G. Liu, 2015. *J. Org. Chem.*, 80, 790.

95. T. Ryu, J. Kim, Y. Park, S. Kim and P. H. Lee, 2013. *Org. Lett.*, 15, 3986.

96. T. Matsuda and Y. Ichioka, 2012. *Org. Biomol. Chem.*, 10, 3175.

97. Y. Hua, H. H. Nguyen, W. R. Scaggs and J. Jeon, 2013. *Org. Lett.*, 15, 3412.

98. M. Onoe, K. Baba, Y. Kim, Y. Kita, M. Tobisu and N. Chatani, 2012. *J. Am. Chem. Soc.*, 134, 19477.

99. Z. Yu and Y. Lan, 2013. *J. Org. Chem.*, 78, 11501.

CHAPTER 4

Rh-Catalyzed Six-Membered Heterocycles Synthesis

Lin Dong, Ji-Rong Huang, Shuai-Shuai Li, Qian-Ru Zhang

Sichuan University, China
dongl@scu.edu.cn

Rhodium-catalyzed C–H activation has witnessed exciting progress over the past decades due to its high efficiency, selectivity, and functional-group tolerance. A number of efficient rhodium-catalyzed reactions for the formation of six-membered heterocycle frameworks has been summarized, involving diversity of recent examples and plausible mechanisms.

1. Introduction

Six-membered ring heterocyclics are some of the important structural motifs that are widely implicated in natural products, biological properties, and medicinal chemistry, such as marketed drug GSK812397 and Rimifon.[1] The synthesis of heterocyclic frameworks has attracted considerable attention because of the unique structures. The demand for environmentally benign and sustainable chemistry has inspired chemists to seek efficient and convenient procedures to construct chemical bonds during the synthesis of heterocyclics.

Rimifon   Quinine   GSK812397

Transition metal-catalyzed organic reactions have been increasingly exploited in the synthesis of a variety of heterocyclic scaffolds, because they can minimize pre-functionalization steps and waste formation via C–H functionalization.[2] Activating the C–H bond to generate reactive organometallic intermediates is an eco-friendly alternative, to conventional catalytic addition method.[3] Such novel approaches provide significant benefits toward the synthesis of various heterocyclics, especially six-membered heterocycles.

Over the past few years, rhodium-catalyzed methods for C–H activation have witnessed drastic progress because of high efficiency, selectivity, and functional group tolerance.[4] The Rh(III)/Rh(I) cycles are widely present in catalysis, stand out in the C–H activation pathway owing to wide range of synthetic utility.[5] These syntheses are highlighted following the discussion of the method they employ.

2. Synthesis of Pyridines

2.1 Rh(I) catalysis

2.1.1 Procedures with [2 + 2 + 2]

[2+2+2] Cycloadditions of nitriles and alkynes catalyzed or mediated by transition metals have been well developed for building pyridines.[6] The pioneering work in this field was using cobalt as the metal catalysts, reported by Yamazaki and co-worker,[7] Vollhardt,[8] and Bönnemann[9] since the first case back to 1973.[7a]

Besides cobalt catalysts, other metal catalysts have also been proved to be efficient for this transformation, such as ruthenium, nickel, titanium, and tantalum complexes. In 1993, Diversi and co-workers found that the [2 + 2 + 2] cycloaddition of alkynes and nitriles could be carried out under the catalysis of $[RhCp'L_n]$, a half-sandwich rhodium(I) complexes (Scheme 1).[10] In this study, they observed an interesting phenomenon by increasing the electron-withdrawing character of the substituents on the Cp' ring. Although reduced reactivity at low temperature was observed, increased chemoselectivity and regioselectivity could be obtained, which tends to produce pyridines rather than benzenes and pyridines, and 2,3,6-isomers rather than benzenes and 2,4,6-isomers, respectively.

In 2006, Tanaka and co-workers reported a chemo- and regio-selective [2 + 2 + 2] cycloaddition of a wide variety of alkynes and nitriles catalyzed by cationic rhodium(I)/modified-BINAP complexes, leading to a series of highly functionalized pyridines under mild reaction conditions (Scheme 2).[11] Taking advantage of this stratagem, not only five-membered ring, but also six- and seven-membered rings could be formed by choosing 1,6-diyne, 1,7-diyne, and 1,8-diyne as the coupling partner separately. The asymmetric version of this reaction,

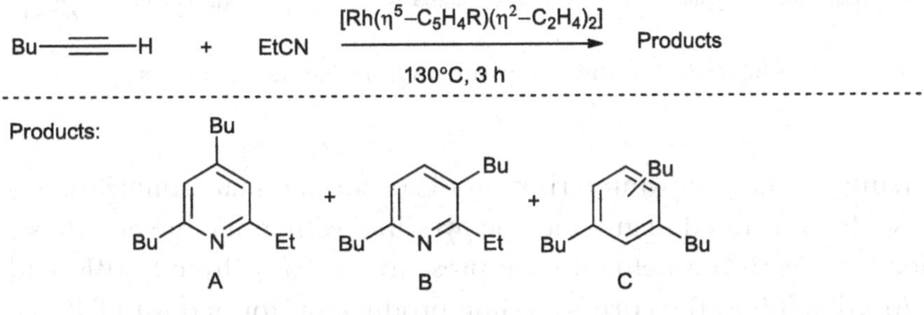

R	T = 100°C/Yield[a]	T = 130°C/Yield[a]
NMe$_2$	26.3(41.1)	21.3(34.5)
^{t}Bu	28.3(10.6)	67.8(21.6)
Me	18.4(7.6)	62.3(19.6)
H	6.6(3.3)	64.5(18.7)
Cl	8.3(7.3)	32.2(20.0)
CF$_3$	9.0(9.2)	65.7(15.5)
COOMe	18.8(5.6)	67.7(19.4)

[a] Combined yields of A+B, the yields of C were given in parentheses.

Scheme 1. Synthesis of pyridines from nitriles and alkynes

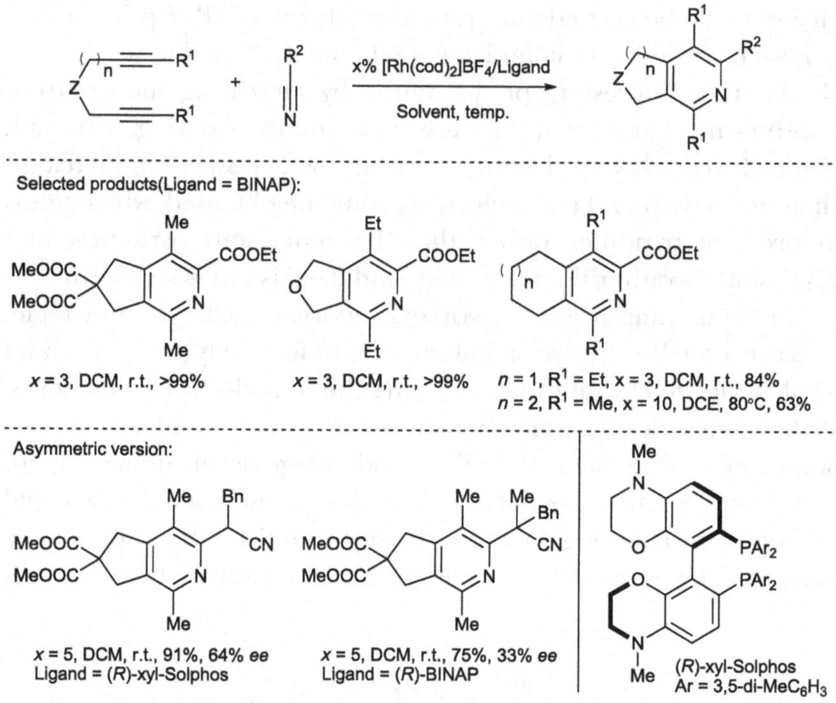

Scheme 2. Synthesis of pyridines from nitriles and diynes

enantioselective desymmetrization of substituted malononitriles, was also demonstrated to give bicyclic pyridines with moderate enantioselectivity. As shown below, in the presence of 5% [Rh(cod)$_2$]BF$_4$ and (R)-xyl-Solphos, the corresponding product was formed with 64% ee at room temperature by reaction of 1,6-diyne with monosubstituted malononitrile. When proceeding this reaction with disubstituted malononitrile, bicyclic pyridine with a quaternary stereocenter was delivered in 75% yield and 33% ee under the catalytic system of 5% [Rh(cod)$_2$]BF$_4$ and (R)-BINAP.

Although metal-catalyzed [2 + 2 + 2] annulation has been well established for the synthesis of pyridines, the nitrogen source is very restricted to nitriles. In 2013, Wan and co-workers found an alternative way for straightforward generation of pyridines by using oximes in the rhodium-catalyzed [2 + 2 + 2] cycloaddition of diynes

(Scheme 3).[12] Under the optimized conditions with catalytic system contained [Rh(cod)$_2$BF$_4$] and 1,1'-Bis(diphenylphosphino)ferrocene (DPPF), pyridine derivatives were formed efficiently in CF$_3$CH$_2$OH. Notably, simple aldehyde also could be used directly in the one-pot reaction with hydroxylamine and diyne.

Based on the mechanism experiments, two possible reaction pathways have been proposed. In path A, Rh(I) catalyst coordinated with both oxime and alkyne moiety, forming five-membered

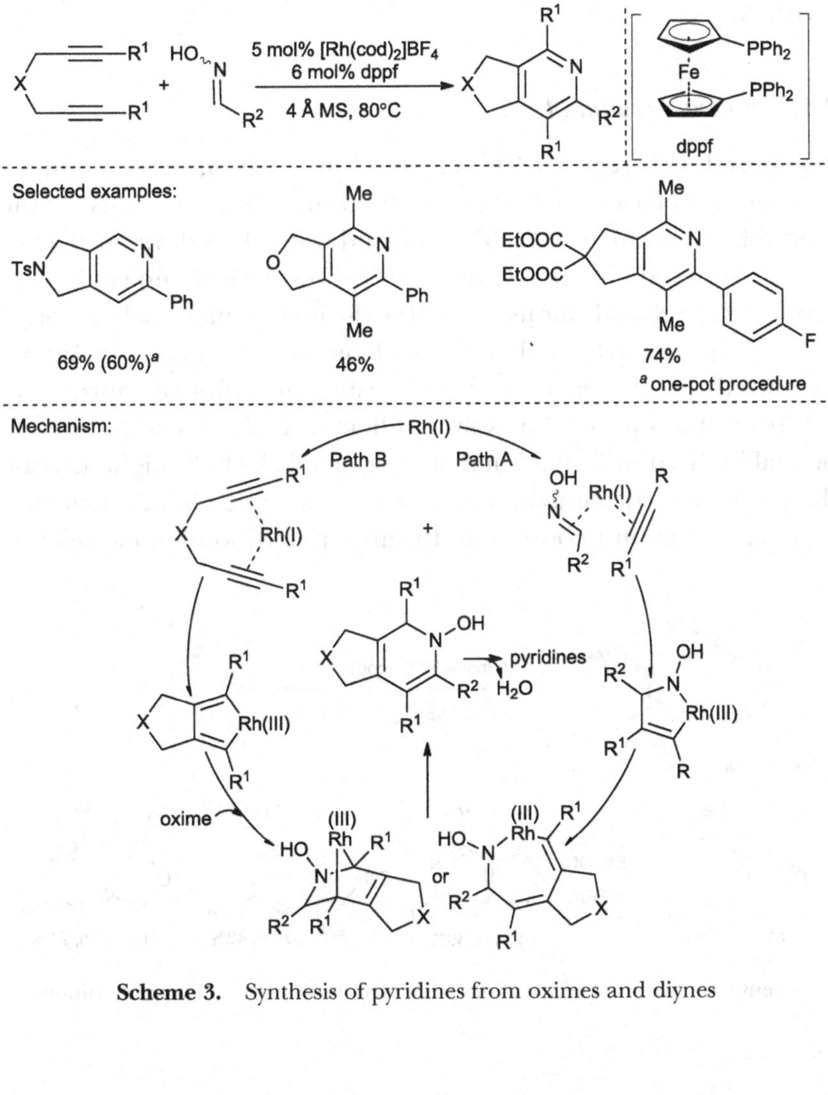

Selected examples:

69% (60%)[a]   46%   74%

[a] one-pot procedure

Mechanism:

Scheme 3. Synthesis of pyridines from oximes and diynes

azametallacycle intermediate. Subsequent intramolecular alkyne insertion afforded seven-membered complex, which could generate pyridines by the following reductive elimination and dehydration. Alternatively, two alkynes of one molecular of diyne delivered metallacyclopentadiene intermediate by coordination reaction with rhodium center, which was followed by a [4 + 2] cycloaddition or an insertion of oxime to give bridged metallacycle or seven-membered ring. Similarly, final product could be obtained by regeneration of the Rh(I) species and dehydration process in this reaction pathway (path B).

2.1.2 Procedures with [4 + 2]

Compared with [2 + 2 + 2] cycloaddition reaction, [4 + 2] annulation seems to have attached more attention among chemists for the synthesis of pyridines. In 2007, Saito and co-workers developed a facile way to afford bicyclic pyridine derivatives *via* intramolecular hetero-[4 + 2] cycloadditions of ω-alkynyl–vinyl oximes (Scheme 4).[13] Cationic rhodium(I) catalyst derived from [RhCl(cod)]$_2$ and AgSbF$_6$ exhibited great reactivity when using hexafluoroisopropanol (HFIP) as the solvent. The solvent effect of HFIP was considered to be vital in this transformation that fluorinated alcohols might increase the preferable conformation of substrates to cycloaddition reactions by polar effect and coordinate to unsaturated bonds to accelerate

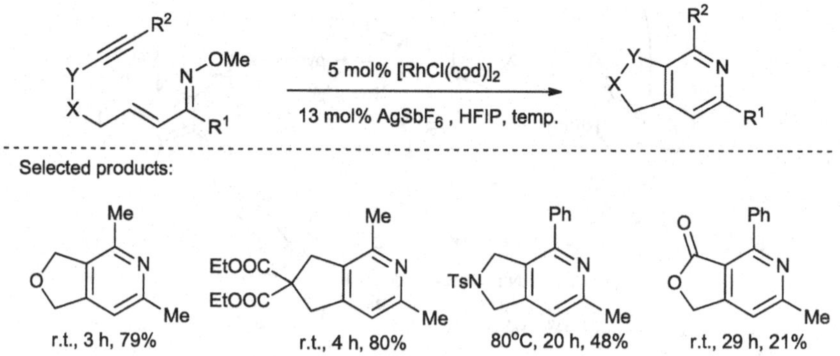

Scheme 4. Synthesis of pyridines from intramolecular [4+2] cycloadditions

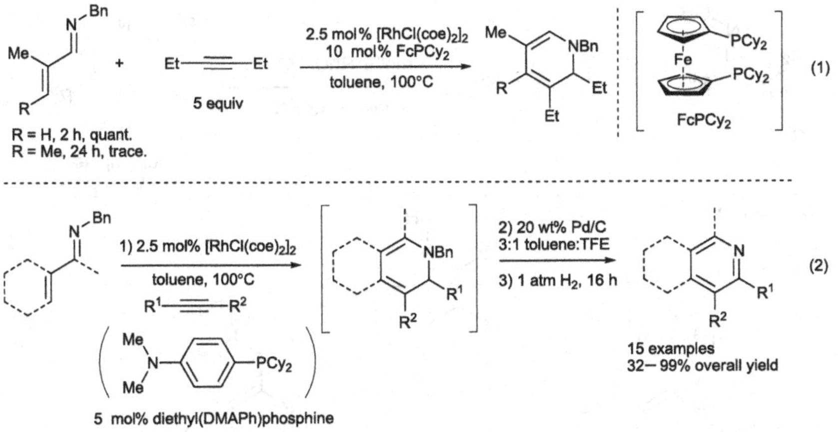

Scheme 5. Synthesis of pyridines from α,β-unsaturated N-benzyl imines and alkynes

the reactivity of cationic Rh(I) species. However, this method is invalid for the formation of tetrahydroisoquinolines.

In the next year, a convenient one-pot C–H alkenylation/electrocyclization/aromatization sequence has been established by Bergman and Ellman (Scheme 5).[4h] Highly substituted pyridines were generated from alkynes and α,β-unsaturated N-benzyl aldimines or ketimines *via* dihydropyridine intermediates. To begin their investigation, the electron-donating (dicyclohexylphosphinyl) ferrocene (FcPCy$_2$) ligand was tested with [RhCl(coe)$_2$]$_2$ to provide dihydropyridine (DHP) in quant yield *via* alkenylation and *in situ* electrocyclization. However, this catalytic system was proved to be ineffective when using aldimines bearing a β-substituent (Scheme 5, Eq. 1). To solve this issue, a new class of ligands for C–H activation was developed and the diethyl(DMAPh)phosphine ligand turned to be more efficient. Using a 2:1 ratio of diethyl(DMAPh)phosphine ligand to [RhCl(coe)$_2$]$_2$, various functional pyridine derivatives could be synthesized with moderate to good yields by *in situ* oxidation of the DHPs to N-benzyl pyridinium intermediate and the following debenzylation process.

α,β-unsaturated ketoximes were also revealed to be good coupling partners with alkynes under the rhodium-catalyzed chelation-assisted C–H activation process. Cheng and co-workers treated α,β-unsaturated

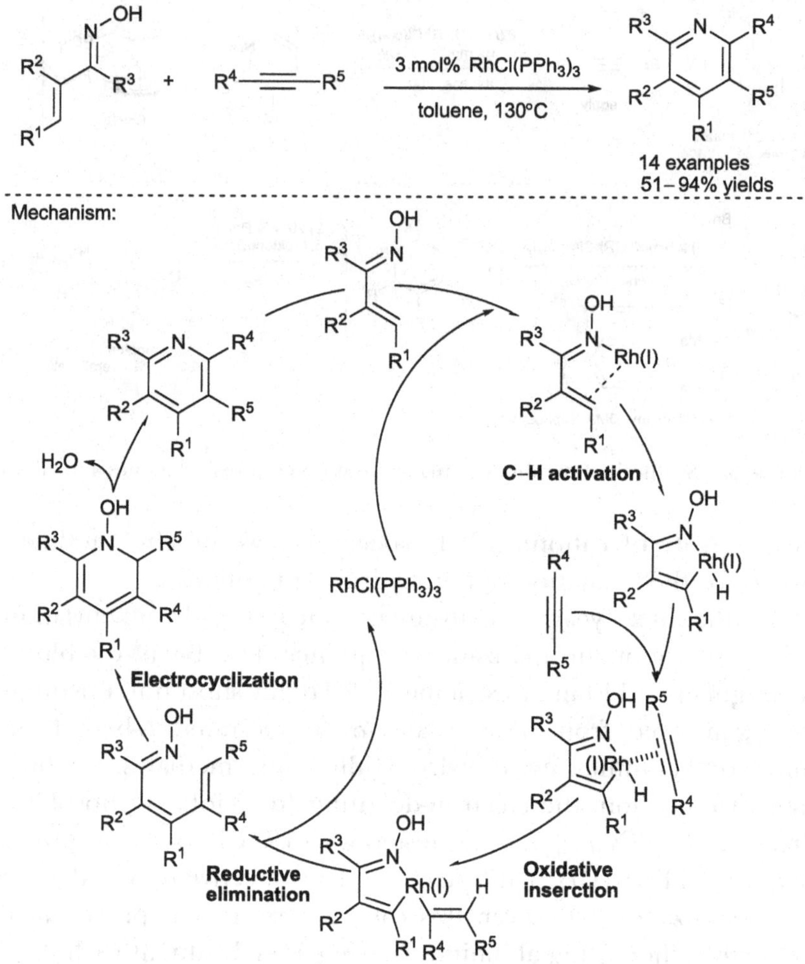

Scheme 6. Synthesis of pyridines from α,β-unsaturated ketoximes and alkynes

ketoximes and alkynes with 3 mol% of RhCl(PPh$_3$)$_3$ in toluene at 130°C to afford corresponding substituted pyridines with moderate to excellent yields (Scheme 6).[14] They thought the first step of this reaction involved coordination of the oxime nitrogen of ketoxime to the metal center, then the alkenyl C–H activation occurred to form the hydrometallacycle. Alkyne insertion followed by reductive

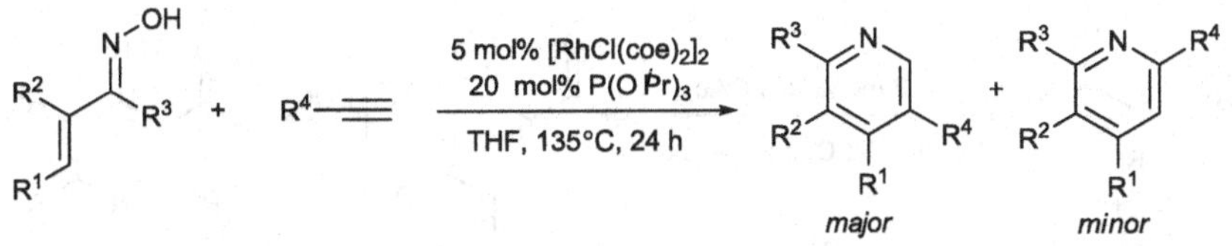

Scheme 7. Synthesis of pyridines from α,β-unsaturated ketoximes and terminal alkynes

elimination gave alkenylated intermediate and regenerated the rhodium catalyst. Under the high temperature reaction conditions, 6π-electrocyclization and elimination of water proceeded to produce the final pyridine derivatives.

Terminal alkynes are usually problematic coupling partners in transition metal-mediated functionalizations because of the serious self-dimerization by path. To continue the study of rhodium-catalyzed pyridine synthesis, Bergman and co-workers found that using inexpensive triisopropyl phosphite as ligand could minimize alkyne homocoupling side reactions.[15] In the presence of 5 mol% of $[RhCl(coe)_2]_2$ and 20 mol% of $P(O^iPr)_3$, functionalized pyridines with moderate to excellent regioselectivities were delivered from α,β-unsaturated ketoximes and terminal alkynes.

2.2 Rh(II) catalysis

In 2008, Davies and co-workers developed a one-pot synthesis of highly functionalized pyridines based on ring expansion of isoxazoles induced by rhodium carbenoid (Scheme 8).[16] The donor and acceptor substituted rhodium carbenoid derived from vinyldiazomethanes and $Rh_2(OAc)_4$ lead to the N–O insertion of isoxazoles.[17] Upon heating, two reaction pathways were proposed to generate 1,4-dihydropyridines before the tautomerization step: (1) Claisen rearrangement or (2) electrocyclic ring opening followed by 6π electrocyclization. The 2,3-Dichloro-5,6-dicyano-1,4-benzoquinone (DDQ) was then added *in situ* to give the oxidative multisubstituted pyridines, in low to good yields. The substrate scope showed that both aromatic

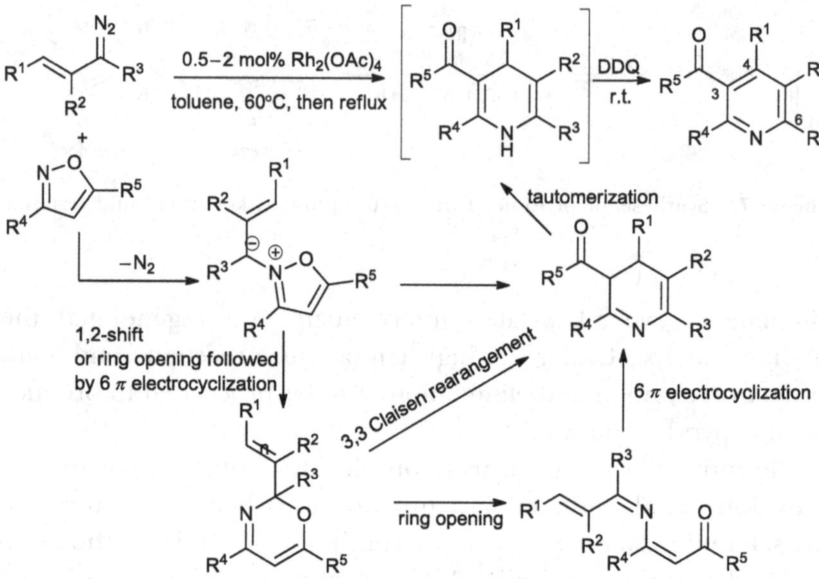

Scheme 8. Synthesis of pyridines based on ring expansion strategy

and non-aromatic substituents at the 4-position were tolerant but either 3- or 6-position was limited to carbonyl moieties.

2.3 Rh(III) catalysis

Pyridines synthesis from α,β-unsaturated oximes and alkynes under Rh(I) catalysis always require high temperature (130°C) and lead to low regioselectivities when using unsymmetrical alkynes as coupling partners. In 2011, Rovis and co-workers have developed a facile construction of pyridine derivatives under the Rh(III) catalysis with mild reaction condition (45°C) (Scheme 9).[18] In this report, it was found that the use of sterically different ligands allowed for complementary selectivities (Cp^t offered the alkyne regioisomer opposite to what is typically observed in Rh(III)-catalyzed C–H activation by using Cp* as the ligand). In the possible reaction mechanism proposed here, the rhodium catalyst coordinated to the basic nitrogen, followed by the formation of five-membered rhodacycle *via* carbonate promoted C–H activation. Regioselective alkyne insertion under the

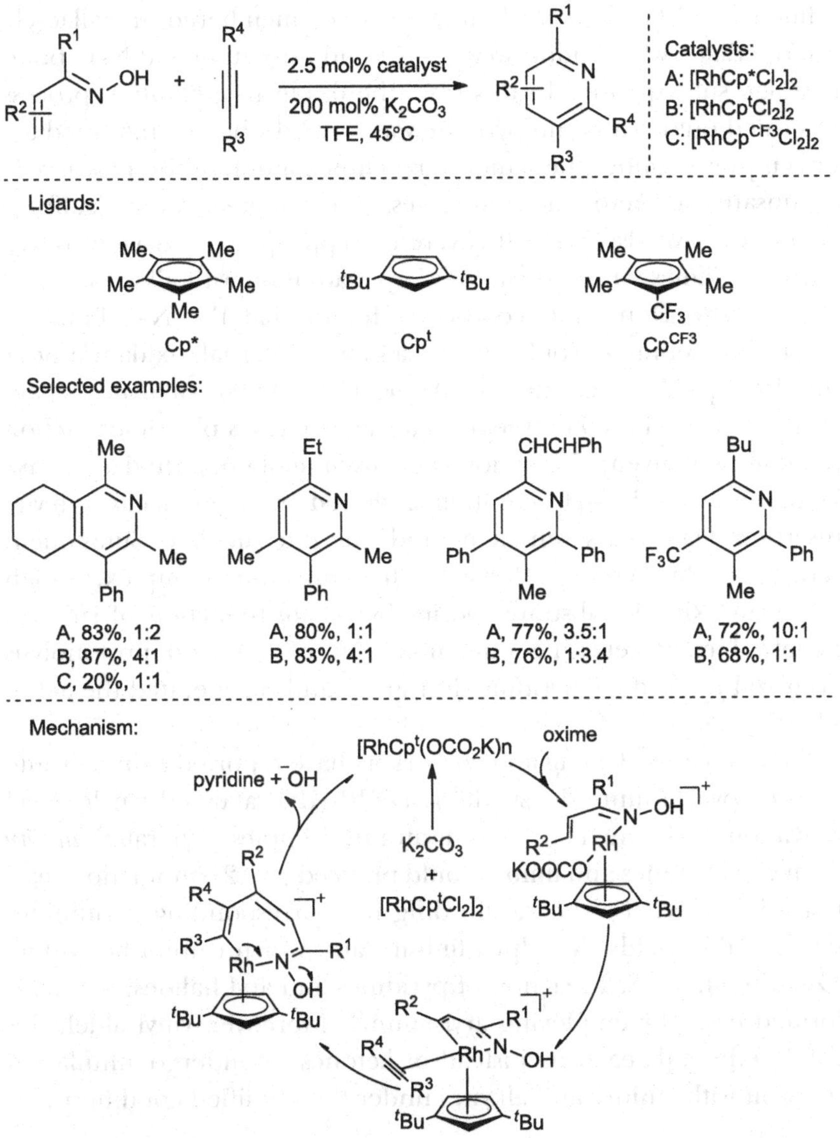
2.5 mol% catalyst
200 mol% K2CO3
TFE, 45°C
Catalysts:
A: [RhCp*Cl2]2
B: [RhCptCl2]2
C: [RhCpCF3Cl2]2
Ligards:
Cp*
Cpt
CpCF3
Selected examples:
A, 83%, 1:2
B, 87%, 4:1
C, 20%, 1:1
A, 80%, 1:1
B, 83%, 4:1
A, 77%, 3.5:1
B, 76%, 1:3.4
A, 72%, 10:1
B, 68%, 1:1
Mechanism:
[RhCpt(OCO2K)n
oxime
pyridine + OH
K2CO3
+
[RhCptCl2]2
KOOCO

Scheme 9. Synthesis of pyridines from α,β-unsaturated oximes and alkynes under the Rh(III) catalysis

influence of Cp ligand then gave seven-membered metallacycle intermediate, which underwent C–N bond formation and N–O bond cleavage subsequently. In this case, the 6π-electrocyclization process was ruled out because no pyridine was formed when 6π intermediate was employed under the standard reaction conditions. Besides acyclic α,β-unsaturated ketoximes, aldimines, and ketimines, cyclic N-sulfonyl ketimines have also been discovered as appropriate nitrogen source in the pyridines synthesis *via* [4+2] cycloaddition with alkynes.

In 2014, Dong and co-workers found that the N–S bond of N-sulfonyl ketimine could act as a novel internal oxidant under the [RhCp*Cl$_2$]$_2$ (2.5 mol%)/AgBF$_4$ (50 mol%) catalytic system (Scheme 10).[19] Based on the mechanism studies, a plausible reaction pathway was given. Formation of a five-membered rhodacycle *via* ketimine assisted C–H activation, followed by regioselective alkyne insertion, lead to a seven-membered ring intermediate. Subsequent reductive elimination afforded Rh(I)-substrate complex, which turned to Rh(III)-substrate species by *in situ* reduction of sulfonyl moiety. Finally, releasing one molecular of SO$_2$ and protonolysis occurred to produce pyridine derivatives and regenerated the active Rh(III) catalyst.

Very recently, Cheng and co-workers have reported a direct route towards pyridinium salts synthesis *via* Rh(III)-catalyzed vinylic C–H activation (Scheme 11).[20] α,β-unsaturated imines generated *in situ* from vinyl ketones and imines could proceed [4 + 2] annulation reaction with internal alkynes, affording the corresponding pyridinium salts in good yields. N-arylpyridinium salts, which cannot be synthesized directly by S$_N$2 reaction of pyridines and aryl halides, were also formed easily by employing aryl amines. Moreover, vinyl aldehydes exhibited good reactivity instead of ketones to undergo annulation reaction with amines and alkynes under the modified conditions.

3. Synthesis of Isoquinoline

Isoquinoline has existed in naturally occurring alkaloids as the key structural backbone and plays very important roles in industrial chemistry, such as dyes, paints, insecticides, as well as antifungals and

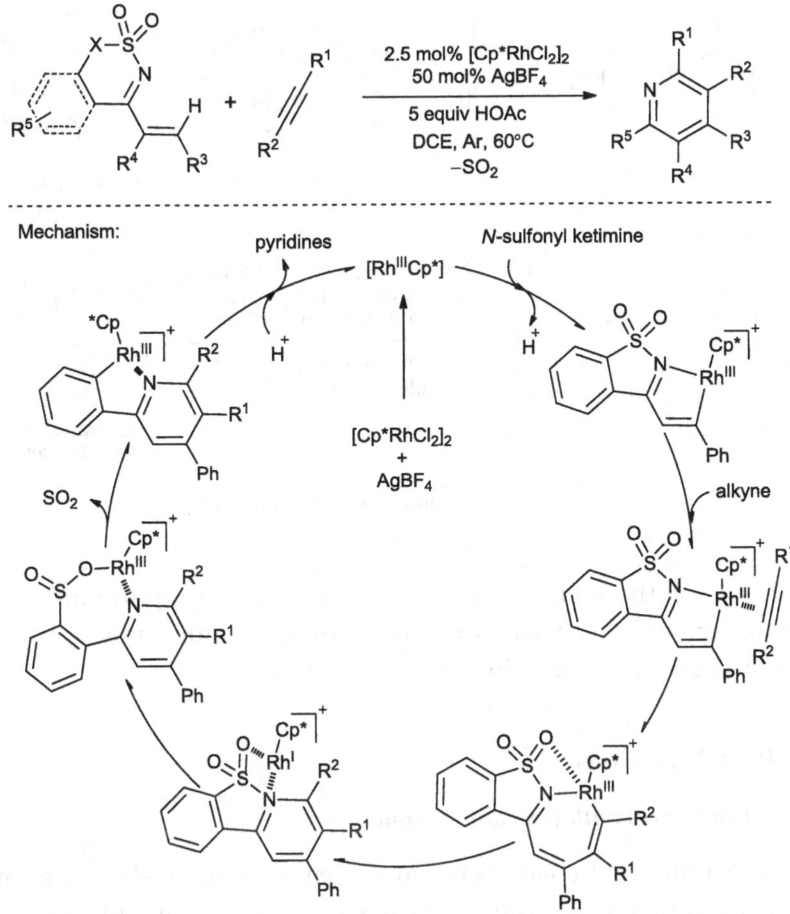

Scheme 10. Synthesis of pyridines from *N*-sulfonyl ketimines and alkynes

pharmaceutically in anesthetics, antihypertension agents, disinfectants, and vasodilators. The synthesis of isoquinolines always among the top topics in organic chemistry, and the classical Bischler–Napieralski reaction, Pomeranz–Fritsch reaction and Pictet–Spengler reaction usually suffer from the requirement of strong acid or prepreparation of complicated phenylethylamines. Transition metal-mediated processes also have been developed in recent decades; however, high temperature, tedious reaction steps, low yield, and limited substrate scope were always accompanied. New and efficient

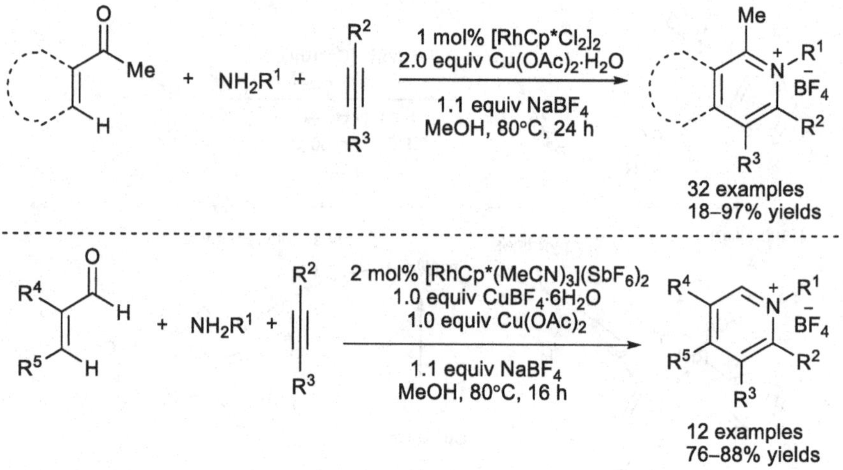

Scheme 11. Synthesis of pyridinium salts

methods for the facile to access functionalized isoquinolines are still in demand, especially ones that could provide high yields in very mild reaction conditions.

3.1 Rh(III) catalysis

3.1.1 *Procedures with external oxidants*

In 2008, Jones and co-workers found that aromatic C–H activation could be achieved by using $[RhCp^*Cl_2]_2$ companion with sodium acetate at room temperature *via* the formation of cyclometalated complexes (Scheme 12).[4i] Insertion of dimethyl-acetylenedicarboxylate (DMAD) and the following oxidative coupling gave the desired isoquinolines and regenerated the catalyst. In this transformation, 3 equiv of $CuCl_2$ is required to complete the oxidative-coupling step and the isoquinoline salts were isolated simply by washing with hexane to remove the excess substrates and alkynes. Besides, all the intermediates following C–H activation, alkyne insertion, and oxidative coupling were fully characterized *via* X-ray analysis.

In the next year, Fagnou and co-workers described a high efficient formation of isoquinoline compounds by treatment of

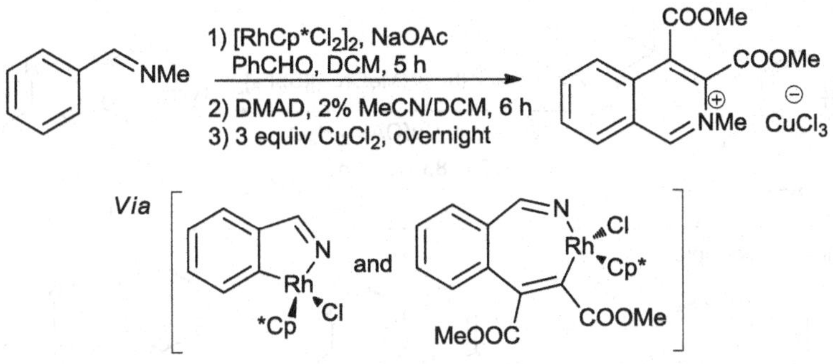

Scheme 12. Synthesis of isoquinoline salts from benzadimines and DMAD

N-tert-butylbenzadimines with internal alkynes under the catalytic system of $[RhCp^*(MeCN)_3][SbF_6]_2$ (2.5 mol%) and $Cu(OAc)_2 \cdot H_2O$ (2.1 equiv) in refluxing dichloroethane (DCE) (Scheme 13).[4j] Both aryl bromide substituent and free phenolic OH group were intact during this reaction, and larger substituent was regioselectively placed at the benzylic position away from the aldimine moiety when unsymmetrical alkynes were employed. During the mechanism study, both electrocyclization/oxidation and the Rh(IV) involved pathways were found not account for the product generation, while Rh(III) species were implicated in each of the bond-breaking/bond-forming steps and C–N reductive elimination procedure.

In 2011, Li and co-workers found that N-substituted 1-aminoiso-quinolines could be successfully synthesized *via* Rh(III)-catalyzed oxidative coupling reactions (Scheme 14).[21] When benzamidines contained different substituted N-phenyl groups were employed to react with diphenyldiacetylene, no direct correlation between the electronic character of phenyl moieties and the yields was observed. N-alkyl benzamidines were next examined and found to be well tol-erant under the standard conditions. Two-fold oxidative coupling products have also been exclusively isolated in some cases, probably because of the steric repulsion assisted the second cyclometalation, which was raised from the o-substituent and the N-phenyl group of the 1:1 oxidative coupling versions.

Although the formation of isoquinolinium salts from C–H activation of N-benzylidenemethylamine, 2-phenylpyridine, and

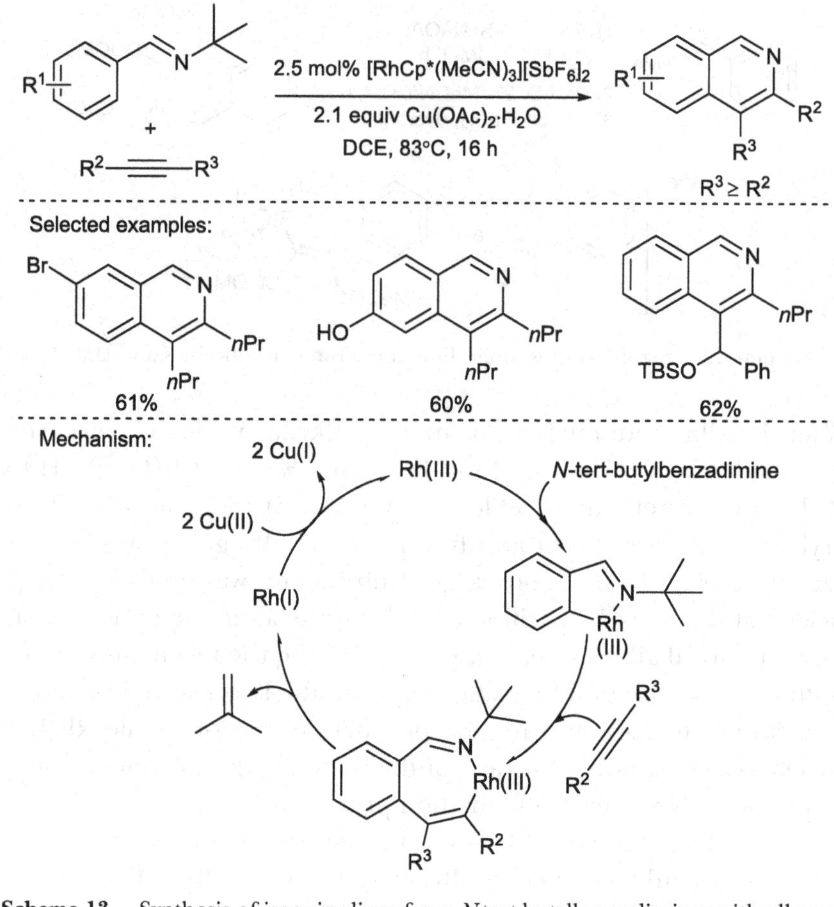

Scheme 13. Synthesis of isoquinolines from *N*-tert-butylbenzadimines with alkynes

benzo[*h*]quinoline with [RhCp*Cl$_2$]$_2$ has been demonstrated by Jones and co-workers.[4i] This methodology was largely limited by requiring stoichiometric amount of [RhCp*Cl$_2$]$_2$ and DMAD as the only coupling partner. The first report for the catalytic synthesis of isoquinolinium salt by Rh(III)-mediated C–H activation and annulation was presented later in 2012 by Cheng and co-workers (Scheme 15).[22] Benzaldehydes, amines, and alkynes were treated with [RhCp*Cl$_2$]$_2$ (2 mol%), AgBF$_4$ (1.0 equiv), and Cu(OAc)$_2$ (1.0 equiv) in tert-amyl alcohol at 110°C to give isoquinoliniums with good yields. AgBF$_4$ was very essential in this transformation for providing BF$_4^-$ as an inert

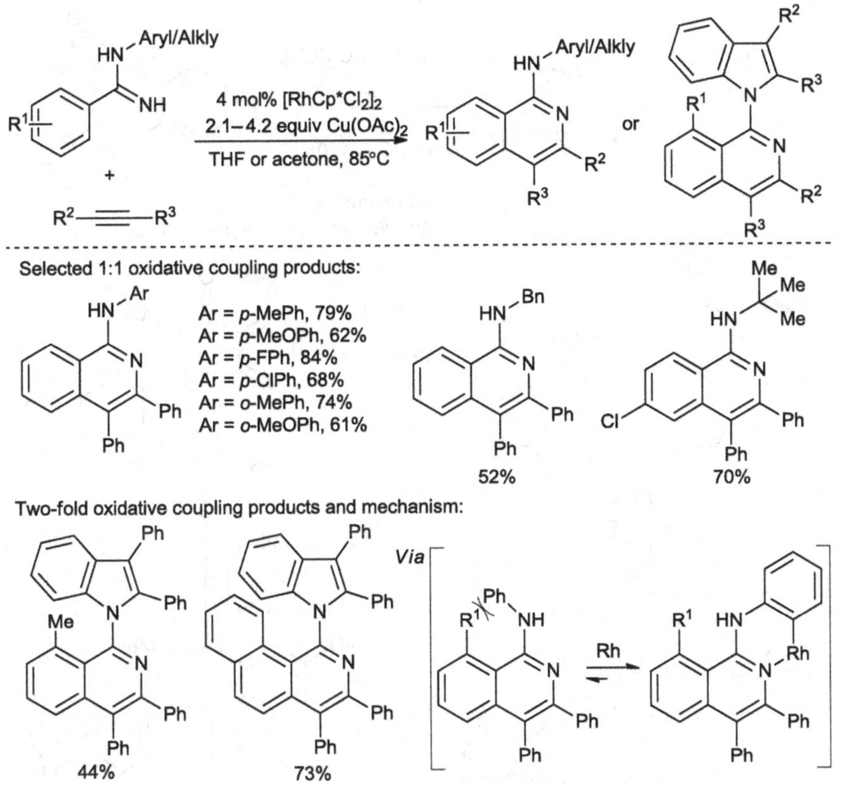

Scheme 14. Synthesis of isoquinolines from substituted benzamidines with alkynes

anion in the final product, activating the catalyst by removing the chloride on the rhodium complex and serving as an oxidant. Interestingly, the yield was decreased dramatically when pre-prepared imine was employed instead of the corresponding benzaldehyde and amine. To further demonstrate the utility of this methodology, total synthesis of the isoquinolinone alkaloid oxychelerythrine was realized within four steps and 62% overall yield was achieved.

In the field of Rh(III)-catalyzed C–H activation, the synthesis of isoquinolinium salts involving the *in situ* generated ketimines is more difficult than aldehyde imines, mainly because of thermodynamically unfavorable formation of ketimines from the reaction of ketones and amines. Based on the previous study on the one-pot

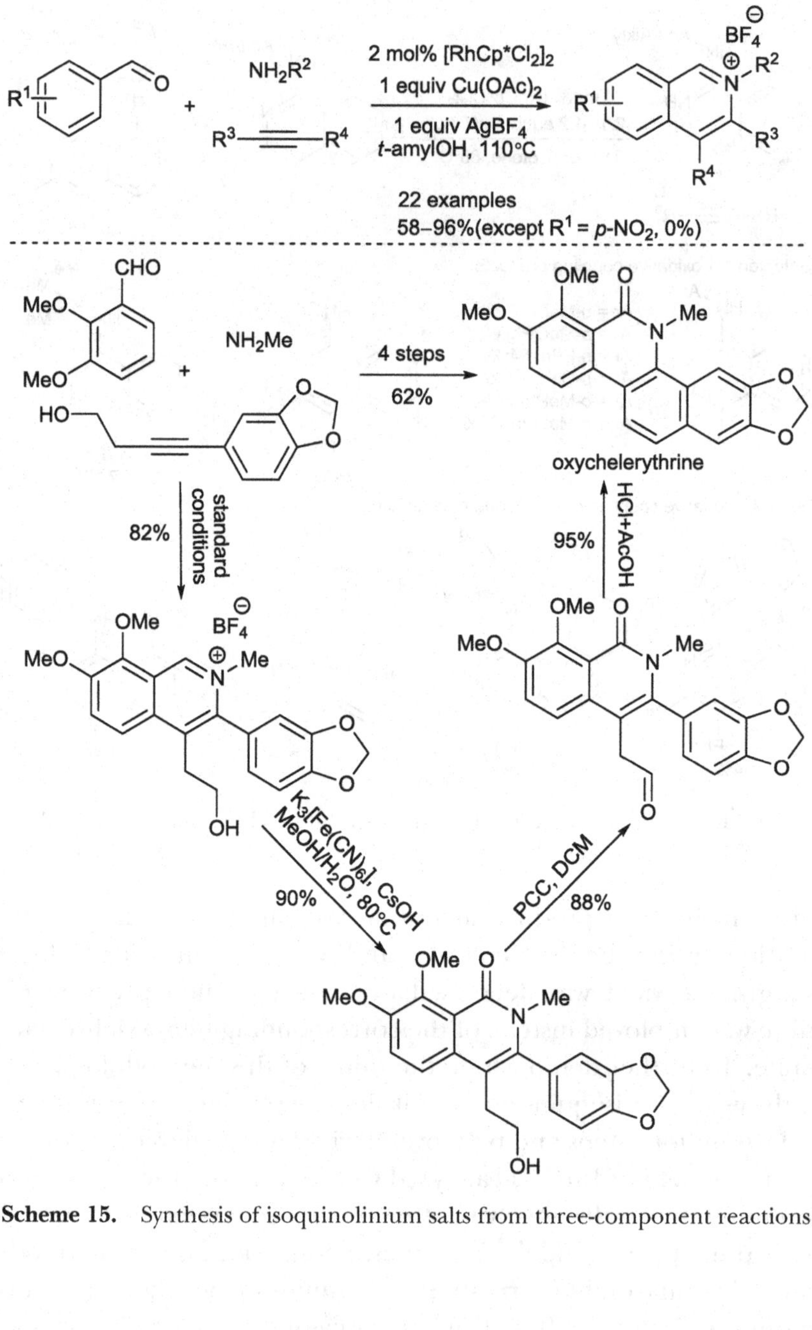

Scheme 15. Synthesis of isoquinolinium salts from three-component reactions

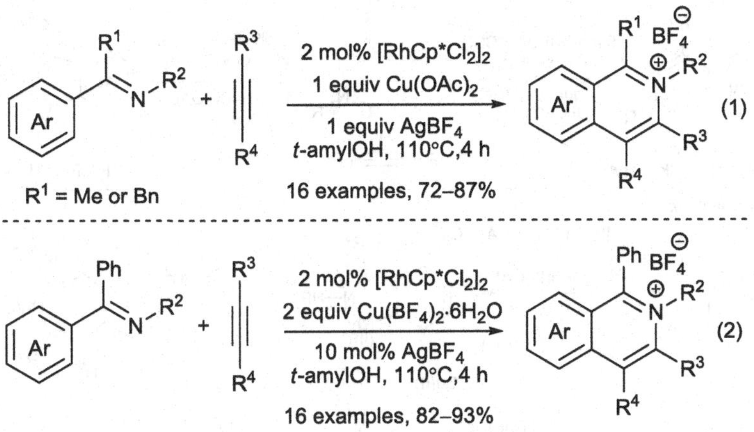

Scheme 16. Synthesis of isoquinolinium salts from ketimines and alkynes

synthesis of isoquinolinium salts from benzaldehydes, amines, and alkynes,[20] Cheng and co-workers found that prepared ketimines could also give 1-substituted isoquinolinium salts under the same conditions (Scheme 16, Eq. 1).[23] When employing benzophenone imines as the coupling partners, the optical reaction condition was modified with two equiv of $Cu(BF_4)_2 \cdot 6H_2O$ to avoid the excess amount of expensive Ag salt (Scheme 16, Eq. 2). Besides, ruthenium complexes were also revealed to be applicable for this conversion.

Ketimines, aldimines, and amidine have been found eligible for the Rh(III)-catalyzed annulation with alkynes in the presence of stoichiometric external oxidant for the construction of isoquinolines or isoquinolinium salts. After that, benzylidenehydrazones also have been demonstrated to be applicable coupling candidates by Li and co-workers in 2014 (Scheme 17).[24] $[RhCp*Cl_2]_2/AgNO_3/Cu(OAc)_2$ catalytic system turned to be efficient in reflux EtOH for the selective C–H activation and cyclization process. During the optimization of the reaction conditions, aziridinylimine was found to exhibit higher reactivity than other 1,1-dialkyl-substituted hydrazones for the synthesis of isoquinoline, while (1-phenylethylidene)hydrazone bearing two free N–H tended to give indene instead.

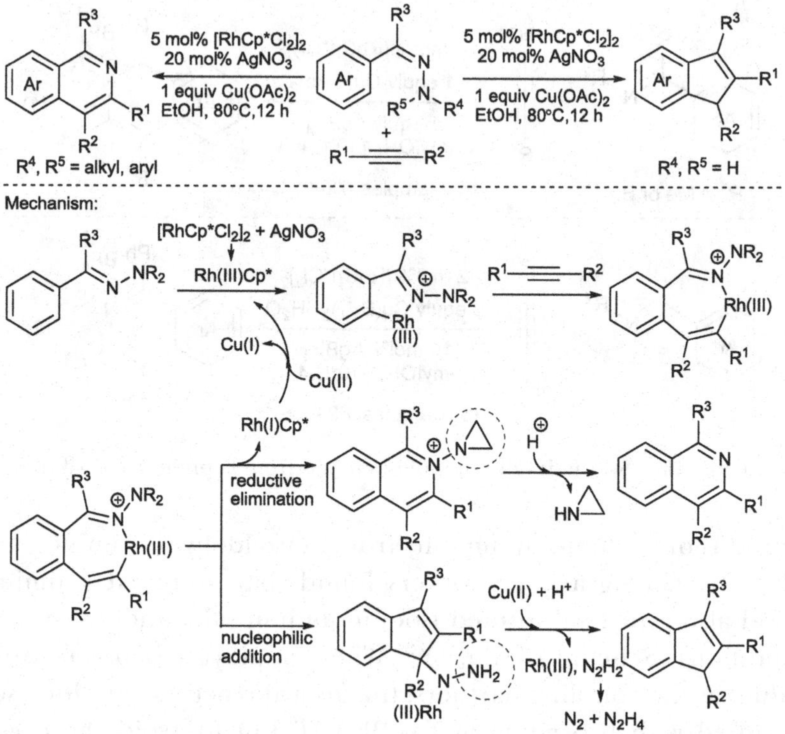

Scheme 17. Synthesis of isoquinolines from substituted hydrazones and alkynes

Two different reaction pathways have been proposed depending on the different substituents of the benzylidenehydrazone after the initial C–H activation and alkyne insertion steps: (1) direct reductive elimination followed by N–N bond cleavage to afford isoquinolines when aziridinylimine was employed or (2) intramolecular addition of C=N bond followed by C–N bond cleavage to provide indenes in the cases with two free N–H bond benzylidenehydrazones.

3.1.2 Redox-neutral procedures

3.1.2.1 Redox-neutral procedures with internal oxidants

C–H bond cleavage and subsequent C–N bond formation sequences under the transition metal-catalysis have emerged as attractive alternatives for the traditional construction of nitrogen contained heterocycles in recent years. However, requiring an external oxidant is very essential to make the catalyst reactive by oxidation after the elimination step. Pioneering study to address this issue was demonstrated by Hartwig and co-workers, using N–O bond embedded in the substrate as a built-in oxidant for the intramolecular palladium-catalyzed indole synthesis.[25] In the field of rhodium catalysis, the novel stratagem of internal oxidant was first illuminated by Guimond and co-worker in the synthesis of isoquinolones.[26]

Later, isoquinolines formation was reported by Chiba and co-workers from the Rh(III)-catalyzed coupling reaction of aryl ketone *O*-acyloxime derivatives and internal alkynes (Scheme 18, Eq. 1).[4n] In this transformation, the typical directing group-assisted C–H activation and subsequent alkyne insertion offered the seven-membered rhodacyclic iminium cation intermediate. Reductive elimination gave *N*-acetoxyisoquinolinium and released Rh(I) species, and the following redox reaction regenerated Rh(III) catalyst and produced the isoquinoline. Alternatively, direct formation of the desired product *via* a concerted redox process was also possible. The iminium cation intermediate with acidic α-proton presented in both reaction pathways was strongly supported by the observation of deuterium incorporation into the methyl moiety in MeOD (Scheme 18, Eq. 2).

Besides, isoxazol-5-ones as well as *O*-acetyloximes could react under the optimized conditions, providing isoquinolines with moderate to good yields *via* decarboxylation process (Scheme 19).

Simple oximine is foreseeably more synthetic, available, and environment friendly compared with its' *O*-acyloxime derivative in the coupling reaction to produce isoquinoline. In 2011, Li and co-workers found that Rh(III)-catalyzed dehydrative C–C and C–N coupling reactions were applicable between oximines and alkynes (Scheme 20, Eq. 1).[27] Donating group decorated either on the aryl moiety of the acetophenone oximes or alkynes could facilitate this transformation, which was contrast to the earlier reported results in the Rh(III)-catalyzed oxidative coupling of amides and alkynes.

To get more information about the reaction mechanism, two important experiments were designed and carried out under the

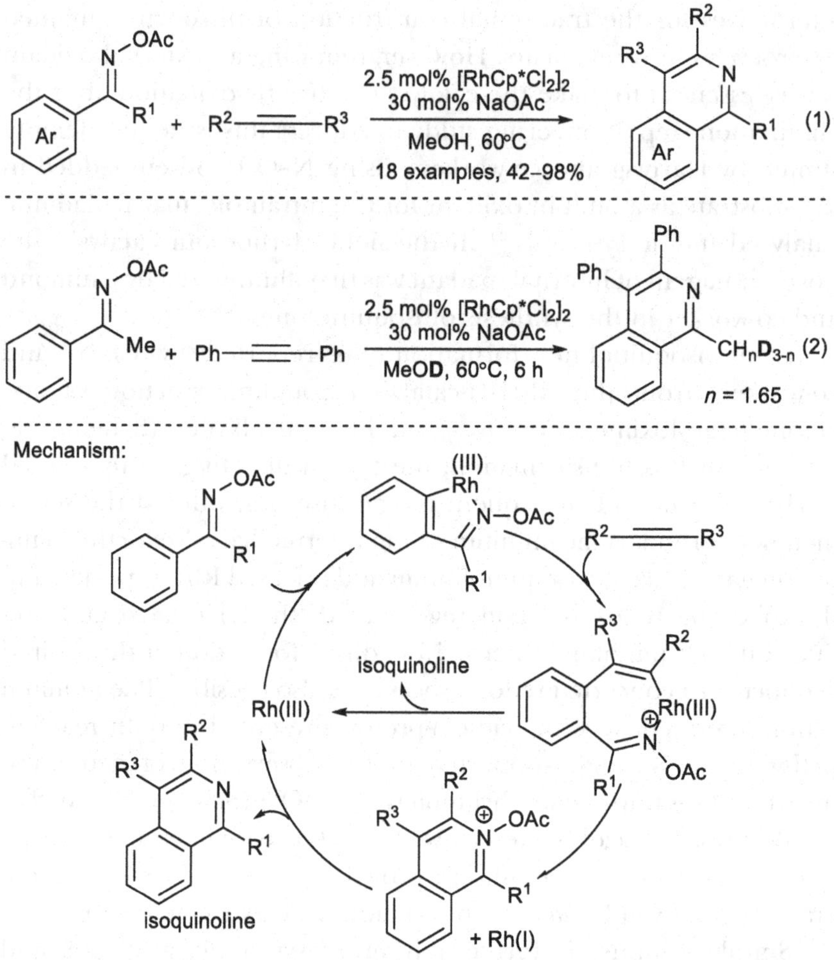

Scheme 18. Synthesis of isoquinolines from O-acyloxime derivatives and alkynes

optimized conditions. The observation that prepared olefin intermediate was totally ineffective to give annulation product exclude
the possibility of electrocyclization (Scheme 20, Eq. 2). On the other
hand, the N-oxide intermediate has previously been considered to
act as the internal oxidant to regenerate the Rh(III) species from
Rh(I) and result in the corresponding isoquinoline. However, this
kind of Rh(III)–Rh(I) catalytic cycle also seemed unlikely because
no isoquinoline was detected by gas chromatography (GC) when

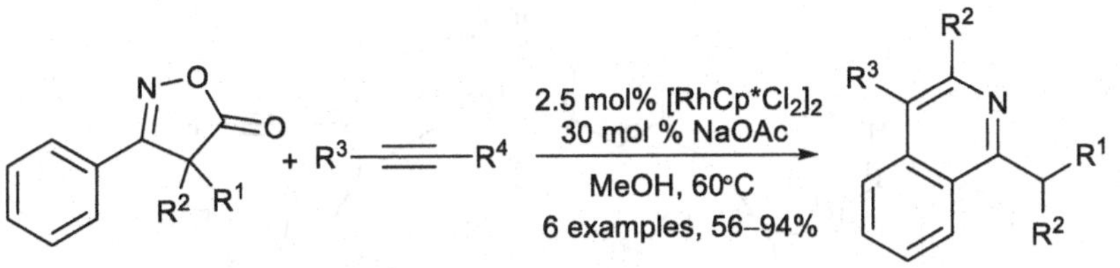

Scheme 19. Synthesis of isoquinolines from isoxazol-5-ones and alkynes

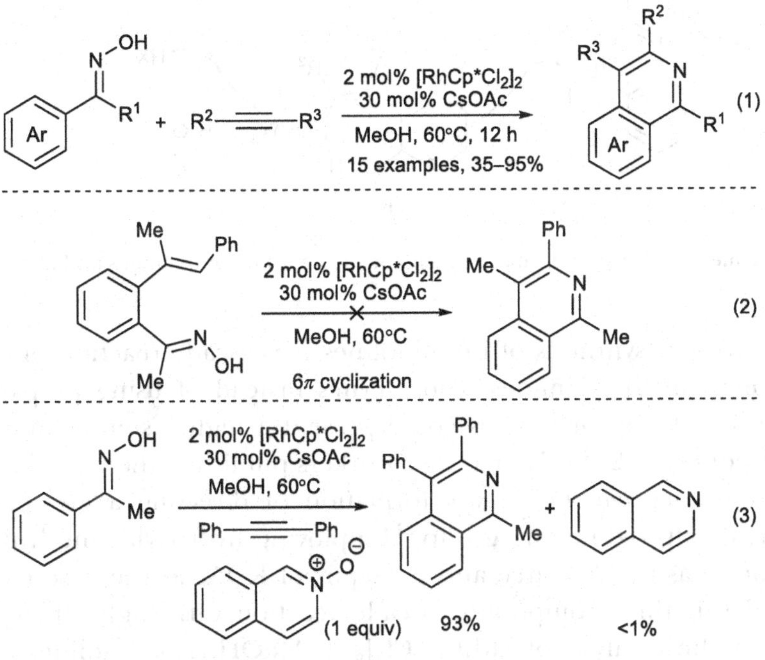

Scheme 20. Synthesis of isoquinolines from oximines and alkynes

isoquinoline *N*-oxide reacted with acetophenone oximine and alkyne in 1:1:1.2 ratio (Scheme 20, Eq. 3). Finally, a Rh(III)–Rh(V) catalytic cycle was regarded as a more preferred process (Scheme 21). In this reaction pathway, acid generated from C–H activation step was considered to promote the formation of Rh(V) species companion with releasing one molecule of water through protonation. Subsequent C–N bond formation *via* reductive elimination afforded desired product and the Rh(III) catalyst.

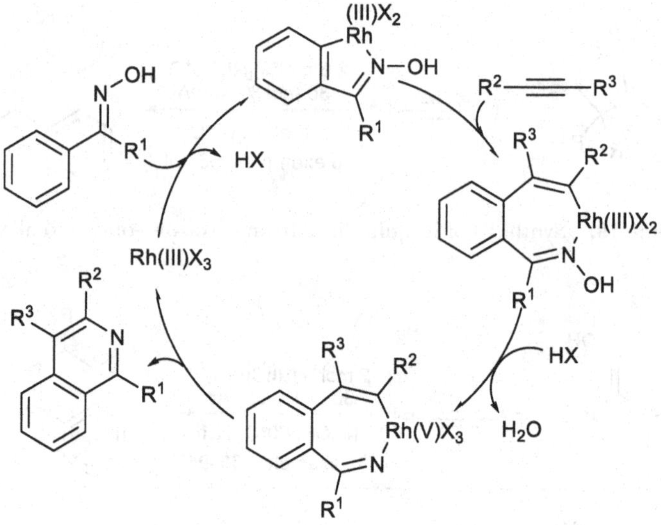

Scheme 21. Proposed catalytic cycle for the coupling of oximines and alkynes

One-pot synthesis of isoquinolines *via* cascade reaction of aryl ketones, hydroxylamines, and alkynes instead of using prepared *O*-acetyl oximes or oximes could present a higher step economic procedure. In 2012, Hua and co-workers published their results on such studies *via* the *in situ* formation of directing a group, or named as pro-directing group.[28] Employing hydroxylamine hydrochloride as the *N*-source and 2.1 equiv of KOAc as the base facilitated this three-compound cascade reaction with high efficiency under the catalysis of [RhCp*Cl$_2$]$_2$ in MeOH. Isoquinolines and heterocycle-fused pyridines had been synthesized with moderate to good yields under the standard conditions (Scheme 22, Eq. 1). In line with Li's previous work on the synthesis of isoquinoline from ketone oximes and alkynes,[27] ketones with EDGs favored this coupling reaction than those with electron withdrawing group (EWGs). Moreover, the yield was declined when the steric hindrance of R^1 was increased, owing to the formation of the inactive oxime isomer in the condensation step that can hardly direct Rh to the ortho site. This one-pot cascade produce was also applicable for the pyridine synthesis under the re-optimized reaction conditions using K$_2$CO$_3$

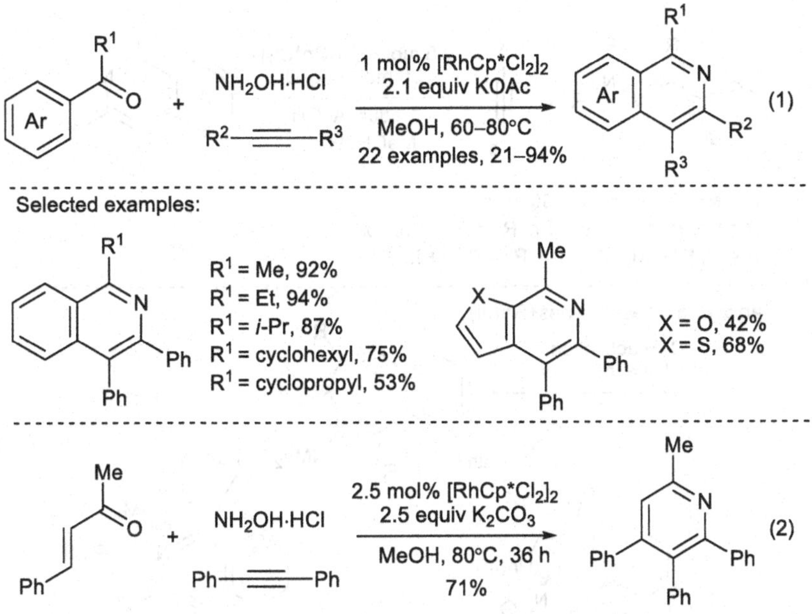

Scheme 22. Synthesis of isoquinolines *via* cascade reaction

as the base and higher catalyst loading as well as prolonged reaction time (Scheme 22, Eq. 2).

Redox-neutral processes in the synthesis of isoquinolines usually involve the N–O bond cleavage to exclude the requirement of external oxidants by using *N*-oxide, oxime, or *N*-methoxyamide as the directing group. A complementary alternative *via* N–N bond cleavage was developed later by Cheng and co-worker in 2013, employing hydrazone as the coupling partner to react with alkynes (Scheme 23).[29] Isoquinoline derivatives with moderate to excellent yields have been obtained from various *N,N*-dimethylhydrazones and alkynes. *N,N*-diphenylacetophenone hydrazone also exhibited satisfied reactivity under the standard conditions, releasing the corresponding isoquinoline along with diphenylamine while unsubstituted acetophenone hydrazone only gave very poor product yield. However, *N,N*-dimethylbenzaldehyde hydrazone was not applicable under the catalytic system with some unclear reason.

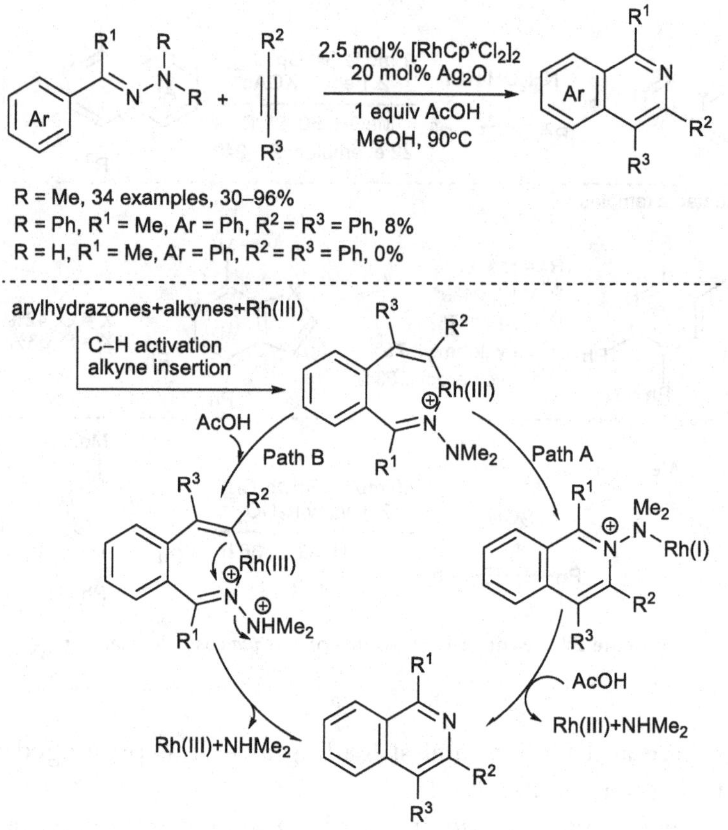

Scheme 23. Synthesis of isoquinolines from hydrazones and alkynes

Two different reaction pathways have been speculated after formation of the seven-membered intermediate *via* C–H activation and alkyne insertion steps. In path A, reductive elimination occurred to give isoquinolinium cation and Rh(I) species, which provided isoquinoline and Rh(III) catalyst by redox process. In path B, intramolecular nucleophilic substitution promoted by acetic acid also could lead to C–N bond formation and N–N bond cleavage, and release Rh(III) species to restart the catalytic cycle. The authors thought that the path B might be more favorable because acetic acid was very crucial in this catalytic system, which made *N,N*-dimethyl group a better leaving group *via* protonation.

Redox-neutral strategy using N–O or N–N motif as both the directing group and internal oxidant could avoid the requirement of external oxidant *via* reductive bond cleavage, but these oxidizing parts usually cannot be incorporated in the final products. By employing ketazines as the coupling partners under the Rh(III)-catalyzed oxidative C–H annulation with alkynes, a higher atom economy strategy was developed by Li and co-workers, and both *N*-atoms of the ketazine were incorporated to the desired isoquinoline products (Scheme 24).[30] In this transformation, ketazine first reacted with alkyne under the catalyst of RhCp*$(H_2O)_3(OTf)_2$ to give one molecule of isoquinoline companion with a Rh(III)-imide complex which resulted from the N–N bond reductive cleavage. Then the Rh(III)-imide intermediate underwent redox coupling pathway in the presence of air and acid, providing the second molecule of product. This simple method was performed under mild conditions and the ketazines could be easily prepared from corresponding ketones and hydrazines.

Although Rh-catalyzed synthesis of isoquinolines *via* C–H activation/cyclization of aromatic imines or oximes with alkynes has

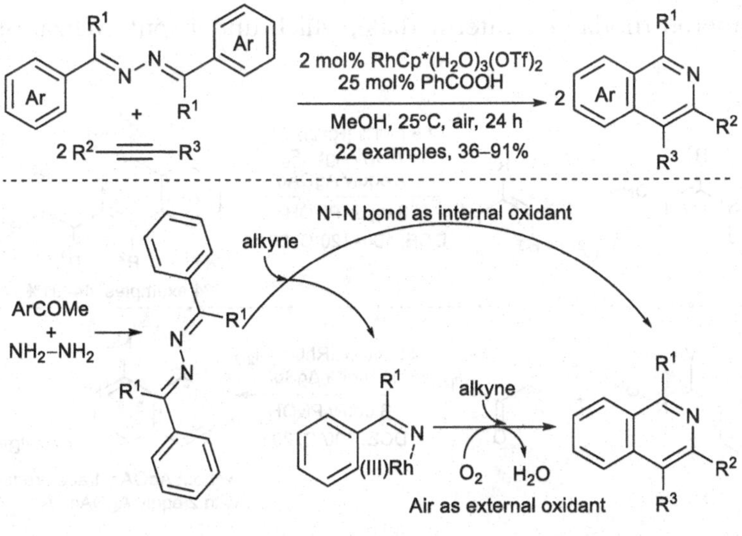

Scheme 24. Synthesis of isoquinolines from ketazines and alkynes

presented a practical process in recent years, the synthesis of biologically important 3-alkyl-substituted isoquinoline cores still remains as a challenge in this field. To solve this issue, Glorius and co-workers developed an efficient Rh(III)-catalyzed redox-neutral approach by using 1,3-dienes as coupling partner instead of alkynes to react with aromatic oxime esters.[31] In this study, isoquinoline motifs were formed *via* a tandem C–H/C–N cyclization/double-bond migration process. Excellent regioselectivities and moderate to good yields were obtained when different aromatic *o*-pivaloyl ketoximes reacted with a variety of electro-deficient dienes. However, when coupling with electron–neutral dienes, higher temperature of 120°C and stoichiometric AgOAc were required to give single oxidative product instead of a mixture contained product and over oxidative product (Scheme 25, Eq. 1). Under the standard reaction conditions, simple acrylate failed to give any product unless AgOAc was added as external oxidant (Scheme 25, Eq. 2). This oxidative Heck reaction of aromatic oxime esters with simple alkenes could provide a complementary method to the redox-neutral strategy in the synthesis of isoquinolines.

Similar as the earlier reported Rh(III)-catalyzed redox-neutral process, C–H activation followed by alkene insertion lead to the seven-membered rhodacycle intermediate, which underwent cyclization to

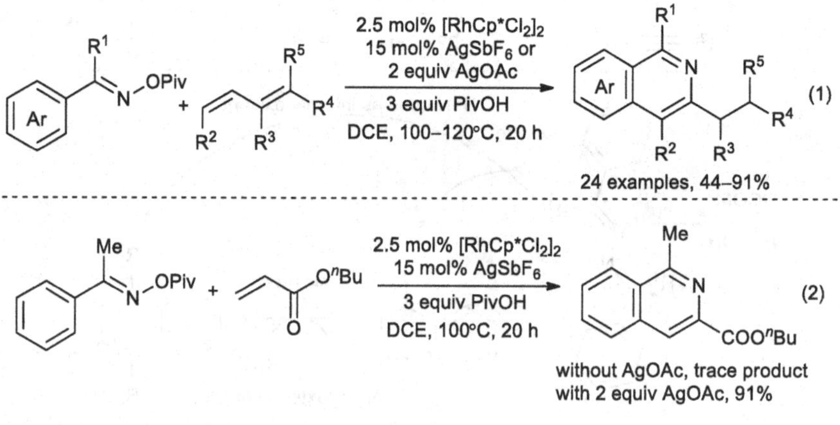

Scheme 25. Synthesis of isoquinolines from oxime esters and 1,3-dienes

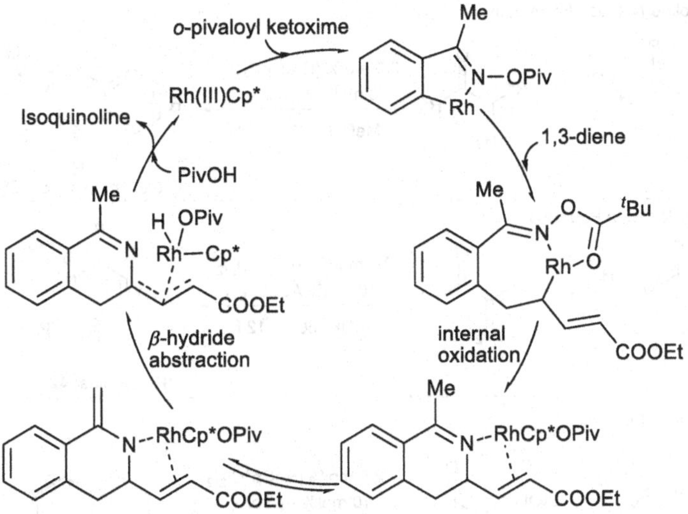

Scheme 26. Proposed catalytic cycle for the coupling of oximines and 1,3-dienes

give complex with double bond coordinated with metal center. Rh-mediated β-hydride abstraction then provided the π-allylmetal hydride compound. Finally, isomerization occurred to produce the corresponding isoquinoline motifs with aromatization as the probable driving force.

3.1.2.2 *Another redox-neutral procedure*

The traditional synthesis of isoquinoline and pyridine N-oxides usually required different oxidants and preparation of their parent heterocycles in advance, and suffered from the potential over oxidation.[32] In 2013, Glorius and co-workers developed a Rh(III)-catalyzed multisubstituted isoquinoline and pyridine N-oxides synthesis by the cyclization of oximes and diazo compounds *via* Sp$_2$ C–H activation (Scheme 27, Eqs. 1 and 2).[33] The tandem C–H activation, cyclization, and condensation sequences occurred in mild reaction conditions and no oxidants were necessary. A broad substrate scope were observed by employing different ketoximes, aldoximes, α,β-unsaturated oximes and diazo compounds. The findings also proved that diazo compounds could act as equivalents of unsymmetrical and electron

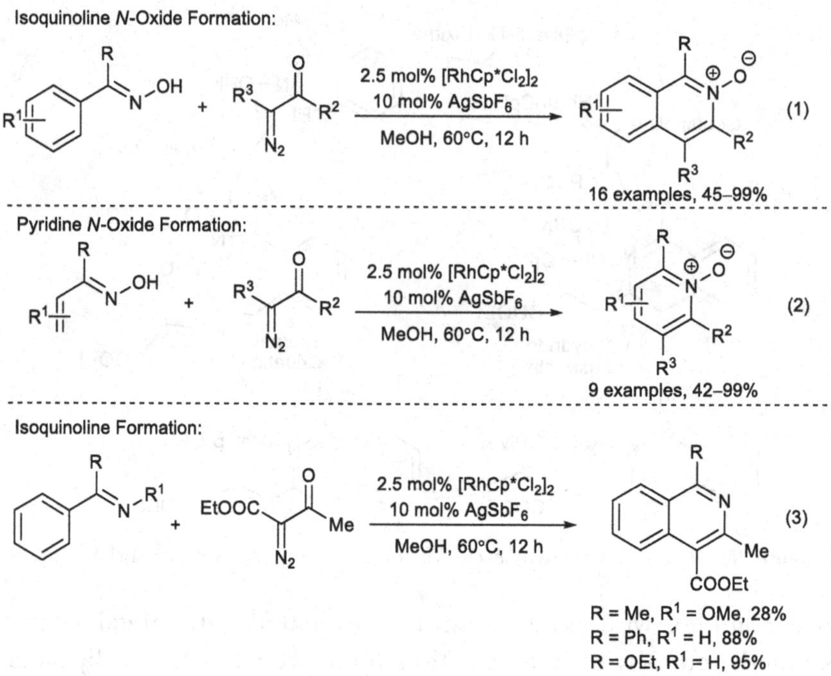

Isoquinoline *N*-Oxide Formation:

(1)

16 examples, 45–99%

Pyridine *N*-Oxide Formation:

(2)

9 examples, 42–99%

Isoquinoline Formation:

(3)

R = Me, R^1 = OMe, 28%
R = Ph, R^1 = H, 88%
R = OEt, R^1 = H, 95%

Scheme 27. Synthesis of isoquinolines from ketoximes and diazos

deficient alkynes in Rh-catalyzed isoquinoline synthesis, providing an alternative method to avoid low regioselectivites and reactivities which have always been obtained by employing the latter coupling partners (Scheme 27, Eq. 3).

The possible reaction mechanism was proposed. The rate-determining C–H activation occurred with the assistance of the oxime group *via* the formation of five-membered rhodacycle, which underwent carbene insertion and sequent protonolysis to give alkylation intermediate. Two reaction pathways were then speculated to be responsible for the construction of the final product: (1) tautomerization followed by 6π electrocyclization and elimination of one molecule of water gave the corresponding isoquinolines and *N*-oxidative ones, or (2) intramolecular nucleophilic attack by the *N* atom of imine on the carbonyl moiety to complete the cyclization step.

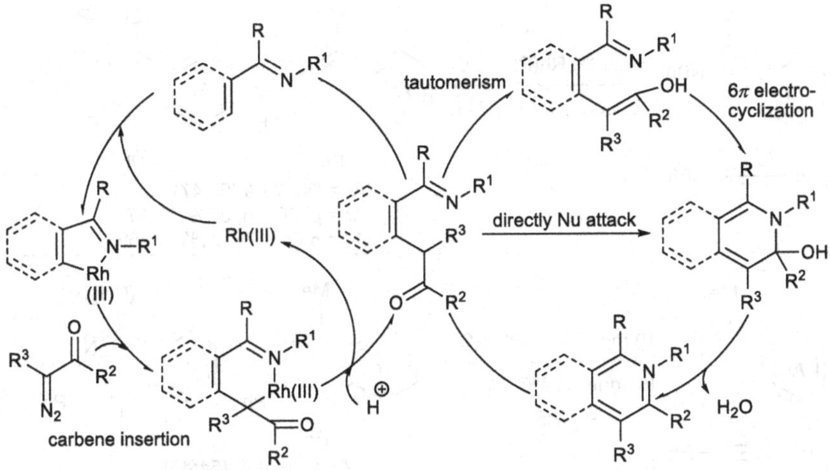

Scheme 28. Proposed catalytic cycle for the coupling of ketoximes and diazos

3.2 Rh(I) catalysis

Unlike the Rh(III)–Rh(I) catalytic system in the synthesis of isoquinolines, regenerating active Rh(III) species usually was not involved in Rh(I)-catalyzed C–H activation/annulation progress. In 2003, Jun and co-workers developed the first Rh(I)-catalyzed formation of isoquinoline derivatives *via* ortho-alkenylation of aromatic ketimines with alkynes (Scheme 29, Eq. 1).[34] Alternatively, one-pot synthesis of isoquinolines also could be achieved by simply mixing aromatic ketones, benzylamines, and alkynes, affording two different isoquinoline products with good yields (Scheme 29, Eq. 2).

Although the exact mechanism was unclear, a plausible explanation for the formation of two kinds of isoquinolines have been given. After the generation of the ortho-alkenylated ketimine *via* Rh(I)-mediated C–H activation, electrocyclic reaction occurred to give the cyclized intermediate under high reaction temperature. Such cyclic complex was not stable and could abstract an acidic α-hydrogen from alkenylated ketimine to form an ion pair composed of iminium and enamide species, which could easily undergo nucleophilic attack to generate 3,4-dihydroisoquinoline and imine (Scheme 30). Under the similar steps, the new formed imine could also provide

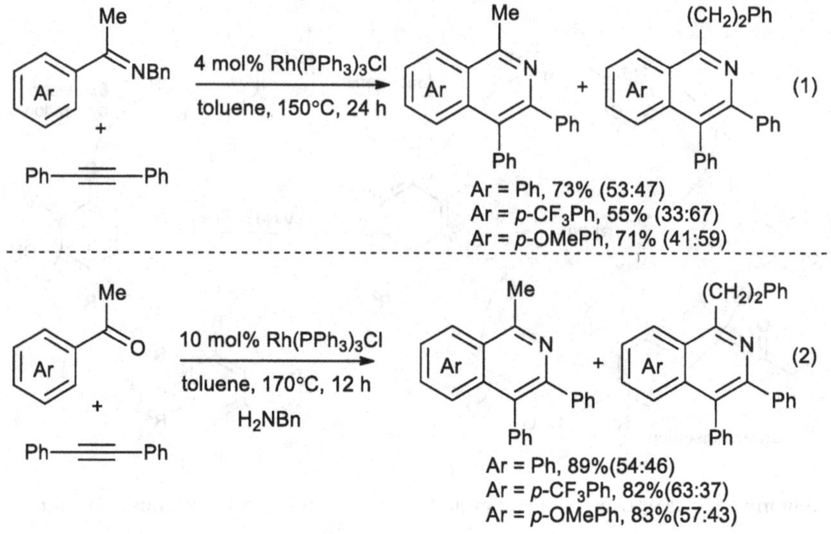

Scheme 29. Synthesis of isoquinolines from ketoximes with alkynes under Rh(I) catalysis

the second molecular of dihydroisoquinoline with a different substitution at the α-position, and both dihydroisoquinolines were subsequently oxidized into the corresponding isoquinoline products separately.

Although Jun's work has presented a high efficient method for the synthesis of isoquinoline compounds under Rh(I) catalysis, the reaction was complicated by formation of two different products with low selectivity. In 2009, Cheng and co-workers developed an alternative Rh(I)-catalyzed isoquinoline synthesis by using aromatic ketoximes instead of ketimines.[35] High reactivity and excellent regioselectivity were observed when ketoximes reacted with symmetrical and unsymmetrical alkynes. Under the optimized conditions, even terminal alkyne could be applicable to give a single isoquinoline product in moderate yield with the phenyl substitution located at C_4 position (Scheme 31, Eq. 1). Tetrahydroquinoline derivatives have also been successfully produced when ketoximes decorated with α-exocyclic double bond or tetrahydroxanthone oximes were employed as the coupling partners (Scheme 31, Eq. 2).

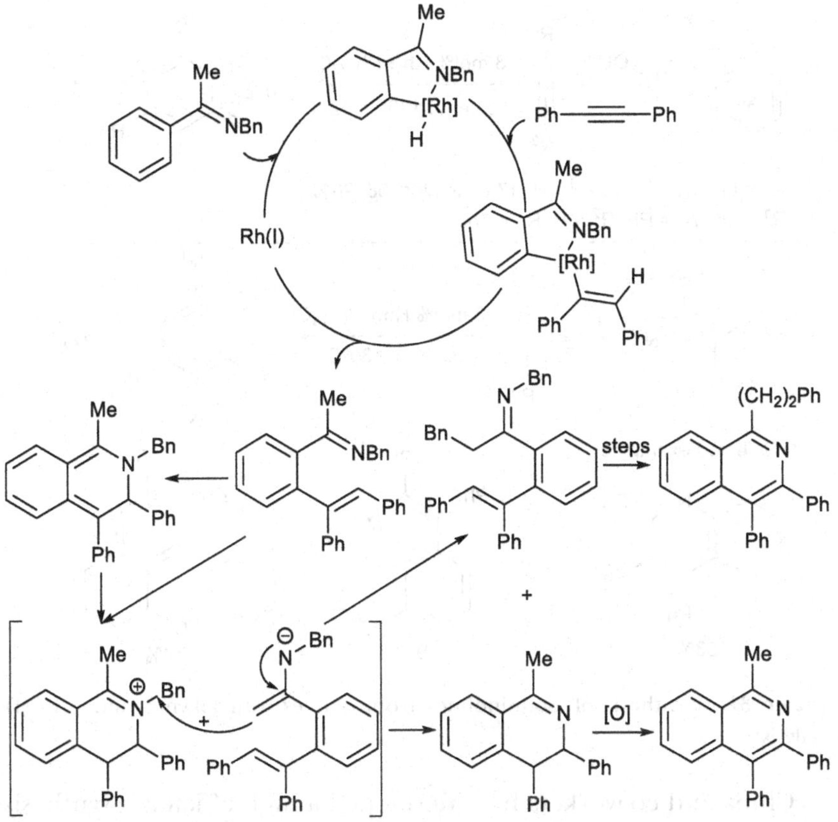

Scheme 30. Proposed catalytic cycle for the coupling of ketimines with alkynes under Rh(I) catalysis

3.3 Rh–Cu bimetallic catalysis

In Rh(III)–Rh(I) catalytic cycle, copper salts were usually employed as oxidants to regenerate the active Rh(III) species after the reductive elimination step. However, copper additives also have played other significant roles more than just oxidizing agent in some Rh(III)–Rh(I) process. Those Rh–Cu bimetallic systems could promote some chemical transformations that are unprecedented with monometallic catalysts, in which one metal catalyst afforded the active intermediate for the next step mediated by another metal catalyst to give the final product.

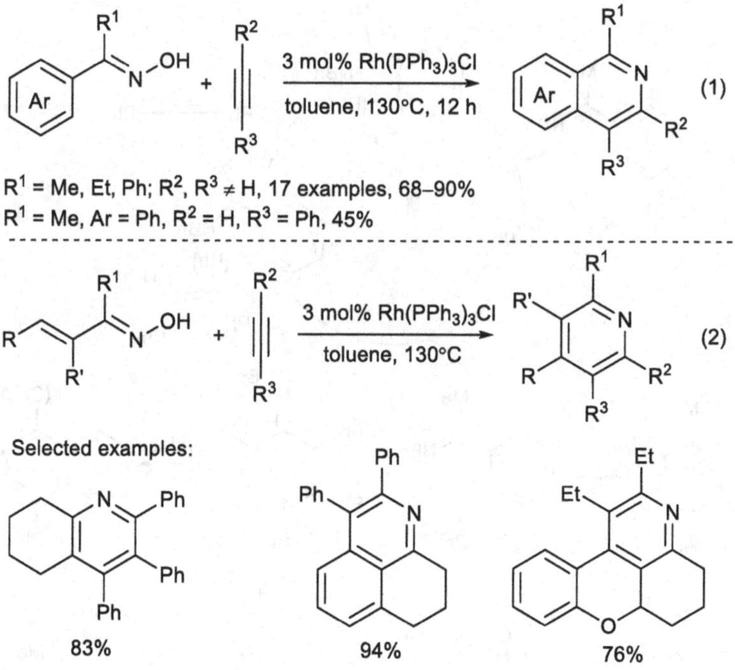
3 mol% Rh(PPh3)3Cl
toluene, 130°C, 12 h
(1)
R1 = Me, Et, Ph; R2, R3 ≠ H, 17 examples, 68–90%
R1 = Me, Ar = Ph, R2 = H, R3 = Ph, 45%
3 mol% Rh(PPh3)3Cl
toluene, 130°C
(2)
Selected examples:
83%
94%
76%

Scheme 31. Synthesis of isoquinolines from oximes with alkynes under Rh(I) catalysis

Chiba and co-workers have developed a high efficiency synthesis of isoquinolines from α-aryl vinyl azides and internal alkynes under an Rh/Cu bimetallic catalytic system in 2011 (Scheme 32).[36] This catalytic synthesis of isoquinoline motifs showed a great generality and a possible mechanism has been proposed based on experiments. First, the reaction solvent of dimethylformamide (DMF) could act as reductant to form Cu(I) species from $Cu(OAc)_2$. Thermal denitrogenative decomposition of vinyl azide afforded 2*H*-azirine, which subsequently underwent Cu(I)-mediated reduction to generate an anion radical intermediate (Cycle 1, path A). Then consecutive C–N bond cleavage was followed to get iminyl copper(II) radical compound, which also resulted from the direct reduction of the vinyl azide instead (Cycle 1, path B). Further reduction and protonation delivered imine and its Cu(II)-complex, both of which were then subjected to the typical Rh(III)–Rh(I) catalytic cycle, producing the

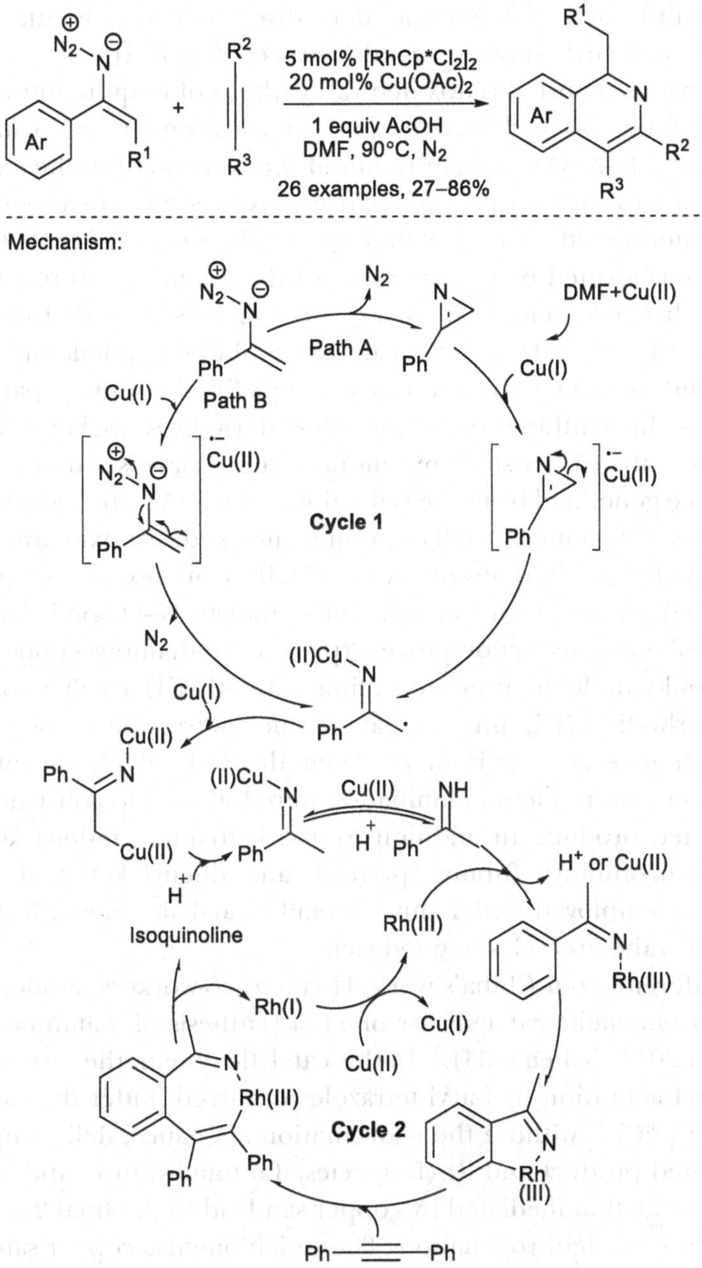

Scheme 32. Synthesis of isoquinolines from α-aryl vinyl azides and alkynes under an Rh/Cu bimetallic catalytic system

final substituted isoquinoline derivatives and regenerating the Rh(III) and Cu(I) species *via* redox process (Cycle 2).

Previous reported redox-neutral synthesis of isoquinolines from aryl ketone *O*-acetyl oximes and alkynes under the catalytic $[RhCp^*Cl_2]_2$–NaOAc system required the N–O bond of the oximes to be *anti* to the aryl moiety.[4n] No reactivity of the corresponding *syn*-isomers could be attributed to the invalid ortho C–H activation, which was assisted by the lone pair of the oxime Sp_2 nitrogen. To solve this problem, Chiba and co-workers found that the $[RhCp^*Cl_2]_2$–Cu(OAc)$_2$ relay catalysts could be applied, and both *anti*- and *syn*-isomers of oximes were qualified coupling partners towards the synthesis of isoquinoline derivatives (Scheme 33).[37] Similar as the earlier study on this bimetallic catalysis, Cu(I) species might be generated from the reduction of Cu(OAc)$_2$ by DMF, which could provide iminyl-Cu(III) intermediate *via* direct oxidative addition (path A). Alternatively, iminyl-Cu(II) complex also might be delivered *via* one electron reduction, hemolytic N–O bond cleavage and a following reduction process (path B). Both iminyl copper species could undergo transmetalation with Rh(III) catalyst to give iminyl rhodium(III) intermediates, which were free to isomerize between *anti*- and *syn*-isomers. Then the C–H activation, alkyne insertion, and reductive elimination were followed to construct the annulated product. In addition to aryl ketoximes, indolyl ketoximes, benzofuranyl, furanyl, pyrrolyl, and thienyl ketoximes also could be employed under this bimetallic catalytic system, forming various azaheterocycles in good yields.

Different from Chiba's work, Hua and co-workers developed a Rh–Cu bimetallic catalysis for one-pot synthesis of 2-aminoquinolines in 2014 (Scheme 34).[38] In this catalytic system, the direct double C–H activation of 1-aryl tetrazoles occurred under the catalyst of $[RhCp^*Cl_2]_2$ without the coordination assistance, delivering the annulated product and Rh(I) species. Tautomerization and subsequent reduction mediated by copper salt lead to the final 2-amino-quinoline *via* denitrogenation. The stoichiometric copper salt also acted as the oxidant to restart the catalytic cycle by regenerating the Rh(III) species.

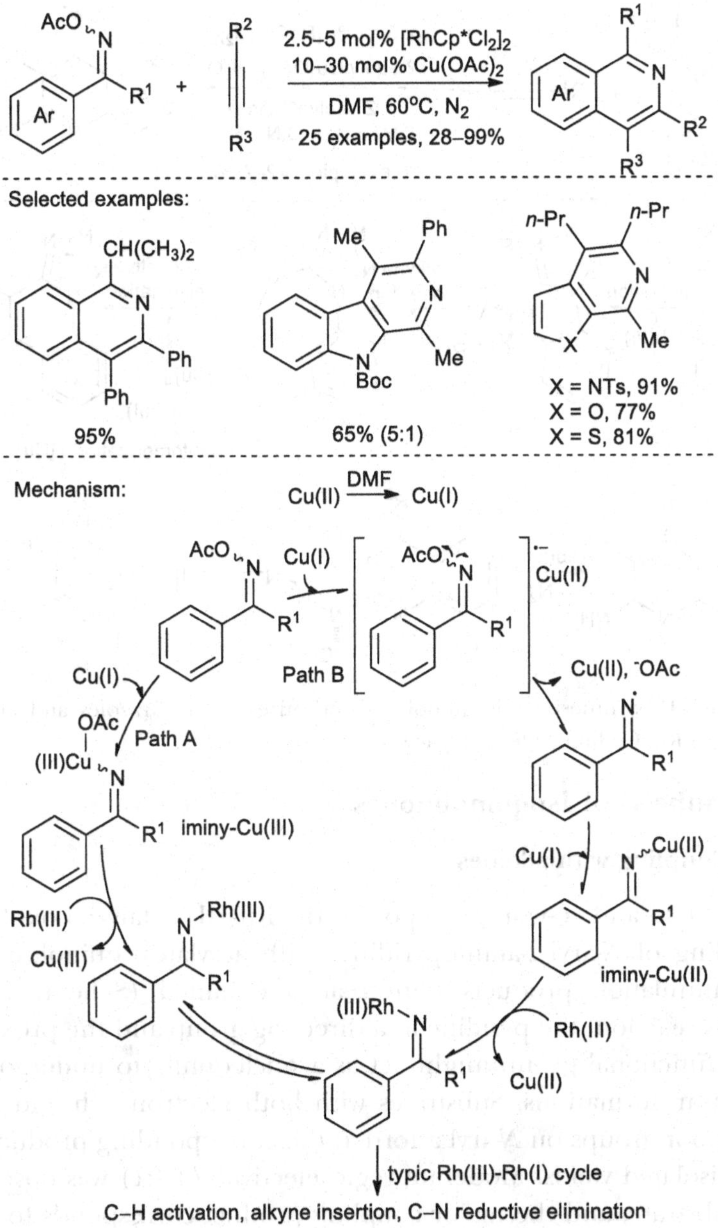
AcO
N
R1
Ar
+
R2
R3
2.5–5 mol% [RhCp*Cl2]2
10–30 mol% Cu(OAc)2
DMF, 60°C, N2
25 examples, 28–99%
R1
N
Ar
R2
R3
Selected examples:
CH(CH3)2
N
Ph
Ph
95%
Me
Ph
N
N
Boc
Me
65% (5:1)
n-Pr
n-Pr
N
X
Me
X = NTs, 91%
X = O, 77%
X = S, 81%
Mechanism:
DMF
Cu(II) → Cu(I)
AcO
N
Cu(I)
R1
Path B
Cu(II)
Cu(II), ⁻OAc
Cu(I)
OAc
Path A
(III)Cu
N
R1
iminy-Cu(III)
N
R1
Cu(I)
N Cu(II)
R1
iminy-Cu(II)
Rh(III)
Cu(III)
N Rh(III)
R1
(III)Rh
N
R1
Rh(III)
Cu(II)
typic Rh(III)-Rh(I) cycle
C–H activation, alkyne insertion, C–N reductive elimination

Scheme 33. Synthesis of isoquinolines from oximes and alkynes under an Rh/Cu bimetallic catalytic system

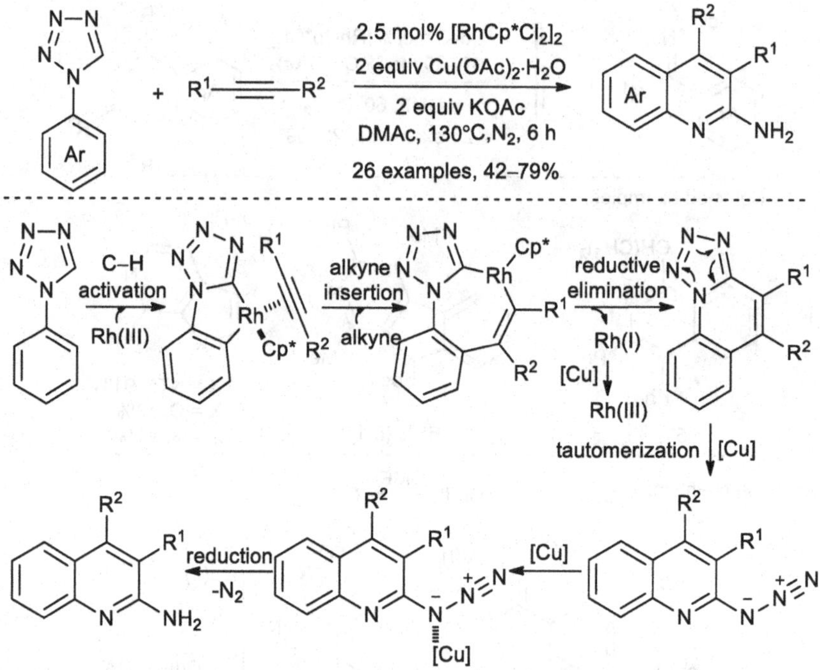

Scheme 34. Synthesis of isoquinolines derivatives from tetrazoles and alkynes under an Rh/Cu bimetallic catalytic system

4. Synthesis of Isoquinolinones

4.1 Coupling with alkenes

In 2010, Li and co-workers reported the Rh(III)-catalyzed oxidative coupling of *N*-aryl-2-aminopyridines with acrylates, quinolones as the annulation products were readily obtained (Scheme 35).[39] In this reaction, the pyridine as a directing group and the proximal N–H functional group might act as a nucleophile to undergo further transformations. Substrates with both electron-rich and electron-poor groups on *N*-aryl afforded the corresponding products in high isolated yields. Moderate regioselectivity (4.4:1) was observed for substrate, and the major coupling product corresponds to C–C coupling at a more hindered position. This observation is most plausibly ascribed to the directing effect of the O-atom and/or the electronic effect of the aryl ring.

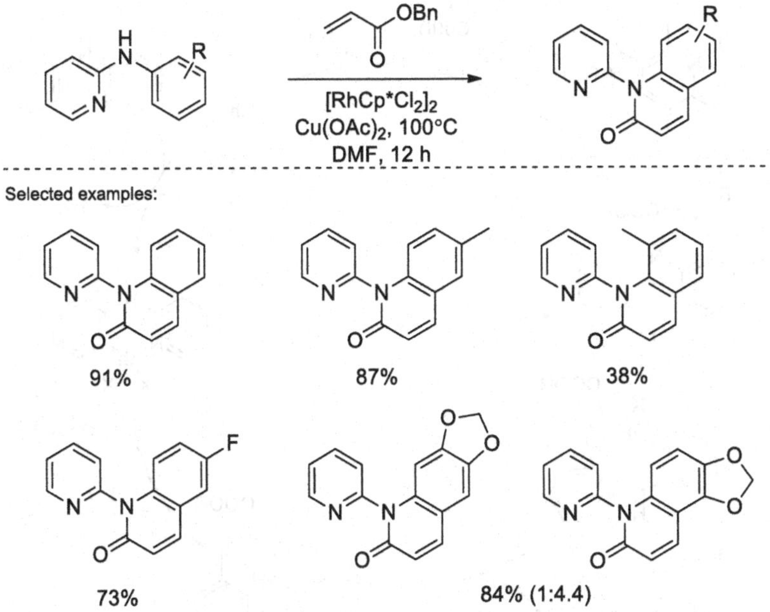

Scheme 35. Synthesis of isoquinolines from *N*-aryl-2-aminopyridines with acrylates

During the cyclometalation process, a six-membered Rh(III) intermediate is provided with a loss of an acid, which then undergoes insertion of an incoming acylate (Scheme 36). Subsequent β-hydride elimination occurs to afford a (metal-bound)trans-olefin, which can isomerize to the *cis*-isomer. Intramolecular amidation and cyclization of this *cis*-intermediate generates the final product.

Later, Glorius and co-workers reported an uniquely efficient Rh(III)-catalyzed and *O*-pivaloyl group directed reaction to yield valuable cyclized lactam products (Scheme 37).[40] The key to success was the use of an *O*-pivaloyl group on the benzamide, which can potentially chelate the Rh center. In this reaction, small amounts of pivalic acid were found to be beneficial, resulting in the selective product. The scope and selectivity of substrates for this reaction are good for acrylates and styrenes. Interestingly, the desired cyclic amide product was given in excellent yield under five bar of ethylene.

In the same year, Fagnou and Guimond and co-workers used benzhydroxamic acids as oxidizing directing groups, which were

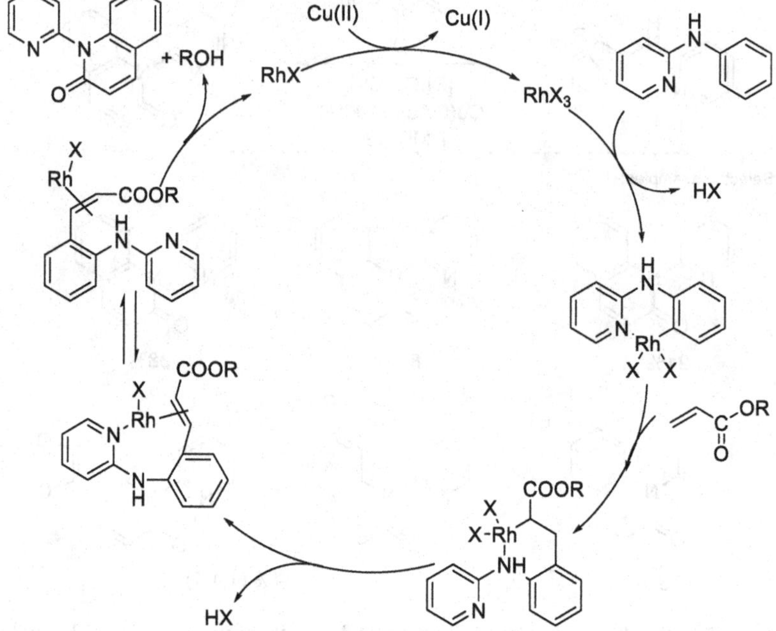

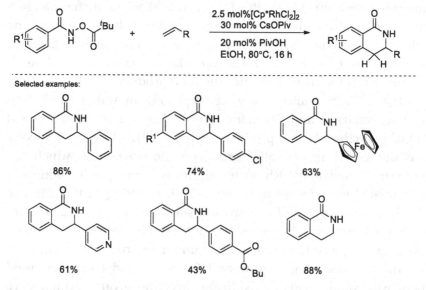

Scheme 36. Proposed catalytic for the coupling of *N*-aryl-2-aminopyridines with acrylates

Scheme 37. Synthesis of cyclized lactams from benzamide derivatives with alkenes

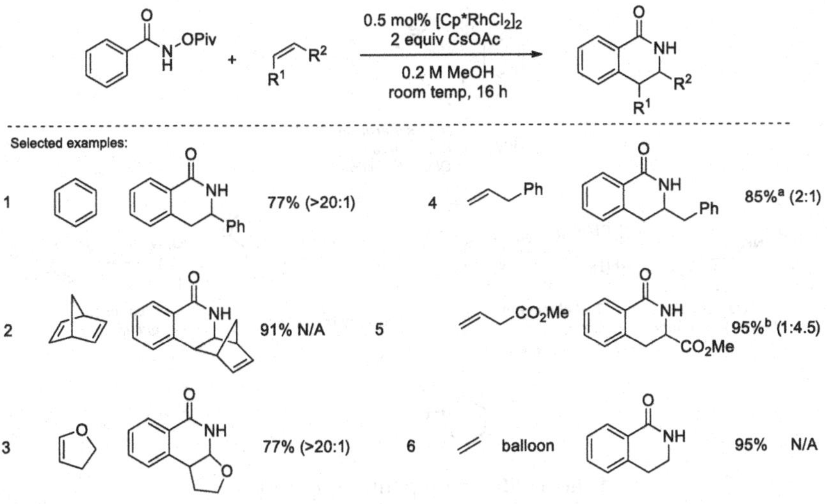

Scheme 38. Synthesis of isoquinolones from benzamide derivatives with alkenes
[a] Isolated yield of both regioisomers. [b] Inseparable mixture of regioisomer.

introduced to Rh(III)-catalyzed redox-neutral isoquinolone synthesis (Scheme 38).[41] The present methodology afforded a wide variety of isoquinolones at room temperature while employing low catalyst loadings (0.5 mol%). However, the inherent challenge with this method was the formation of the C(sp3)–N bond, avoiding the well-documented β-hydride elimination which furnished Heck-type products. To do so, the energy for C–N bond forming/N–O bond cleaving must be lower than β-hydride elimination.

Later, Xia reported the study about Rh(III)-catalyzed C–H functionalizations of benzamide derivatives with olefin by Density functional theory (DFT) calculations which proved the different pathways controlled by the N–OR internal oxidants (Scheme 39).[42] When using N–OMe as internal oxidant, the intermediate is unstable, therefore the olefination occurs easily *via* a β-hydride elimination/reductive elimination (RE) sequence to generate the Rh(I) intermediate, which is then oxidized to the active Rh(III) *via* MeOH elimination from the N–OMe reduction in the presence of HOAc. However, if a seven-membered rhodacycle intermediate containing an N–OPiv moiety, the coordination of the acyloxy carbonyl oxygen stabilizes this intermediate and increases the barrier of the olefination pathway.

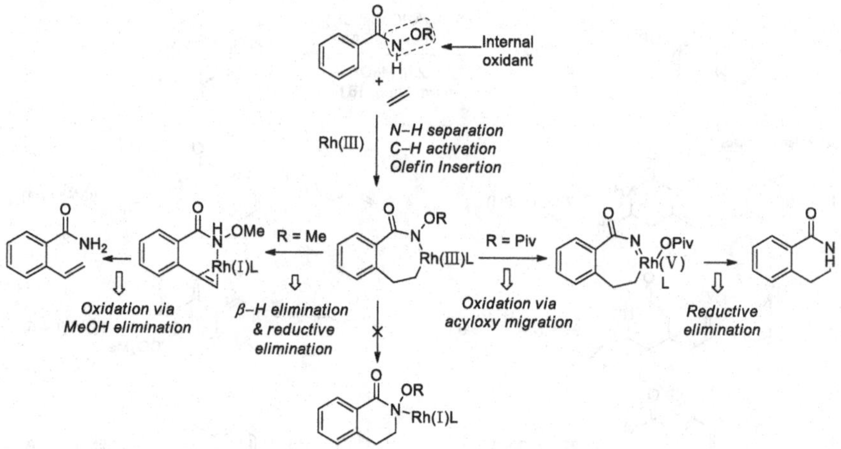

Scheme 39. Competition reactions

Though allenes have been widely applied in transition metal-catalyzed reactions, the application of this synthetically attractive functional group in C–H activation reactions remains rare. In 2012, Glorius reported the Rh(III)-catalyzed coupling reaction with allenes in the presence of a substoichiometric amount of CsOAc, which delivered some interesting heterocycles (Scheme 40).[43] In the reactions, good regioselectivity favoring activation of the less hindered C–H bond was observed of *meta*-substituted substrate. Because of steric interference, the ortho substituent retarded the reaction. Interestingly, heterocycles were very well tolerated, leading to some valuable products. Electron-rich heterocycles like furan and thiophene were suitable substrates giving the corresponding products in excellent yields. What is more, under slightly modified conditions, nicotinic provided the corresponding product in good yield.

In standard cases, the coordination of five-membered intermediate and allene was governed by electronic factors, then the C–C bond formed with the central carbon atom of the allene moiety (Scheme 41). The seven-membered rhodacycle intermediate might be slowly isomerized to other allylic species. However, the reductive elimination occurred preferentially at the less hindered carbon to afford the product with an exocyclic double bond. The active Rh(III) catalyst was regenerated by the cleavage of the N–O bond.

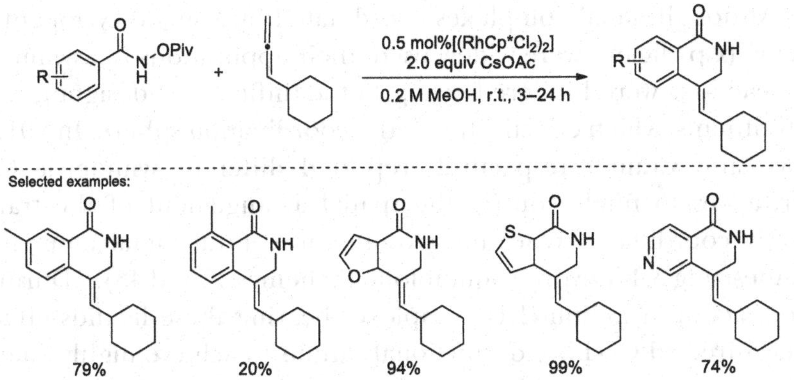

Scheme 40. Synthesis of isoquinolinones from benzamide derivatives with allenes

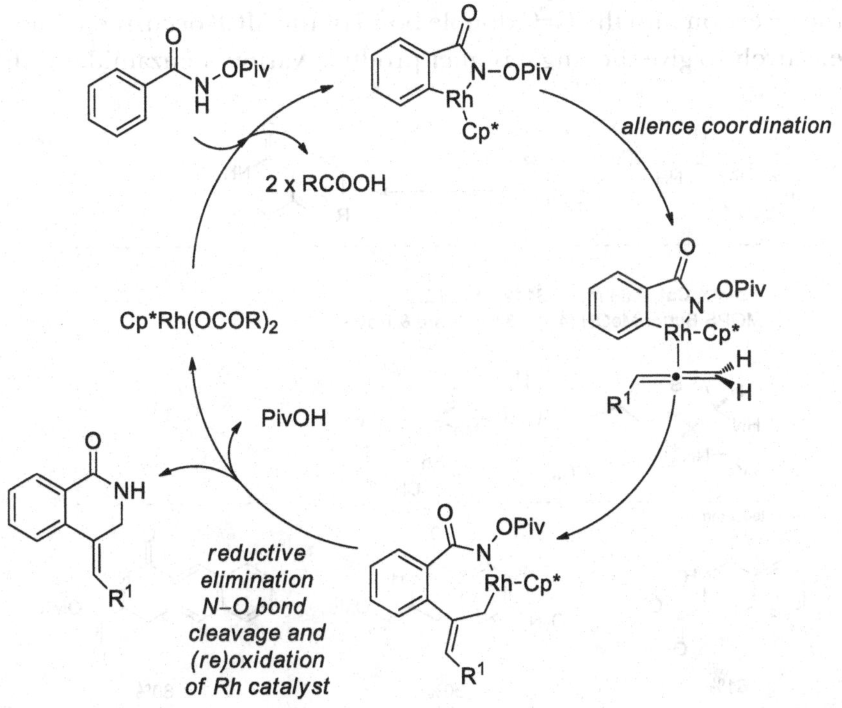

Scheme 41. Proposed catalytic cycle for the coupling of benzamide derivatives with allenes

Although metal complexes coordinated by a single cyclopentadienyl (Cp) ligand were widely used, their applications to asymmetric reactions were hindered owing to the difficulty of designing Cp substituents, which effectively bias the coordination sphere. In 2012, Rovis and Cramer respectively reported different symmetric Cp derivatives to finely control the spatial arrangement of the transiently coordinated reactants around the central metal atom for synthesizing 3,4-dihydroisoquinolone (Schemes 42 and 43).[44] During the reaction, rhodium(III) complexes bearing these ligands effectively directed C–H bond functionalizations to achieve highly enantioselective coupling with benzhydroxamic acids and olefins.

In 2013, Wu and co-workers opened up a new avenue in Rh(III)-catalyzed C–H activation/cycloaddition of N-pivaloyloxybenzamide and methylenecyclopropanes (MCPs) for the construction of privileged spiro dihydroisoquinolinones (Scheme 44).[45] It is noted that the insertion into the C=C double bond of the MCP occurred regioselectively to give the single isomer product. Various benzamides with

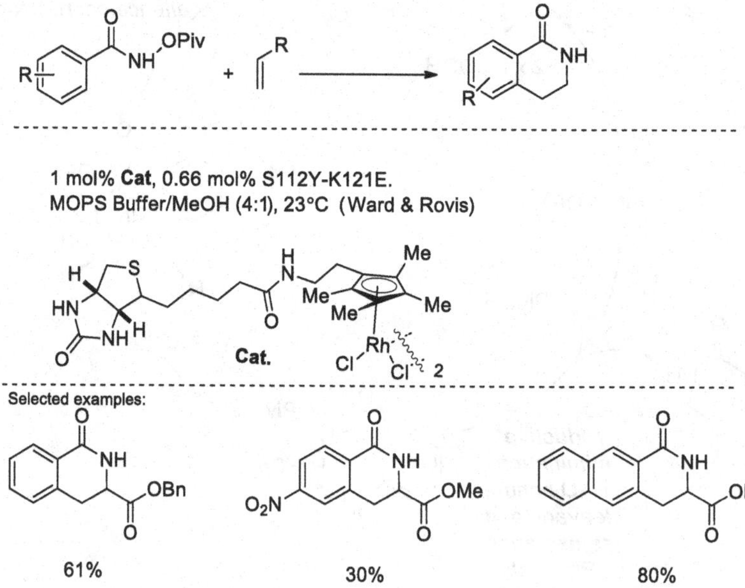

1 mol% **Cat**, 0.66 mol% S112Y-K121E.
MOPS Buffer/MeOH (4:1), 23°C (Ward & Rovis)

Selected examples:

61%   30%   80%

Scheme 42. Synthesis of dihydroisoquinolones under novel rhodium(III) complexes

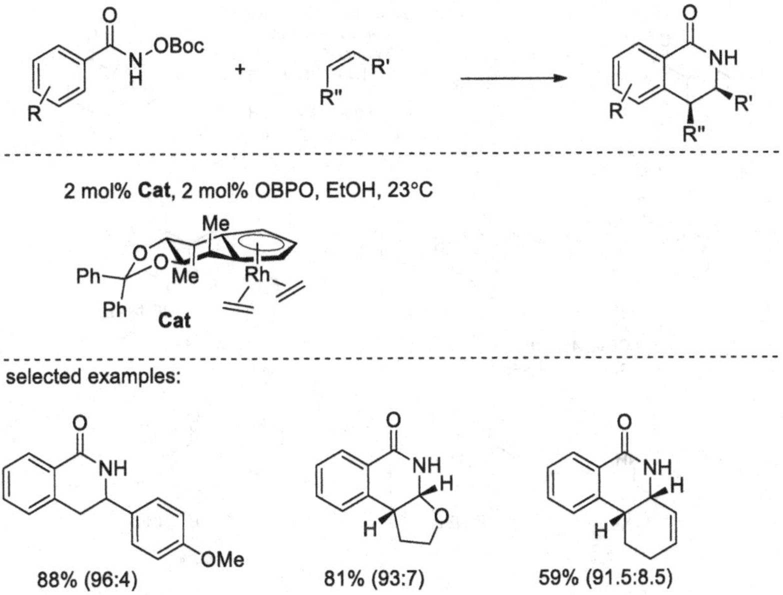

2 mol% **Cat**, 2 mol% OBPO, EtOH, 23°C

selected examples:

88% (96:4)   81% (93:7)   59% (91.5:8.5)

Scheme 43. Synthesis of dihydroisoquinolones under fixed rhodium(III) complexes

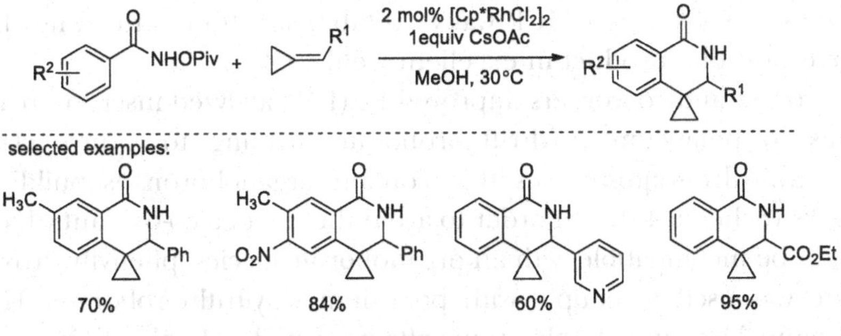

2 mol% [Cp*RhCl₂]₂
1equiv CsOAc

MeOH, 30°C

selected examples:

70%   84%   60%   95%

Scheme 44. Synthesis of dihydroisoquinolones from benzamides and methylenecyclopropanes

valuable functional groups could react smoothly to afford corresponding spiro dihydroisoquinolinones in moderate to excellent yields. When *meta*-substituted benzamides were applied, sole products were obtained in good yields because of favoring activation of the less hindered C–H bond. Notably, heterocyclic pyridine derivative was applicable under slightly modified conditions.

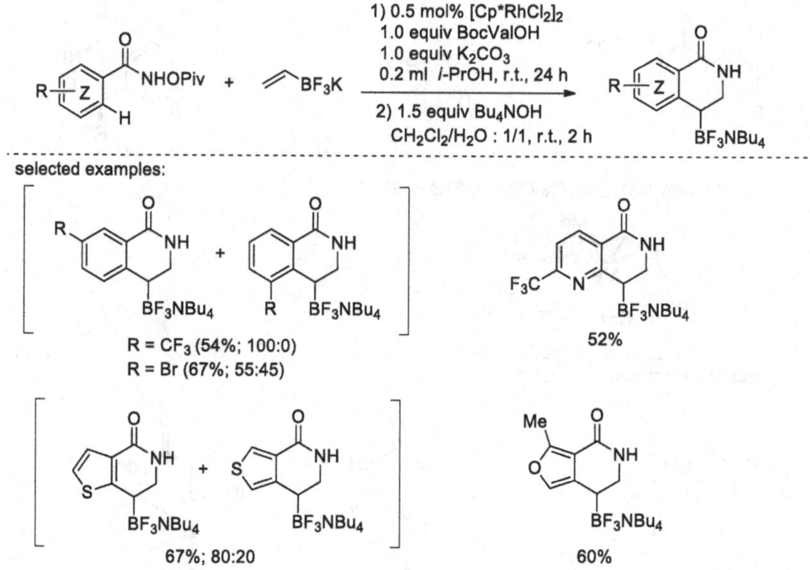

Scheme 45. Synthesis of dihydroisoquinolones from benzamides and potassium-vinyl trifluoroborates

The ester substituted MCP was a suitable substrate, affording the corresponding product in excellent yield (95%).

Presset and co-workers improved Rh(III)-catalyzed insertion reactions to potassium vinyltrifluoroborate, leading to a new class of tetrahydroisoquinolones that contain organoboron as building blocks (Scheme 45).[46] In order to avoid the use of a co-oxidant which might be incompatible with an organoboron species, phenylhydroxamate was used to couple with potassium vinyltrifluoroborate. The substantial and unique electronic effect seemed to be the major contributing factor to the unusual regiochemistry of the olefin insertion product. The regioselectivity of the C–H activation is strongly influenced by steric factors. In addition, pyridine, thiophene, and furan derivatives were also applicable for this reaction providing good yields.

Davis and co-workers designed simple alkenes tethered to the benzamide system on the *meta*-position to complete coupling in 2013 (Scheme 46).[47] Under the catalysis of [Cp*Rh(III)], no reaction occurred even if the more sterically accessible C–H bond was

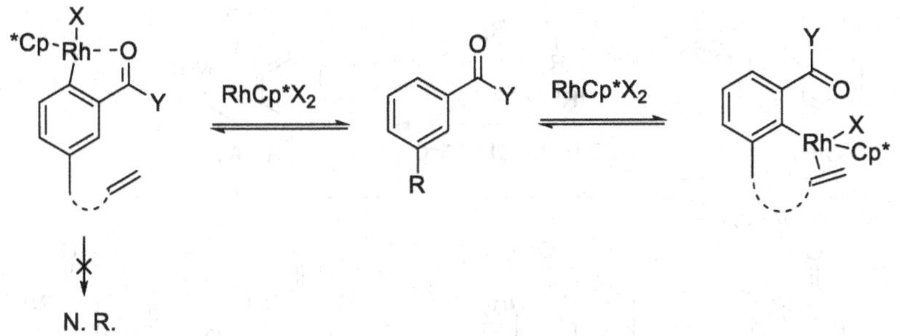

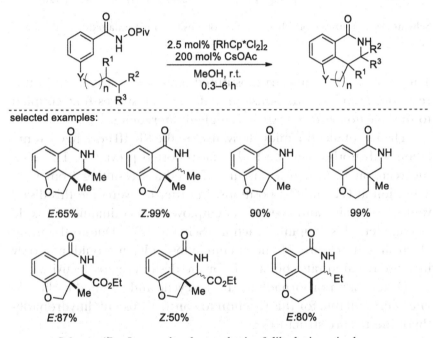

Scheme 46. Proposed catalytic cycle for the *meta*-position coupling

selected examples:

E:65%   Z:99%   90%   99%

E:87%   Z:50%   E:80%

Scheme 47. Intramolecular synthesis of dihydroisoquinolones

first activated. The lack of an appropriate coupling partner induced the intramolecular coupling reaction with high regioselectivity.

Through examining the impact of olefin geometry in the reaction, alkenes subjected to the reaction conditions provided a single isomer (Scheme 47). It indicated that stereochemistry of the product was dependent on the geometry of the starting olefin.

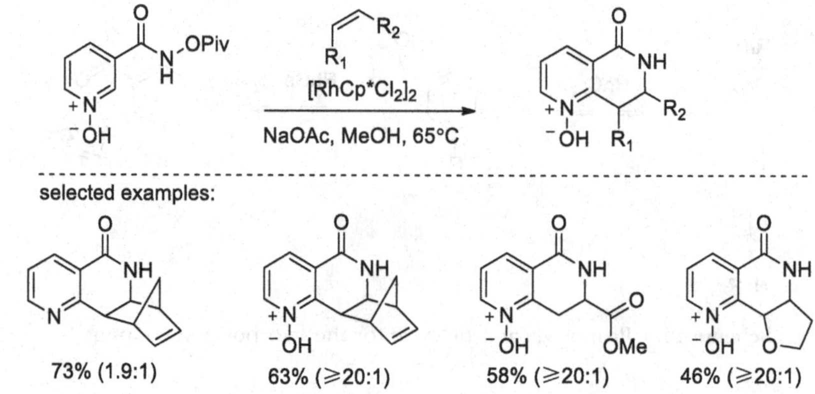

Scheme 48. Synthesis of dihydroisoquinolones from nicotinamide *N*-oxides and olefines

The mono-, 1,1-di-, and tri-substituted olefins were tolerated in this reaction. *E* and *Z* of α, β-unsaturated esters all successfully coupled to the reaction and only gave a single diastereomer.

The use of nicotinamide derivatives in the Rh(III)-catalyzed annulation with norbornadiene had been reported previously. However, the reaction furnished a mixture of products resulting from C–H activation at C-2 and C-4 positions. In order to overcome this disadvantage, Huckins and co-workers employed a nicotinamide *N*-oxide to complete this coupling reaction (Scheme 48).[48] Due to the more electron rich nature of nicotinamide *N*-oxide, it would not only increase reaction rates but also favor the desired regioselectivity.

Other than norbornadiene, both acyclic and strained cyclic alkenes were suitable for this reaction to construct useful dihydronaphthyridinone core structures.

4.2 Coupling with alkynes

In 2010, Rovis reported the reaction that utilized Rh(III) catalysts in the presence of Cu(II) oxidants to catalyze *N*-methyl benzamide and diphenylacetylene (Scheme 49).[49] The reaction is tolerant of aryl bromides and unprotected aldehydes and other substituents on nitrogen are also tolerated.

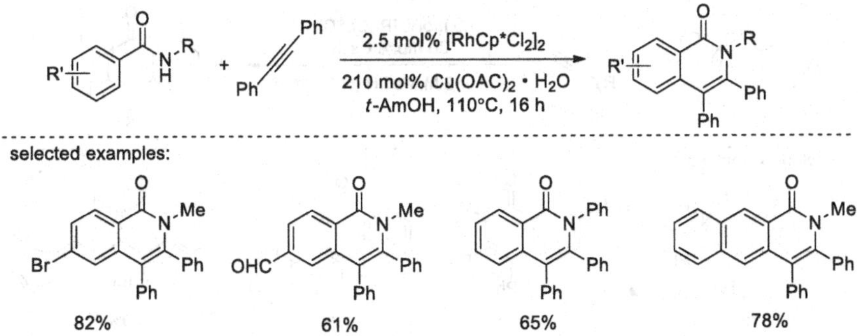

selected examples:

82%	61%	65%	78%

Scheme 49. Synthesis of isoquinolones from benzamides and alkynes

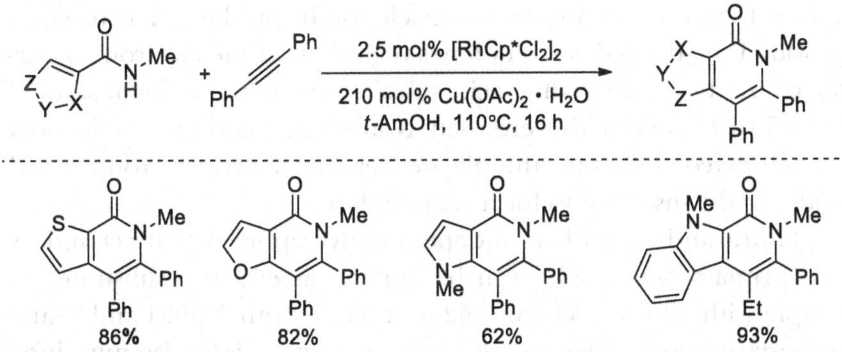

86%	82%	62%	93%

Scheme 50. Synthesis of isoquinolones from heteroaryl carboxamides and alkynes

It is worth mentioning that heteroaryl carboxamides were suitable for these reaction conditions (Scheme 50). The 3-position furan and pyrrole carboxamides furnished a single product resulting from C–H activation at the more activated 2-position. Thiophene and indolyl carboxamides afforded the corresponding products in excellent yield.

Almost at the same time, Guimond and co-workers reported the synthesis of isoquinolone motif *via* Rh(III)-catalyzed annulation of benzhydroxamic acids with alkynes (Scheme 51).[26] The process is external oxidant-free which just used the installed N–O bond as internal oxidant for C–N bond formation. Using a catalytic amount of base

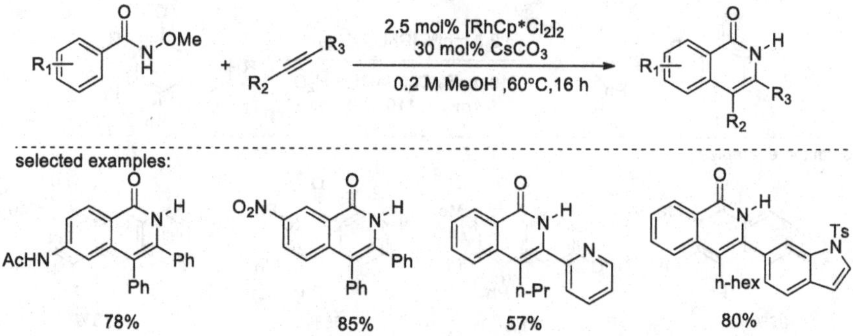

Scheme 51. Synthesis of isoquinolones from benzhydroxamic acids and alkynes

and methanol as solvent gave high yields of the products. The reaction provided the desired isoquinolones regardless of the electron character of the benzhydroxamic acid substituents. When *meta*-substituted benzohydroxamic acids were used as substrates, good regioselectivity was observed. Of note, the alkyne bearing a pyridyl group could undergo the insertion to form isoquinolone.

Miura and co-workers independently reported that secondary and primary benzamides can be applied as effective substrates to couple with alkyne (Schemes 52 and 53).[50] Both *N*-alkyl and *N*-aryl secondary benzamides can be used to synthesis of isoquinolone motif *via* the method. The oxidative coupling of primary benzamide with diphenylacetylene proceeded smoothly accompanied by regioselective CH bond cleavage, giving unexpected 1:2 coupling tetracyclic structures which exhibit a broad range of interesting biological and optoelectronic properties.

Except arylamides, acrylamides can also achieve oxidative coupling with alkynes that was reported by Li. Using 0.5 mol% loading of $[RhCp*Cl_2]_2$ with $Cu(OAc)_2$ as an oxidant, an efficient and selective method to synthesis 2-pyridones *via* oxidative insertion of alkynes into acrylamides had been put forward (Scheme 54).[51] Various acrylamides with internal alkynes were examined in the reaction. A thiophenyl-substituted alkyne could also be applied but with lower reactivity. Alkyl- and aryl-substituted unsymmetrical alkynes participated in this reaction affording a mixture of regioisomers (7:1 ratio)

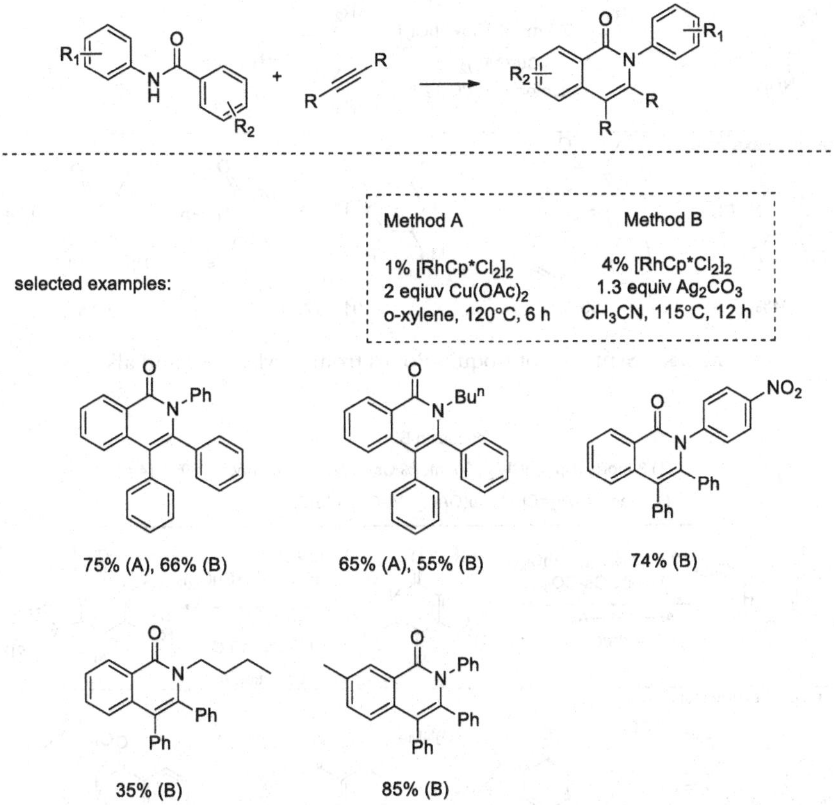

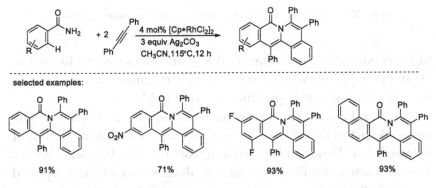

Scheme 52. Synthesis of isoquinolones from secondary benzamides and alkynes

Scheme 53. Synthesis of isoquinolones from primary benzamides and alkynes

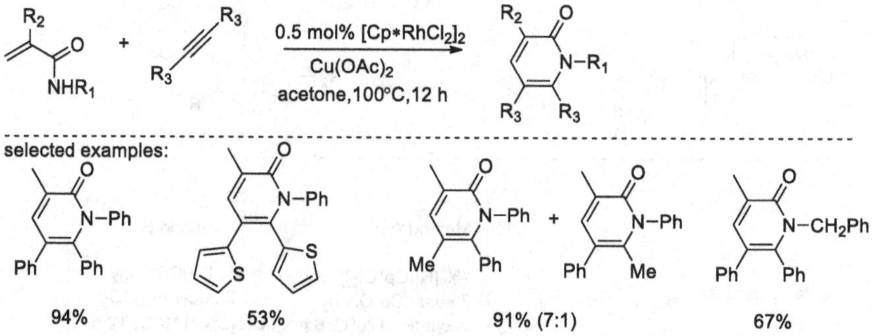

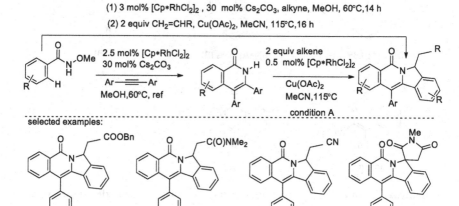

Scheme 54. Synthesis of isoquinolones from acrylamides and alkynes

Scheme 55. Synthesis of isoquinolones *via* cascade coupling reactions

in excellent yield. Furthermore, the *N*-substituent is not limited to an aryl group, and *N*-benzyl-substituted acrylamide was also suitable for this reaction to give pyridone in good yield.

In 2011, Li and co-workers considered the nitrogen in isoquinolones could act as an efficient directing group to achieve activation of proximal C–H bonds and subsequent oxidative coupling with alkynes. The functional similarities between alkynes and alkenes, applied activated alkenes to react with isoquinolone (Scheme 55).[52]

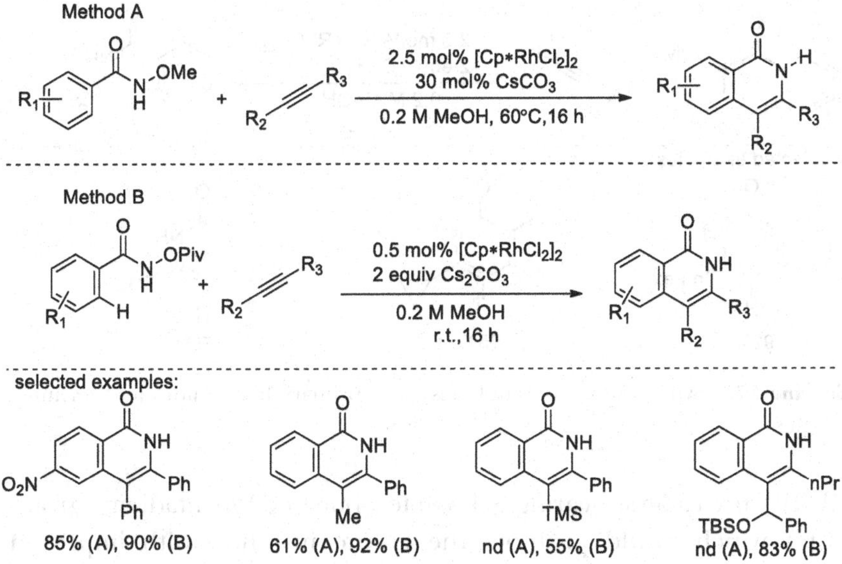

Method A

	2.5 mol% [Cp*RhCl₂]₂	
	30 mol% CsCO₃	
	0.2 M MeOH, 60°C,16 h	

Method B

	0.5 mol% [Cp*RhCl₂]₂	
	2 equiv Cs₂CO₃	
	0.2 M MeOH	
	r.t.,16 h	

selected examples:

85% (A), 90% (B)   61% (A), 92% (B)   nd (A), 55% (B)   nd (A), 83% (B)

Scheme 56.   Synthesis of isoquinolones from primary benzamides and alkynes

In 2011, Guimond and Fagnou improved a more reactive internal oxidant/directing group for the synthesis of a wide variety of isoquinolones at mild conditions while low catalyst loadings (0.5 mol%) were employed. Moreover, in contrast to the first generation system (N–OMe), the new pivaloyl system was more suitable for the synthesis of various isoquinolone (Scheme 56).[41] The alkyl–aryl disubstituted alkynes gave higher yields and better regioselectivity than previous report. Interestingly, a TMS-protected alkyne was also tolerated, which can be utilized to further functionalization. Additionally, dialkyl-substituted alkynes were well tolerated for the pivaloyl system.

In addition, the terminal alkynes were employed to afford the desired mono-substituted isoquinolones in moderate to high yields and the terminal end was usually located at the 4-position (Scheme 57). Alkyl-substituted terminal alkynes and trimethylsilylacetylene were suitable alkynes yielding the predictable isoquinolones.

Rh(III)-catalyzed C–H activation has been used for the synthesis of a variety of heterocycles. However, limited examples about ligand development had been reported. Hyster and Rovis synthesized a

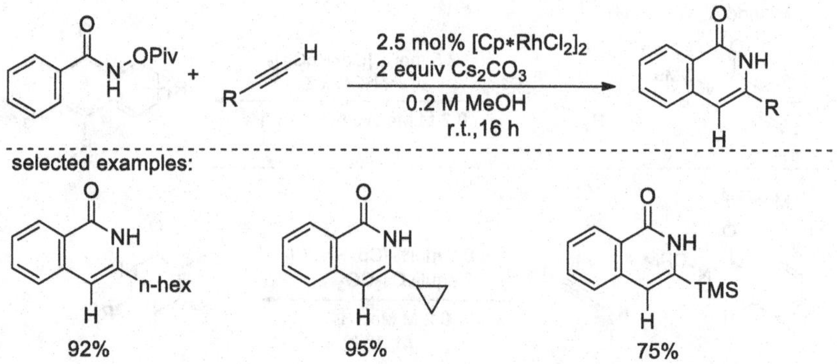

Scheme 57. Synthesis of isoquinolones from primary benzamides and terminal alkynes

Rh(III) pre-catalyst bearing a 1,3-di-tert-butyl cyclopentadienyl group (Cp^t) which would promote the alkyne insertion with improved regioselectivity and no loss in yield (Scheme 58).[53] Extensive substitution on the acrylamide was suitable for pyridone formation. Alkyl N-substituted acrylamides were tolerated affording the corresponding products with high yields and high regioselectivity. Enynes were employed in this reaction with high levels of selectivity to afford the alkene products which could be used for further functionalization.

In 2011, Li and co-workers reported chelation-assisted rhodium(III)-catalyzed direct C–H functionalization of π-deficient pyridines with alkynes through two-fold C–H activation to synthesis quinolone motif (Scheme 59).[4a] Isonicotinamide and two equivalent of diphenylacetylene using [RhCp*Cl$_2$]$_2$ (4 mol%) as a catalyst and Cu(OAc)$_2$ as an oxidant gave quinolone product, with two-fold C–H activation of the pyridine ring at the two- and three-positions. Isonicotina-mides bearing either electron-rich or -poor groups in the N-phenyl ring were suitable for the reaction giving corresponding products in high yield under the standard conditions. The heteroaryl and aliphatic alkynes were tolerated to afford corresponding quinolone. The catalytic reaction was successfully extended to 2,4-bipyridine, 4-(1-methylimidazol-2-yl)pyridine and 4-(1-phenylimidazol-2-yl) pyridine, as a result two-fold and four-fold C–H activation products were isolated, respectively.

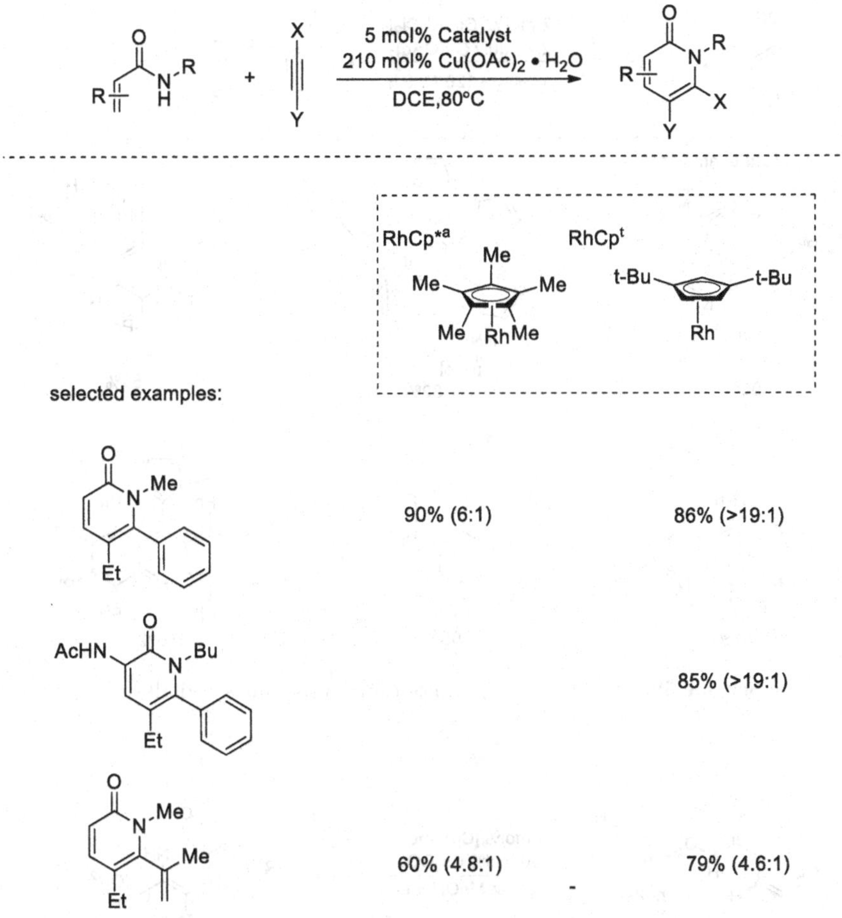
5 mol% Catalyst
210 mol% Cu(OAc)2 • H2O
DCE,80°C
selected examples:
RhCp*a
RhCpt
90% (6:1)
86% (>19:1)
85% (>19:1)
60% (4.8:1)
79% (4.6:1)

Scheme 58. Synthesis of isoquinolones from acrylamides and alkynes

In 2012, Xu and Park developed an intramolecular reaction of alkyne-tethered hydroxamic esters to synthesis hydroxyalkyl-substituted isoquinolone/2-pyridone derivatives, which could be readily transformed into indolizidine scaffolds (Scheme 60).[54] The intramolecular N–O bond played the role as *N*-methoxy/pivaloyloxy in Fagnou's rhodium-catalyzed intermolecular C–H activation. Moreover, this intramolecular rhodium-catalyzed C–H/N–O bond functionalization reaction provided isoquinolones with reverse regioselectivity compared to the reported intermolecular version,

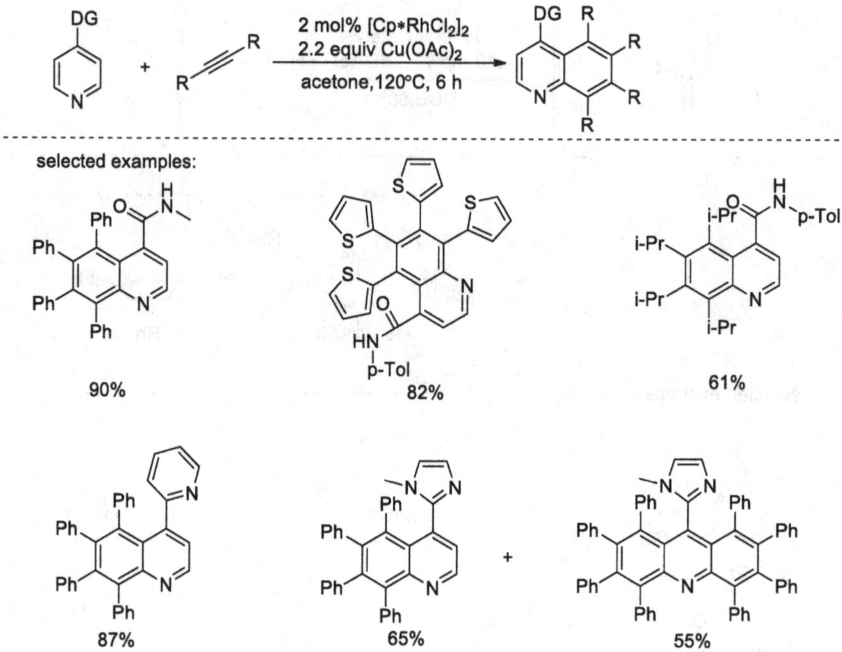

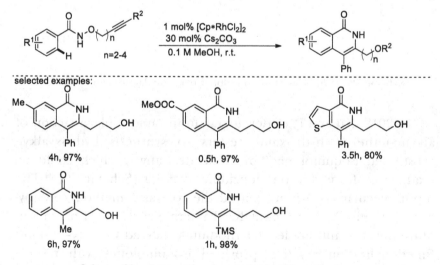

Scheme 59. Synthesis of quinolones from pyridines and alkynes

Scheme 60. Intramolecular synthesis of isoquinolones

which offered a great synthetic potential for the construction of benzoindolizidine and indolizidine frameworks. Various substrates proceeded smoothly to furnish isoquinolones and heteroaryl analogs in good to excellent yields. *Meta*-substituted benzamides afforded the corresponding isoquinolones in excellent yields with high regioselectivity. Heteroaryl carboxamides were tolerated in the reaction under slightly changed condition. TMS-protected alkyne turned out to be efficient and afforded product in excellent yield.

Additionally, the reaction extended to *N*-(pent-4-yn-1-yloxy) acrylamides and gave the corresponding products (Scheme 61). In the presence of [(Cp*RhCl$_2$)$_2$] (2.5 mol%) and CsOAc (30 mol%) the acrylamide at both α and β positions were tolerated well and afforded the 2-pyridone in excellent yield.

Wang and co-workers developed a mild, short, and efficient method for the formation of bench-stable 3-isoquinolone MIDA boronates under the catalysis of Rh(III) (Scheme 62).[55] The annulated isoquinolone MIDA boronates were capable of efficient Suzuki–Miyaura couplings to deliver interesting heterocycle compounds. The steric interaction of the boron motif with the aryl ring was more pronounced than that with the metal center, which governed the regioselectivity of this reaction. Both electron-donating and -withdrawing properties on the aromatic ring were well tolerated, affording the corresponding products in moderate to excellent yields. In addition, several heterocyclic derivatives, e.g. thiophene and indole proceeded smoothly to deliver the cyclization products

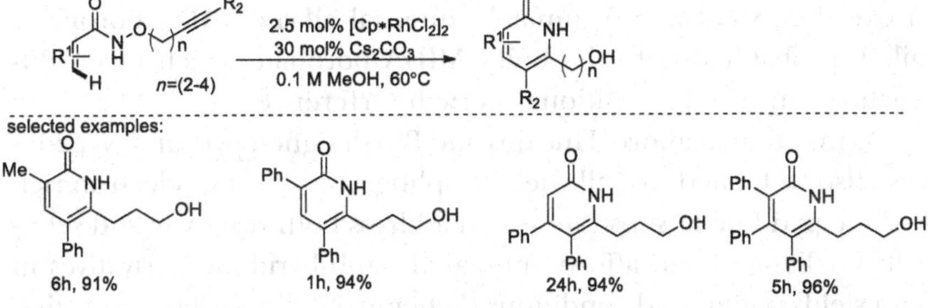

Scheme 61. Intramolecular synthesis of isoquinolones from acrylamide derivatives

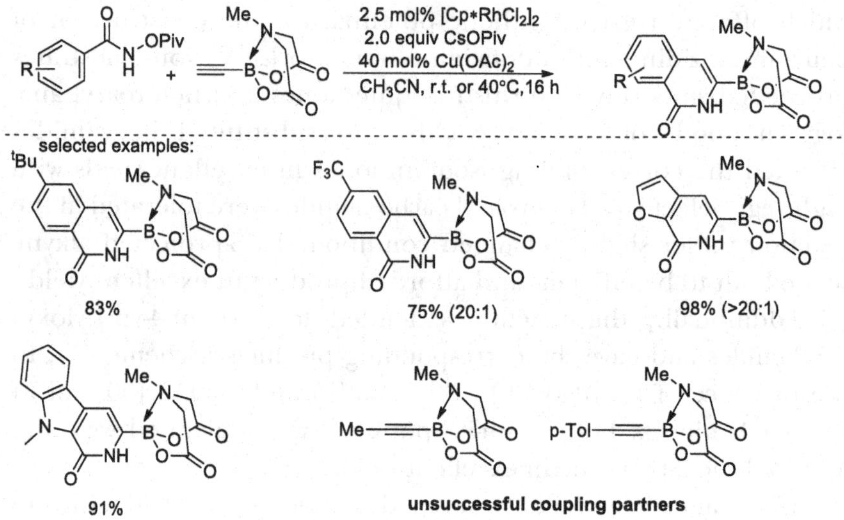

Scheme 62. Synthesis of bench-stable 3-isoquinolone MIDA boronates

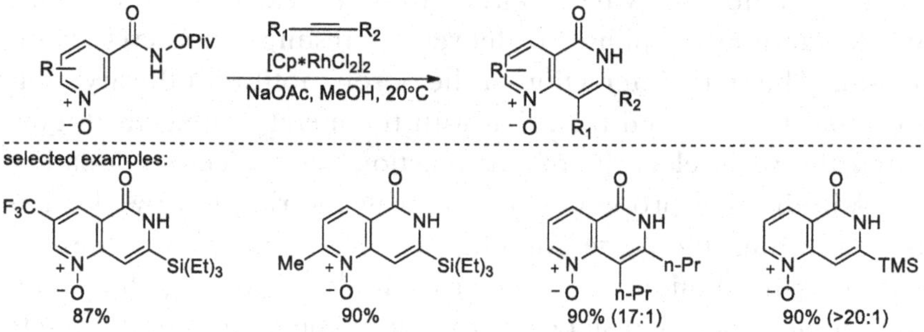

Scheme 63. Synthesis of isoquinolones from pyridine N-oxide and alkynes

in excellent yields. Unfortunately, internal alkyne MIDA boronates failed, probably due to the bulky MIDA boronate which made this reaction sensitive to additional steric interference.

As mentioned above, Huckins and Bio designed pyridine N-oxides was also tolerated to alkynes coupling. Employing electron-rich N-oxide pyridine as substrate would address both reactivity and selectivity challenges, and afford annulated naphthyridone derivatives in high yields under mild conditions (Scheme 63).[48] A variety of substituents on the five-position and six-position of nicotinamide core were well tolerated in the reaction. Dialkyl-substituted internal alkynes were also acceptable substrates.

5. Synthesis of Isocoumarins

In 2007, Miura reported two methods to afford the corresponding isocoumarin derivatives under rhodium/copper catalyst system through direct oxidative coupling of benzoic acids with internal alkynes (Scheme 64).[4r,56] The reaction proceeded efficiently forming no wastes except for water. Dialkylacetylenes proceeded efficiently similar as diphenylacetylene to produce 3,4-dialkylisocoumarins in good yields. When unsymmetrical alkylphenylacetylenes used as coupling partners, 4-alkyl-3-phenylisocoumarins were predominantly formed. Electron-rich and electron-deficient benzoic acids were found proceeded smoothly to afford corresponding isocoumarins.

A plausible mechanism for the reaction had been illustrated (Scheme 65). The carboxylate oxygen coordinated to $Rh(III)X_3$ giving a rhodium(III) benzoate, subsequently formed a rhodacycle

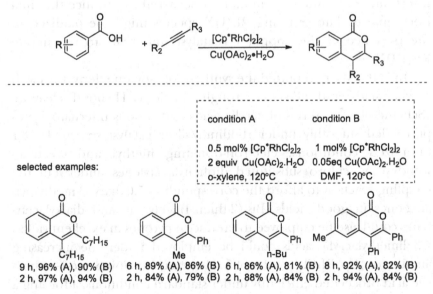

Scheme 64. Synthesis of isocoumarins from benzoic acids and alkynes

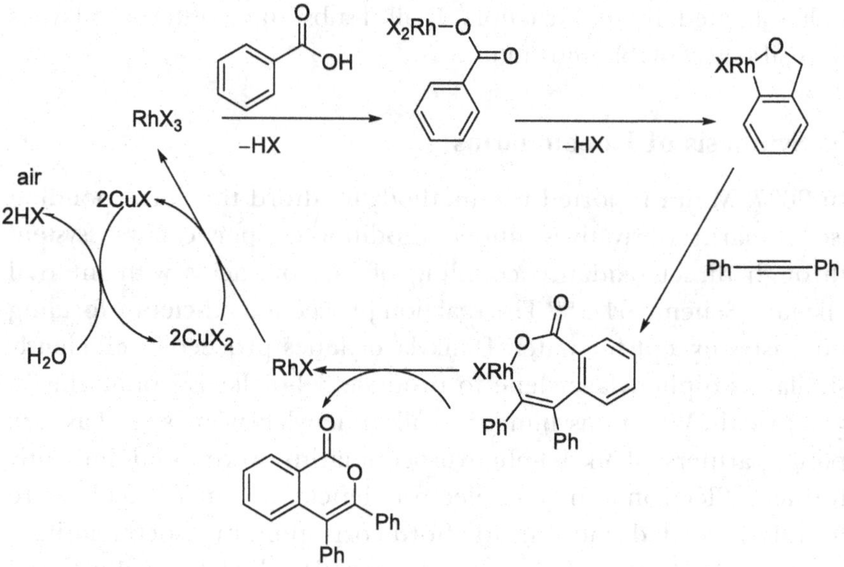

Scheme 65. Proposed catalytic cycle for the coupling reactions of benzoic acids and alkynes

intermediate through ortho C–H bond activation, then alkyne insertion, and reductive elimination occurred to produce the final isocoumarin. The resulting Rh(I)X species might be oxidized in the presence of the copper co-catalyst under air to regenerate Rh(III)X$_3$.

In 2009, Miura reported the synthesis of α-pyrone derivatives with acrylic acids and alkynes through vinylic C–H bond cleavage (Scheme 66).[57] A series of acrylic acids and various internal alkynes proceeded smoothly under rhodium/silver catalyst system. Under optimized conditions, electron-donating methyl and electron-withdrawing chloro substituted diphenylacetylenes underwent the coupling reaction to afford the corresponding 5,6-diaryl-3-methylpyran-2-ones in good yields. Bis-(2-thienyl)acetylene and dialkylacetylenes could also be employed to produce α-pyrones in excellent yields. 2,3-dimethylacrylic acids could be employed under the increasing amount of [Cp*RhCl$_2$]$_2$ to 2 mol%. However, 1-cyclohexene-1-carboxylic acid proceeded efficiently under standard conditions, affording a bicyclic product in 76% yield.

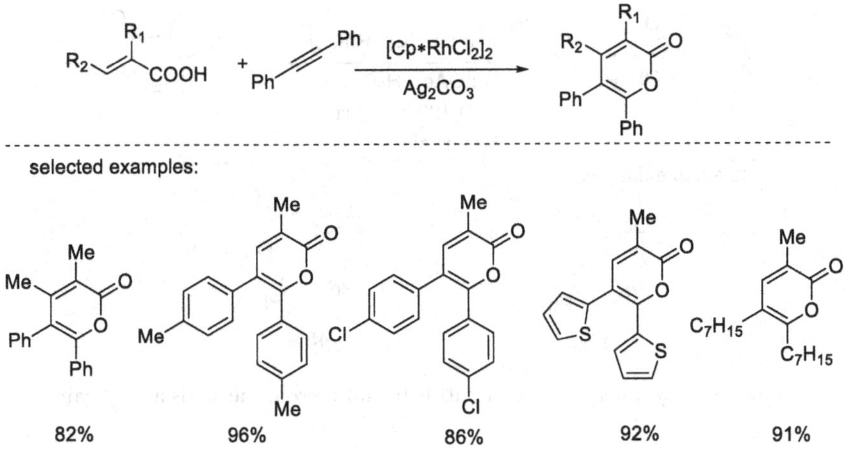

selected examples:

82%　　96%　　86%　　92%　　91%

Scheme 66. Synthesis of isocoumarins from acrylic acids and alkynes

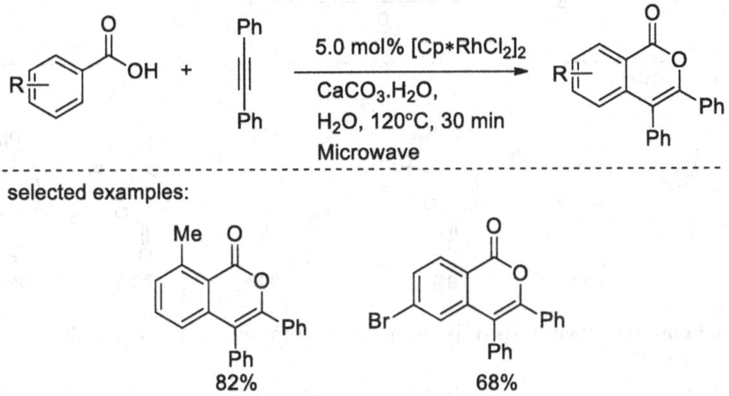

selected examples:

82%　　68%

Scheme 67. Synthesis of isocoumarins from acrylic aids and alkynes under microwave

In 2013, Yi and co-workers reported a "greener" route to form isocoumarins derivatives *via* a water-mediated and Rh/Cu-catalyzed coupling reaction of benzoic acids with alkynes under microwave-assisted heating procedures (Scheme 67).[58] This method is simple, efficient and fast, which could be extended to the synthesis of a variety of biologically important heterocyclic compounds. The reaction proceeded smoothly regardless of the electron-donating or -withdrawing character of the benzoic acids.

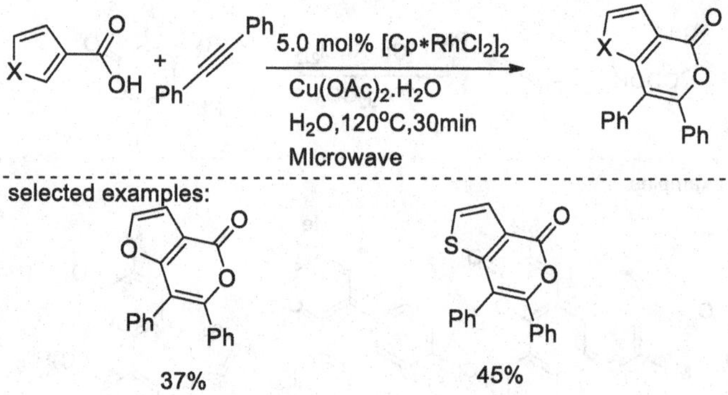

Scheme 68. Synthesis of isocoumarins from heteroarene acids and alkynes

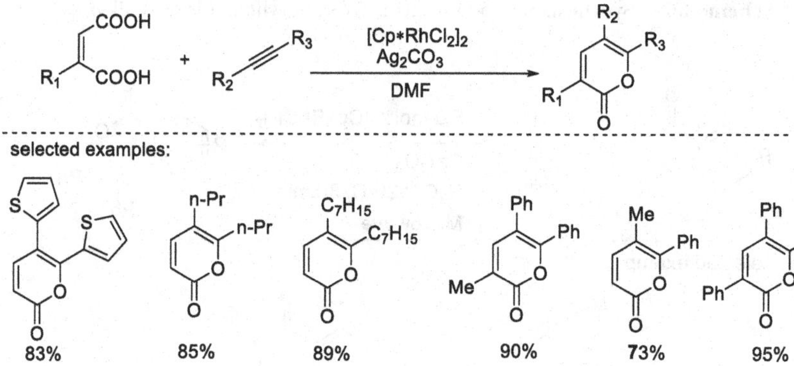

Scheme 69. Synthesis of isocoumarins from maleic acids and alkynes

Additionally, heteroarene acids were also suitable for this coupling reaction affording the corresponding products (Scheme 68).

Satoh and co-workers reported the rhodium-catalyzed decarboxylative coupling of maleic acid in place of acrylic acid for the synthesis of α-pyrones (Scheme 69).[59] Similar conditions, which were previously reported by using 1 mol% of [Cp*RhCl$_2$]$_2$ and 1 equiv of Ag$_2$CO$_3$ in DMF at 120°C, were used to promote maleic acids reacted with alkynes, as a result the decarboxylation corresponding α-pyrones were produced in excellent yields. Various internal alkynes and maleic acids were effective for the coupling reaction. Di-(2-thienyl)acetylene, 4-octyne, and 8-hexadecyne also proceeded smoothly to give 5,6-dithieny and 5,6-dialkyl-α-pyrones in good yields. Unsymmetrical alkylphenylacetylenes coupled with maleic acids to form 5-alkyl-6-phenyl-α-pyrones predominantly. 2-methylmaleic acid and 2-phenylmaleic acid were also employed in the reaction to furnish 3,5,6-trisubstituted-α-pyrones in excellent yields.

A possible mechanism for the coupling of maleic acids with alkyne is illustrated in Scheme 70. Coordination of the carboxyl oxygen atoms coordinated to a [Cp*Rh(III)X$_2$] species giving a rhodium (III) dicarboxylate **A**, then through decarboxylating to form a five-membered rhodacycle **B**, followed by alkyne insertion to give **C**, and reductive elimination took place to produce α-pyrone. The resulting Cp*Rh(I) species might be oxidized in the presence of Ag$_2$CO$_3$ to regenerate active [Cp*Rh(III)X$_2$].

Interestingly, 1,3-diynes could also be used to couple with maleic acid to furnish 6H,6′H-[2,2′-bipyran]-6,6′-dione frameworks (Scheme 71).

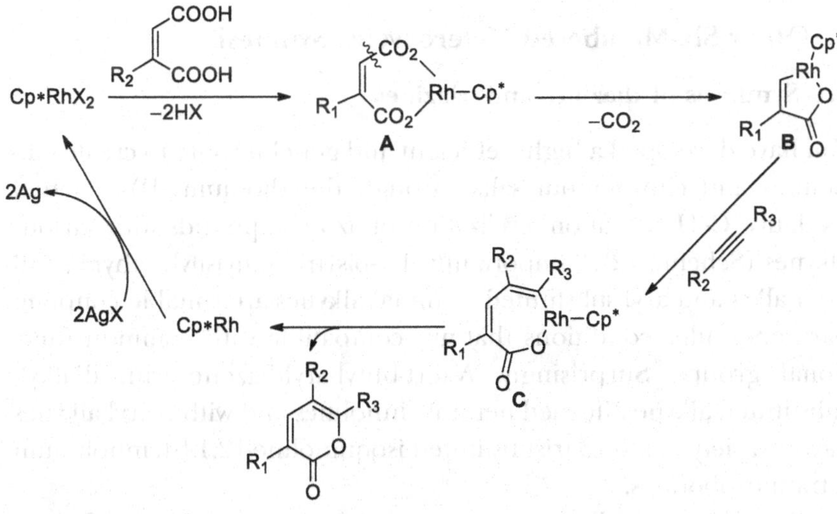

Scheme 70. Proposed catalytic cycle for the coupling reactions of maleic acids and alkynes

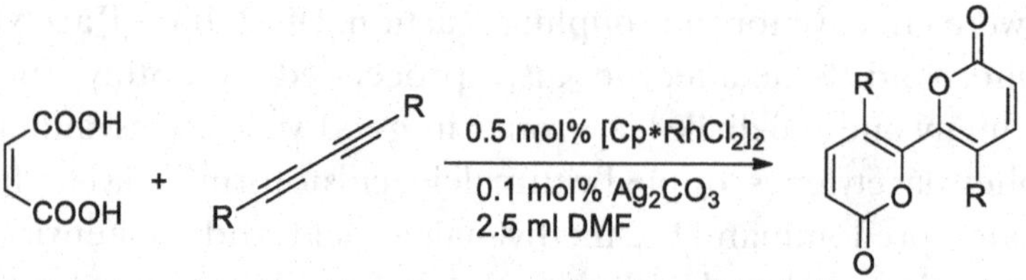

Scheme 71. Synthesis of isocoumarins from maleic acids and 1,3-diynes

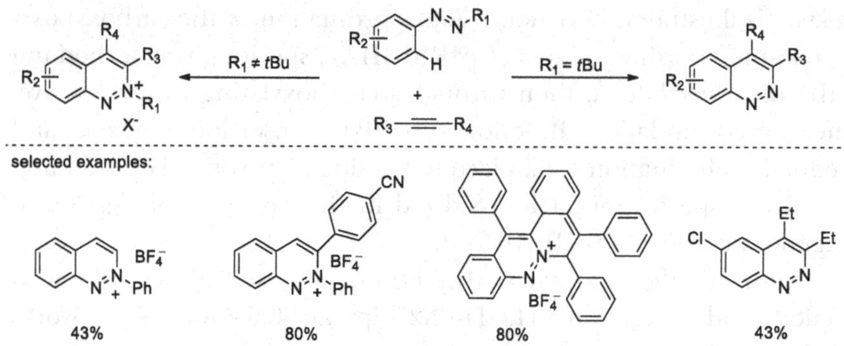

selected examples:

43% 80% 80% 43%

Scheme 72. Synthesis of cinnoline derivatives from azos and alkynes

6. Other Six-Membered Heterocycles Synthesis

6.1 Synthesis of diazines and triazines

You have developed a highly efficient and general route to create cinnolines and cinnolinium salts through the rhodium(III)-catalyzed oxidative C–H activation/cyclization of azo compounds with various alkynes (Scheme 72).[60] Importantly, 1,2-bis(trimethylsilyl)-ethyne, and even alkyl and aryl-substituted terminal alkynes are suitable coupling partners under conditions that are compatible with common functional groups. Surprisingly, N-tert-butyl-aryldiazene with dialkyl-substituted alkyne afforded neutral cinnolines and with diaryl alkynes gave a variety of 5,6,13-trisubstituted isoquinolino-[2,1-b]cinnolinium tetrafluoroborates.

Bergman and Ellman also reported formal [3 + 3] annulations of aromatic azides to give phenazines when azobenzenes were employed as the substrate. Various azobenzenes including

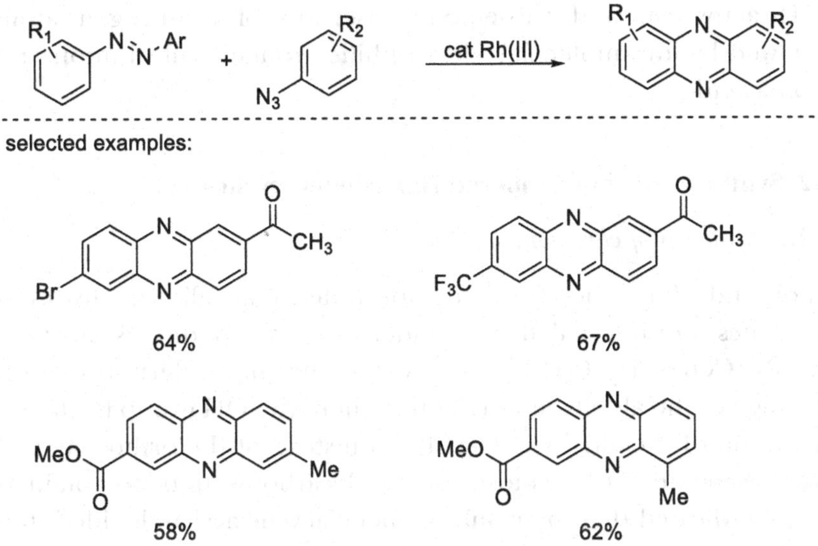

selected examples:

64%

67%

58%

62%

Scheme 73. Synthesis of phenazines from azobenzenes and azides

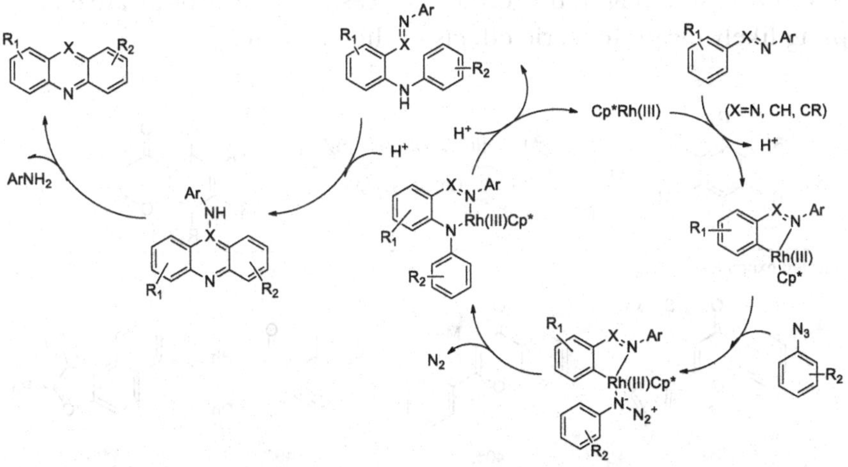

Scheme 74. Proposed catalytic cycle for the coupling reactions of azobenzenes and azides

unsymmetrical disubstituted derivatives can be introduced into ortho C–H amination (Scheme 73).[61] A mechanism that account for the catalytic transformations is proposed in Scheme 74. A five-membered rhodacycle was formed from imine chelation-assisted

C–H activation and subsequent insertion of a nitrogen atom followed by intramolecular electrophilic aromatic substitution and aromatization.

6.2 Synthesis of six-membered ring *o*-heterocyclics

6.2.1 *Synthesis of chromone product*

Satoh and Miura successfully demonstrated that salicylaldehydes as substrates with internal alkynes under oxidative conditions catalyzed by $[Rh(COD)Cl]_2/C_5H_2Ph_4$ to provide chromone derivatives with cleavage of the aldehyde C–H bond, where the OH group facilitates activation of the aldehyde C–H bond instead of the ortho Caryl–H bond (Scheme 75).[4b] In most cases no decarbonylation was obtained which indicated that the resulting metallacyclic acyl-aryloxide intermediate is resistant to any decarbonylation likely because of the chelation effect. In a unique example, a substituted benzofuran was found as a side reaction product as a result of decarbonylative coupling likely owing to steric effects of the aryl ring.

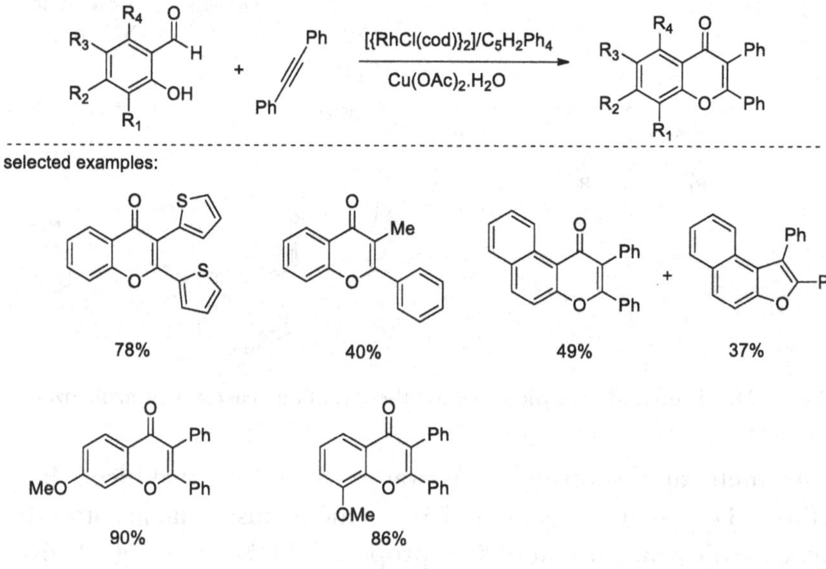

Scheme 75. Synthesis of chromones from salicylaldehydes and alkynes

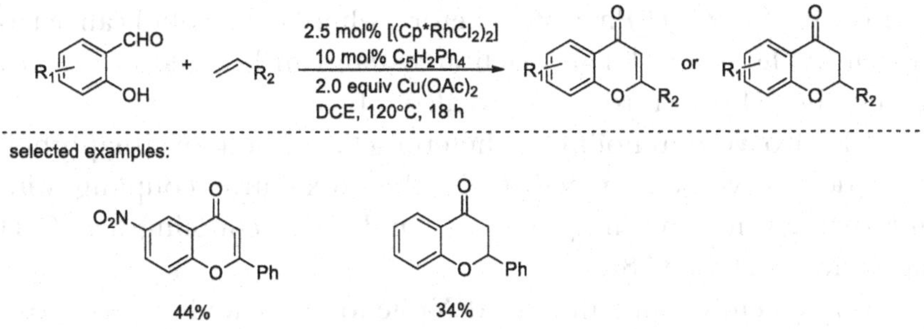

selected examples:

44%   34%

Scheme 76. Synthesis of chromones from salicylaldehydes and alkynes

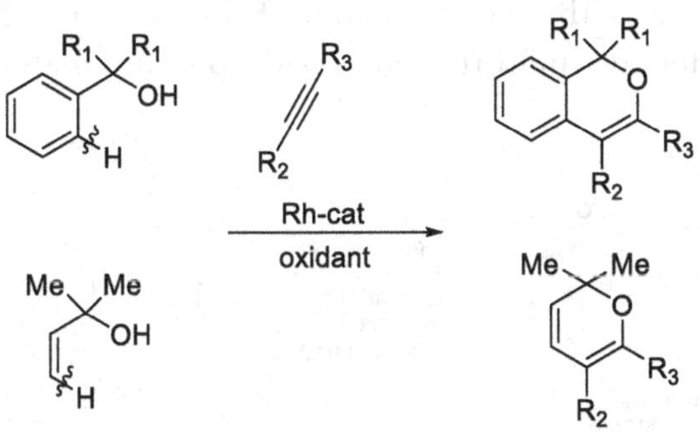

Scheme 77. Synthesis of isochromenes from tertiary alcohols with alkynes

Lately, Glorius reported a Rh(III)-catalyzed regioselective DHR of aldehyde C–H bonds with various classes of olefins (Scheme 76).[62] The most electron-withdrawing substrate, such as 5-nitrosalicylalde-hyde, selectively gave corresponding product in 44% yield. In addition, only flavanone as substrate was afforded from DHR of the aldehyde C–H bond and subsequent Michael addition when Ag_2CO_3 was used as the oxidant.

6.2.2 *Synthesis of ether product*

Satoh and Miura successfully demonstrated that the oxidative coupling of tertiary alcohols with alkynes can be performed in the presence of Rh(III) and a copper salt to provide the corresponding isochromene derivatives (Scheme 77),[63] which has been reported by

$[Rh(COD)Cl]_2/C_5H_2Ph_4$ system before, where the alcohol can act as a removable group.[64] The reaction system can be used to directly access the 2H-pyran from an allyl alcohol.

1-Hydroxylisoquinoline, a heterocyclic variant of 1-naphthol, may deliver versatile reactivity in their oxidative coupling with internal alkynes resulting in C–C and C–O coupling *via* C–H activation (Scheme 78).[65]

Very recently, Wang described *N*-phenoxyacetamide internal oxidants to the coupling reaction with cyclopropenes *via* a rhodium(III)-catalyzed C–H activation, which represents the first example of using cyclopropene as a three-carbon unit (Scheme 79).[66] The reaction proceeds under very mild reaction conditions and features very high

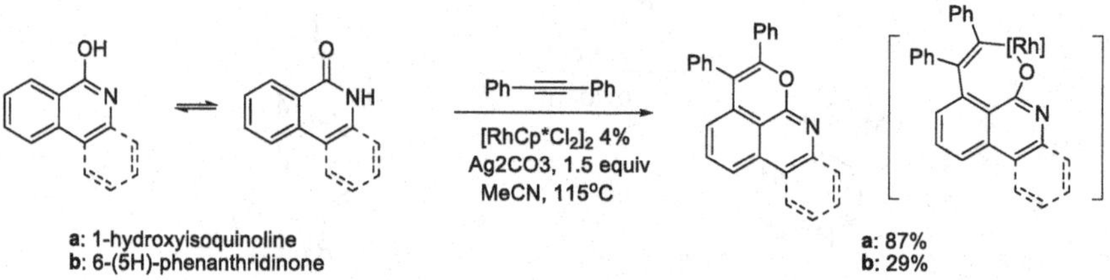

a: 1-hydroxyisoquinoline
b: 6-(5H)-phenanthridinone

a: 87%
b: 29%

Scheme 78. Oxidative coupling between 1-naphthols and alkynes

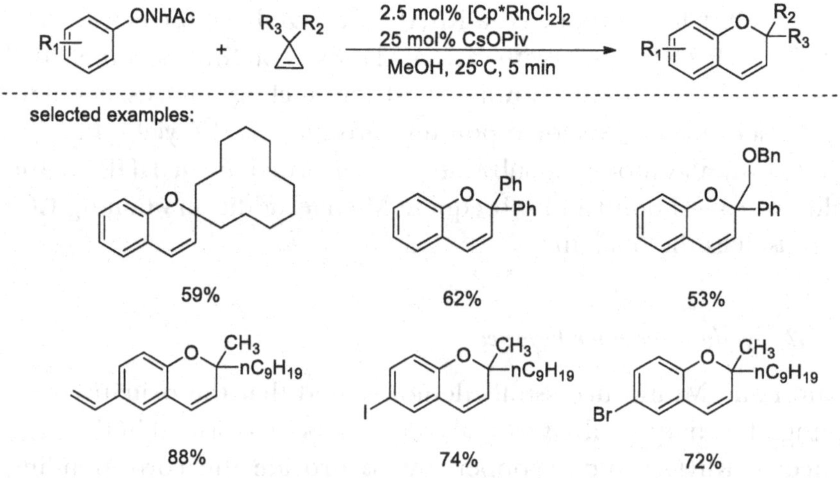

selected examples:

59%   62%   53%

88%   74%   72%

Scheme 79. Oxidative coupling between *N*-phenoxyacetamides and cyclopropenes

efficiency towards a wide range of substrates due to the highly strained ring system of cyclopropenes.

6.3 Synthesis of sultams product

Cramer successfully developed acylated sulfonamides as suitable directing groups for rhodium(III)-catalyzed C–H bond activations. The cyclometalated intermediates readily react with a broad range of internal alkynes and open a rapid and general access to aryl sultams (Scheme 80).[67]

The internal directing group competition of tosyl benzamide was carried out. A mixture of the quinolone and its detosylated congener were obtained, which confirms that sulfonamides have an attenuated directing group character which might be attributed to a different hybridization (sp_3-hybridized sulfur atom versus sp_2^- hybridized carbon atom) and a less favorable alignment for the cyclometalation (Scheme 81).

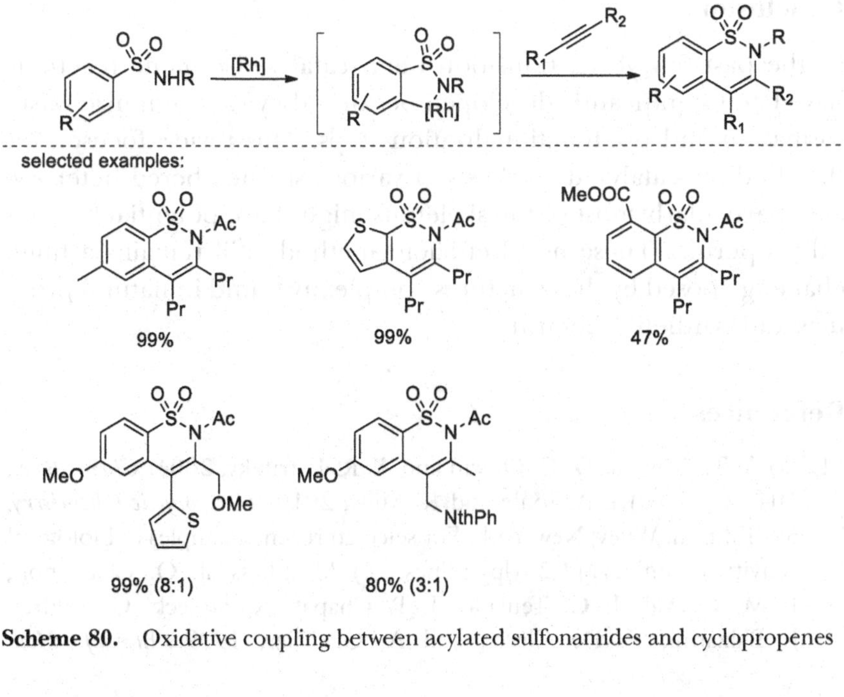

selected examples:

| 99% | 99% | 47% |

| 99% (8:1) | 80% (3:1) |

Scheme 80. Oxidative coupling between acylated sulfonamides and cyclopropenes

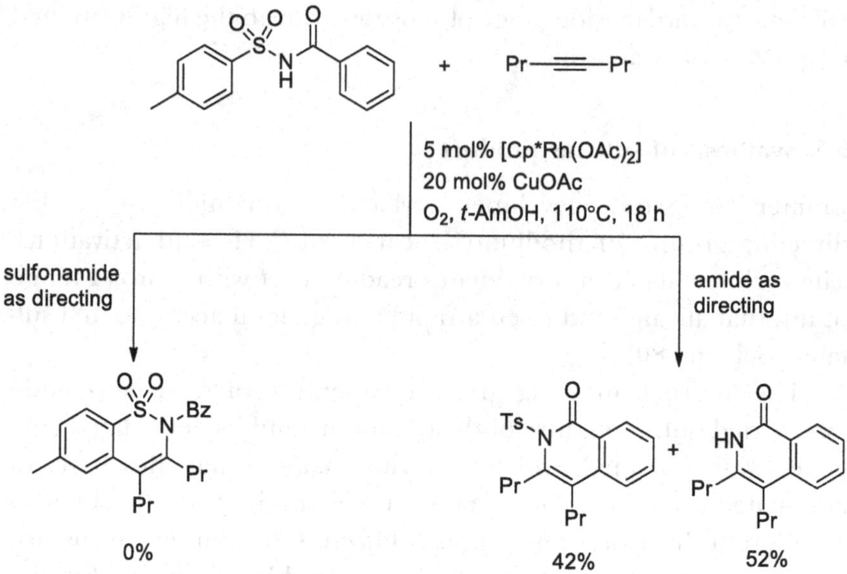

Scheme 81. Competition reactions

Conclusion

In the past few years, transition metal-catalyzed organic reactions have been significantly developed because they can minimize waste formation and pre-functionalization steps. This review focused on the rhodium-catalyzed synthesis of various six-membered heterocycles, particularly most of the skeletons might have potential biological properties. These novel efficient methods still remains a huge challenge posed by the structures complexity found in natural products and synthetic chemistry.

References

1. (a) A. T. Balaban, D. C. Oniciu and A. R. Katritzky, 2004. *Chem. Rev.*, 104, 2777. (b) J. A. Joule and K. Mills, 2010. *Heterocyclic Chemistry*, 5th Edition. Wiley, New York. For selected recent examples of biological activity of imidazo[1,2-α]pyridines: (c) M. Lhassani, O. Chavignon, J. M. Chezal, J. C. Teulade, J. P. Chapat, R. Snoeck, G. Andrei, J. Balzarini, E. De Clercq and A. Gueiffier, 1999. *Eur. J. Med.*

Chem., 34, 271. (d) K. C. Rupert, J. R. Henry, J. H. Dodd, S. A. Wadsworth, D. E. Cavender, G. C. Olini, B. Fahmy and J. Siekierka, 2003. *Bioorg. Med. Chem. Lett.*, 13, 347. (e) Y. Rival, G. Grassy and G. Michel, 1992. *Chem. Pharm. Bull.*, 40, 1170. (f) Y. Abe, H. Kayakiri, S. Satoh, T. Inoue, Y. Sawada, K. Imai, N. Inamura, M. Asano, C. Hatori, A. Katayama, T. Oku and H. Tanaka, 1998. *J. Med. Chem.*, 41, 564. For selected recent examples of biological activity of pyrido[1,2-α]benzimidazoles: (g) M. Hammad, A. Mequid, M. E. Ananni and N. Shafik, 1987. *Egypt. J. Chem.*, 29, 5401. (h) M. Hranjec, I. Piantanida, M. Kralj, L. Šuman, K. Pavelí and G. Karminski-Zamola, 2008. *J. Med. Chem.*, 51, 4899. (i) S. K. Kotovskaya, Z. M. Baskakova, V. N. Charushin, O. N. Chupakhin, E. F. Belanov, N. I. Bormotov, S. M. Balakhnin and O. A. Serova, 2005. *Pharm. Chem. J.*, 39, 574. (j) H. L. Koo and H. L. Dupont, 2010. *Curr. Opin. Gastroenterol.*, 26, 17. GSK812397: (k) K. Gudmundsson and S. D. Boggs, 2006. *PCT Int. Appl. WO 2006026703, CAN*, 144, 274. (l) S. Boggs, V. I. Elitzin, K. Gudmundsson, M. T. Martin and M. J. Sharp, 2009. *Org. Process Res. Dev.*, 13, 781.

2. (a) D. A. Colby, R. G. Bergman and J. A. Ellman, 2010. *Chem. Rev.*, 110, 624. (b) K. S. Masters, T. R. M. Rauws, A. K. Yadav, W. A. Herrebout, B. V. D. Veken and B. U. W. Maes, 2011. *Chem. Eur. J.*, 17, 6315. (c) H.-G. Wang, Y. Wang, C.-L. Peng, J.-C. Zhang and Q. Zhu, 2010. *J. Am. Chem. Soc.*, 132, 13217. (d) Y. Nakao, K.-S. Kanyiva, S. Oda and T. Hiyama, 2006. *J. Am. Chem. Soc.*, 128, 8146. (e) C. S. Yeung and V. M. Dong, 2011. *Chem. Rev.*, 111, 1215. (f) O. Daugulis, 2010. *Top. Curr. Chem.*, 292, 57. (g) L, Ackermann, 2010. *Chem. Commun.*, 46, 4866. (h) K. Fagnou, 2010. *Top. Curr. Chem.*, 292, 35.

3. (a) K. R. Roesch and R. C. J. Larock, 2002. *Org. Chem.*, 67, 86. (b) K. Godula and D. Sames, 2006. *Science*, 312, 67. (c) T. Uto, M. Shimizu, K. Ueura, H. Tsurugi, T. Satoh and M. J. Miura, 2008. *Org. Chem.*, 73, 298. (d) X. Chen, K. M. Engle, D.-H. Wang and J.-Q. Yu, 2009. *Angew. Chem., Int. Ed.*, 48, 5094. (e) L. Ackermann, R. Vicente and A. R. Kapdi, 2009. *Angew. Chem., Int. Ed.*, 48, 9792. (f) C.-L. Sun, B.-J. Li and Z.-J. Shi, 2010. *Chem. Commun.*, 110, 1147. (g) G.-Y. Song, F. Wang and X.-W. Li, 2012. *Chem. Soc. Rev.*, 41, 3651.

4. For selected examples on Rh(III) as catalyst: (a) G.-Y. Song, X. Gong and X.-W. Li, 2011. *J. Org. Chem.*, 76, 7583. (b) M. Shimizu, H. Tsurugi, T. Satoh and M. Miura, 2008. *Chem. Asian J.*, 3, 881.

(c) M. Imai, M. Tanaka, S. Nagumo, N. Kawahara and H. Suemune, 2007. *J. Org. Chem.*, 72, 2543. (d) R. T. Stemmler and C. Bolm, 2007. *Adv. Synth. Catal.*, 349, 1185. (e) S. Mochida, M. Shimizu, K. Hirano, T. Satoh and M. Miura, 2010. *Chem. Asian. J.*, 5, 847. (f) T. Uto, M. Shimizu, K. Ueura, H. Tsurugi, T. Satoh and M. Miura, 2008. *J. Org. Chem.*, 73, 298. (g) T. Fukutani, N. Umeda, K. Hirano, T. Satoh and M. Miura, 2009. *Chem. Commun.*, 5141. (h) D. A. Colby, R. G. Bergman and J. A. Ellman, 2008. *J. Am. Chem. Soc.*, 130, 3645. (i) L. Li, W. W. Brennessel and W. D. Jones, 2008. *J. Am. Chem. Soc.*, 130, 12414. (j) N. Guimond and K. Fagnou, 2009. *J. Am. Chem. Soc.*, 131, 12050. (k) D. R. Stuart, M. Bertrand-Laperle, K. M. N. Burgess and K. Fagnou, 2008. *J. Am. Chem. Soc.*, 130, 16474. (l) N. Umeda, H. Tsurugi, T. Satoh and M. Miura, 2008. *Angew. Chem., Int. Ed.*, 47, 4019. (m) K. Muralirajan, K. Parthasarathy and C.-H. Cheng, 2011. *Angew. Chem., Int. Ed.*, 50, 4169. (n) P. C. Too, Y.-F. Wang and S. Chiba, 2010. *Org. Lett.*, 12, 5688. (o) D. R. Stuart, P. Alsabeh, M. Kuhn and K. Fagnou, 2010. *J. Am. Chem. Soc.*, 132, 18326. (p) B.-J. Li, H.-Y. Wang, Q.-L. Zhu and Z.-J. Shi, 2012. *Angew. Chem., Int. Ed.*, 51, 3948. (q) N. Umeda, K. Hirano, T. Satoh, N. Shibata, H. Sato and M. Miura, 2011. *J. Org. Chem.*, 76, 13. (r) K. Ueura, T. Satoh and M. Miura, 2007. *Org. Lett.*, 9, 1407. (s) J.-R. Huang, L. Qin, Y.-Q. Zhu, Q. Song and L. Dong, 2015. *Chem. Commun.*, 51, 2844. (t) J.-R. Huang, Q. Song, Y.-Q. Zhu, L. Qin, Z.-Y. Qian and L. Dong, 2014. *Chem. Eur. J.*, 20, 16882. (u) L. Dong, C.-H. Qu, J.-R. Huang, W. Zhang, Q.-R. Zhang and J.-G. Deng, 2013. *Chem. Eur. J.*, 19, 16537. For selected examples on Rh(I) as catalyst: (v) J. C. Lewis, R. G. Bergman and J. A. Ellman, 2008. *Acc. Chem. Res.*, 41, 1013. (w) K.-L. Tan, R. G. Bergman and J. A. Ellman, 2002. *J. Am. Chem. Soc.*, 124, 3202. (x) S. H. Wiedemann, J. C. Lewis, J. A. Ellman and R. G. Bergman, 2006. *J. Am. Chem. Soc.*, 128, 2452.
5. (a) J.-R. Huang, Q.-R. Zhang, C.-H. Qu, X.-H. Sun, L. Dong and Y.-C. Chen, 2013. *Org. Lett.*, 15, 1878. (b) L. Dong, J.-R. Huang, C.-H. Qu, Q.-R. Zhang, W. Zhang, B. Han and C. Peng, 2013. *Org. Biomol. Chem.*, 11, 6142. (c) T. Satoh and M. Miura, 2010. *Chem.-Eur. J.*, 16, 11212. (d) D. A. Colby, A. S. Tsai, R. G. Bergman and J. A. Ellman, 2012. *Acc. Chem. Res.*, 45, 814. (d) F. W. Patureau, J. Wencel-Delord and F. Glorius, 2012. *Aldrichimica Acta*, 45, 31.
6. J. A. Varela and C. Saá, 2003. *Chem. Rev.*, 103, 3787.

7. (a) Y. Wakatsuki and H. Yamazaki, 1973. *Tetrahedron Lett.*, 3383.
(b) Y. Wakatsuki and H. Yamazaki, 1985. *Bull. Chem. Soc. Jpn.*, 58, 2715.
8. K. P. C. Vollhardt, 1984. *Angew. Chem.*, *Int. Ed. Engl.*, 23, 539.
9. H. Bönnemann, W. Brijoux, R. Brinkmann, W. Meurers and R. Mynott, 1984. *J. Organomet. Chem.*, 272, 231.
10. P. Diversi, L. Ermini, G. Ingrosso and A. Lucherini, 1993. *J. Organomet. Chem.*, 447, 291.
11. K. Tanaka, N, Suzuki and G. Nishida, 2006. *Eur. J. Org. Chem.*, 3917.
12. F. Xu, C. Wang, D. Wang, X. Li and B. Wang, 2013. *Chem. Eur. J.*, 19, 2252.
13. A. Saito, M. Hironaga, S. Oda and Y. Hanzawa, 2007. *Tetrahedron Lett.*, 48, 6852.
14. K. Parthasarathy, M. Jeganmohan and C.-H. Cheng, 2008. *Org. Lett.*, 10, 325.
15. R. M. Martin, R. G. Bergman and J. A. Ellman, 2012. *J. Org. Chem.*, 77, 2501.
16. J. R. Manning and H. M. L. Davies, 2008. *J. AM. Chem. Soc.*, 130, 8602.
17. J. R. Manning and H. M. L. Davies, 2008. *Tetrahedron*, 64, 6901.
18. T. K. Hyster and T. Rovis, 2011. *Chem. Commun.*, 47, 11846.
19. Q.-R. Zhang, J.-R. Huang, W. Zhang and L. Dong, 2014. *Org. Lett.*, 16, 684.
20. C.-Z. Luo, J. Jayakumar, P. Gandeepan, Y.-C. Wu and C.-H. Cheng, 2015. *Org. Lett.*, 17, 924.
21. X. Wei, M. Zhao, Z. Du and X. Li, 2011. *Org. Lett.*, 13, 4636.
22. J. Jayakumar, K. Parthasarathy and C.-H. Cheng, 2012. *Angew. Chem.*, *Int. Ed.*, 51, 197.
23. N. Senthilkumar, P. Gandeepan, J. Jayakumar and C.-H. Cheng, 2014. *Chem. Commun.*, 50, 3106.
24. X.-C. Huang, X.-H. Yang, R.-J. Song and J.-H. Li, 2014. *J. Org. Chem.*, 79, 1025.
25. Y. Tan, J. F. Hartwig, 2010. *J. Am. Chem. Soc.*, 132, 3676.
26. N. Guimond, C. Gouliaras and K. Fagnou, 2010. *J. Am. Chem. Soc.*, 132, 6908.
27. X. Zhang, D. Chen, M. Zhao, J. Zhao, A. Jia and X. Li, 2011. *Adv. Synth. Catal.*, 353, 719.
28. L. Zheng, J. Ju, Y. Bin and R. Hua, 2012. *J. Org. Chem.*, 77, 5794.

29. S.-C. Chuang, P. Gandeepan and C.-H. Cheng, 2013. *Org. Lett.*, 15, 5750.

30. W. Han, G. Zhang, G. Li and H. Huang, 2014. *Org. Lett.*, 16, 3532.

31. D. Zhao, F. Lieb and F. Glorius, 2014. *Chem. Sci.*, 5, 2869.

32. J. A. Bull, J. J. Mousseau, G. Pelletier and A. B. Charette, 2012. *Chem. Rev.*, 112, 2642.

33. Z. Shi, D. C. Koester, M. Boultadakis-Arapinis and F. Glorius, 2013. *J. Am. Chem. Soc.*, 135, 12204.

34. S.-G. Lim, J. H. Lee, C. W. Moon, J.-B. Hong and C.-H. Jun, 2003. *Org. Lett.*, 5, 2759.

35. K. Parthasarathy and C.-H. Cheng, 2009. *J. Org. Chem.*, 74, 9359.

36. Y.-F. Wang, K. K. Toh, J.-Y. Lee and S. Chiba, 2011. *Angew. Chem., Int. Ed.*, 50, 5927.

37. P. C. Too, S. H. Chua, S. H. Wong and S. Chiba, 2011. *J. Org. Chem.*, 76, 6159.

38. L. Zhang, L. Zheng, B. Guo and R. Hua, 2014. *J. Org. Chem.*, 79, 11541.

39. J.-L. Chen, G.-Y. Song, C.-L. Pan and X.-W. Li, 2010. *Org. Lett.*, 12, 5426.

40. S. Rakshit, C. Grohmann, T. Besset and F. Glorius, 2011. *J. Am. Chem. Soc.*, 133, 2350.

41. N. Guimond, S. I. Gorelsky and K. Fagnou, 2011. *J. Am. Chem. Soc.*, 133, 6449.

42. L. Xu, Q. Zhu, G.-P. Huang, B. Cheng and Y.-Z. Xia, 2012. *J. Org. Chem,*. 77, 3017.

43. H.-G. Wang and F. Glorius, 2012. *Angew. Chem., Int. Ed.*, 51, 7318.

44. (a) T. K. Hyster, L. Knörr, T. R. Ward and T. Rovis, 2012. *Science*, 338, 501. (b) B.-H. Ye and N. Cramer, 2012. *Science*, 338, 505.

45. S.-L. Cui, Y. Zhang and Q.-F. Wu, 2013. *Chem. Sci.*, 4, 3421.

46. M. Presset, D.l Oehlrich, F. Rombouts and G. A. Molander, 2013. *Org. Lett.*, 15, 1528.

47. T. A. Davis, T. K. Hyster and T. Rovis, 2013. *Angew. Chem., Int. Ed.*, 52, 14181.

48. J. R. Huckins, E. A. Bercot, O. R. Thiel, T.-L. Hwang and M. M. Bio, 2013. *J. Am. Chem. Soc.*, 135, 14492.

49. T. K. Hyster and T. Rovis, 2010. *J. Am. Chem. Soc.*, 132, 10565.

50. (a) G.-Y. Song, D. Chen, C.-L. Pan, R. H. Crabtree and X.-W. Li, 2010. *J. Org. Chem.*, 75, 7487. (b) S. Mochida, N. Umeda, K. Hirano, T. Satoh and M. Miura, 2010. *Chem. Lett.*, 39, 744.

51. Y. Su, M. Zhao, K.-L, Han, G.-Y. Song and X.-W. Li, 2010. *Org. Lett.*, 12, 5462.

52. F. Wang, G.-Y. Song, Z.-Y. Du and X.-W. Li, 2011. *J. Org. Chem.*, 76, 2926.

53. T. K. Hyster and T. Rovis, 2011. *Chem. Sci.*, 2, 1606.

54. X.-X. Xu, Y. Liu and C.-M. Park, 2012. *Angew. Chem., Int. Ed.*, 51, 9372.

55. H.-G. Wang, C. Grohmann, C. Nimphius and F. Glorius, 2012. *J. Am. Chem. Soc.*, 134, 19592.

56. K. Ueura, T. Satoh and M. Miura, 2007. *J. Org. Chem.*, 72, 5362.

57. S. Mochida, K. Hirano, T. Satoh and M. Miura, 2009. *J. Org. Chem.*, 74, 6295.

58. Q. Li, Y.-N. Yan, X.-W. Wang, B.-W. Gong, X.-B. Tang, J.-J. Shi, H. E. Xu and W. Yi, 2013. *RSC Adv.*, 3, 23402.

59. M. Itoh, M. Shimizu, K. Hirano, T. Satoh and M. Miura, 2013. *J. Org. Chem.*, 78, 11427.

60. D.-B. Zhao, Q. Wu, X.-L. Huang, F.-J. Song, T.-Y. Lv and J.-S. You, 2013. *Chem. Eur. J.*, 19, 6239.

61. Y.-J. Lian, J. R. Hummel, R. G. Bergman and J. A. Ellman, 2013. *J. Am. Chem. Soc.*, 135, 12548.

62. Z. Shi, N. Schröder, F. Glorius, 2012. *Angew. Chem., Int. Ed.*, 51, 8092.

63. K. Morimoto, K. Hirano, T. Satoh and M. Miura, 2011. *J. Org. Chem.*, 76, 9548.

64. T. Uto, M. Shimizu, K. Ueura, H. Tsurugi, T. Satoh and M. Miura, 2008. *J. Org. Chem.*, 73, 298.

65. G. Song, D. Chen, C.-L. Pan, R. H. Crabtree and X. Li, 2010. *J. Org. Chem.*, 75, 7487.

66. H. Zhang, K. Wang, B. Wang, H. Yi, F. Hu, C. Li, Y. Zhang and J. Wang, 2014. *Angew. Chem., Int. Ed.*, 53, 13234.

67. M. V. Pham, B.-H. Ye and N. Cramer, 2012. *Angew. Chem., Int. Ed.*, 51, 10610.

CHAPTER 5

Palladium-Catalyzed Synthesis
of Five-Membered Heterocycles

Hai-Tao Zhu, Zheng-Hui Guan

Northwest University, China
guanzhh@nwu.edu.cn

1. Introduction

Five-membered heterocycles are important synthetic targets as a result of their occurrence in numerous biologically active molecules, their important roles in diverse living processes, and their utility as versatile intermediates (Scheme 1).[1] As a consequence, numerous efforts are focused on the development of concise and efficient methods for the construction of these heterocycles. To date, transition metal-catalyzed cyclizations have proven to be powerful tool in heterocyclic chemistry. Compared with other transition metals (Au, Pt, Ru, Rh, etc.), palladium metal is probably the most versatile and widely used for the synthesis of five-membered heterocycles in organic chemistry.[2] Considering the importance of both the topics in organic synthesis, here we have summarized the main achievements in this field.

2. Palladium-Catalyzed Synthesis of Five-Membered Oxygen-Containing Heterocycles

2.1 Palladium-catalyzed reactions of alkenes

2.1.1 *Palladium-catalyzed tandem cyclizations of alkenes*

The intramolecular tandem cyclization of palladium π-olefin complexes is an effective method for the construction of five-membered

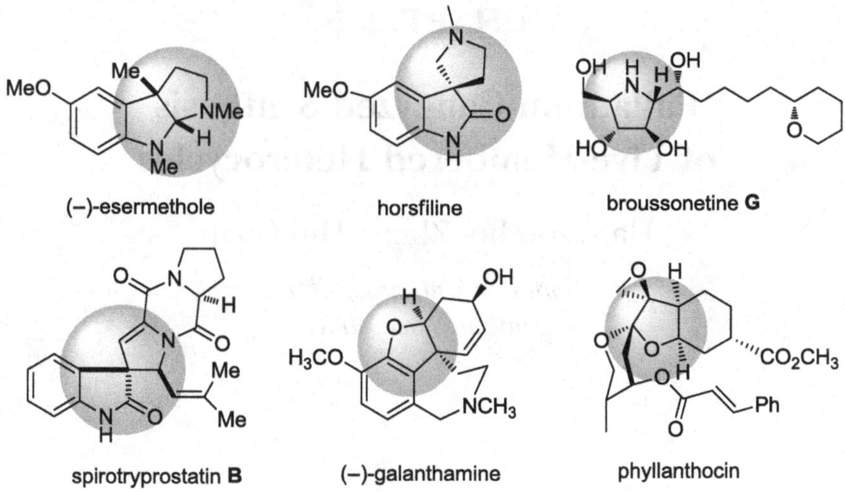
(−)-esermethole
horsfiline
broussonetine G
spirotryprostatin B
(−)-galanthamine
phyllanthocin

Scheme 1. Several important heterocyclic compounds

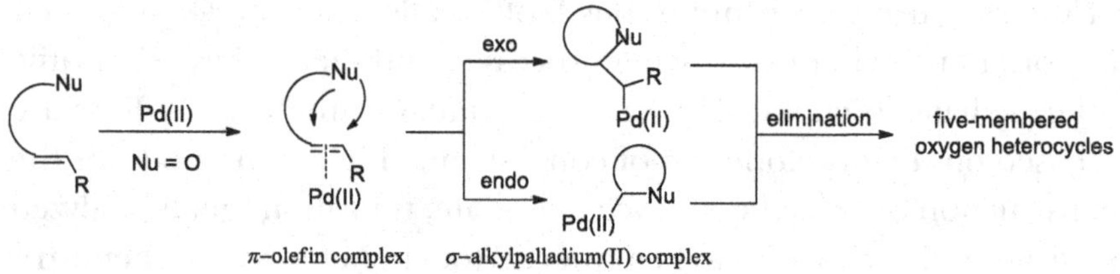
Pd(II)
Nu = O
exo
endo
elimination
five-membered oxygen heterocycles
π–olefin complex
σ–alkylpalladium(II) complex

Scheme 2. Plausible mechanism of intramolecular cyclization of palladium π-olefin complexes

oxygen heterocycles. This chemistry is normally believed to proceed through a stepwise mechanism including the complexation of the olefin with a Pd(II) salt, intramolecular nucleophilic attack on the π-olefin complexes to give a σ-alkylpalladium(II) species, β-hydride elimination of palladium (Scheme 2). In this process, generated Pd(0) needs to be oxidized to Pd(II) for the catalytic cycle. Oxidants commonly used are O_2, $CuCl_2$, benzoquinone, dimethyl sulfoxide (DMSO), and so on.

Oxygen nucleophiles at an appropriate distance from the C–C double bond can readily attack palladium π-olefin complexes to cyclize to a wide variety of five-membered oxygen heterocycles. In this cyclization, phenols, alcohols, ketones, and acids can be used as nucleophiles.

In the 1970s, Hosokawa and co-workers made a detailed study on the intramolecular oxypalladation of a variety of allylic phenols to synthesize benzofurans.[3] 2-Allylic phenols 1 underwent 5-exo-trig cyclization in the presence of stoichiometric or catalytic amounts of $PdCl_2(PhCN)_2$ to afford benzofurans 2 in moderate yields (Scheme 3).

Later, Uozumi and co-workers reported the enantioselective cyclization of 2-(2,3-dimethyl-2-butenyl)phenol 3 for the synthesis of optical 2,3-dihydro-benzofuran derivative (S)-4 (Scheme 4). Attribute

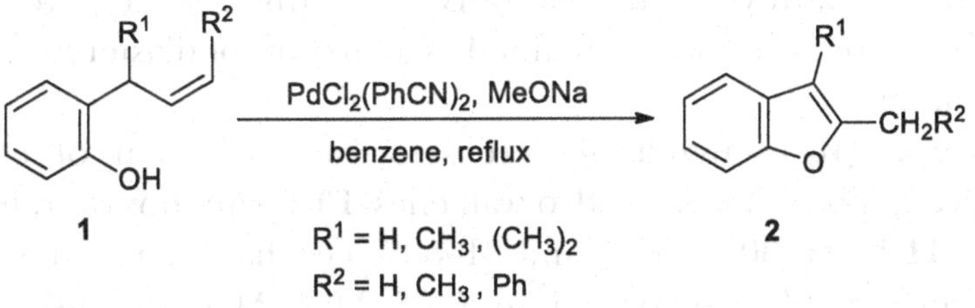

Scheme 3. Synthesis of benzofurans from allylic phenols

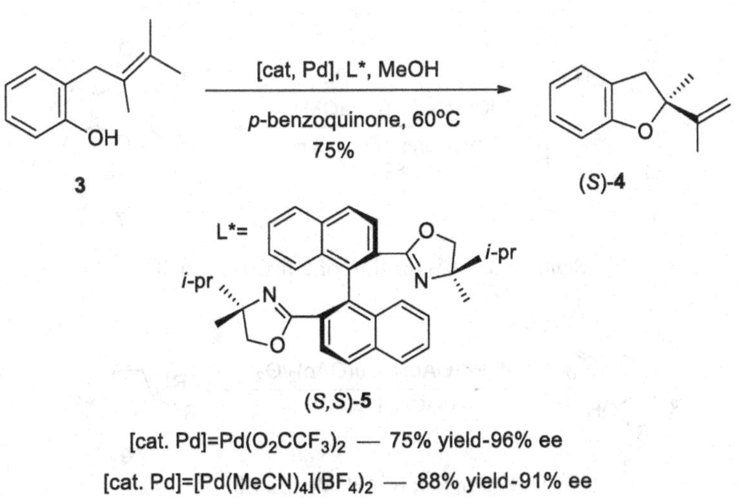

Scheme 4. Synthesis of optical 2,3-dihydrobenzofuran

to the coordination between the chiral ligand and $Pd(O_2CCF_3)_2$ catalyst, this system showed good yields and high enantioselectivity.[4a] Subsequently, they also obtained better results using 5 mol% $[Pd(MeCN)_4](BF_4)_2$.[4b]

The palladium(II)-catalyzed cyclization of alkenes with intramolecular alcohol group is a versatile way to produce furans, dihydrofurans and tetrahydrofurans. The cyclization of 5-hydroxyuracil **6** with a stoichiometric amount of $PdCl_2(PhCN)_2$ in benzene in the presence of NaOMe could afford furan derivative **7** in 82% yield (Scheme 5).[5]

The γ,δ-unsaturated alcohols **8** reacted with a catalytic amount of $Pd(OAc)_2$ plus catalytic amounts of $Cu(OAc)_2$ and O_2 as a reoxidant to generate 2-vinyltetrahydrofurans **9**. In this reaction, the five-membered products were obtained as a mixture of diastereoisomers (Scheme 6).[6]

The Pd(II)-catalyzed diastereoselective cyclization of enantiomerically pure (2S,3S)-hydroxyalkenes **10** to methoxytetrahydrofurans **11** in 71–80% yields and 81–88% ee's has been achieved in the presence of catalytic amounts of $PdCl_2(MeCN)_2$ and CuCl/ $CuCl_2$ under O_2 (Scheme 7).[7]

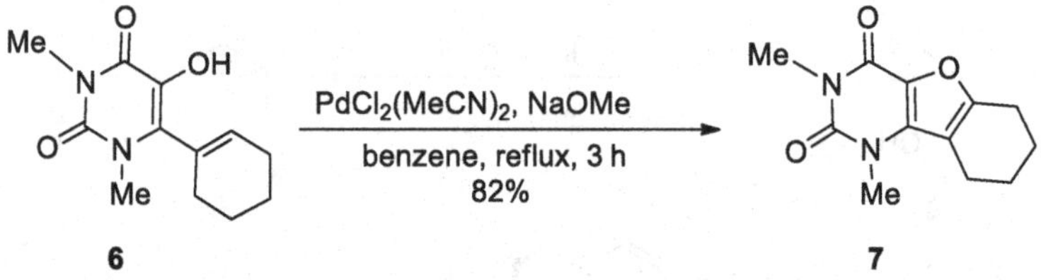

Scheme 5.   Cyclization of 5-hydroxyuracil

R^1=Ph, Me; R^2= H, Me, Et, Ph

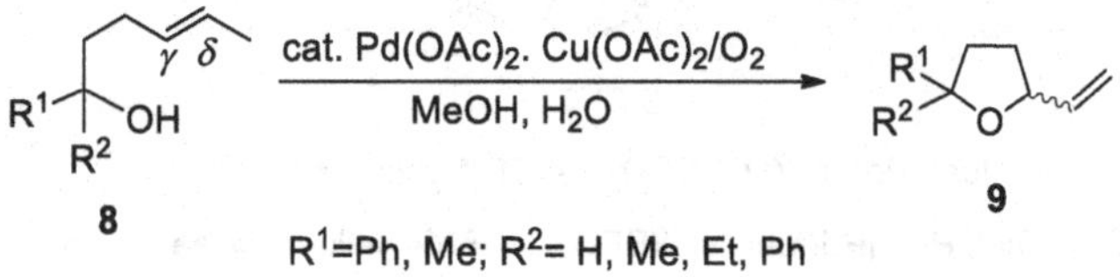

Scheme 6.   Cyclization of γ, δ-unsaturated alcohols

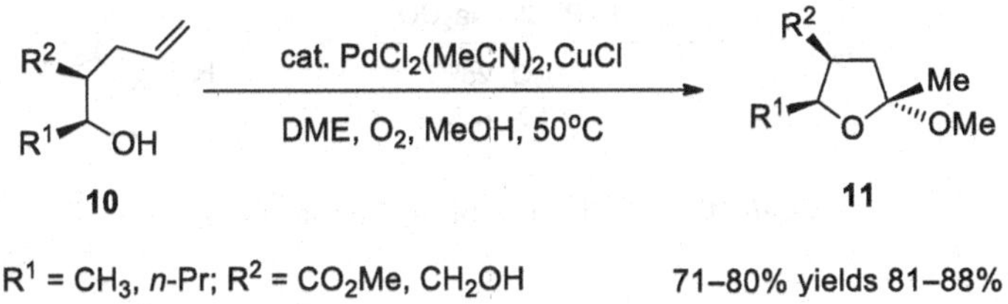
cat. PdCl2(MeCN)2,CuCl
DME, O2, MeOH, 50°C
10
11
R1 = CH3, n-Pr; R2 = CO2Me, CH2OH
71–80% yields 81–88% ees'

Scheme 7. Diastereoselective cyclization of (2S,3S)-hydroxyalkenes

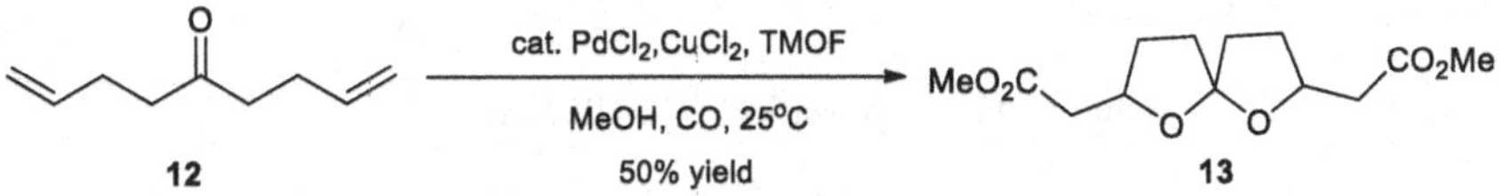
cat. PdCl2,CuCl2, TMOF
MeOH, CO, 25°C
50% yield
12
13

Scheme 8. Cyclization of dienones

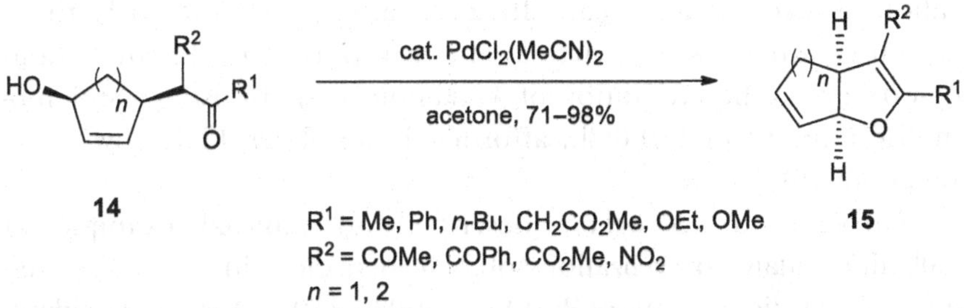
cat. PdCl2(MeCN)2
acetone, 71–98%
14
15
R1 = Me, Ph, n-Bu, CH2CO2Me, OEt, OMe
R2 = COMe, COPh, CO2Me, NO2
n = 1, 2

Scheme 9. Cyclization of cis-1,4-disubstituted cyclopentenones

The palladium (II)-catalyzed intramolecular C–O bond formation of alkenes with an oxygen atom from the carbonyl group of an alkenone as a nucleophile can also give five-membered oxygen heterocycles. The $PdCl_2$-catalyzed double cyclization of dienones **12** in the presence of $CuCl_2$, CO, methanol, and trimethylorthoformate (TMOF) afforded symmetric spiroacetals **13** in moderate yields (Scheme 8).[8]

In addition, the bicyclic dihydrofurans **15** are prepared from *cis*-1,4-disubstituted cyclopentenones **14** in 71–98% yields using $PdCl_2(MeCN)_2$ as a catalyst. This process was stepwise, which involved cyclic oxypalladation and palladium hydroxide elimination (Scheme 9).[9]

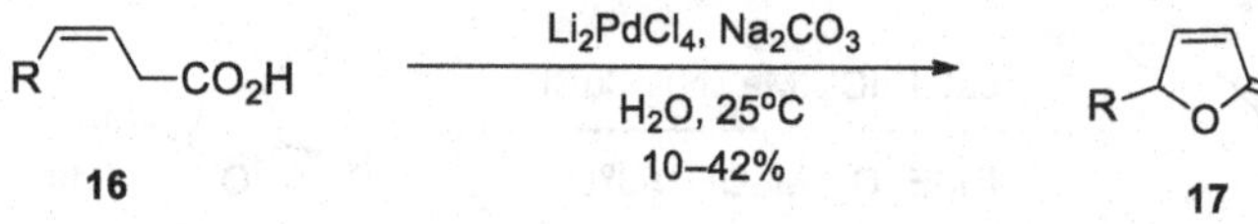

R⌒⌒CO$_2$H $\xrightarrow[\substack{H_2O,\ 25^\circ C \\ 10\text{–}42\%}]{Li_2PdCl_4,\ Na_2CO_3}$ R⌒O⌒O

16   **17**

Scheme 10.   Cyclization of 3-alkenoic acids

18 — $\xrightarrow[\substack{DMSO,\ O_2 \\ 78\%}]{cat.\ Pd(OAc)_2,\ NaOAc}$ — 19

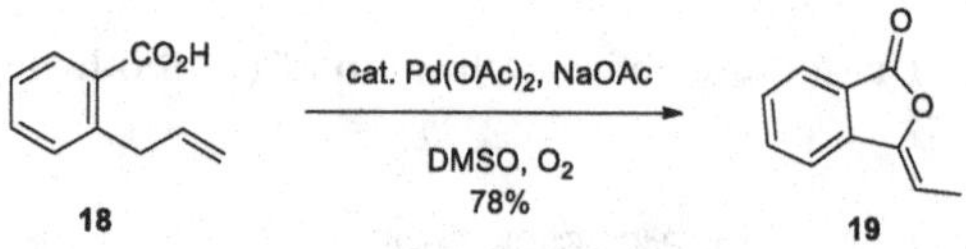

Scheme 11.   Cyclization of 2-allylbenzoic acid

Lactones, which occur widely in nature, play an important role in biological activities molecules. The palladium (II)-catalyzed cyclization of alkenoic acids can efficiently afford unsaturated lactones in one synthetic step. Early attempts reported by Kasahara and co-workers to the cyclization of 3-alkenoic acids **16** using a stoichiometric amount of Li$_2$PdCl$_4$, afforded butenolides **17** in low yields (Scheme 10).[10]

Larock and co-workers described an unusual example of palladium-catalyzed cyclizations of 2-allylbenzoic acid **18** in the presence of catalytic amounts of Pd(OAc)$_2$ and O$_2$/DMSO as the reoxidant. The 5-exo-trig cyclization product **19** was obtained involving subsequent double-bond isomerization into conjugation (Scheme 11).[11]

2.1.2 *Palladium-catalyzed intramolecular allylic alkylations of alkenes*

Lactones are common subunits found in a variety of natural products and biologically active compounds.[12] They can also be used as building blocks in organic synthesis. Thus their synthesis has received a tremendous amount of attention from the synthetic community and many strategies have been developed to them. Intramolecular Pd-catalyzed allylic substitution reactions which involve the attack of carbanions at an allylic metal intermediate have proved to be a powerful strategy to generate five-membered heterocycles such as γ-lactones and are conveniently functional group tolerant, alternative to the

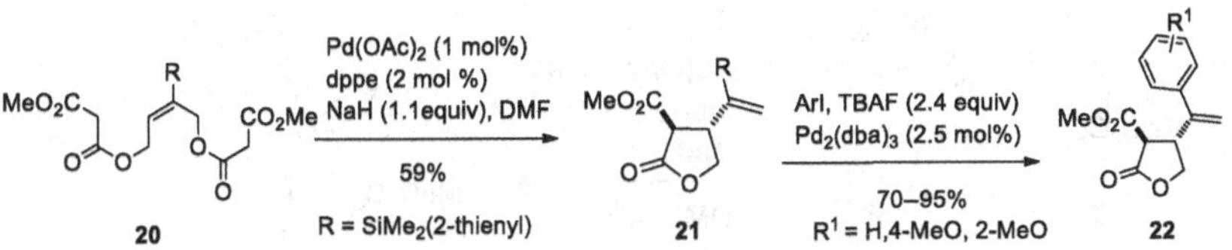

Scheme 12. Cyclization of the unsaturated esters

classical lactonization reaction.[13] In addition, much attention to Pd-catalyzed intramolecular asymmetric allylic alkylation reaction has been paid.

Pd-catalyzed intramolecular allylation reactions wherein the nucleophilic and the electrophilic partners are tethered by an ester moiety and can readily generate butyrolactones. Many investigations reported by Poli and co-workers to the cyclization of the unsaturated ester **20** bearing trialkylsilyl substituents in the presence of $Pd(OAc)_2$, 1,2-bis(diphenylphosphino)ethane (DPPE) as a ligand and NaH as a base, afforded the expected γ-lactones **21** with single *trans* diastereoselectivity.[14] In this cyclization, the trialkylsilyl substituents served as important roles to force displacement of the acetoxy group vicinal to the silicon atom. Furthemore, the trialkylsilyl-substituted lactones could be derived with various aryl iodides to produce the expected 4-(α-styryl) γ-lactones **22** in good yields through the palladium-catalyzed coupling reaction (Scheme 12).

Subsequently, they successfully applied the Pd-catalyzed intramolecular allylation of esters to synthesize two analogs of picropodophyllin, which showed good cytotoxic activity. The palladium-catalyzed two-step allylation/Hiyama coupling sequence involving ester **23** and iodide **24** afforded the desired styryl γ-lactone **26**. Further functional transformation including a key aldolization and an electrophilic aromatic substitution, gave the two target analogs (Scheme 13).[15]

2.1.3 *Palladium-catalyzed cyclocarbonylations of alkenes*

Palladium-catalyzed alkoxycarbonylative intramolecular cyclization of allylic alcohols has become a powerful tool for the synthesis of five-membered lactones.

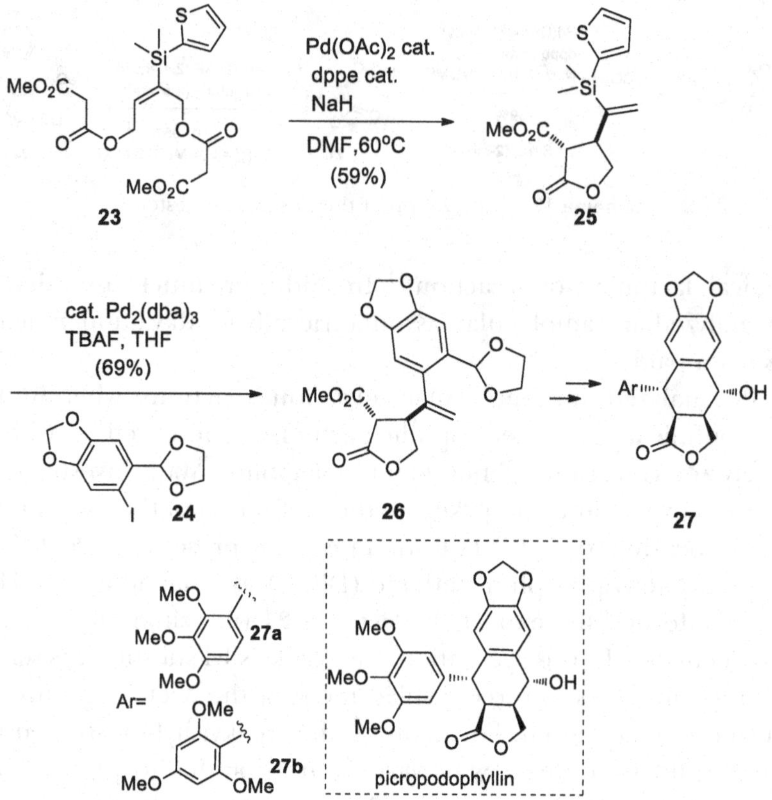

Scheme 13.   Synthesis of two analogs of picropodophyllin

In 1984, Semmelhack and co-workers developed an effective and concise method for the constructure of polycyclic lactones *via* a palladium-catalyzed tandem carbonylation (Scheme 14).[16] Interestingly, this reaction proceeded in a stereoselective manner.

Subsequently, Alper and Leonard described a palladium-catalyzed carbonylation of alkenols into five-membered lactones.[17] By carrying out the carbonylation in the presence of (-)-BPPM (2S,4S)-tert-butyl 4-(diphenylphosphino)-2-((diphenylphosphino)methyl)pyrrolidine-1-carboxylate, the reaction proceeded in an enantioselective manner (Scheme 15).

In 1999, Zhang and co-workers developed the first highly enantioselective cyclocarbonylation of β,γ-substituted allylic alcohols lacking

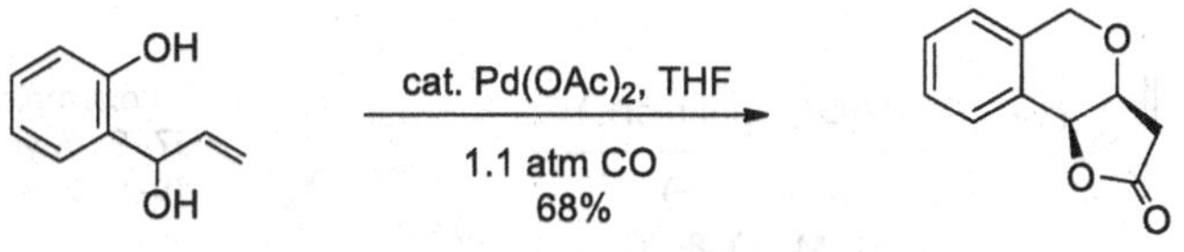

Scheme 14. Synthesis of polycyclic lactones

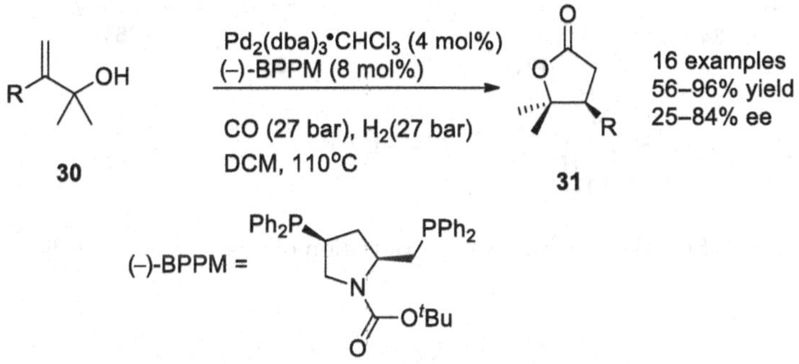

Scheme 15. Enantioselective synthesis of five-membered lactones

dialkyl substituents at the α-position.[18] The so-called BICP ligand played a significant role on this palladium-catalyzed asymmetric carbonylation (Scheme 16).

Tamaru and co-workers reported the dialkoxycarbonylation of 3-butenols **36**, to provide γ-butyrolactone-2-acetic acid esters **37** in moderate yields under 1 bar of CO (Scheme 17a). In this catalyst system, 1–50 mol% of $PdCl_2$ and 3 equiv of $CuCl_2$ were needed, and also propylene oxide and ethyl orthoacetate were required as additives.[19] Interestingly, the palladium-catalyzed decarboxylative carbonylation of 3-vinyl-1-oxo-2,6- dioxacyclohexanes **38** to 2-vinyl-γ-butyrolactones **39** was reported by the same group (Scheme 17b).[20]

At the same period, Yoshida and co-workers demonstrated that the carbonylation reaction of diols **40** with $PdCl_2$ as the catalyst and $CuCl_2$ as the reoxidant could readily synthesize *cis*-3-Hydroxytetrahydrofuran acetic acid lactones **41** in good yields under 1 bar of CO at room temperature (Scheme 18).[21]

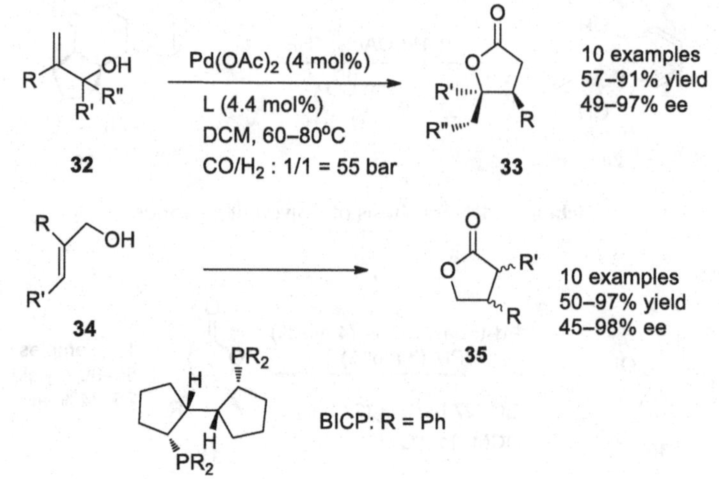

Scheme 16. Enantioselective cyclocarbonylation of β,γ-substituted allylic alcohols

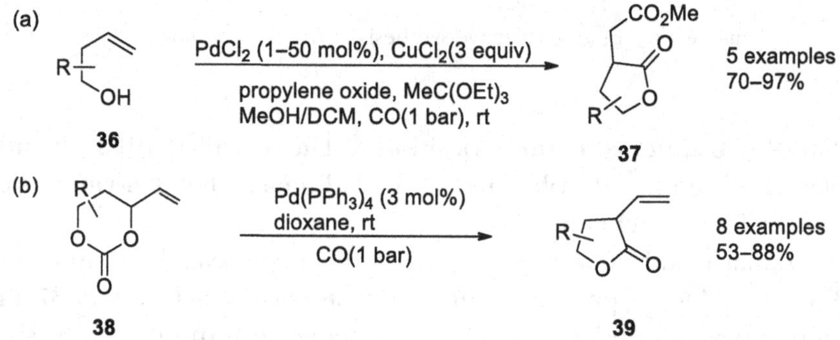

Scheme 17. Palladium-catalyzed dialkoxycarbonylation and decarboxylative carbonylation

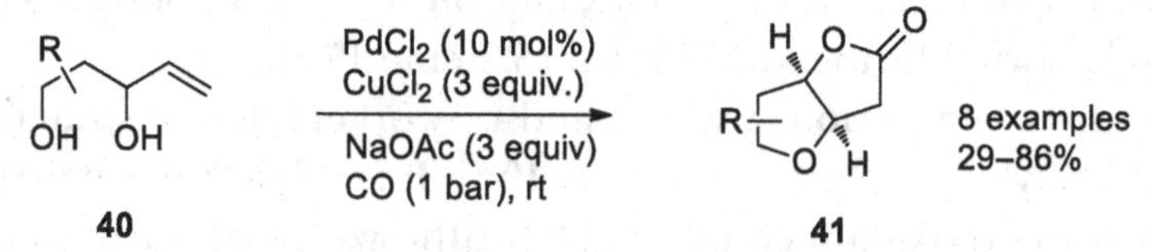

Scheme 18. Carbonylation of diols

2.1.4 *Palladium-catalyzed intramolecular heck reactions of alkenes*

Due to their diverse and efficient catalytic performance, palladium-catalyzed coupling reactions have become a powerful tool to forge new C–C or C–heteroatom bonds for rapid access to complex and useful heterocycles. The importance of palladium catalysis was underlined by the 2010 Nobel Price to R. Heck for his pioneering works in this field. In light of the importance and usefulness of intramolecular Heck reaction, several main achievements on the constructure of five-membered heterocycles in this area were summarized.

The accepted mechanism of the Heck reaction involves oxidative addition, carbon-palladium insertion, and β-hydride elimination.

Larock has reported early, an intramolecular Heck cyclization. The reaction of *o*-iodoaryl allyl ethers **42** in the presence of catalytic amounts of $Pd(OAc)_2$, Na_2CO_3, HCO_2Na, and n-Bu_4NCl afforded benzofurans **43** in reasonably good isolated yields under mild conditions (Scheme 19).[22] To improve the overall yields of benzofurans, HCO_2Na should be added. The role of HCO_2Na presumably reduced any π-allylpalladium intermediates formed by carbon–oxygen insertion back to $Pd(0)$, then to reenter the desired catalytic cycle.

In 2003, Lautens and co-workers reported an intramolecular Heck reaction with dihydronaphthalene substrates in the presence of $Pd_2(dba)_3/HP(t\text{-}Bu)_3 \bullet BF_4$ and DABCO for the synthesis of five-membered oxygen-containing fused cycles. For dibromo substrates **44**, two step Heck reactions are accessible by adding an external olefin to afford more highly functionalized polycycles **46** (Scheme 20).[23]

The enantioselective intramolecular Heck reactions have emerged as one of the most powerful methods for the formation of oxygen heterocycles. Early example of the asymmetric construction

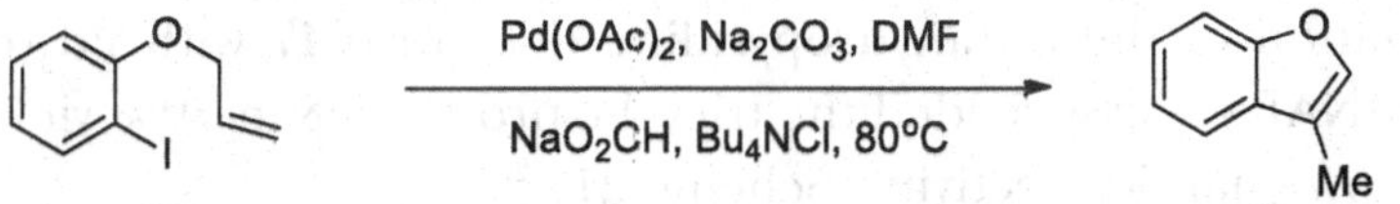

Scheme 19. Intramolecular Heck cyclization of o-iodoaryl allyl ethers

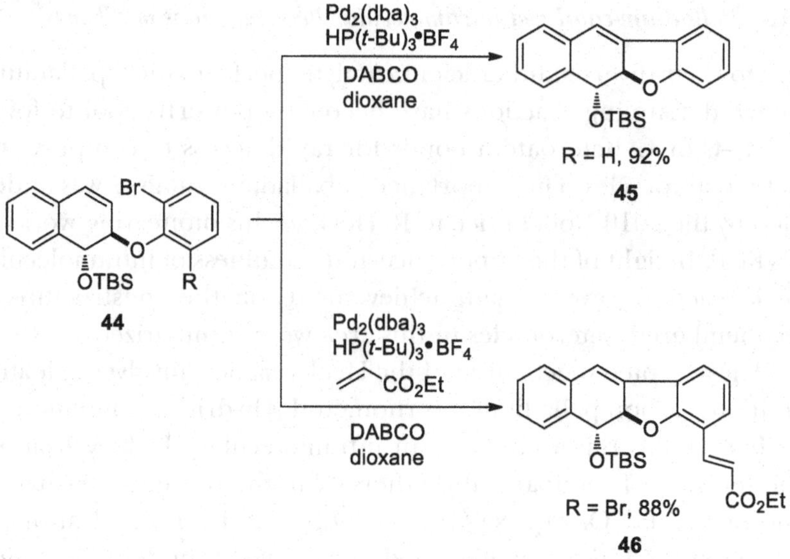

Scheme 20. Synthesis of five membered oxygen-containing fused cycles

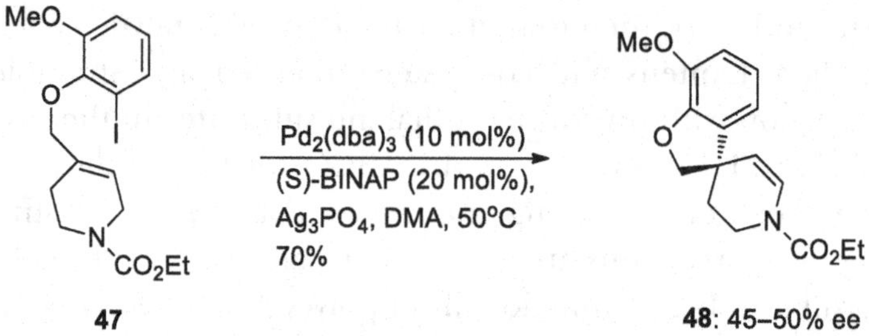

Scheme 21. Cyclization of the tetrahydropyridines

of a spirodihydrobenzofuran framework by an enantioselective Heck reaction has been described by Cheng and co-workers. The cyclization of the tetrahydropyridine derivative **47** with $Pd_2(dba)_3$-(S)-BINAP catalyst yielded the tricyclic product **48** in 70% yield, but with low enantioselectivity (Scheme 21).[24]

Diaz and co-workers found a tandem asymmetric Heck cyclization-hydride capture reaction. Upon reaction with $Pd(OAc)_2$, (R)-BINAP,

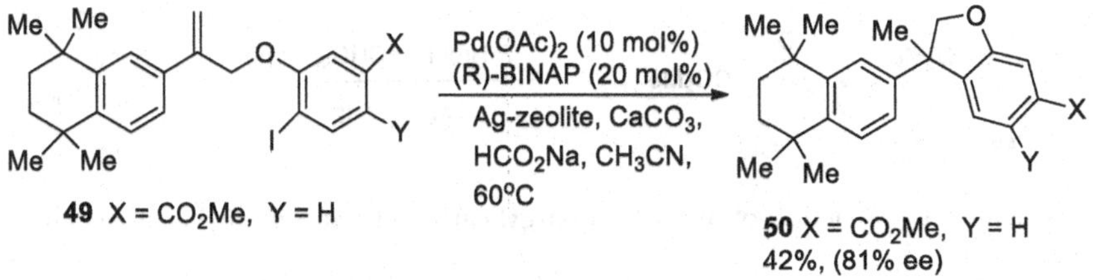
Me
Me
Me
Me
O
X
I
Y
49 X = CO2Me, Y = H
Pd(OAc)2 (10 mol%)
(R)-BINAP (20 mol%)
Ag-zeolite, CaCO3,
HCO2Na, CH3CN,
60°C
Me
Me
Me
O
Me
Me
X
Y
50 X = CO2Me, Y = H
42%, (81% ee)

Scheme 22. Tandem asymmetric Heck cyclization hydride capture reaction

Ag-exchanged zeolite, $CaCO_3$, and sodium formate in acetonitrile, aryl iodide **49** produced dihydrobenzofuran **50** in moderate yield and good enantioselectivity (Scheme 22).[25]

2.2 Palladium-catalyzed reactions of alkynes

2.2.1 *Palladium-catalyzed tandem cyclizations of alkynes*

The intramolecular tandem cyclization of palladium π-alkyne complexes could effectively yield five-membered oxygen heterocycles. For example, the first Pd(II)-catalyzed cyclization of alkynes was reported by Utimoto. Using 2-methoxy-3-alkyn-1-ols **51** as subtrates, the expected furans **52** were obtained in good yields (Scheme 23).[26]

Subsequently, Tsuji and co-workers developed a palladium-catalyzed cyclization of propargyl carbonates with soft carbonucleophiles under neutral conditions. Reaction of methyl propargyl carbonate with methyl acetoacetate in the presence of $Pd_2(dba)_3$•$CHCl_3$/dppe catalyst gave 4,5-dihydrofuran in good yield (Scheme 24).[27]

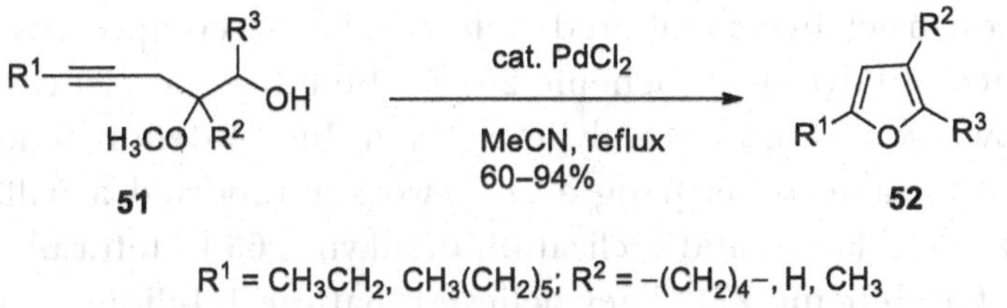
R1
R3
OH
H3CO R2
51
cat. PdCl2
MeCN, reflux
60–94%
R2
R1
O
R3
52
R1 = CH3CH2, CH3(CH2)5; R2 = –(CH2)4–, H, CH3

Scheme 23. The first Pd(II)-catalyzed cyclization of alkynes

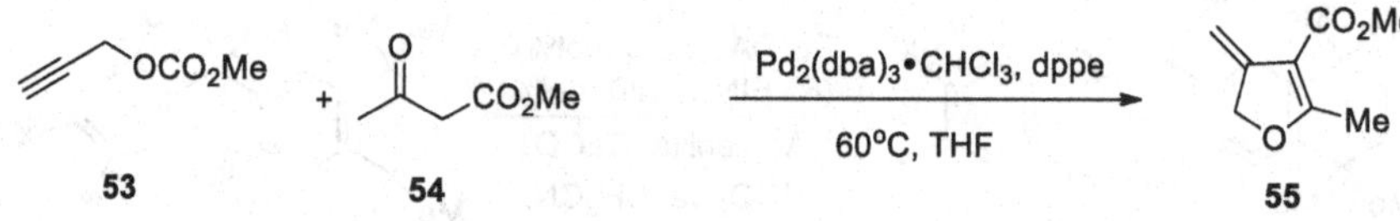

Scheme 24. Reaction of methyl propargyl carbonate with methyl acetoacetate

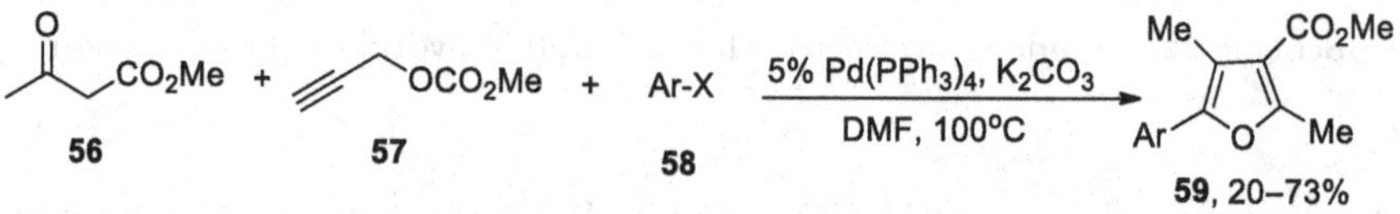

Scheme 25. Pd(0)-catalyzed one-pot three-component cyclization reaction

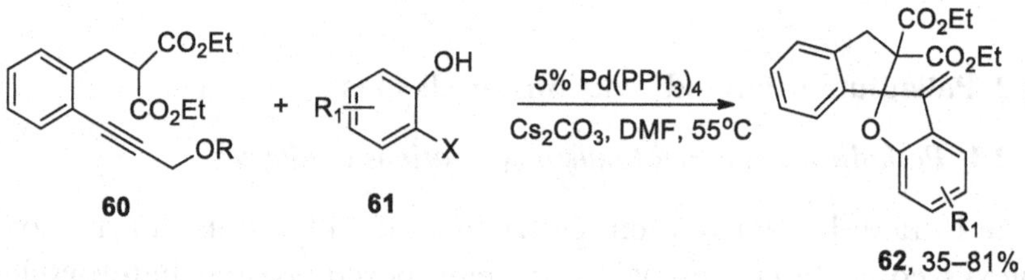

Scheme 26. Reaction of the allenes generated *in situ* propargylic compounds

Based on the above researches, Liang and co-workers reported an efficient one-pot three-component cyclization reaction catalyzed by palladium(0). The coupling- cyclization reactions of β-keto ester **56**, primary propargylic carbonate **57**, and aryl halides **58** presented various tetrasubstituted furans **59** with excellent regioselectivities in the presence of Pd(PPh$_3$)$_4$ as catalyst and K$_2$CO$_3$ as base (Scheme 25).[28]

Recently, they investigated the reaction of the allenes generated *in situ* propargylic compounds **60** with 2-halophenols **61** to successfully construct five-membered heterocyclic **62** compounds in the presence of Pd(PPh$_3$)$_4$ (Scheme 26).[29] The Pd(0) catalyst is simultaneously involved in two catalytic cycles in this tandem reaction.

At the same time, Jiang and co-worker reported a palladium-catalyzed oxidation and cyclization of alkynes **63** to tetrasubstituted furans **65** (Scheme 27). They believed that the 1,4-dione formed in the presence of Pd(OAc)$_2$ and O$_2$ was a key intermediate **64**.[30]

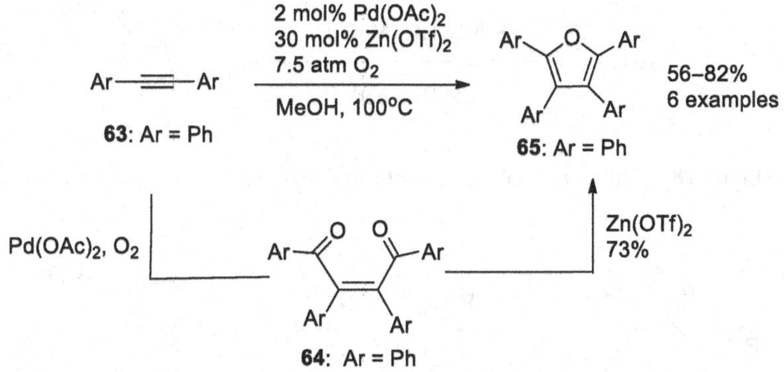

Scheme 27. Palladium-catalyzed oxidation and cyclization of alkynes

2.2.2 *Palladium-catalyzed cyclocarbonylations of alkynes*

Early examples on the palladium-catalyzed cyclocarbonylation of alkynes bearing an oxygen nucleophile were reported by Nogi and Tsuji. The propargyl alcohols **66** were subjected to the carbonylative condition in methanol to yield the corresponding five-membered γ-lactones **67** (Scheme 28).[31]

Changing the catalysts for carbonylation, Kato and co-workers described the preparation of cyclic β-alkoxyacrylates **69** in good yields from simple alkynols **68**. The best catalyst system was $PdCl_2(MeCN)_2/$ p-benzoquinone in methanol at 0°C under a CO atmosphere (Scheme 29).[32]

An asymmetric palladium-catalyzed carbonylation of alkynes were described by the same group. The cyclic alkynediol **70** reacted with a Pd(II) catalyst and chiral bisoxazoline to afford bicyclic alkoxyacrylate **71** in good yield with moderate enantioselectivity (Scheme 30).[33]

2.3 Palladium-catalyzed reactions of allenes

2.3.1 *Palladium-catalyzed tandem cyclizations of allenes*

Because of the unique structure with a 1,2-diene moiety, allenes show distinctive reactivity and serve as potential precursors for the synthesis of highly complex heterocycles in organic synthesis. In 1993, Walkup

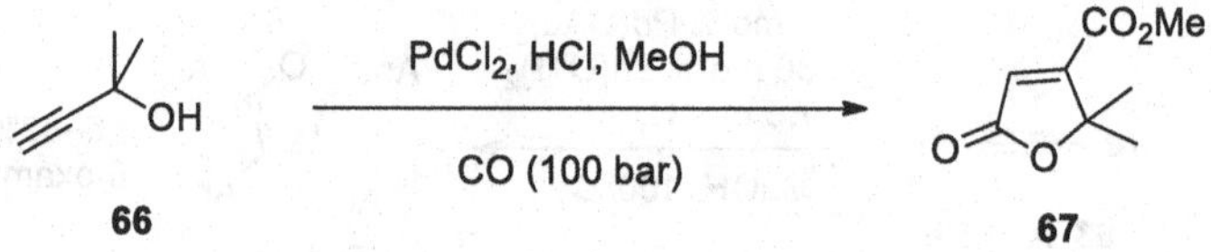

Scheme 28. Palladium-catalyzed cyclocarbonylation of propargyl alcohols

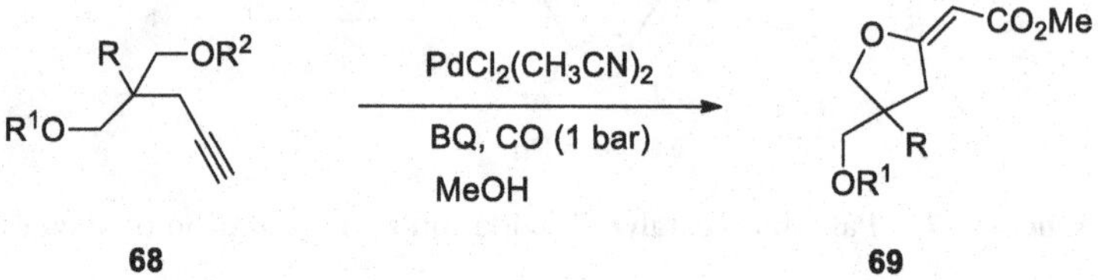

Scheme 29. Synthesis of cyclic β-alkoxyacrylates

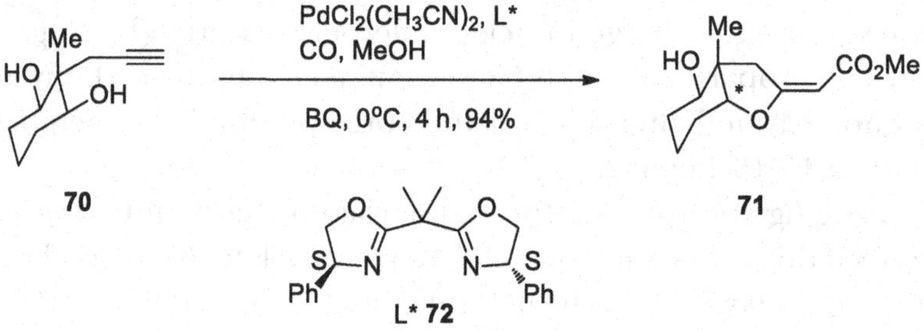

Scheme 30. Asymmetric palladium-catalyzed carbonylation of alkynes

and co-workers developed the cyclization-coupling reactions of γ-hydroxyallenes **72** and **74**, giving alkenyl-substituted tetrahydrofurans **73** and 2-furanones **75**, respectively, in the presence of Pd(PPh$_3$)$_4$ and K$_2$CO$_3$ as base. This transition showed moderate yields and modest *cis/trans* selectivity. Aryl or alkenylpalladium(II) halides are believed to be the key intermediates (Scheme 31).[34]

Later, a Pd-catalyzed cyclization reaction of 2,3-allenoic acids **76** with allylic halides **77** was reported by Ma and co-workers, affording β-allyl polysubstituted butenolides **78** in moderate to excellent yields (Scheme 32).[35]

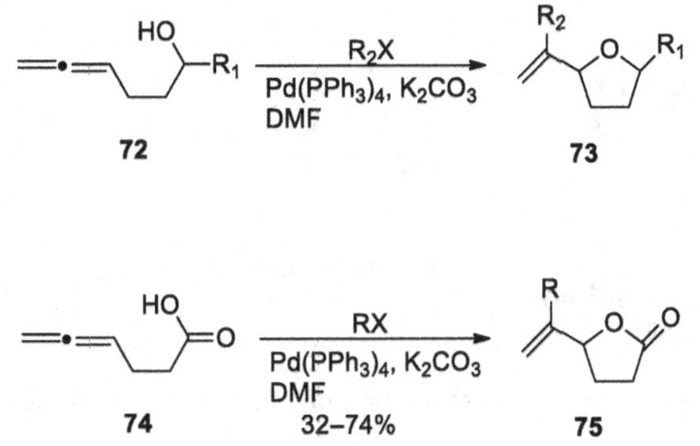

R = Ph, p-MeOC$_6$H$_4$, p-MeC$_6$H$_4$, 1-and 2-Naphthyl,
PhHC=CH, p-HalC$_6$H$_4$
X = Br or I

Scheme 31. Cyclization-coupling reactions of γ-hydroxyallenes

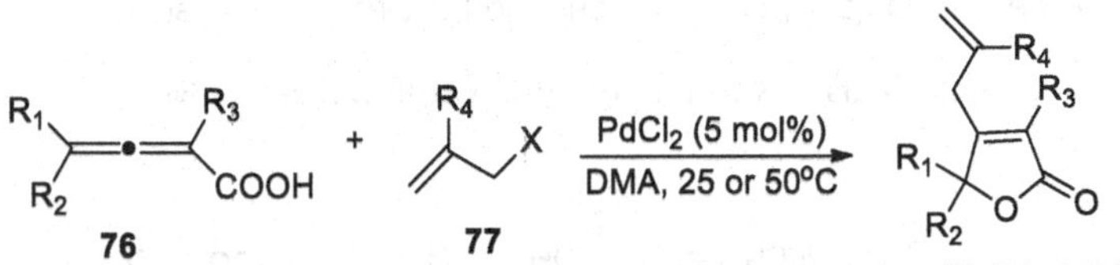

78 (52–95%)

R$_1$ = aryl, alkyl
R$_2$ = H, alkyl
R$_3$ = alkyl

R$_4$ = H, Ph, Me, Br

X = Cl, Br

Scheme 32. Pd-catalyzed cyclization of 2,3-allenoic acids with allylic halides

Using $PdCl_2$ as catalyst, 2,3-allenols **79** reacted with allylic bromides to give polysubstituted 2,5-dihydrofurans **80** having a modifiable allylic group in good yields (Scheme 33).[36]

2.3.2 *Palladium-catalyzed cyclocarbonylations of allenes*

In the mid-1980s, Walkup and Park found that the Pd-catalyzed carbonylative cyclization of allenes bearing a hydroxyl or silyl group could synthesize tetrahydrofuran derivatives **82** and **83** (Scheme 34).[37]

Using $PdCl_2$, $CuCl_2$, and CO, a mixture of *cis* and *trans* isomers was formed by a 5-exo-trig cyclization. Subsequently, when 4,5-hexadienal

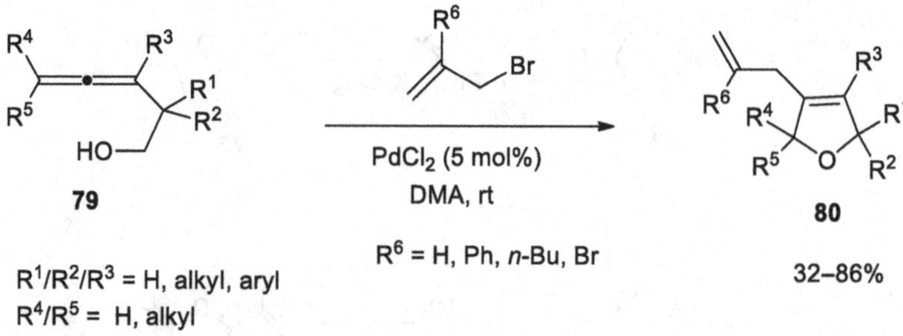

Scheme 33.   Pd-catalyzed cyclization of 2,3-allenols with allylic bromides

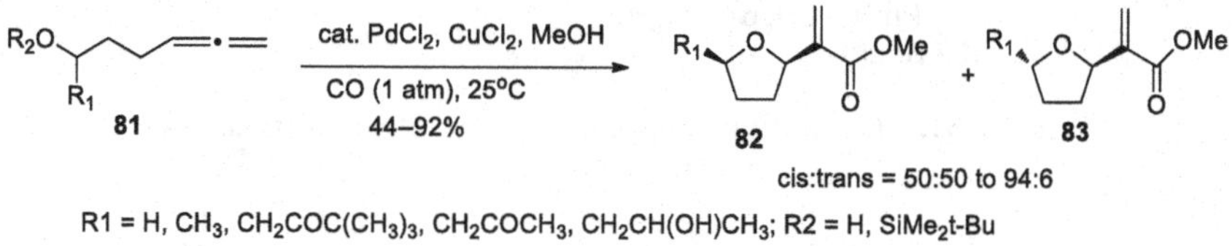

$R1 = H, CH_3, CH_2COC(CH_3)_3, CH_2COCH_3, CH_2CH(OH)CH_3; R2 = H, SiMe_2t-Bu$

Scheme 34.   Synthesis of tetrahydrofuran derivatives

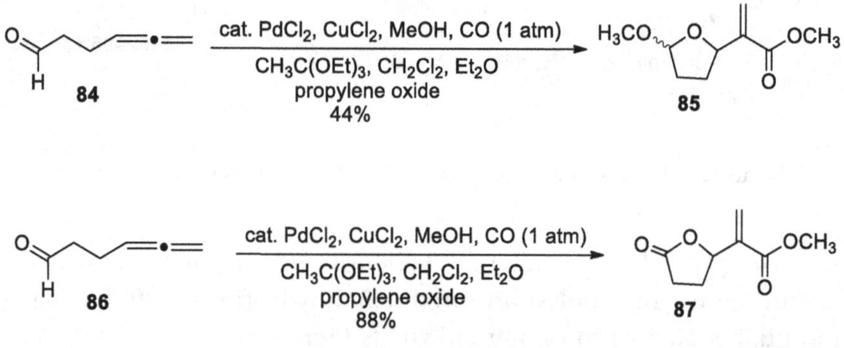

Scheme 35.   Intramolecular carbonylative oxypalladation for the synthesis of furans and furanones

84 and 4,5-hexadienoic acid 86 were subjected to the reaction conditions of intramolecular carbonylative oxypalladation, furan derivative 85 and furanone 87 were also formed in good yields (Scheme 35).[38] In these reactions, propylene oxide served as an acid trap and triethyl orthoacetate was used as a water scavenger.

3. Palladium-Catalyzed Synthesis of Five-Membered Nitrogen-Containing Heterocycles

3.1 Palladium-catalyzed reactions of alkenes

3.1.1 Palladium-catalyzed tandem cyclizations of alkenes

The intramolecular palladium-catalyzed tandem cyclization of nitrogen nucleophiles with alkenes has proven to be an effective method for the synthesis of five-membered nitrogen heterocyclic compounds. The Pd-catalyzed intramolecular C–N bond formation was first studied by Hegedus and co-workers. They found that in the presence of a stoichiometric amount of Pd, the cyclization of *o*-allylic anilines **88** could afford 2-methylindoles **89** in good yields under mild conditions (Scheme 36).[39] This reaction tolerated various functional groups on the benzene ring.

Interestingly, using oxime esters as substrates, Narasaka and co-workers developed a Pd-catalyzed cyclizations for the synthesis of pyrroles. They found that the reaction of O-pentafluorobenzoyl oxime **90** showed better result than the one of O-sulfoxime **91** using triethylamine as base and Pd(OAc)$_2$ as catalyst after heating (Scheme 37). The high yield of pyrrole was considered to be duo to the high reactivity of O-pentafluorobenzoyl oxime in palladium-catalyzed cyclization and the promotion of trimethylsilyl chloride.[40]

Various tosylamides have been widely employed to produce five-membered nitrogen heterocycles under standard catalytic Pd cyclization conditions.

Early examples on the cyclization of olefinic tosylamides **93** were reported by Hegedus and McKearin. Using PdCl$_2$(CH$_3$CN)$_2$ as catalyst,

$$R \underset{NH_2}{\overset{}{\bigcirc\!\!\!-\!\!\!\bigcirc}}\!\!\diagup\!\!\diagdown \quad \xrightarrow[\text{2) Et}_3\text{N}]{\text{1) PdCl}_2\text{(MeCN)}_2\text{, THF}} \quad R \overset{}{\bigcirc\!\!\!-\!\!\!\bigcirc}\!\!\overset{}{\underset{H}{N}}\!\!-\text{Me}$$

88   74–84%   **89**

R = H, 5-CO$_2$Et, 6-OMe

Scheme 36. Cyclization of o-allylic anilines

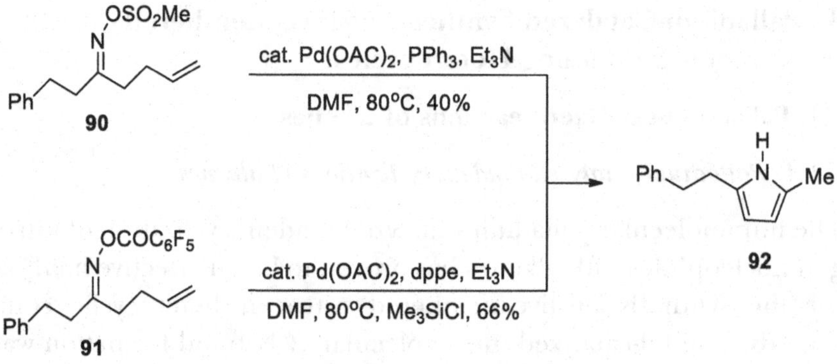

Scheme 37.   Pd-catalyzed cyclization of o-pentafluorobenzoyl oximes

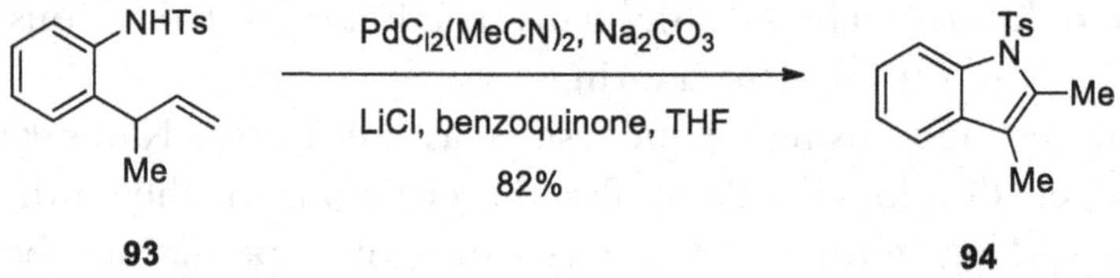

Scheme 38.   Pd π-olefin cyclization of olefinic tosylamides

indoles **94** were formed exclusively through a Pd π-olefin cyclization (Scheme 38).[41]

Larock and co-worker reported that in the presence of $Pd(OAc)_2$ plus DMSO and molecular oxygen as a reoxidant, the cyclization of acyclic and cyclic olefinic tosylamides (**95** and **96**) could give the corresponding five-membered ring products containing an allylic nitrogen moiety (Scheme 39). This process was believed to be proceeding by way of a π-allylpalladium intermediate rather than Pd π-olefin chemistry.[42]

Recently, Guan and co-worker reported a $Cu(TFA)_2$-catalyzed oxidative tandem cyclization/1,2-alkyl migration of enamino amides **99** for the synthesis of pyrrolin-4-ones **100** (Scheme 40).[43] The reaction showed the excellent tolerance of functional groups. Subsequently, they developed a palladium-catalyzed intramolecular oxidative cyclization reaction. The tertiary enamines **101** in the presence of $Pd(OAc)_2$ or $PdCl_2$, $Cu(OAc)_2$ as oxidant and trifluoroacetic acid could be

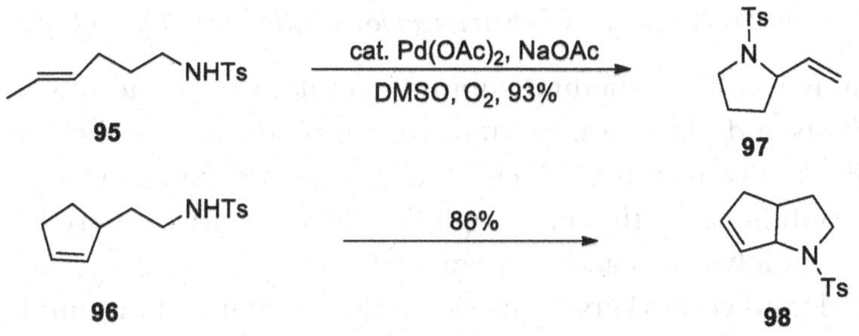

Scheme 39. π-Allylpalladium cyclization of olefinic tosylamides

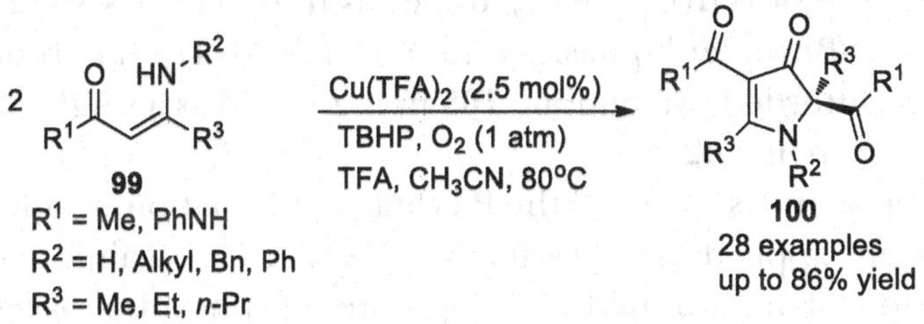

Scheme 40. Tandem cyclization of enamino amides

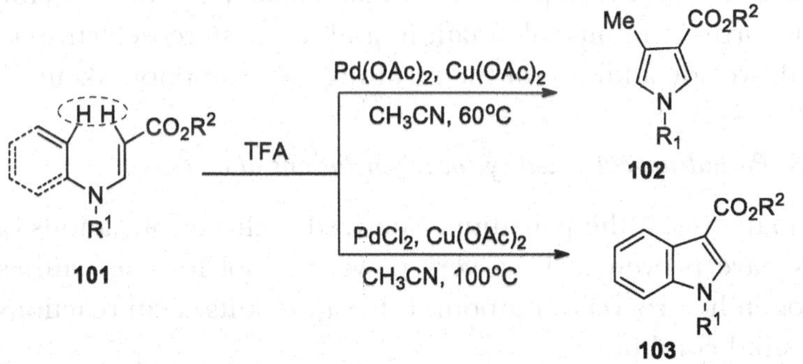

Scheme 41. Palladium-catalyzed intramolecular oxidative cyclization of tertiary enamines

cyclized to a variety of substituted 1,3,4-trisubstituted pyrroles **102** and 1,3-disubstituted indoles **103**. Palladium-catalyzed vinyl C–H activation of a tertiary enamine sequence of intramolecular oxidative cyclization was achieved by employing trifluoroacetic acid as an additive (Scheme 41).[44]

3.1.2 Palladium-catalyzed intramolecular allylic alkylations of alkenes

Lactams is a key structural subunit prevalent in numerous natural products and plays an important role in pharmaceutical chemistry.

Pd-catalyzed intramolecular allylation reactions wherein the nucleophilic and the electrophilic partners are tethered by an amide moiety can readily generate γ-lactams.

Poli and co-workers developed a Pd(0)-catalyzed intramolecular allylic alkylations for the synthesis of pyrrolidone derivatives under mild conditions. The cyclization of the unsaturated amide **104**, in the presence of $[Pd(C_3H_5)Cl]_2/dppe]$ as the catalytic system, KOH, and n-Bu$_4$NBr under biphasic conditions (DCM/H$_2$O) at room temperature, afforded pyrrolidone **105** in high yield as a single diastereoisomer (Scheme 42).[45]

Later on, the strategy of the Pd-catalyzed allylation of amides was successfully applied to the formal synthesis of (±)-α-kainic acid. The cyclization of unsaturated starting material, phosphonoacetamide **106**, quantitatively presented the *trans* pyrrolidone intermediate **107** under the optimal conditions (Scheme 43).[46] Further functionalization to yield the advanced precursor of kainic acid, included a Horner–Wadsworth–Emmons olefination and a diastereoselective conjugated hydride addition of the resulting electron-poor alkene.

3.1.3 Palladium-catalyzed cyclocarbonylations of alkenes

For many years, the palladium-catalyzed cyclocarbonylations of alkenes have proven to be a very powerful tool for the synthesis of nitrogen heterocyclic compounds because of its clean reactions and very mild conditions.

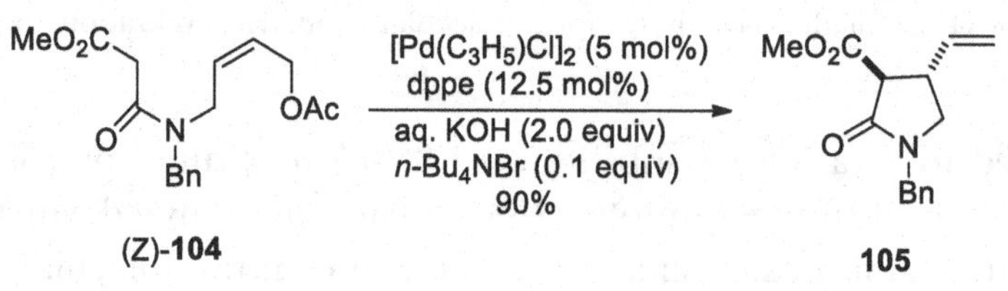

Scheme 42. Pd(0)-catalyzed intramolecular allylic alkylations of unsaturated amide

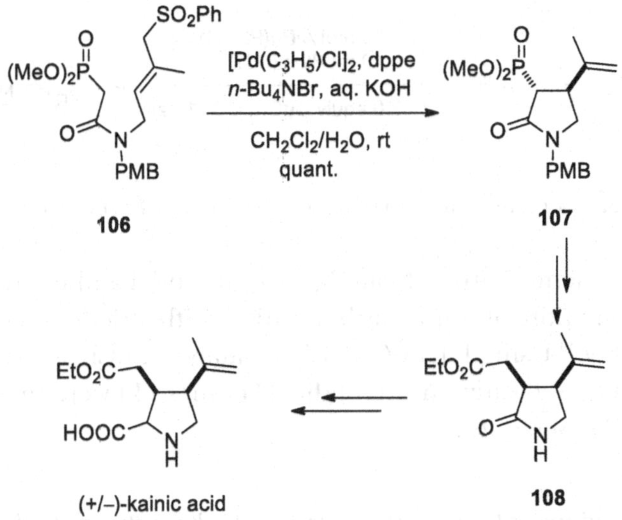

Scheme 43. Application of Pd-catalyzed allylation of amides

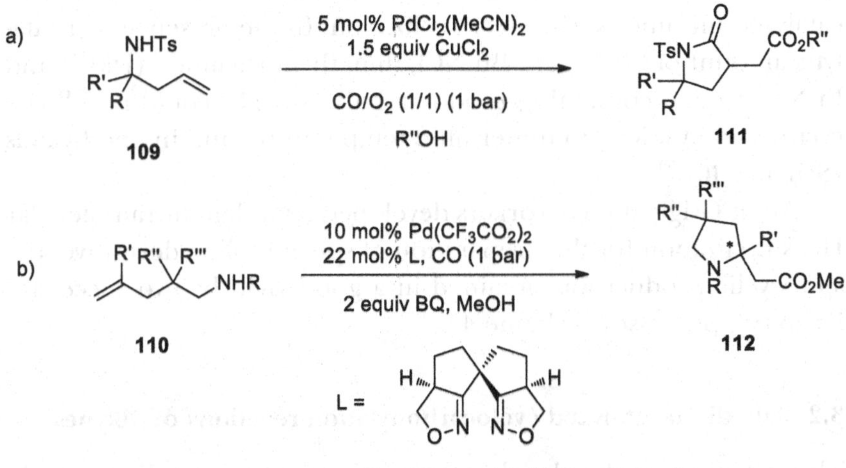

Scheme 44. Palladium-catalyzed cyclocarbonylation of *N*-tosylhomoallylamines

In the presence of palladium and copper salts, the reactions of *N*-tosylhomoallylamines **109** were carried out at 1 bar of CO to furnish 2-pyrrolidones **111** as products (Scheme 44a).[47] Meanwhile, the enantioselective palladium-catalyzed cyclocarbonylation reaction was also realized by application of spiro bis(isoxazoline) as chiral

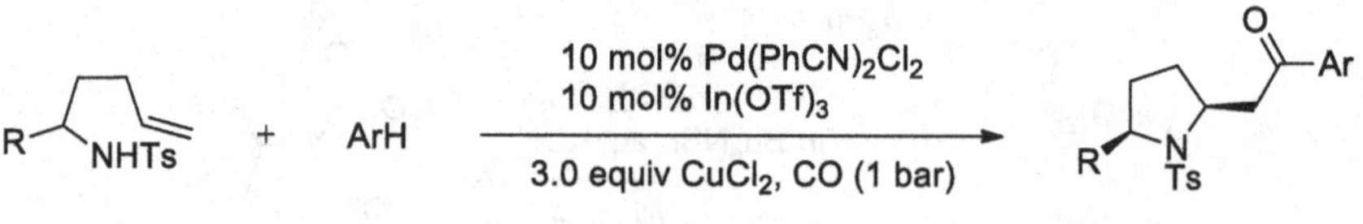

Scheme 45. Combination of aminocarbonylation with Friedel–Crafts acylation

ligands (Scheme 44b).[48] Notably, Cernak and Lambert developed the combination of aminocarbonylation with Friedel–Crafts acylation in 2009.[49] Using $Pd(PhCN)_2Cl_2$ as catalyst, $CuCl_2$ as oxidant and indium salts as additive, α-pyrrolidinyl ketones **114** were produced in good yields (Scheme 45).

3.1.4 Palladium-catalyzed intramolecular heck reactions of alkenes

As early as 1987, Larock and co-workers reported the preparation of five-membered nitrogen heterocycles, such as indoles, by palladium-catalyzed intramolecular Heck cyclization. In the presence of a catalytic amount of $Pd(OAc)_2$, Bu_4NCl, dimethylformamide (DMF) and Et_3N, nitrogen containing *o*-iodo aryl alkenes **115** could rapidly be cyclized to indoles **116** under mild temperatures and in good yields (Scheme 46).[50]

After, Grigg and co-workers developed a tandem intramolecular Heck cyclization for the preparation of spiroindoline derivative, the spirocyclic product was obtained in a good yield by two successive 5-exo-trig processes (Scheme 47).[51]

3.2 Palladium-catalyzed cyclocarbonylation reactions of alkynes

The palladium-catalyzed cyclocarbonylation reactions of alkynes could effectively yield five-membered oxygen heterocycles. For example, Arcadi and co-workers reported the palladium-catalyzed cyclocarbonylation of 2-alkynyl trifluoroacetanilides **121** with aryl halides and vinyl triflates for the synthesis of 2-substituted 3-acylindoles **122** in good yields (Scheme 48a).[52] The mechanism presumably involved the oxidative addition of the organohalide to active Pd(0) species,

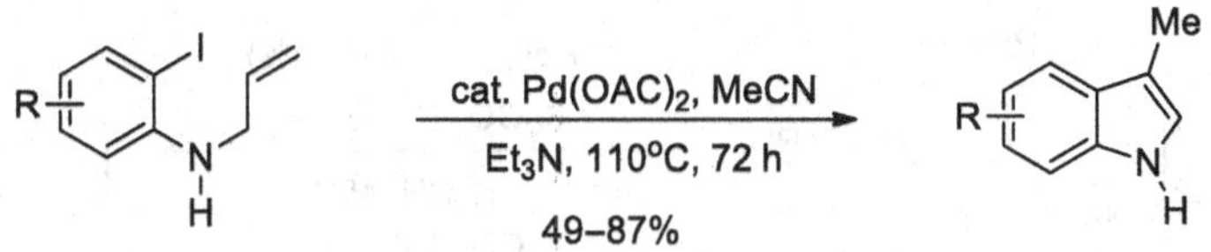

115 116

R = CH_3, CH_3O, CH_3CH_2O,

Scheme 46. Palladium-catalyzed intramolecular Heck cyclization of *N*-allyl phenylamine

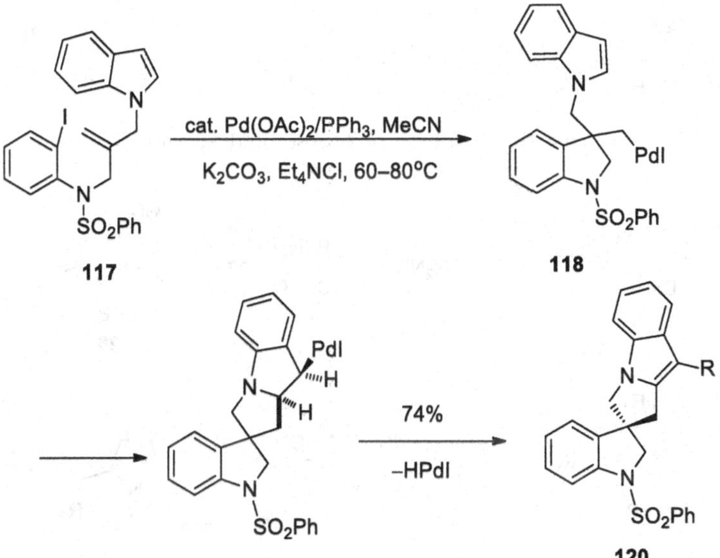

117 118 119 120

Scheme 47. Synthesis of spiroindolines

the coordination and insertion of CO, the intramolecular nucleophilic cyclization of the nitrogen atom with Pd-activated triple bond and the reductive elimination of Pd(II) species. Interestingly, 2-(1-alkynyl)benzenamines **123** were cyclized to 3-(halomethylene) indolin-2-ones **124** in the presence of PdX_2 and CuX_2 (X = Br, Cl) (Scheme 48b). In this reaction, the active Pd(II) species can be regenerated by the oxidation of Pd(0) with CuX_2 to join a new catalytic cycle.

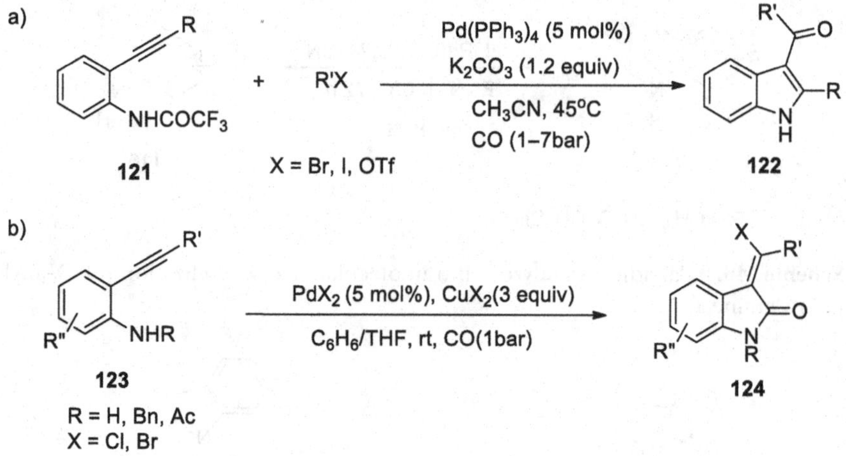
a)
R
NHCOCF3
121
+ R'X
X = Br, I, OTf
Pd(PPh3)4 (5 mol%)
K2CO3 (1.2 equiv)
CH3CN, 45°C
CO (1–7bar)
R'
O
R
N
H
122
b)
R'
R''
NHR
123
R = H, Bn, Ac
X = Cl, Br
PdX2 (5 mol%), CuX2(3 equiv)
C6H6/THF, rt, CO(1bar)
X
R'
O
R''
N
R
124

Scheme 48. Synthesis of 2-substituted 3-acylindoles

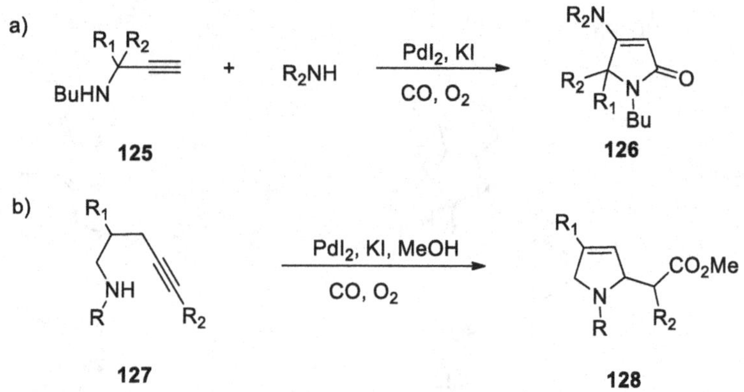
a)
R1 R2
BuHN
125
+ R2NH
PdI2, KI
CO, O2
R2N
R2
R1
N
Bu
O
126
b)
R1
NH
R
R2
127
PdI2, KI, MeOH
CO, O2
R1
CO2Me
N
R
R2
128

Scheme 49. Cyclocarbonylations of 2-ynylamines and (Z)-(2-en-4-ynyl)-amines

Later on, when Gabriele and co-worker carried out the cyclocarbonylations reaction of 2-ynylamines **125** in the presence of PdI_2 under oxidative conditions, they successfully produced pyrrole-2(5*H*)-ones **126** in moderate yields (Scheme 49a).[53] Using similar reaction conditions, the same authors also described the palladium-catalyzed cyclocarbonylations of (*Z*)-(2-en-4-ynyl)-amines **127** to the corresponding pyrroles **128** (Scheme 49b).[54]

3.3 Palladium-catalyzed tandem cyclization reactions of allenes

Allenes having a nitrogen functionality as nucleophile are extremely useful precursors for the synthesis of five-membered nitrogen heterocycles through an intramolecular palladium-catalyzed hydroaminations. In 1998, Yamamoto and co-worker developed an intramolecular hydroamination of aminoallenes in the presence of $[(\eta^3\text{-}C_3H_5)PdCl]_2$ (Scheme 50). The aminoallenes **129** reacted readily in a 5-exo-trig cyclization to afford the corresponding vinyl-substituted dihydropyrroles **130**.[55]

In 2003, Ma's group reported a Pd(II)-catalyzed coupling–cyclization of α-amino allenes with allylic halides. The reaction of 2,3-allenyl amines **131** with allylic bromide in a similar manner afforded 2,5-dihydropyrroles **132** in good to excellent yields. When the chiral substrate, (S)-2,3-allenyl amine **133**, was subjected to the standard conditions, optically active (S)-2,5-dihydropyrrole **134** was obtained without obvious loss of enantiopurity (Scheme 51).[56]

4. Palladium-Catalyzed Synthesis of Other Five-Membered Heterocycles Cyclocarbonylations

As early as 1994, Perry and co-workers developed a palladium-catalyzed carbonylative reaction for the synthesis of oxazolines. Substrates, aryl halides or triflates **135**, coupled with amino alcohols to present the intermediate aryl amides **136**, were activated by $SOCl_2$ or PTSA (*p*-toluene sulphonic acid) to produce oxazolines **137** in good yields (Scheme 52).[57]

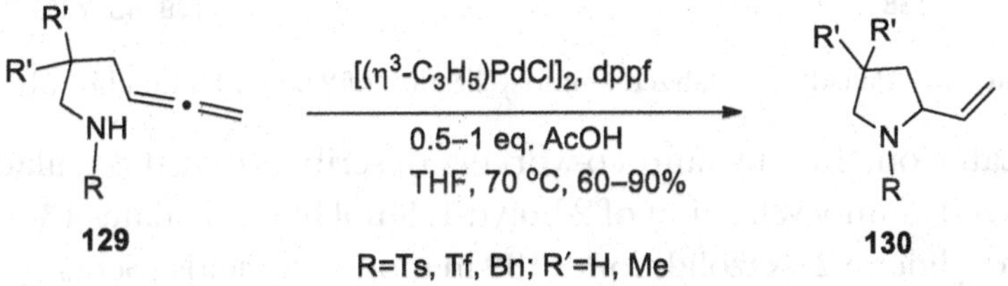

Scheme 50. Palladium-catalyzed intramolecular hydroamination of aminoallenes

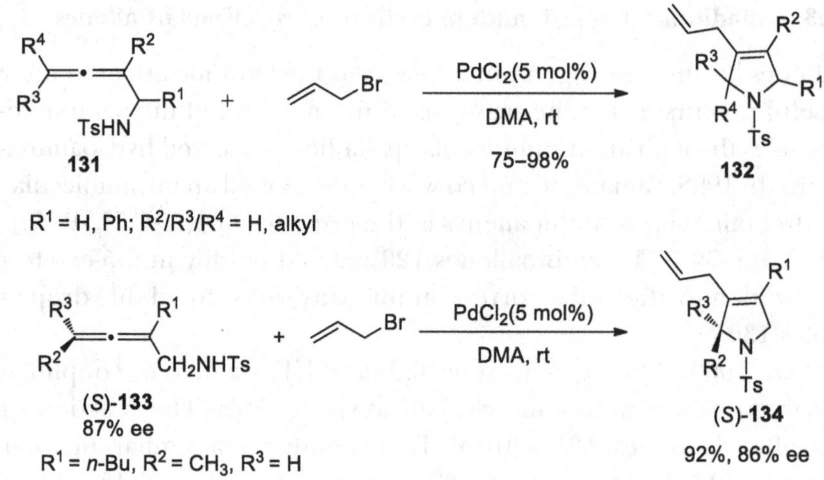
PdCl2(5 mol%)
DMA, rt
75–98%
131
132
R1 = H, Ph; R2/R3/R4 = H, alkyl
PdCl2(5 mol%)
DMA, rt
(S)-133
87% ee
R1 = n-Bu, R2 = CH3, R3 = H
(S)-134
92%, 86% ee

Scheme 51. Pd(II)-catalyzed couplingcyclization of α-amino allenes with allylic halides

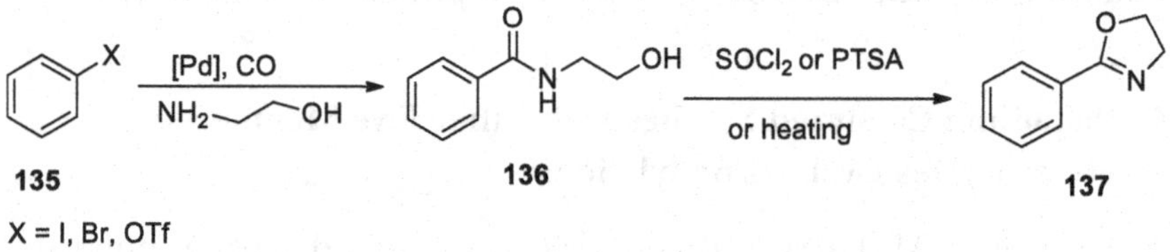
[Pd], CO
SOCl2 or PTSA
or heating
135
136
137
X = I, Br, OTf

Scheme 52. Synthesis of oxazolines

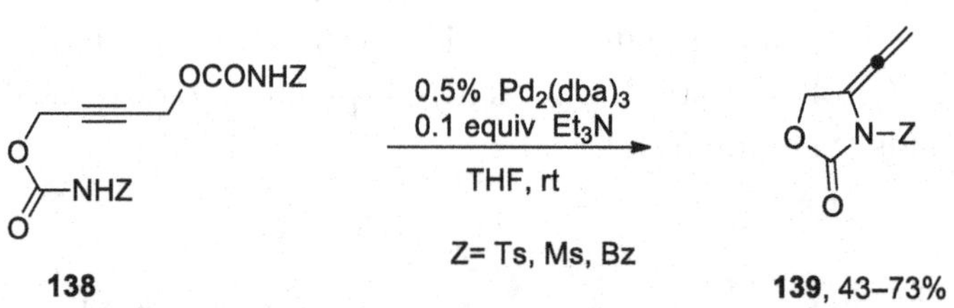
0.5% Pd2(dba)3
0.1 equiv Et3N
THF, rt
Z= Ts, Ms, Bz
138
139, 43–73%

Scheme 53. Palladium-catalyzed aminocyclization of 2-butyn-1,4-diol biscarbamates

Later on, Tamaru and co-workers described that the palladium-catalyzed aminocyclization of 2-butyn-1,4-diol biscarbamates **138** led to 4-ethenylidene-2-oxazolidinones **139** in moderate yields (Scheme 53).[58]

At the same period, Young and DeVita reported a novel approach to the synthesis of oxadiazoles. The one-pot reaction of aryl iodides

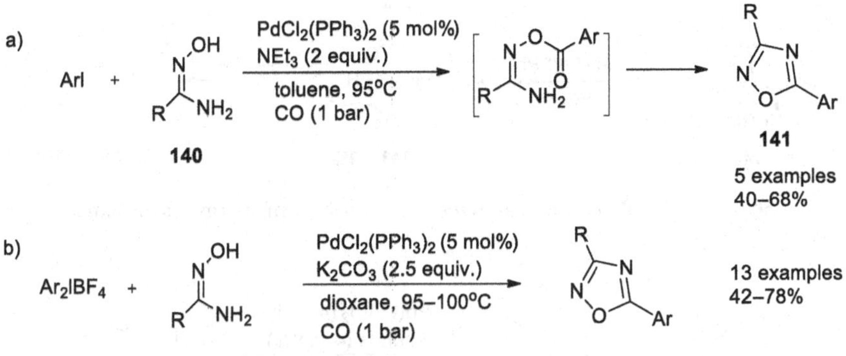

a)

ArI + (structure **140**) $\xrightarrow[\text{toluene, 95°C}]{\begin{array}{c}\text{PdCl}_2(\text{PPh}_3)_2 \text{ (5 mol\%)}\\ \text{NEt}_3 \text{ (2 equiv.)}\\ \text{CO (1 bar)}\end{array}}$ [intermediate] $\longrightarrow$ **141**

5 examples
40–68%

b)

Ar_2IBF_4 + (structure) $\xrightarrow[\text{dioxane, 95–100°C}]{\begin{array}{c}\text{PdCl}_2(\text{PPh}_3)_2 \text{ (5 mol\%)}\\ \text{K}_2\text{CO}_3 \text{ (2.5 equiv.)}\\ \text{CO (1 bar)}\end{array}}$ (product)

13 examples
42–78%

Scheme 54. Synthesis of oxadiazoles

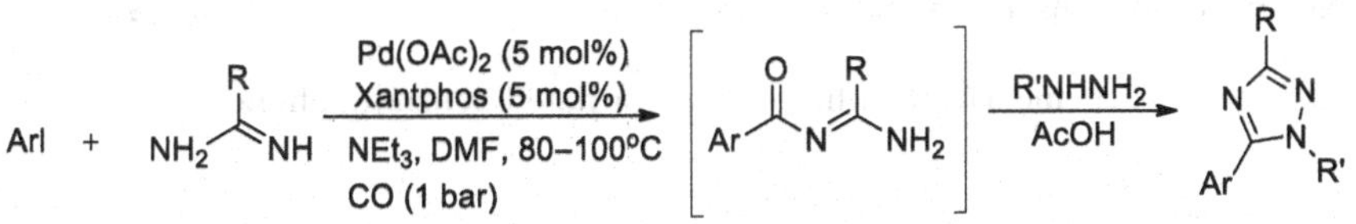

ArI + (structure) $\xrightarrow[\text{NEt}_3, \text{DMF, 80–100°C}]{\begin{array}{c}\text{Pd(OAc)}_2 \text{ (5 mol\%)}\\ \text{Xantphos (5 mol\%)}\\ \text{CO (1 bar)}\end{array}}$ [intermediate] $\xrightarrow[\text{AcOH}]{\text{R'NHNH}_2}$ (product)

Scheme 55. Synthesis of 1,2,4-triazoles

with amidoximes **140** catalyzed by $\text{PdCl}_2(\text{PPh}_3)_2$, afforded the expected oxadiazoles **141** in moderate yields under 1 bar of CO (Scheme 54a).[59] The catalytic system had the excellent tolerance of both electron-withdrawing and electron-donating substituents. Afterwards, Zhou and Chen demonstrated a similar carbonylation reaction with diaryliodonium salts as starting materials (Scheme 54b).[60]

Noteworthily, Staben and Blaquiere recently discovered a four-component carbonylation reaction for the synthesis of 1,2,4-triazoles under mild conditions and low CO pressure. This method showed the good tolerance of substrate (Scheme 55).[61]

Ohno and co-workers found that the reaction of propargyl carbonate **142** with $\text{Pd(PPh}_3)_4$ (5 mol%) could result in the desired bicyclic tetrahydrofuran **143** under mild condition (Scheme 56).[62] The cyclic product was applied to synthesize the pachastrissamine, which displays remarkable cytotoxic activity against several tumor cell lines.

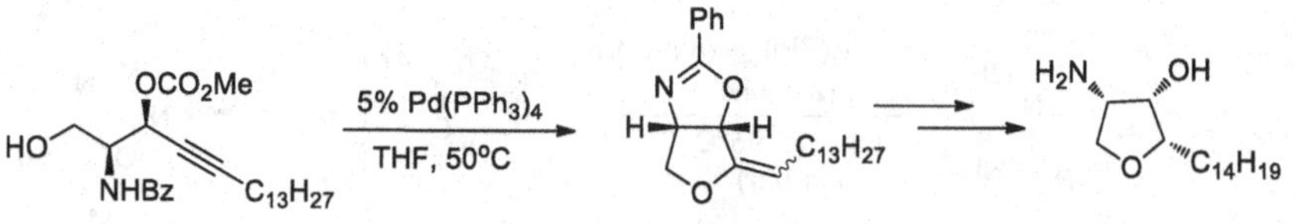

142 **143**, 69% Pachastrissamine

Scheme 56. Palladium-catalyzed cyclization of propargyl carbonates

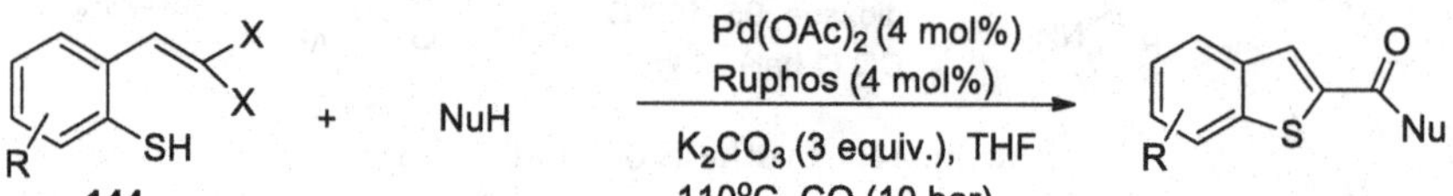

144 **145**

X = Br, Cl
NuH = Alcohols, phenol, and amines

15 examples
24–73%

Scheme 57. Synthesis of carbonylated benzothiophenes

In 2011, Alper developed a facile and selective palladium-catalyzed domino synthesis of carbonylated benzothiophenes derivatives **145** from 2-gem-dihalovinylthiophenols **144** in moderate yields (Scheme 57).[63] This protocol was believed involving an intramolecular C–S coupling/intermolecular carbonylation cascade sequence.

5. Summary

We have summarized the number of contributions of palladium-catalyzed cyclization of unsaturated hydrocarbon systems for the synthesis of five-membered heterocycles over the past decades. Palladium-catalyzed reactions have become a very powerful tool in this area due to the excellent tolerance of functional groups and the advantage of avoiding protection group. In short, it is apparent from a synthetic perspective that the application of palladium catalysis to five-membered heterocycle chemistry plays an important role on the synthesis of this important class of compounds. Despite the impressive results obtained, various effective and concise processes are still desired, and that it is significant to improve the scope of reactions and mildness of the conditions for palladium chemistry.

References

1. (a) S. M. Kupchan, E. J. LaVoie, A. R. Branfman, B. Y. Fei, W. M. Bright and R. F. Bryan, 1977. *J. Am. Chem. Soc.*, 99, 3199. (b) B. M. Trost, D. B. Horne and M. J. Woltering, 2003. *Angew. Chem. Int. Ed. Engl.*, 42, 5987. (c) C. Marti and E. M. Carreira, 2005. *J. Am. Chem. Soc.*, 127, 11505. (d) X. D. Hu, Y. Q. Tu, E. Zhang, S. Gao, S. Wang, A. Wang, C. A. Fan and M. Wang, 2006. *Org. Lett.* 8, 1823. (e) B. M. Trost and M. K. Brennan, 2006. *Org. Lett.*, 8, 2027. (f) V. J. Reddy and C. J. Douglas, 2010. *Org. Lett.*, 12, 952.

2. (a) L. A. Agrofoglio, I. Gillaizeau and Y. Saito, 2003. *Chem. Rev.*, 103, 1875. (b) E. Negishi and L. Anastasia, 2003. *Chem. Rev.*, 103, 1979. (c) G. Zeni and R. C. Larock, 2004. *Chem. Rev.*, 104, 2285. (d) G. Zeni and R. C. Larock, 2006. *Chem. Rev.* 106, 4644. (e) L. N. Guo, X. H. Duan and Y. M. Liang, 2011. *Acc. Chem. Res.*, 44, 111. (f) X.-F. Wu, H. Neumann, and M. Beller, 2013. *Chem. Rev.*, 113, 1.

3. T. Hosokawa, H. Ohkata and I. Moritani, 1975. *Bull. Chem. Soc. Jpn.*, 48, 1533.

4. (a) Y. Uozumi, K. Kato and T. Hayashi, 1997. *J. Am. Chem. Soc.* 119, 5063; (b) Y. Uozumi, H. Kyota, K. Kato, M. Ogasawara and T. Hayashi, 1999. *J. Org. Chem.*, 64, 1620.

5. K. C. Majundar, U. Das, U. K. Kundu and A. Bandyopadhyay, 2001. *Tetrahedron*, 57, 7003.

6. T. Hosokawa, M. Hirata, S.-I. Murahashi and A. Sonoda, 1976. *Tetrahedron Lett.*, 17, 1821.

7. T. Hosokawa, F. Nakajima, S. Iwasa and S.-I. Murahashi, 1990. *Chem. Lett.*, 19, 1387.

8. J. S. Yadav, E. S. Rao, V. S. Rao and B. M. Choudary, 1990. *Tetrahedron Lett.*, 31, 2491.

9. A. Tenaglia and F. Kammerer, 1996. *Synlett*, 6, 576.

10. A. Kasahara, T. Izumi, K. Sato, M. Maemura and T. Hayasaka, 1977. *Bull. Chem. Soc. Jpn.*, 50, 1899.

11. R. C. Larock and T. R. Hightower, 1993. *J. Org. Chem.*, 58, 5298.

12. B. M. Trost, 1989. *Angew. Chem., Int. Ed. Engl.*, 28, 1173.

13. C. Kammerer, G. Prestat, D. Madec and G. Poli, 2014. *Acc. Chem. Res.*, 47, 3439.

14. M. Vitale, G. Prestat, D. Lopes, D. Madec and G. Poli, 2006. *Synlett*, 2231.

15. M. Vitale, G. Prestat, D. Lopes, D. Madec, C. Kammerer, G. Poli and L. Girnita, 2008. *J. Org. Chem.*, 73, 5795.

16. M. F. Semmelhack, C. Bodurow and M. Baum, 1984. *Tetrahedron Lett.*, 25, 3171.

17. W. -Y. Yu, C. Bensimon and H. Alper, 1997. *Chem. Eur. J.*, 3, 417.

18. P. Cao and X. Zhang, 1999. *J. Am. Chem. Soc.*, 121, 7708.

19. Y. Tamaru, M. Hojo and Z. Yoshida, 1987. *Tetrahedron Lett.*, 28, 325.

20. Y. Tamaru, T. Bando, M. Hojo and Z. Yoshida, 1987. *Tetrahedron Lett.*, 28, 3497.

21. Y. Tamaru, T. Kobayashi, S. Kawamura, H. Ochiai, M. Hojo and Z. Yoshida, 1985. *Tetrahedron Lett.*, 26, 3207.

22. R. C. Larock and D. E. Stinn, 1988. *Tetrahedron Lett.*, 29, 4687.

23. M. Lautens and Y.-Q. Fang, 2003. *Org. Lett.* 5, 3679.

24. C.-Y. Cheng, J.-P. Liou and M.-J. Lee, 1997. *Tetrahedron Lett.*, 38, 4571.

25. P. Diaz, F. Gendre, L. Stella and B. Charpentier, 1998. *Tetrahedron*, 54, 4579.

26. K. Utimoto, 1983. *Pure Appl. Chem.*, 55, 1845.

27. J. Tsuji, H. Watanabe, I. Minami and I. Shimizu, 1985. *J. Am. Chem. Soc.*, 107, 2196.

28. X.-H. Duan, X.-Y. Liu, L.-N. Guo, M.-N. Liao, W.-M. Liu and Y.-M. Liang, 2005. *J. Org. Chem.*, 70, 6980–6983.

29. H.-P. Bi, X.-Y. Liu, F.-R. Gou, L.-N. Guo, X.-H. Duan, X.-Z. Shu and Y.-M. Liang, 2007. *Angew. Chem., Int. Ed.*, 46, 7068.

30. A.-Z. Wang, H.-F. Jiang and Q.-X. Xu, 2009. *Synlett*, 6, 929.

31. T. Nogi and J. Tsuji, 1969. *Tetrahedron*, 25, 4099.

32. K. Kato, A. Nishimura, Y. Yamamoto and H. Akita, 2001. *Tetrahedron Lett.*, 42, 4203.

33. (a) K. Kato, M. Tanaka, Y. Yamamoto and H. Akita, 2002. *Tetrahedron Lett.*, 43, 1511. (b) K. Kato, A. Nishimura, Y. Yamamoto and H. Akita, 2002. *Tetrahedron Lett.*, 43, 643.

34. R. D. Walkup, L. Guan, M. D. Mosher, S. W. Kim and Y. S. Kim, 1993. *Synlett*, 88.

35. S. Ma and Z. Yu, 2003. *J. Org. Chem.*, 68, 6149.

36. S. Ma and W. Gao, 2000. *Tetrahedron Lett.*, 41, 8933.

37. R. D. Walkup and G. Park, 1987. *Tetrahedron Lett.*, 28, 1023.

38. R. D. Walkup and M. D. Mosher, 1993. *Tetrahedron*, 49, 9285.

39. L. S. Hegedus, G. F. Allen and E. L. Waterman, 1976. *J. Am. Chem. Soc.*, 98, 2674.

40. H. Tsutsui, M. Kitamura and K. Narasaka, 2002. *Bull. Chem. Soc. Jpn.* 75, 1451.

41. L. S. Hegedus and J. M. McKearin, 1982. *J. Am. Chem. Soc.*, 104, 2444.

42. R. C. Larock, T. R. Hightower, L. A. Hasvold and K. P. Peterson, 1996. *J. Org. Chem.*, 61, 3584.

43. Z.-J. Zhang, Z.-H. Ren, Y.-Y. Wang and Z.-H. Guan, 2013. *Org. Lett.*, 15, 4822.

44. X.-L. Lian, Z.-H. Ren, Y.-Y. Wang and Z.-H. Guan, 2014. *Org. Lett.*, 16, 3360.

45. D. Madec, G. Prestat, E. Martini, P. Fristrup, G. Poli and P.-O. Norrby, 2005. *Org. Lett.*, 7, 995.

46. M. Bui The Thuong, S. Sottocornola, G. Prestat, G. Broggini, D. Madec and G. Poli, 2007. *Synlett*, 10, 1521.

47. T. Mizutani, Y. Ukaji and K. Inomata, 2003. *Bull. Chem. Soc. Jpn.*, 76, 1251.

48. (a) T. Shinohara, M. A. Arai, K. Wakita, T. Arai and H. Sasai, 2003. *Tetrahedron Lett.*, 44, 711. (b) C. Granito, L. Troisi, and L. Ronzini, 2004. *Heterocycles*, 63, 1027.

49. T. A. Cernak and T. H. Lambert, 2009. *J. Am. Chem. Soc.*, 131, 3124.

50. R. C. Larock and S. Babu, 1987. *Tetrahedron Lett.*, 28, 5291.

51. R. Grigg, P. Fretwell, C. Meerholtz and V. Sridharan, 1994. *Tetrahedron*, 50, 359.

52. A. Arcadi, S. Cacchi, V. Carnicelli and F. Marinelli, 1994. *Tetrahedron*, 50, 437.

53. B. Gabriele, P. Plastina, G. Salerno and M. Costa, 2005. *Synlett*, 935.

54. (a) B. Gabriele, G. Salerno, A. Fazio and F. B. Campana, 2002. *Chem. Commun.*, 1408. (b) B. Gabriele, G. Salerno, A. Fazio and L. Veltri, 2006. *Adv. Synth. Catal.*, 348, 2212.

55. M. Meguro and Y. Yamamoto, 1998. *Tetrahedron Lett.*, 39, 5421.

56. S. Ma, F. Yu and W. Gao, 2003. *J. Org. Chem.*, 68, 5943.

57. R. J. Perry and B. D. Wilson, 1994. *Macromolecules*, 27, 40.

58. M. Kimura, Y. Wakamiya, Y. Horino and Y. Tamaru, 1997. *Tetrahedron Lett.*, 38, 3963.

59. J. R. Young and R. J. DeVita, 1998. *Tetrahedron Lett.*, 39, 3931.

60. T. Zhou and Z.-C. Chen, 2002. *Synth. Commun.*, 32, 887.

61. S. T. Staben and N. Blaquiere, 2010. *Angew. Chem., Int. Ed.*, 49, 325.

62. S. Inuki, Y. Yoshimitsu, S. Oishi, N. Fujii and H. Ohno, 2010. *J. Org. Chem.*, 75, 3831.

63. F. Zeng and H. Alper, 2011. *Org. Lett.*, 13, 2868.

CHAPTER 6

Pd-Catalyzed Six-Membered Heterocycles Synthesis

Jiangang Mao[*], Weiliang Bao[†,‡]

[*]Department of Chemistry, Zhejiang University,
Hangzhou, Zhejiang, P. R. China
School of Metallurgy and Chemical Engineering, Jiangxi University
of Science and Technology, Ganzhou, Jiangxi, P. R. China
[†]Department of Chemistry, Zhejiang University,
Hangzhou, Zhejiang, P. R. China
[‡]wlbao@zju.edu.cn

1. Introduction

Over the past 20 years, organo palladium chemistry has achieved the explosive growth and found widespread use in organic synthesis.[1] It is widely recognized that palladium is the most versatile metal in facilitating unique transformations, many of which are not readily achieved using classical techniques or with other transition metal catalysts. Another important feature is that palladium-catalyzed reactions proceed under mild reaction conditions and tolerate broad functional groups in many cases. Therefore, the synthesis of many important and biologically active heterocyclic compounds *via* palladium-catalyzed reactions has attracted much attention from organic chemists.[2] This chapter reviews the methodologies of palladium catalysis in the synthesis of six-membered heterocyclic compounds.

2. Synthesis of Six-Membered Nitrogen Containing Heterocycles

2.1 Palladium-catalyzed cycloamination to six-membered N-heterocycles

2.1.1 *Addition of nitrogen atom to triple bond*

The intramolecular regio and stereoselective addition of a nucleophile to an unsaturated bond is a most effective strategy for carbo and heterocyclic ring construction. The cyclization of alkynes containing proximate nucleophilic centers promoted by organopalladium complexes is a straightforward approach to synthesize functionalized heterocycles.

Larock and co-workers developed a synthesis of 3,4-disubstituted isoquinolines *via* palladium-catalyzed cross-coupling of *N-tert*-butyl-*o*-(1-alkynyl)-benzaldimines with aryl, allylic, benzylic, alkynyl, and vinylic halides in 2001 (Scheme 1).[3] This general process avoided the problem of regioselectivity that exists in the synthesis of isoquinolines by the iminoannulation of internal alkynes.[4] The reaction requires an aryl group on the end of the acetylene furthest from the imine functionality in many cases. Electron-rich and *o*-substituted aryl halides give lower yields.

Recently, Chowdhury and co-workers have applied intramolecular cyclizations of terminal alkynes for the preparation of (*E*)-3-arylidene-3,4-dihydro-2*H*-1,4-benzoxazines (Scheme 2).[5] Electron-rich aryl halides give lower yields, whereas electron-deficiency aryl iodides facilitated the reactions compared to electron-rich ones. Heteroaryl

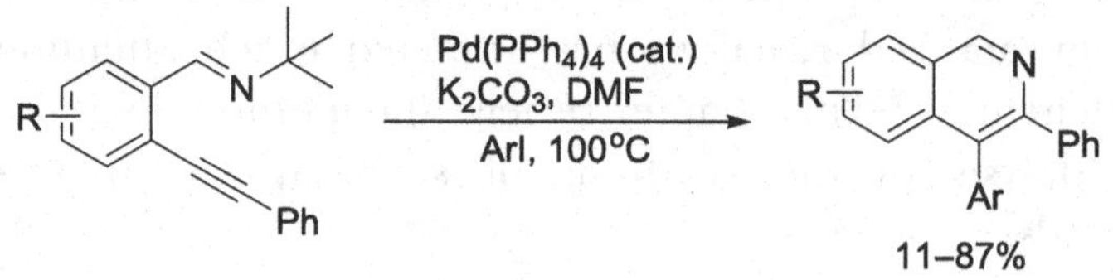

Scheme 1. Synthesis of 3,4-disubstituted isoquinolines

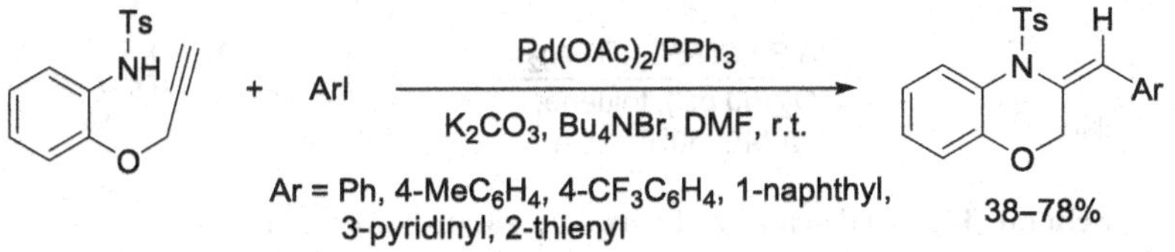

Ts
⟨ ⟩—NH + ArI —— Pd(OAc)₂/PPh₃ ——→ ⟨ ⟩
O   K₂CO₃, Bu₄NBr, DMF, r.t.   O

Ar = Ph, 4-MeC₆H₄, 4-CF₃C₆H₄, 1-naphthyl,
3-pyridinyl, 2-thienyl

38–78%

Scheme 2. Synthesis of (*E*)-3-arylidene-3,4-dihydro-2*H*-1,4-benzoxazines

iodides afforded better product yields compared to simple aryl iodides. The complete regio and stereoselectivity avoid the formation of seven-membered ring products *via* 7-*endo-dig* mode or of products with Z-stereochemistry.

2.1.2 *Addition of nitrogen atom to double bond*

Pd(II)-catalyzed oxidative cyclization of alkenes is an efficient route to five-membered heterocycles. However, few general methods have been reported for the synthesis of six-membered *N*-heterocycles. Recently, Stahl and co-workers described an efficient Pd-(DMSO)₂(TFA)₂ catalyst system, which facilitated intramolecular Pd (II)-catalyzed aerobic oxidative amination of alkenes to prepare six-membered *N*-heterocycles with up to 92% yield (Scheme 3).[6] However, the reaction was performed under 60 psi of O₂.

In 2013, Zhang and co-workers described a Pd(OAc)₂-catalyzed oxidative cyclization for the preparation of 2-substituted 1,2,3,4-tetrahydroquinoline and 1,2,3,4-tetrahydroquinoxaline derivatives with acetoxy functionality in the presence of PhI(OAc)₂ as oxidant (Scheme 4).[7]

Beifuss and co-workers reported a simple procedure for the synthesis of 1,2,3,4-tetrahydroquinoxaline *via* reductive cyclization of ω-nitroalkene using 60 mol% of Pd(OAc)₂, 120 mol% of 1,10-phenanthroline and CO (5 bar) at 140°C in 53% yield (Scheme 5).[8] They also described that this reductive cyclization system can be simply used for the synthesis of 3-isopropyl-3,4-dihydro-2H-1,4-benzoxazine from 3,3-dimethylallyl-2-nitrophenyl ether in 50% yield.

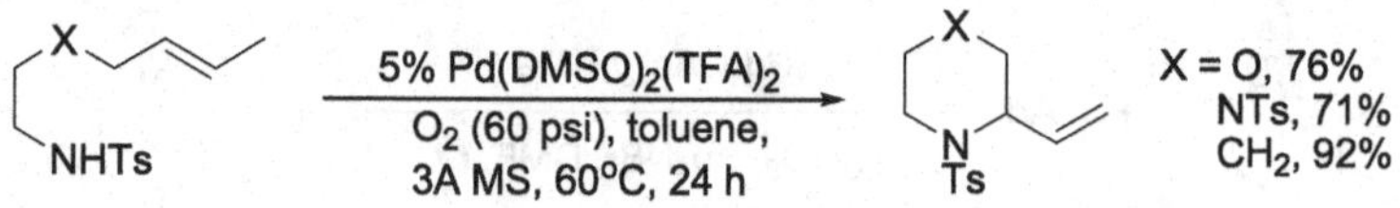

$$\text{(structure: }X\text{ chain with NHTs)}\quad\xrightarrow[\substack{\text{O}_2\text{ (60 psi), toluene,}\\3A\text{ MS, 60°C, 24 h}}]{5\%\text{ Pd(DMSO)}_2(\text{TFA})_2}\quad\text{(cyclized product N–Ts)}$$

X = O, 76%
NTs, 71%
CH$_2$, 92%

Scheme 3. Pd(II)-catalyzed aerobic oxidative cyclization of alkenes

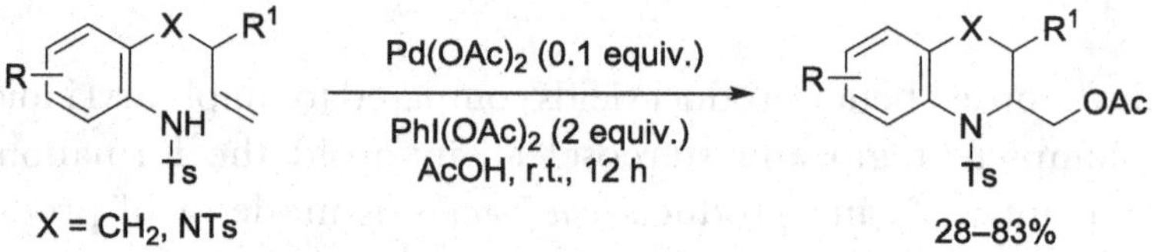

$$\text{(aryl NH–Ts, }X,R^1\text{ vinyl)}\quad\xrightarrow[\substack{\text{PhI(OAc)}_2\text{ (2 equiv.)}\\\text{AcOH, r.t., 12 h}}]{\text{Pd(OAc)}_2\text{ (0.1 equiv.)}}\quad\text{(bicyclic N–Ts, }R^1,\text{ OAc)}$$

X = CH$_2$, NTs

28–83%

Scheme 4. Pd(II)-catalyzed oxidative cyclizations

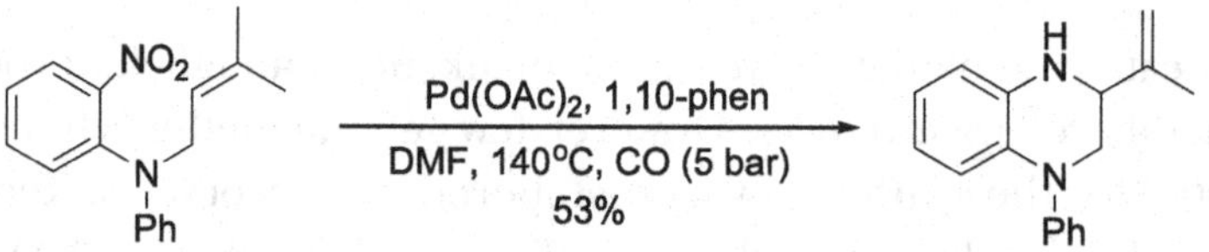

$$\text{(aryl NO}_2,\text{ N–Ph)}\quad\xrightarrow[\substack{\text{DMF, 140°C, CO (5 bar)}\\53\%}]{\text{Pd(OAc)}_2,\ 1,10\text{-phen}}\quad\text{(bicyclic N–H, N–Ph)}$$

Scheme 5. Pd(II)-catalyzed reductive cyclizations

2.1.3 Addition of nitrogen atom to π-allyl intermediates

Yamamoto and co-workers developed a palladium-catalyzed intramolecular asymmetric hydroamination of aminoalkynes with up to 92% yield and 90% ee, respectively. The intramolecular asymmetric hydroamination process is believed to undergo a key step of addition of nitrogen atom to π-allyl intermediates (Scheme 6).[9] When catalyst system was consisting of Pd(0)-methyl norphos (or tolyl renorphos)-benzoic acid, the yield and ee of the reaction increased to 95%, respectively.[10]

Ohno and co-workers reported a palladium-catalyzed domino cyclization of propargyl bromides for the construction of bicyclic heterocycle 2,7-diazabicyclo[4.3.0]non-5-enes in the presence of NaH in MeOH. The reaction also underwent addition of nitrogen atom to π-allyl intermediate (Scheme 7).[11]

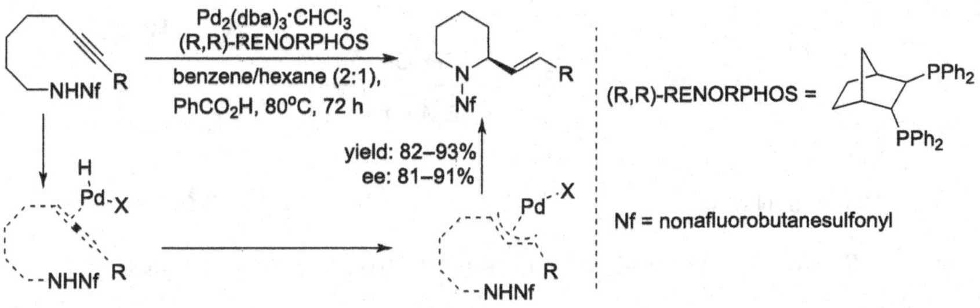

Scheme 6. Pd-catalyzed intramolecular asymmetric hydroamination of aminoalkynes

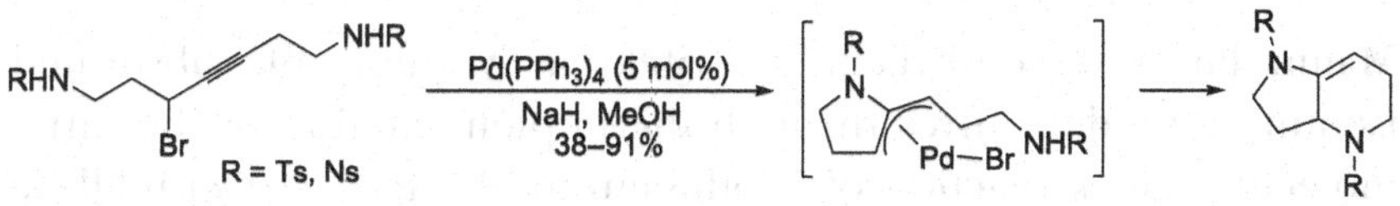

Scheme 7. Domino cyclization of propargyl bromides through π-allyl intermediates

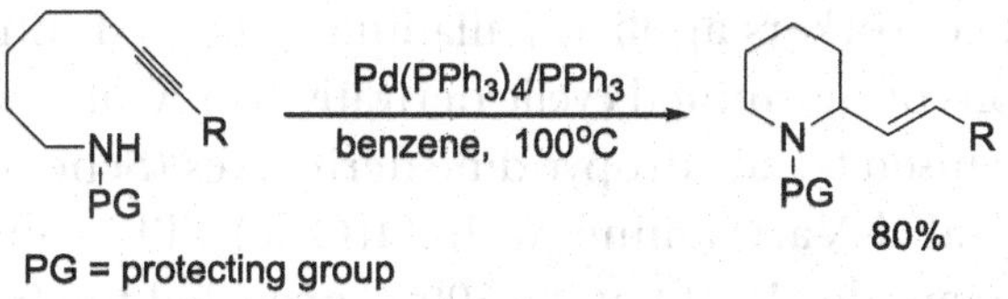

Scheme 8. Domino cyclization of propargyl bromides through α-allyl intermediates

Intramolecular hydroamination of alkynes tethered with amino group catalyzed by $Pd(PPh_3)_4/PPh_3$ under neutral conditions was found by Yamamoto and co-workers in 2005 (Scheme 8).[12] The reaction did not need carboxylic acid as a co-catalyst.

2.1.4 *Addition of nitrogen atom to allenes*

Ma and co-workers reported a palladium-catalyzed coupling-cyclization of β-amino allenes with allylic halides leading to 1,2,3,6-tetrahydropyridines (Scheme 9).[13] The reaction underwent a π-allyl palladium intermediate, the author thought that the current transformation most likely proceeded *via* a Pd(II)-catalyzed pathway.

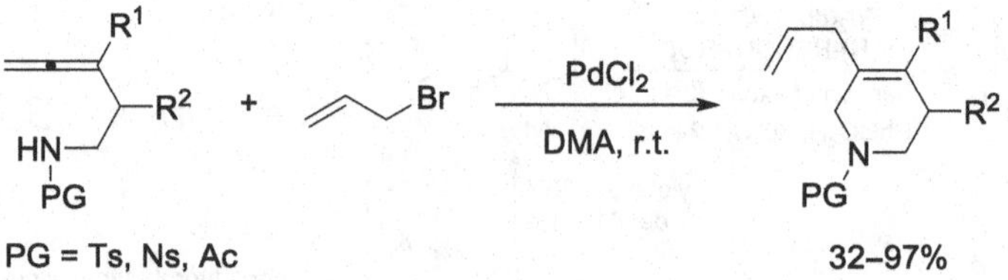

PG = Ts, Ns, Ac   32–97%

Scheme 9. Pd-catalyzed addition of nitrogen atom to allenes

2.2 Heck-type cyclization

Majumdar and co-workers described a synthesis of substituted isoquinolone derivatives through the implementation of the intramolecular Heck reaction of 2-iodobenzamides tethered with allylic group under ligand free conditions in excellent yields (Scheme 10).[14] The reaction proceeded on the unactivated allylic system without the necessity of using any ligand.

Ray and co-workers applied palladium-catalyzed intramolecular Heck reactions of substituted cyclic derivatives of *N*-allyl-*N*-aryl amines to synthesize fused tetrahydropyridine derivatives (Scheme 11).[15] The reaction of *N*-allyl-*N*-aryl amines with $Pd(OAc)_2/PPh_3$ and Cs_2CO_3 in dimethylformamide (DMF) at 90–100°C provided the fused tetrahydropyridines in excellent yields by 6-*exo*-trig cyclization.

Broggini and co-workers described an intramolecular Heck reaction involving σ-alkylpalladium intermediates for the preparation of dihydroisoquinolinones (Scheme 12).[16] The authors isolated σ-alkylpalladium intermediates, the stability of the σ-alkylpalladium complexes is probably a consequence of the strong constraint resulting from the bridged junction that hampers the cisoid conformation essential for β-hydride elimination.

The rapid synthesis of large organic compound collections by the intramolecular Heck reaction on the solid phase is a promising strategy for the discovery of new pharmaceutical leads. Goff and co-workers reported a solid phase synthesis of highly substituted 1-(*2H*)-soquinolinones bearing peptoid side chains (Scheme 13).[17]

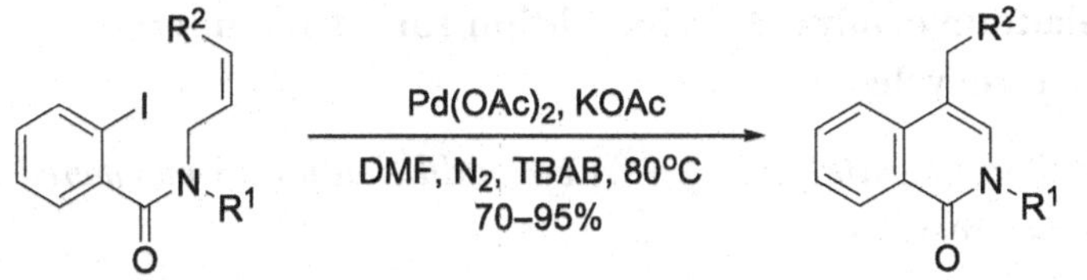

R^1 = Ph, C$_6$H$_4$Cl, C$_{10}$H$_7$, 6-aminocoumarin, *N*-methyl-6-aminoquinolone
R^2 = H, Me

Scheme 10. Intramolecular Heck cyclizations of aryl halides

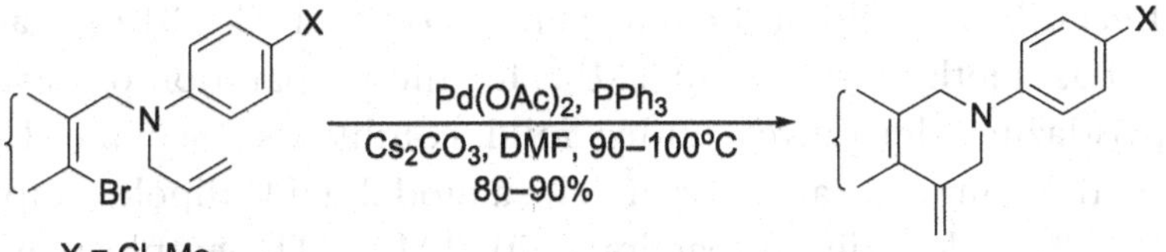

X = Cl, Me

Scheme 11. Intramolecular Heck cyclizations of vinylic halides

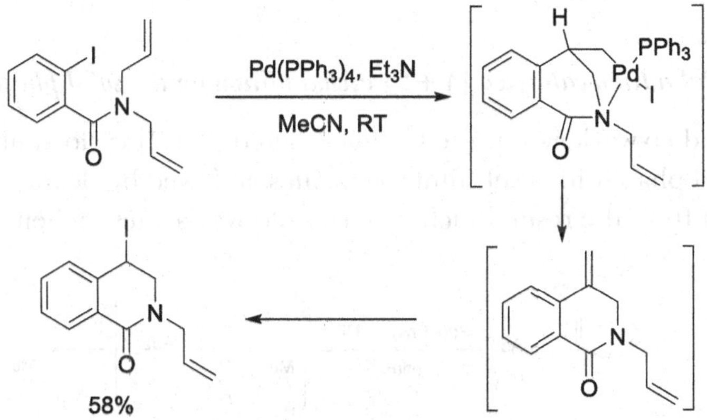

58%
2-allyl-4-iodo-3,4-dihydroisoquinolin-1(2*H*)-one

Scheme 12. Intramolecular Heck cyclizations of vinylic halides

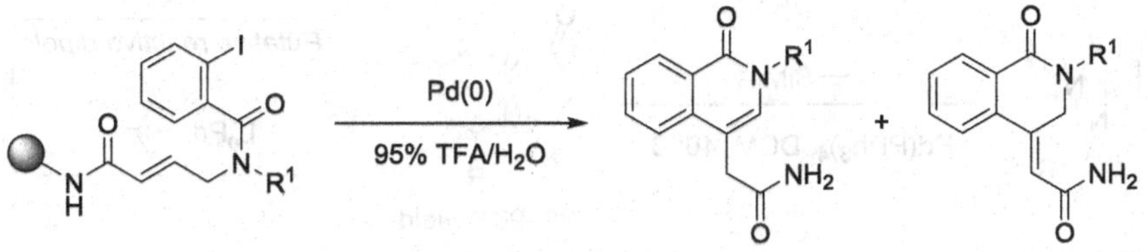

Scheme 13. Intramolecular Heck reaction on the solid phase

2.3 Palladium-catalyzed cycloaddition for six-membered *N*-heterocycles

2.3.1 *Palladium-catalyzed [3 + 3] cycloaddition for six-membered N-heterocycles*

Kemmitt and co-workers reported a [3 + 3] cycloaddition of trimethyle-nemethane (TMM) to activated aziridines for the synthesis of piperidines through palladium TMM cycloaddition as early as 1989 (Scheme 14).[18]

Recently, Hayashi and co-workers reported a [3 + 3] cycloaddition of azomethine imines to TMMs for the preparation of hexahydropyridazine derivatives under mild conditions (Scheme 15).[19] Azomethine imines have proven to be a model stable dipole template for [3 + 3] cycloaddition reactions with TMMs. The employment of substituted TMM precursor highlights the difference of this system from reported [3 + 2] cycloaddition of TMMs under palladium catalysis.[20]

2.3.2 *Palladium-catalyzed [4 + 2] cycloaddition on the solid phase*

Kaur and co-workers reported a Pd-catalyzed [4 + 2] cycloaddition on the solid phase with 4-substituted urazines, followed by cleavage of the product from the resin, which gave triazolopyridazines (Scheme 16).[21]

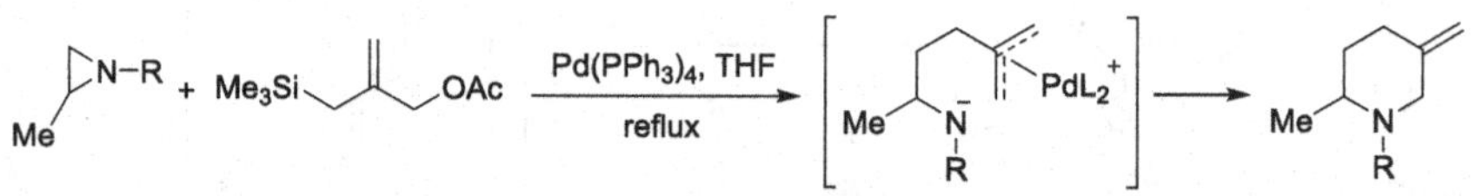

Scheme 14. Pd-catalyzed [3 + 3] cycloaddition of azomethine imines to TMM

20–90% yield

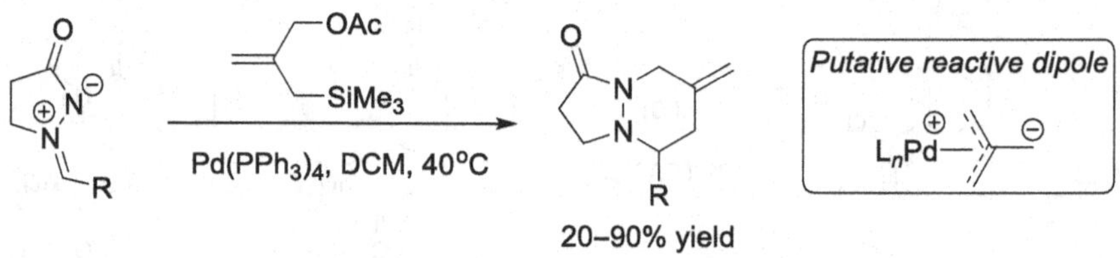

Scheme 15. Pd-catalyzed [3 + 3] cycloaddition of azomethine imines to TMM

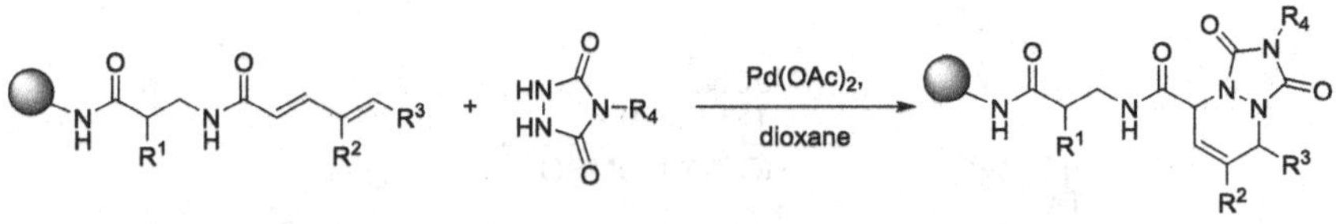

Scheme 16. Pd-catalyzed [3 + 3] cycloaddition of azomethine imines to TMM

2.3.3 Palladium-catalyzed cycloaddition of enynes for six-membered N-heterocycles

Mikami and co-workers developed a highly enantioselective method for the preparation of quinoline bearing a quaternary carbon center or a spiro ring *via* ene-type cyclization of 1,7-enynes catalyzed by a cationic BINAP-Palladium(II) complex (Scheme 17).[22] The formation of these quinolines can be easily followed by the variation of coloration of the reaction, the yield and ee are up to 99%.

Liang and co-workers transformed 1,6-enyne carbonates into 3-vinylidene-1-tosylpyridines in the presence of PdI_2 as the catalyst (Scheme 18),[23] however, 3-vinylidene-1-tosylpyrrolidines were obtained when $Pd(dba)_2$ was used as the catalyst. The reaction underwent an allenylpalladium intermediate from facile decarboxylation, which would be subsequently trapped by an olefin in an intramolecular fashion to form the putative products. The divergent cyclizations provide a versatile cascade reaction for the synthesis of heterocyclic allenes.

2.3.4 Palladium-catalyzed cycloaddition of allenes for six-membered N-heterocycles

A diastereoselective arylative carbocyclization of pro-nucleophile-linked allenes with halides to provide spirocyclic lactam products with moderate to high diastereoselectivities and good yields under Pd(0) catalysis is realized to prepare one heterocyclic ring (Scheme 19).[24]

2.3.5 Palladium-catalyzed hetero-Diels–Alder reactions

Maison and co-workers described a Pd-catalyzed hetero-Diels–Alder reaction *via* a domino procedure including a Heck reaction and a spontaneous Diels–Alder process to six-membered *N*-heterocycle. (Scheme 20).[25]

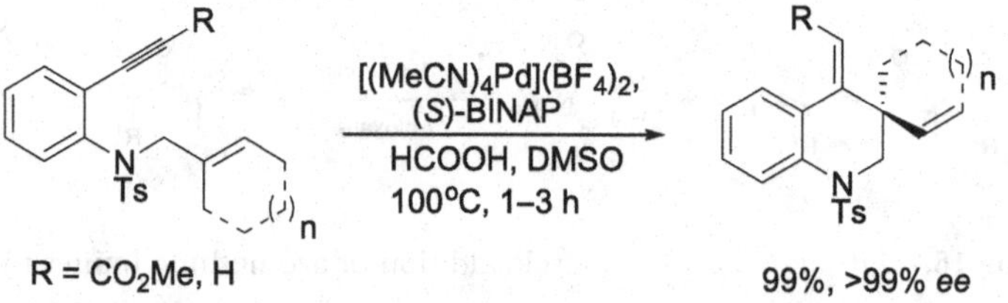

Scheme 17. Pd-catalyzed cycloaddition of 1,7-enynes for six-membered N-heterocycles

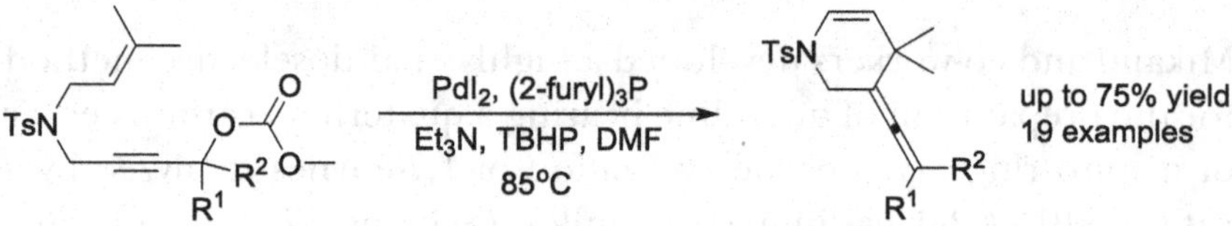

Scheme 18. Pd-catalyzed cycloaddition of 1,6-enyne carbonates for six-membered N-heterocycles

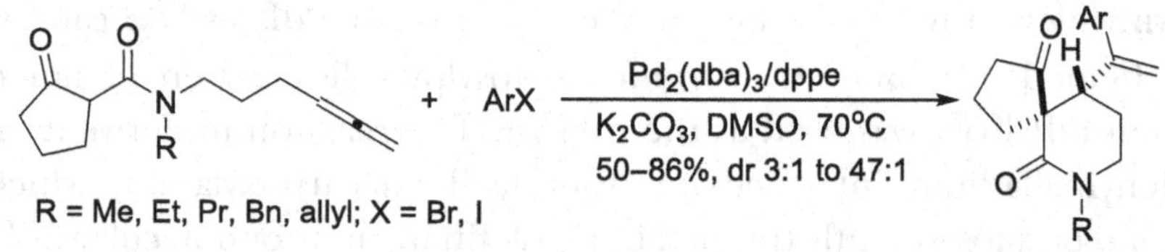

Scheme 19. Palladium catalyzed arylative allene carbocyclization cascades

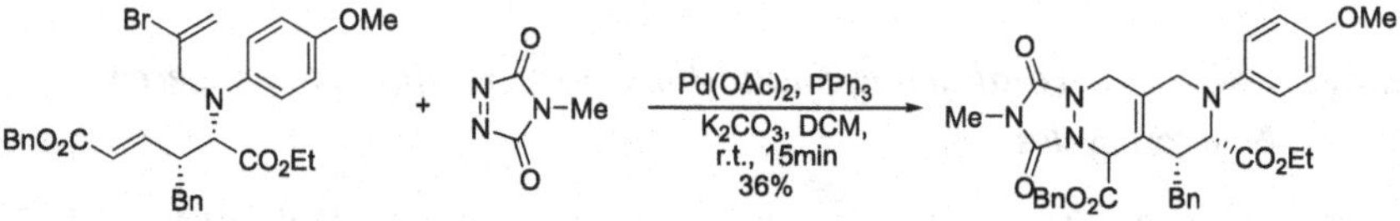

Scheme 20. Palladium-catalyzed hetero-Diels–Alder reactions

2.4 Palladium-catalyzed annulation reaction for six-membered N-heterocycles

2.4.1 Annulation by halo imines

Larock and co-workers reported a Pd-catalyzed iminoannulation of halo imines with terminal and internal acetylenes for the synthesis of

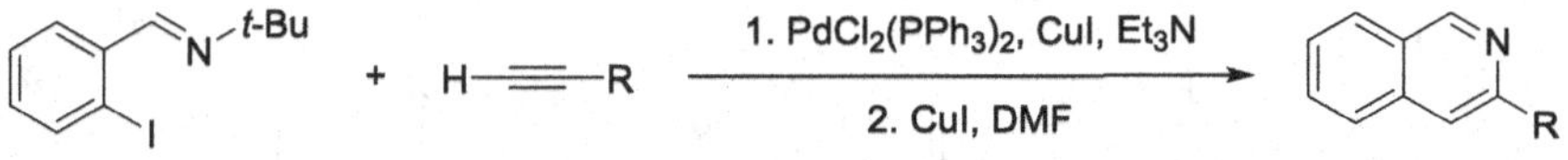

Scheme 21. Pd-catalyzed annulation reaction by halo imines

isoquinolines and pyridines. The reaction underwent a really domino procedure including a Heck-type/Sonogashira reaction and a spontaneous nucleophilic addition to six-membered N-heterocycle (Scheme 21).[26]

2.4.2 *Annulation by other halo amines bearing carbanions*

In order to establish the synthetic utility of palladacycles, Malinakova and co-workers prepared a stable racemic benzannulated azapalladacycle featuring a palladium-bonded sp^3-hybridized stereogenic carbon, which was then converted into a series of racemic 2,3,4-trisubstituted 1,2-dihydroquinolines *via* a regioselective insertion of activated alkynes (RC≡CCOOEt) (Scheme 22).[27]

2.4.3 *Palladium-catalyzed carboamination reactions*

Malinakova and co-workers reported a Pd-catalyzed asymmetric carboamination of aryl bromide with N^1-aryl-N^2-allyl-1,2-diamine for the synthesis of *cis*-2,6-disubstituted piperazines. N^1-aryl-N^2- allyl-1,2-diamine was generated from readily available amino acid precursors. The products are obtained in 14–20:1 *dr*, with >97% ee (Scheme 23).[28]

2.4.4 *Palladium-catalyzed annulation via cyclopropane ring expansion*

Tsuritani and co-workers described a Pd-catalyzed annulation *via* cyclopropane ring expansion to construct 3,4-dihydro-2(1H)-quinolinone derivatives (Scheme 24).[29] The reaction tolerates broad functional groups such as ester, cyano, ether, and ketone groups. This method is also applicable to synthesize 5,6-dihydrobenzimidazo[1,2-a]quinoline.

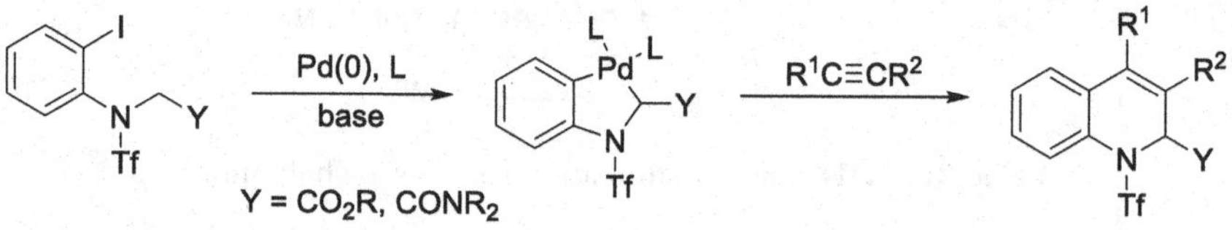

Scheme 22. Pd-catalyzed annulation reaction by halo amines bearing carbanions

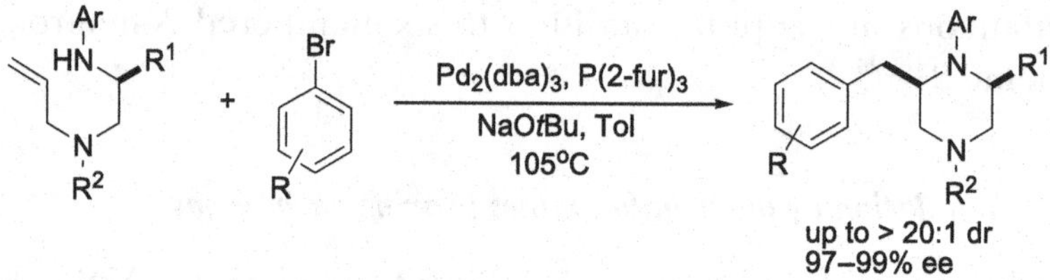

Scheme 23. Pd-catalyzed carboamination reactions

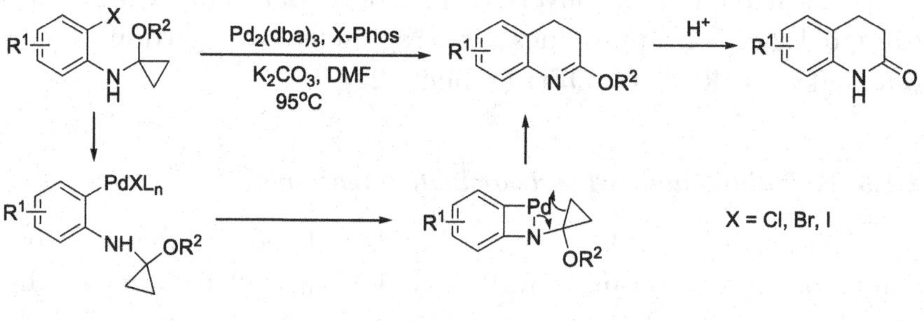

Scheme 24. Pd-catalyzed annulation by alkane substrates

2.4.5 *Palladium-catalyzed annulation by other halo amines bearing carbanions*

Solé and co-workers reported that two different and competitive reaction pathways, involving the enolate arylation and the nucleophilic attack at the carbonyl can be facilitated starting from (2-haloanilino) ketones, which show structure-dependent behavior. Thus, treatment of γ-(2-iodoanilino) ketone with $PdCl_2(PPh_3)_2$ and Cs_2CO_3 afforded the α-arylation compound **1** (Scheme 25).[30] In contrast, under the same reaction conditions, β-(2-iodoanilino) ketone afforded

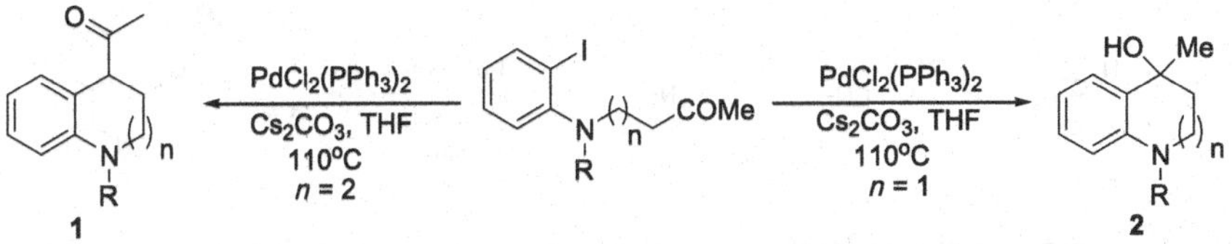

Scheme 25. Pd-catalyzed annulation by other halo amines bearing carbanions

exclusively alcohols **2**, as a result of the addition of the palladium intermediate to the ketone carbonyl group.

2.4.6 *Palladium-catalyzed three component annulation of carbodiimide, isocyanide, and amine/phosphite*

Wu and co-workers reported a $Pd(OAc)_2$-catalyzed three component reaction of carbodiimide, isocyanide, and amine to synthesis quinazolino[3,2-*a*]quinazolines in good yields without ligand (Scheme 26).[31] Multi-bonds are formed in one-pot through nucleophilic attack, isocyanide insertion, and C–N coupling during the reaction process. They also reported a $Pd(OAc)_2/PCy_3$-catalyzed three component reaction of carbodiimide, isocyanide, and amine for the preparation of quinazolin-4(3H)-imines (Scheme 27).[32]

Recently, Wu and co-workers reported a Pd-catalyzed three component reaction of carbodiimide, isocyanide, and phosphite to synthesize 4-imino-3,4-dihydroquinazolin-2-ylphosphonates in moderate to good yields. Three bonds are formed in a one-pot procedure and the tandem process includes nucleophilic attack, isocyanide insertion, and C–N coupling (Scheme 28).[33]

2.4.7 *Palladium-catalyzed two component annulation of amine/ amide/imine and isocyanide*

Jiang and co-workers reported a palladium-catalyzed one-pot cyclization reaction of aryl halides with isocyanides to construct 6-aminophenanthridines in good to excellent yields (up to 93%) (Scheme 29),[34]

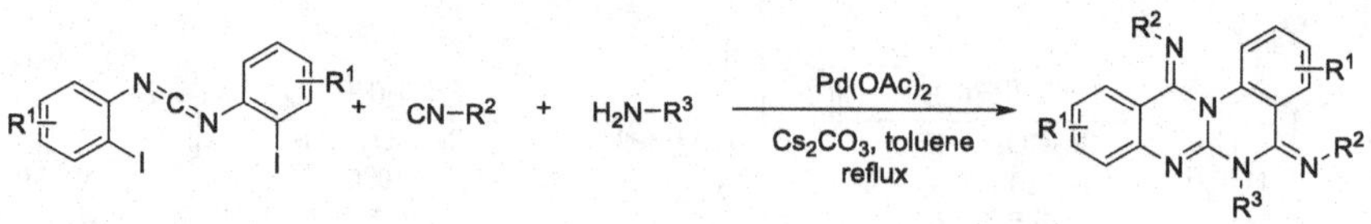

Scheme 26.　Pd(OAc)$_2$-catalyzed three component annulation

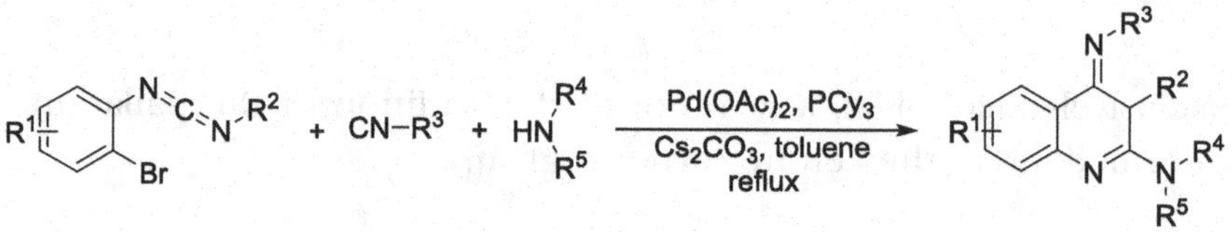

Scheme 27.　Pd(OAc)$_2$/PCy$_3$-catalyzed three component annulation

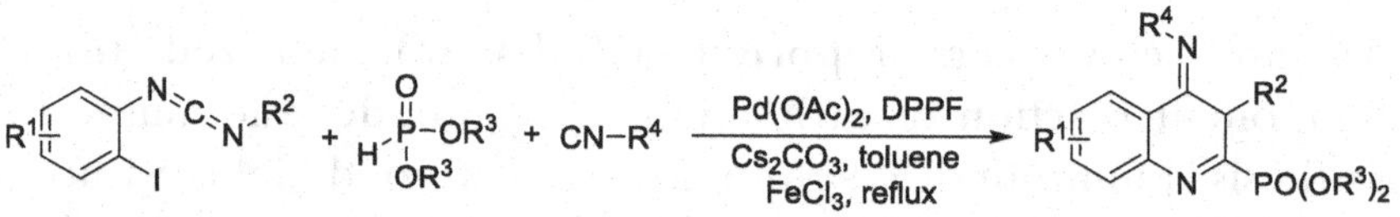

Scheme 28.　Pd(OAc)$_2$/DPPF-catalyzed three-component annulation

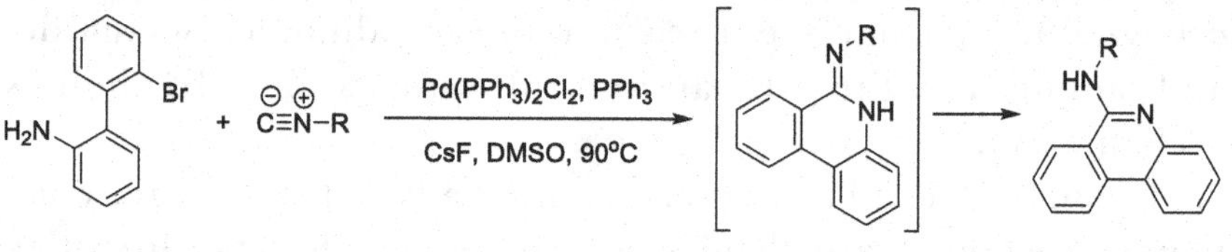

Scheme 29.　Pd-catalyzed two component annulation of amine and isocyanide

which is an important unit in numerous biologically active natural products and medicinally significant compounds.

　　Chauhan and co-workers reported a novel ligand free palladium-catalyzed cascade reaction for the synthesis of highly diverse isoquinolin-1 (2*H*)-one derivatives from isocyanide and amide. The reaction proceeds through tandem isocyanide insertion with intra-molecular cyclization followed by a Mazurciewitcz–Ganesan type sequence to provide isoquinoline-1(2*H*)-one derivatives in moder-ate to good yields (Scheme 30).[35]

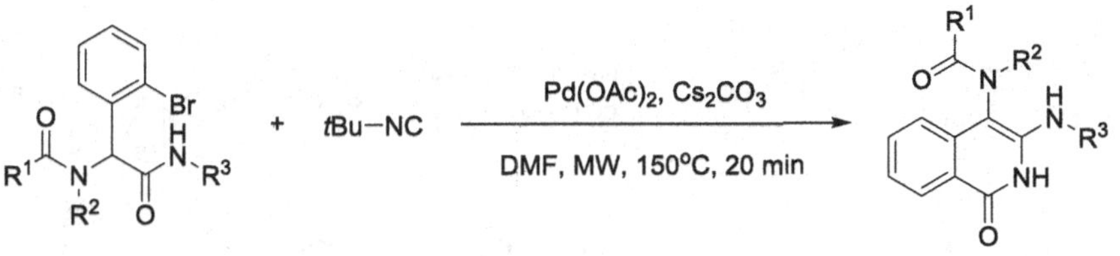

Scheme 30.   Pd-catalyzed two-component annulation of amide and isocyanide

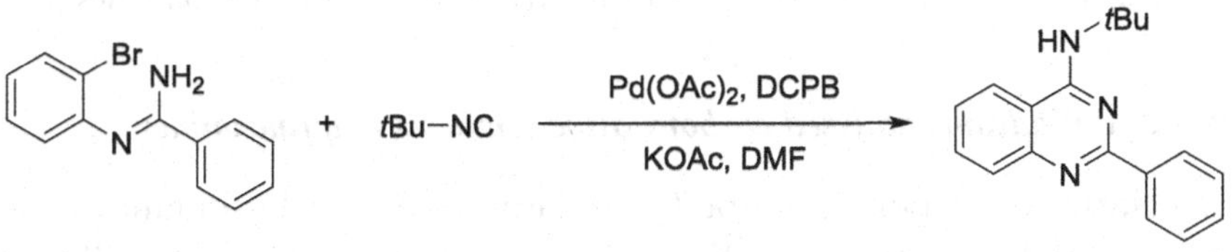

Scheme 31.   Pd-catalyzed two-component annulation of imine and isocyanide

Orru and co-workers developed an efficient palladium-catalyzed intramolecular imidoylation of *N*-(2-bromoaryl)amidines for the synthesis of 4-aminoquinazolines (Scheme 31).[36] Various substituents are tolerated on the amidine and the isocyanide, providing efficient access to a broad range of diversely substituted 4-aminoquinazolines of significant pharmaceutical interest.

2.5 Palladium-catalyzed carbonylation for six-membered *N*-heterocycles

2.5.1 *Palladium-catalyzed carbonylative synthesis of quinolinones*

For the preparation of various quinoline derivatives, the palladium-catalyzed carbonylative coupling of 2-haloanilines with terminal alkynes offers straightforward access. Haddad and co-workers applied 10 mol% of PdCl$_2$ (DPPF) in Et$_2$NH under 20 bar of CO at 120°C in the synthesis of BILN 2061 derivatives (Scheme 32a).[37] Ye and Alper reported that palladium-catalyzed cyclocarbonylation reaction of *o*-iodoanilines with allenes and CO in 1-butyl-3- methylimidazolium hexafluorophosphate afforded similar quinoline derivatives in moderate to excellent yields under a low pressure (5 bar) of CO (Scheme 32b).[38]

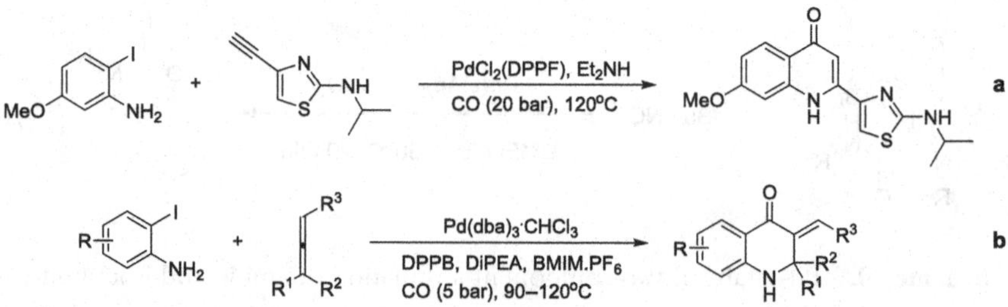

Scheme 32.   Palladium-catalyzed carbonylative synthesis of quinolinones

2.5.2 *Palladium-catalyzed carbonylative synthesis of quinazolines*

Larksarp and Alper developed a palladium acetate-bidentate phosphine catalyst system for the cyclocarbonylation of *o*-iodoanilines with heterocumulenes (Scheme 33a).[39] The nature of the substrates and the electrophilicity of the carbon center of the carbodiimide, as well as the stability of the ketenimine, influenced the product yields of this reaction. Urea-type intermediates are believed to be generated first *in situ* from the reaction of *o*-iodoanilines with heterocumulenes. Recently, Zhu and co-workers reported a palladium catalyzed intramolecular C–H carboxamidation of *N*-arylamidines to the corresponding quinazolines. The reactions were carried out in the presence of 1.0 equiv of CuO as oxidant under atmospheric pressure of CO and provided diversified 2-aryl(alkyl)-quinazolin-4(3H)-ones in reasonable to good yields (Scheme 33b).[40]

2.5.3 *Palladium-catalyzed carbonylative synthesis of pyridinones*

Knight and co-workers developed a palladium catalyzed decarboxylative carbonylation of 5-vinyloxazolidin-2-ones in 2000. 3,6-Dihydro-1H-pyridin-2-ones were obtained in good yields from the corresponding 5-vinyloxazolidin-2-ones (Scheme 34).[41]

2.5.4 *Palladium-catalyzed carbonylative synthesis of isoquinolinones*

Alper and co-workers developed a new route to ring-fused substituted oxazolo and pyrazoloisoquinolinones through a one-pot carboxamidation/aldol-type condensation reaction sequence (Scheme 35a).[42]

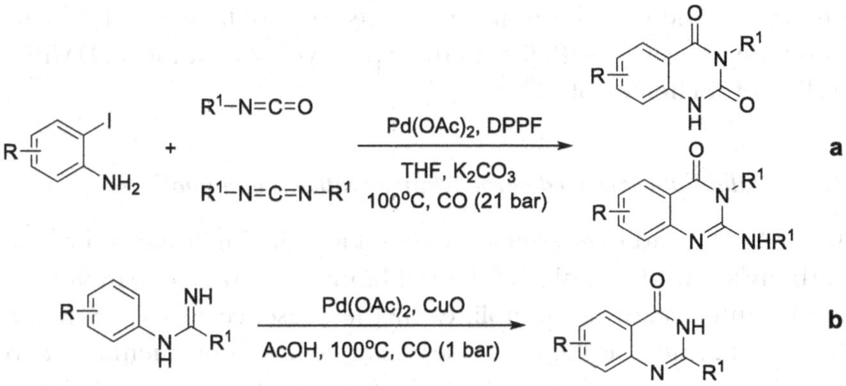

Scheme 33. Palladium-catalyzed carbonylative synthesis of quinazolines

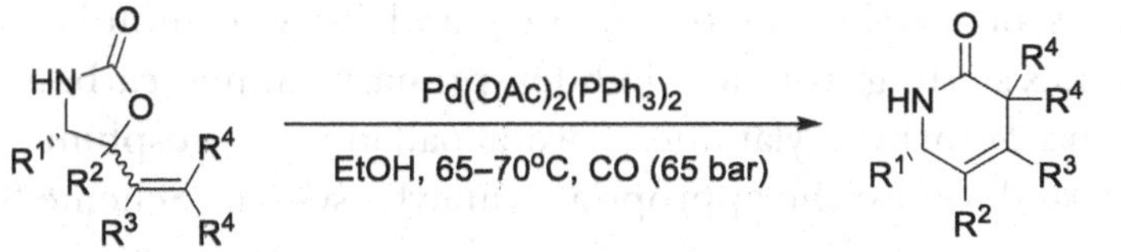

Scheme 34. Palladium-catalyzed carbonylative synthesis of pyridinones

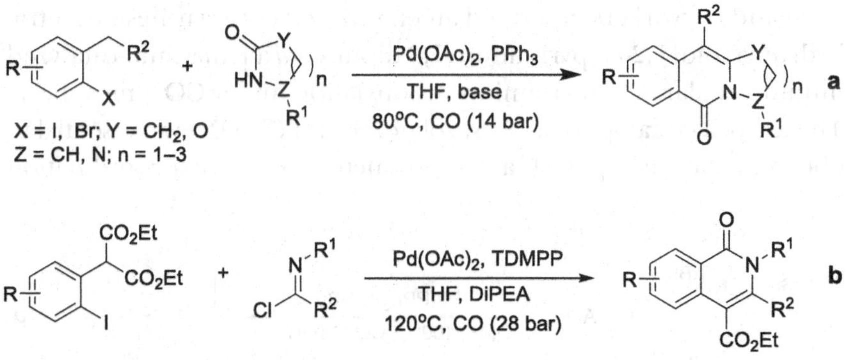

Scheme 35. Palladium-catalyzed carbonylative synthesis of isoquinolinones

A range of ring-fused oxazolo and pyrazoloisoquinolinones were obtained from a variety of active methylenes. The isoquinolinone products contain different functional groups that can be further functionalized. Isoquinolinones can also be prepared *via* palladium-catalyzed carbonylation of diethyl(2-iodoaryl)-malonates and imidoyl chlorides. The corresponding products were obtained with fair to good yields using tris(2,6-dimethoxyphenyl) phosphine (TDMPP) as ligand (Scheme 35b).[43]

2.5.5 *Palladium-catalyzed carbonylative synthesis of quinolines*

In 2002, Dai and co-workers developed a palladium-catalyzed cyclo-carbonylation of *o*-(1-alkynyl)benzaldimines for the preparation of 3-substituted 4-aroylisoquinolines in the presence of CO (Scheme 36a).[44] This methodology provides a simple and convenient route to isoquinolines containing aryl, alkyl, or vinyl substituents at C-3 and an aroyl group at C-4 of the isoquinoline ring in good yields. Rossi and co-workers reported a palladium-catalyzed carbonylation for the synthesis of 2-aryl-4-aminoquinolines and 2-aryl-4-amino[1,8]naph-thyridines starting from aryl iodides, primary amines, carbon mono-xide, and 2-ethynyl-arylamines. Several palladium/phosphine systems were tested to find the appropriate catalytic system (Scheme 36b).[45]

2.5.6 *Palladium-catalyzed carbonylative synthesis of alkenylureas*

Sasai and co-workers described an enantioselective synthesis of tetra-hydropyrrolo[1,2-*c*]- pyrimidine-1,3-diones *via* a palladium-catalyzed intramolecular oxidative aminocarbonylation under CO atmosphere. The use of a chiral spiro bis(isoxazoline) ligand (SPRIX) was essential to obtain the desired optically active products. The peculiar coordination

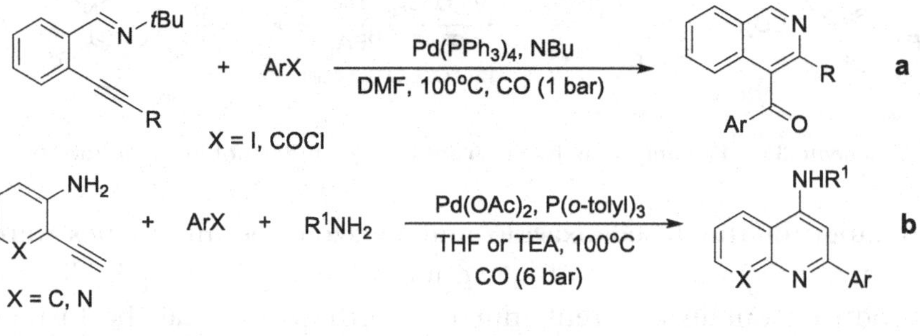

Scheme 36. Palladium-catalyzed carbonylative synthesis of isoquinolinones

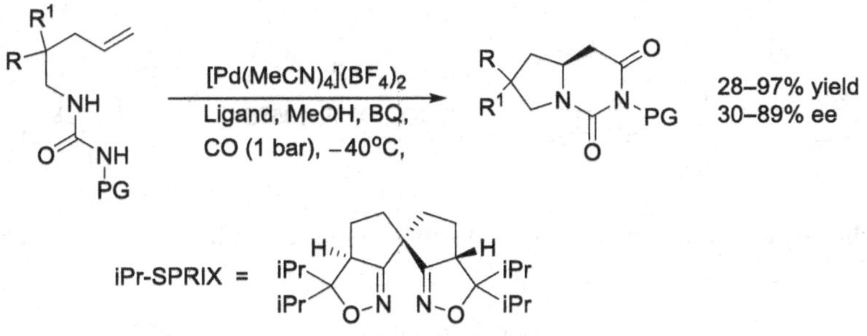

Scheme 37. Palladium-catalyzed carbonylative synthesis of alkenylureas

ability of SPRIX originates from two structural characteristics: low σ-donor ability of the isoxazoline coordination site and rigidity of the spiro skeleton (Scheme 37).[46]

2.6 Norbornene mediated synthesis of six-membered N-heterocycles

In 2004, Catellani and co-workers reported a new coupling methodology for the synthesis of condensed nitrogen heterocycles. Using Pd(OAc)$_2$/tri-2-furylphosphine (TFP) and norbornene as a co-catalyst system in DMF at 105°C, the coupling of *o*-bromobenzamide with *o*-iodotoluene afforded the corresponding phenanthridinone in 86% yield (Scheme 38).[47]

2.7 C–H activation-based methodology

2.7.1 *Intramolecular double C–H activation*

Bao and co-workers reported an atom-economical, concise and efficient methodology to synthesize the imidazole or benzimidazole fused isoquinoline polyheteroaromatic compounds *via* a double C–H activation process (Scheme 39).[48]

2.7.2 *Intramolecular C–H activation of halo amine compounds*

Garratt and co-workers reported a palladium-catalyzed intramolecular C–H activation of aryl bromides to synthesize six-membered annulated

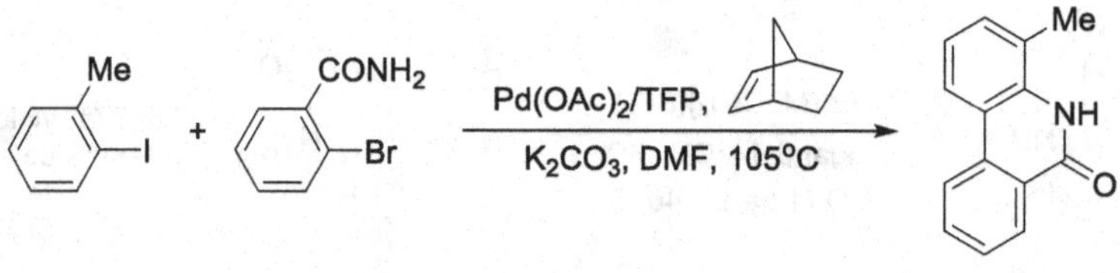

Scheme 38. Norbornene mediated synthesis of six-membered *N*-heterocycles

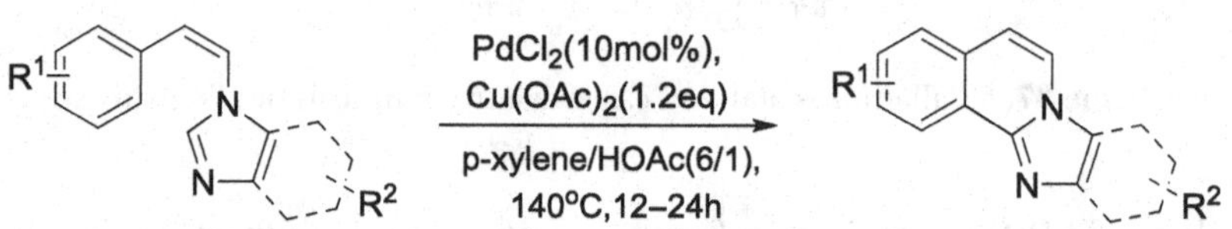

Scheme 39. Intramolecular double C–H activation

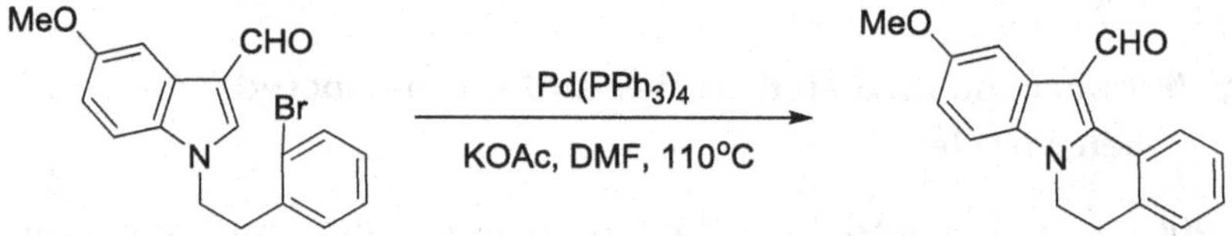

Scheme 40. Intramolecular C–H activation of halo amine compounds

51–98%

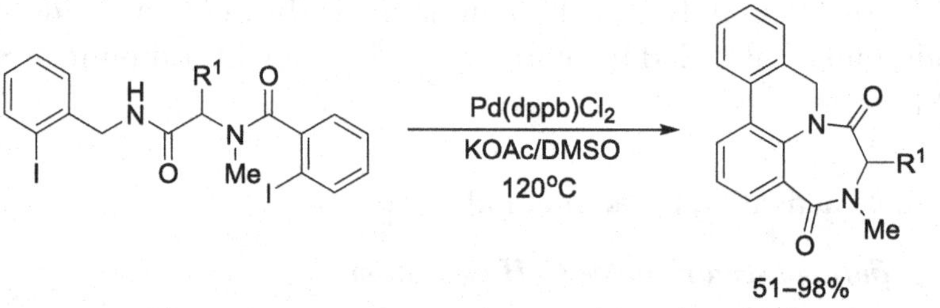

Scheme 41. Intramolecular C–H activation of halo amine compounds

indoles for probing the active site of the receptor for melatonin (Scheme 40).[49] The coupling reaction was catalyzed by $Pd(PPh_3)_4$ in the presence of KOAc as base in dimethylacetamide (DMA).

Zhu and co-workers reported a one-pot synthesis of polyhetero-cycles by a palladium-catalyzed intramolecular N-arylation/C–H acti-vation/aryl–aryl bond-forming domino process (Scheme 41).[50] The

1,4-benzodiazepine-2,5-dione derivatives can be prepared in good to excellent yields by the double cyclization of diiodo compounds. The ready accessibility of the starting material and the generality of this process clearly indicated its potentials in the diversity-oriented synthesis of this family of compounds.

2.7.3 Intramolecular C–H activation of nitrogen containing compounds

Fülöp and co-workers developed a Pd(II)-catalyzed intramolecular oxidative cyclization of tosyl-protected *cis*- and *trans*-*N*-allyl-2-amino-cyclohexanecarboxamides (Scheme 42).[51] The regio and diastereoselective synthesis of cyclohexane-fused pyrimidin-4-ones was achieved under the marked solvent effect. The novel Pd(II)-mediated domino oxidation/oxidative amination reaction proceed *via* a *cis*-aminopalladation mechanism.

Dai and coworkers reported a Pd(OAc)₂-dppf catalyzed intramolecular direct arylation to prepare the 5,6-dihydrophenanthridine derivative in 71% yield along with the minor inseparable debromination by-product in the presence of K_2CO_3 and DMA at 120°C (Scheme 43).[52]

2.7.4 Intermolecular C–H activation of halo and amine compounds

Wang and co-workers described an atom-economical cascade reaction through palladium-catalyzed dual C–H activation for the synthesis of phenanthridinones (Scheme 44).[53] This reaction sequence

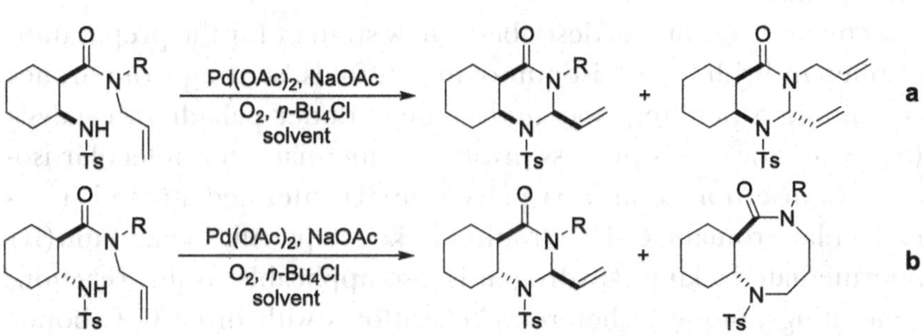

Scheme 42. Intramolecular C–H activation of nitrogen containing compounds

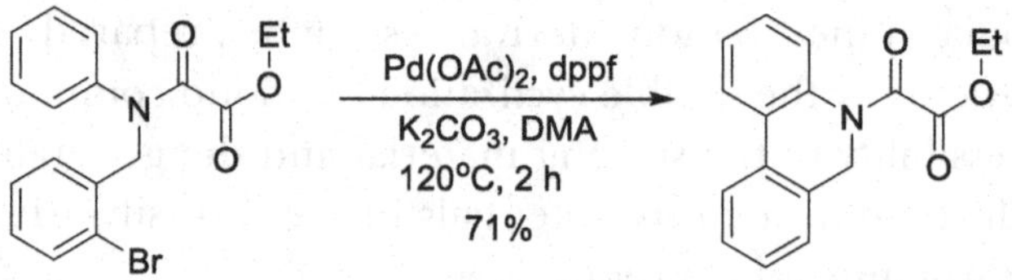

Scheme 43. Intramolecular dual C–H activation of nitrogen containing compounds

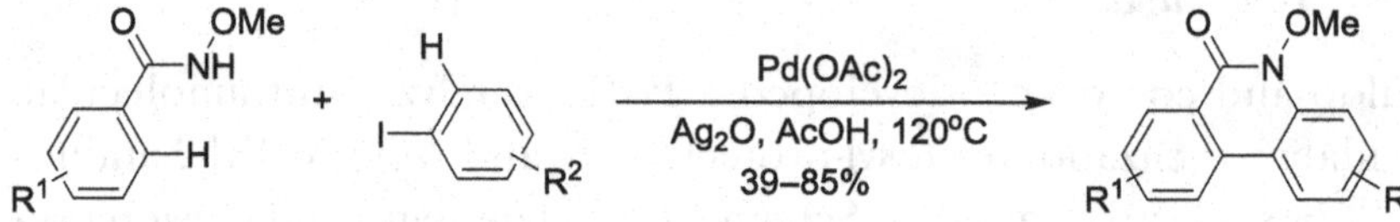

Scheme 44. Intermolecular C–H activation of benzamides and halides

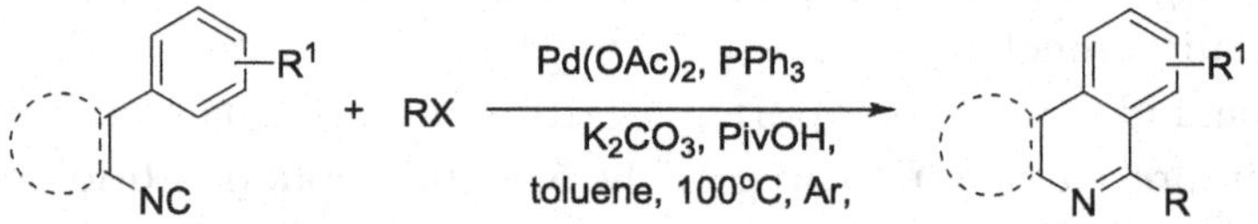

Scheme 45. Intermolecular C–H activation of isocyanides and halides

involves the rupture of two C–H bonds, one C–I bond, and one N–H bond, as well as the formation of one C–C bond and one C–N bond. The Pd(II)–Pd(IV)–Pd(II) and Pd(II)–Pd(0)–Pd(II) catalytic cycles operate simultaneously in the reaction. The replacement of benzamides with non-MeO-substituted benzamides did not afford satisfactory results.

Zhu and co-workers described a new strategy for the preparation of phenanthridine and isoquinoline scaffolds, starting from arenes containing a pending isocyanide moiety under palladium catalysis (Scheme 45).[54] This process involves sequential intermolecular isocyanide insertion to an aryl palladium(II) intermediate and intramolecular aromatic C–H activation as key steps. Alkyl palladium(II) intermediate lacking β-hydrogen is also applicable to this reaction, generating unique bisheterocyclic scaffolds with three C–C bonds being formed consecutively.

2.7.5 *Intermolecular C–H activation of nitrogen containing compounds*

Bao and co-workers developed a convenient one-pot synthesis of pyrimido[1,6-a]indol-1(2H)-one derivatives through a nucleophilic addition/Cu-catalyzed *N*-arylation/Pd-catalyzed C–H activation in a sequential process (Scheme 46).[55] The reaction of easily prepared *ortho*-gem-dibromovinyl isocyanates with *N*-alkyl-anilines gave the desired indole derivatives in moderate to good yields.

Wang and co-workers developed a tandem catalytic system that provides rapid access to vinyl-substituted 5,6-dihydropyridin-2(1*H*)-ones and 3,4-dihydroisoquinolin-1(2*H*)-ones under mild reaction conditions (Scheme 47).[56] A variety of aromatic amides and alkenyl amides bearing diverse substituents are compatible with the reaction conditions, delivering the cyclization product with high regio and stereoselectivities. The alkene effect is found to be the key factor for the success of the reaction.

Zhu and co-workers developed an efficient approach for the synthesis of nitrogen heterocycles containing a cyclic amidine moiety (Scheme 48).[57] The process involves palladium-catalyzed C(sp^2)–H activation and isocyanide insertion starting from readily accessible

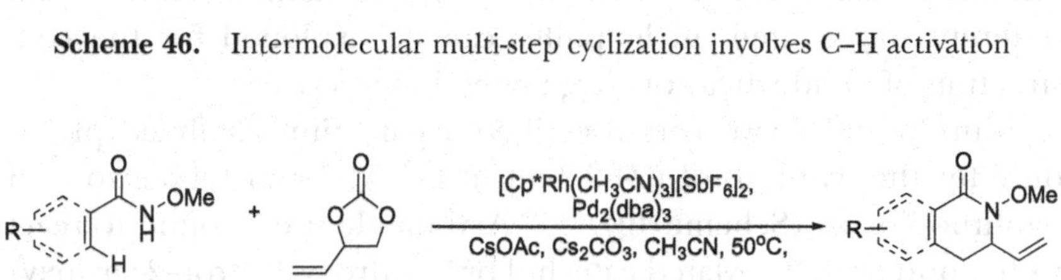

R^1 = H, Cl, Br, Meo

R^2 = H, CF$_3$; R^3 = Me, Bn

R^4 = H, Me, MeO, Cl, Br,CF$_3$O,

1) CuI, DMEDA, K$_2$CO$_3$, toluene, 120°C

2) Pd(dppf)Cl$_2$, KOAc, toluene, 120°C

16 examples, up to 87% yield

Scheme 46. Intermolecular multi-step cyclization involves C–H activation

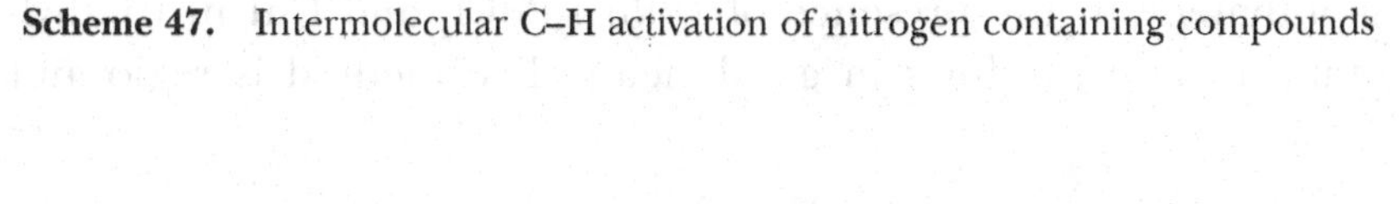

[Cp*Rh(CH$_3$CN)$_3$][SbF$_6$]$_2$, Pd$_2$(dba)$_3$

CsOAc, Cs$_2$CO$_3$, CH$_3$CN, 50°C,

Scheme 47. Intermolecular C–H activation of nitrogen containing compounds

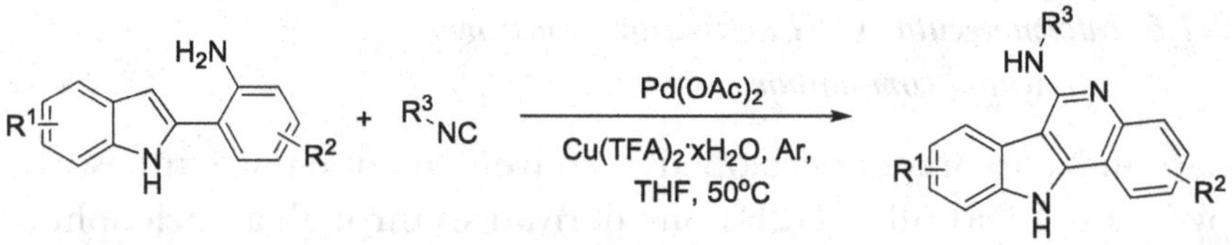

Scheme 48.   Intermolecular C–H activation of nitrogen containing compounds

ortho-heteroarene-substituted aniline derivatives under neutral and mild conditions. A broad range of indole and pyrrole substrates and isocyanides are tolerated.

3. Synthesis of Six-Membered Oxygen Containing Heterocycles

Six-membered *O*-heterocycles are probably one of the most common structural motifs spread across natural products, from simple glucose to structurally complex metabolites and marine natural products. Due to the remarkably rich array of functionalities and chiral centers that these heterocycles can incorporate, their regio and stereoselective synthesis has become a continuous challenge for organic chemists.

3.1 Addition of oxygen atom to C=C bond

3.1.1 *Addition of oxygen atom to triple bond*

The electrophilic activation of carbon–carbon triple bonds have also proved to be a highly efficient entry in the preparation of six-membered oxygen containing heterocycles. Pd-catalyzed intra and intermolecular cyclizations of precursors containing alkynes and different oxygenated nucleophiles can be exploited for the construction of a wide array of oxygenated heterocycles.

Kundu and co-workers dsecribed a palladium-catalyzed procedure for the synthesis of (Z)-2,3-dihydro-2-(ylidene)-1,4-benzo- and naphthodioxins (Scheme 49).[58a,58b] Aryl halides were found to react with mono-prop-2-ynylated catechol or 2-hydroxy-3-(prop-2-ynyloxy) naphthalene in the presence of (PPh$_3$)$_2$PdCl$_2$ and CuI in triethylamine to give products in good yields. The method is regio and

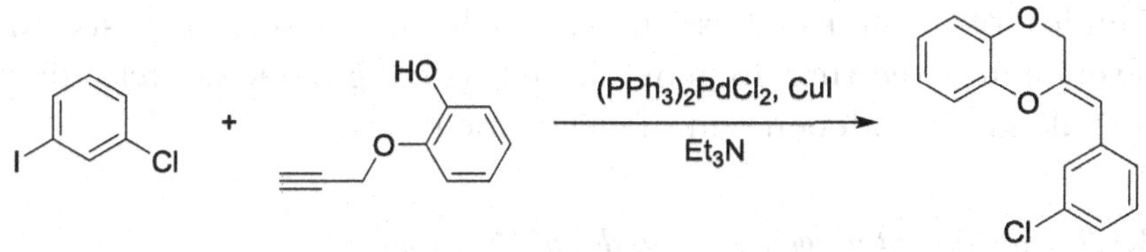

Scheme 49. Pd-catalyzed six-membered *O*-heterocycles *via* addition of oxygen to triple bond

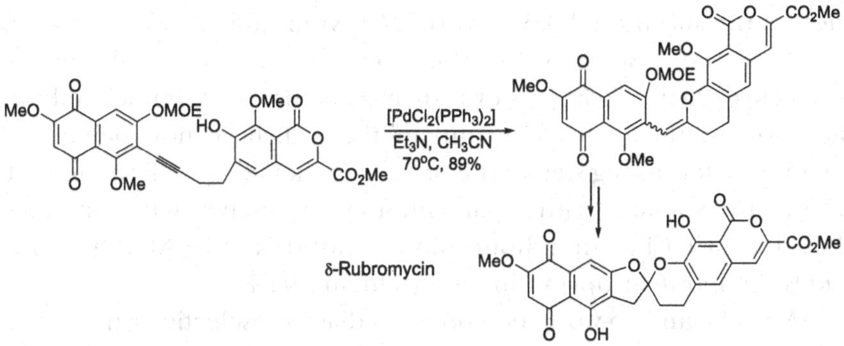

Scheme 50. The total synthesis of δ-Rubromycin

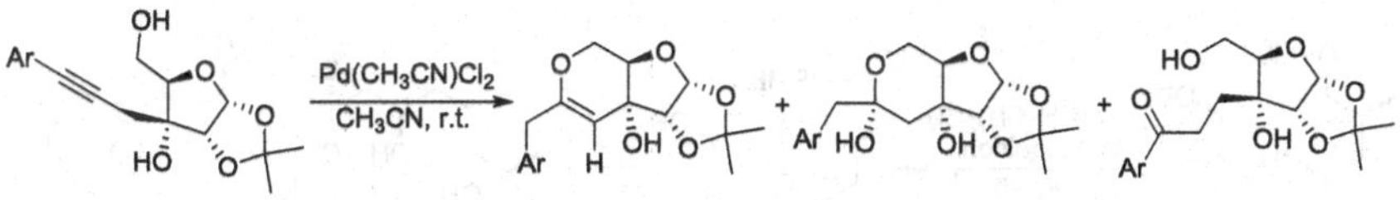

Scheme 51. Pd-catalyzed six-membered *O*-heterocycles *via* addition of oxygen to triple bond

stereoselective and also amenable to bis-heteroannulation. Recently, the reaction was applied to the total synthesis of δ-Rubromycin containing a bis-benzannulated 5,6-spiroketal moiety by Li and co-workers in a 2.7% overall yield (Scheme 50).[58c]

Ramana and co-workers described a Pd-mediated cycloisomerization of 3-*C*-propargyl-ribo- and allopyranose derivatives (Scheme 51),[59] the authors investigated the reaction in detail to understand the influence of electronic factors on the regioselectivity (6-exo- versus

7-endo) of alkynol cycloisomerization leading either to a six- or seven-membered ring. In general, the 6-exo-dig mode of cyclization is facile and is independent of electronic factors.

3.1.2 *Addition of oxygen atom to double bond via the π-allylpalladium cations*

Uenishi and co-workers developed an asymmetric total synthesis of (−)-laulimalide through Pd-catalyzed stereospecific ring construction of the substituted 3,6-dihydro[2H]pyran units (Scheme 52).[60]

Trost and co-workers described an intramolecular highly enantioselective allylic alkylations of hydroxyl alkenes *via* the nucleophilic addition of the hydroxyl group to the π-allylic cation being the enantio-determining step of the process (Scheme 53).[61] Treatment of the hydroxy alkene with a palladium catalyst generated by mixing $Pd_2(dba)_3.CHCl_3$, chiral diphosphine, and Et_3N in DCM at 0°C provides the tetrahydropyran in 80% yield and 94% ee.

Menche and co-workers reported a diastereoselective synthesis of functionalized tetrahydropyrans *via* a palladium-catalyzed domino *oxa*-Michael/Tsuji–Trost reaction with up to 72% yield and 82% ee

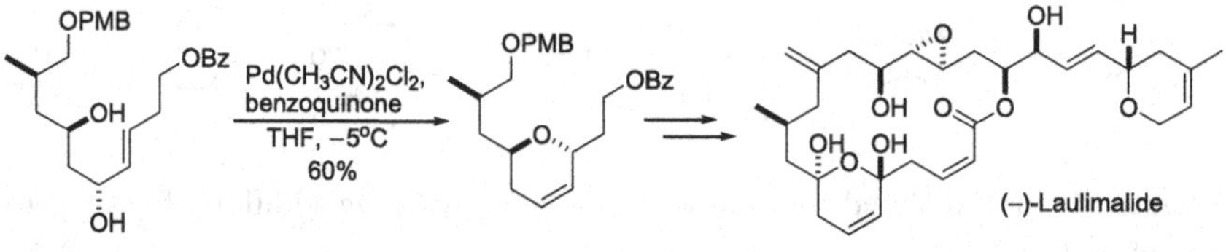

Scheme 52. Pd-catalyzed six-membered *O*-heterocycles *via* addition of oxygen to double bond

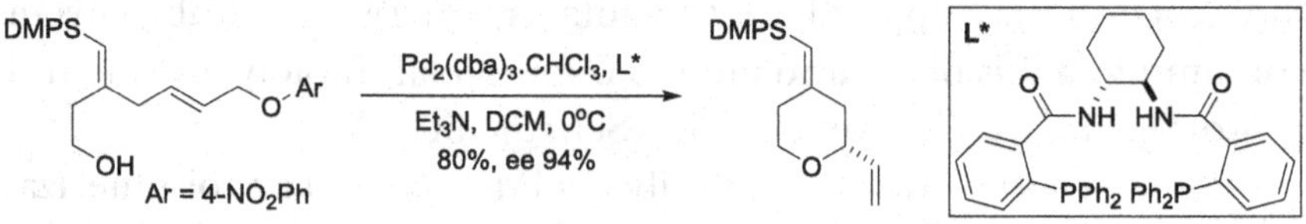

Scheme 53. Pd-catalyzed six-membered *O*-heterocycles *via* addition of oxygen to double bond

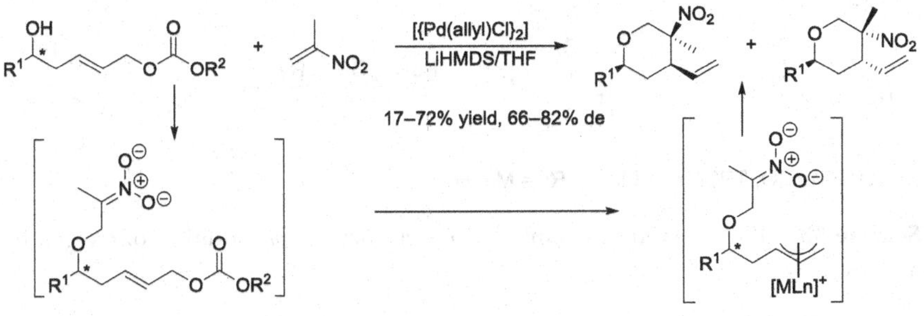

Scheme 54. Pd-catalyzed six-membered *O*-heterocycles *via* addition of oxygen to double bond

(Scheme 54).[62] The process underwent an enolate intermediate, which furnished a π-allyl complex. This π-allyl complex was subsequently trapped in an intramolecular fashion through an allylic substitution reaction to afford polysubstituted tetrahydropyran as a mixture of two diastereomers.

3.1.3 *Addition of oxygen atom to allenes*

Menche and co-workers described a heterocyclization/cross-coupling domino reaction of β, γ-allendiols and α-allenic esters for the synthesis of functionalized buta-1,3-dienyldihydropyrans (Scheme 55).[63] The chemo- and regio-controlled palladium-catalyzed methodology provides access to enantiopure six-membered heterocycles, 3,6-dihydro-pyrans that bear a buta-1,3-dienyl moiety, through cross-coupling reactions of two different allenes.

3.2 Heck-type cyclization

Frech and co-workers reported a Mizoroki–Heck reactions catalyzed by dichloro {bis[1-(dicyclohexylphosphanyl)piperidine]}palladium for the synthesis of 2*H*-Chromen-2-ones (Scheme 56).[64] The [(P{(NC$_5$H$_{10}$) (C$_6$H$_{11}$)$_2$})$_2$-Pd(Cl)$_2$] is a highly active Heck catalyst with excellent functional group tolerance and readily prepared in quantitative yield from the reaction of [Pd(cod)(Cl)$_2$] with two equivalents of 1-(dicyclohexylphosphanyl)piperidine.

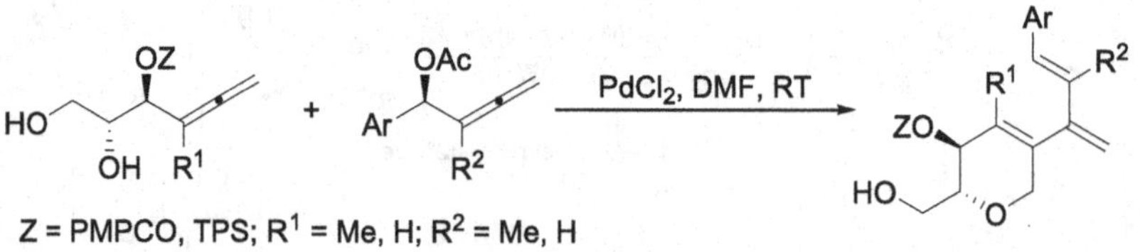

Z = PMPCO, TPS; R^1 = Me, H; R^2 = Me, H

Scheme 55. Pd-catalyzed six-membered *O*-heterocycles *via* addition of oxygen to allenes

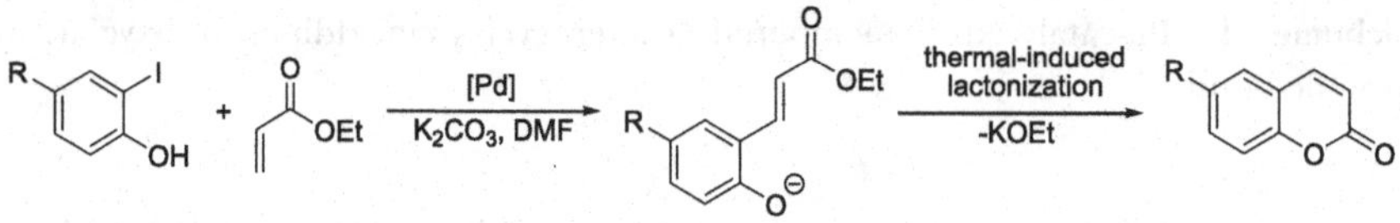

Scheme 56. Heck-type cyclization for six-membered *O*-heterocycles

R = H, OMe; X = CH$_2$, O

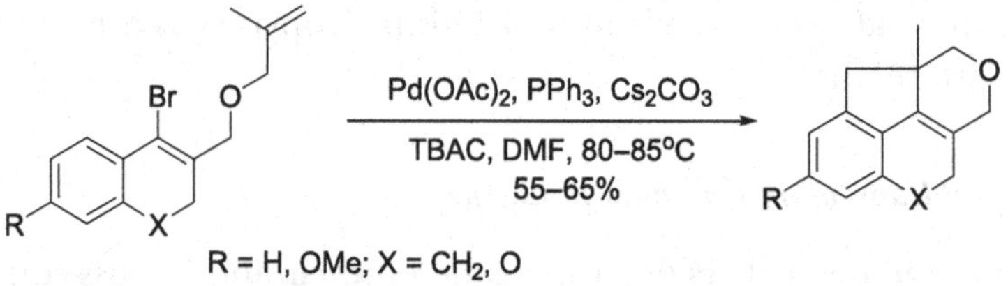

Scheme 57. Heck-type cyclization for six-membered *O*-heterocycles

Ray and co-workers described an efficient and convenient method for the synthesis of fused pyran rings *via* intramolecular palladium-catalyzed cyclization followed by β–H elimination or C–H activation (Scheme 57).[65] It is also possible to utilize this method to synthesis benzopyran systems.

3.3 Palladium-catalyzed cycloaddition for six-membered *O*-heterocycles

3.3.1 *Palladium-catalyzed cycloaddition of enynes for six-membered O-heterocycles*

Mikami and co-workers developed a palladium-catalyzed carbocyclization of 1,6-enynes leading to six-membered *O*-heterocycles by water

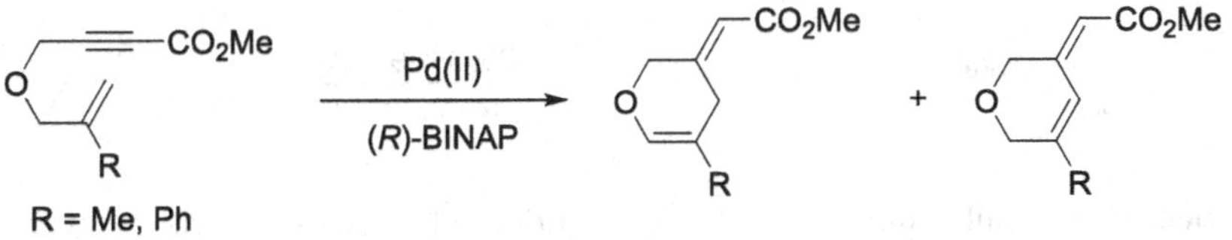

Scheme 58. Palladium-catalyzed six-membered ring formation from 1,6-enynes

originated hydride addition. Mechanistic studies show that generation of palladium hydride species comes from water (Scheme 58).[66] Subsequently, Yamamoto and co-workers also reported similar reaction and investigated the effects of substrate structure on the cyclization of enynes and enediynes.[67]

3.3.2 *Palladium-catalyzed cycloaddition of alkynes for six-membered N-heterocycles*

Tanaka and co-workers developed a palladium-catalyzed annulation reaction to furnish 4,5,6-trisubstituted 2*H*-pyran-2-ones in the presence of triethylamine (Scheme 59),[68] starting from internal alkynes and β-chloro-α, β-unsaturated esters. Treatment of methyl (Z)-3-chloro-2- heptenoate with Pd(PPh)$_4$ generated [(Z)-1-butyl-2-methoxycarbonylethenyl]chlorobis(triphenyl- phosphine)palladium *via* oxidative addition, which gave the corresponding 2*H*-pyran-2-one upon addition of 4-octyne.

Trost and co-workers reported an effective system combined palladium catalysts with formic acid for the catalytic *ortho* vinylation of phenols with activated alkynes (Scheme 60).[69] This catalyst system provided a new atom economic synthesis of coumarins through the simple addition of phenols to alkynoate esters. The scope of the reaction with respect to the phenol and the alkynoates is defined. For unsymmetrical aromatic substrates, generally good regioselectivity that reflects the HOMO coefficients was achieved.

Pal and co-workers reported a regioselective approach to thieno-pyranones through the coupling of bromo or iodo substituted thiophenecarboxylic acid with terminal alkynes (Scheme 61).[70] The best process for the preparation of 4,5-disubstituted derivatives involved

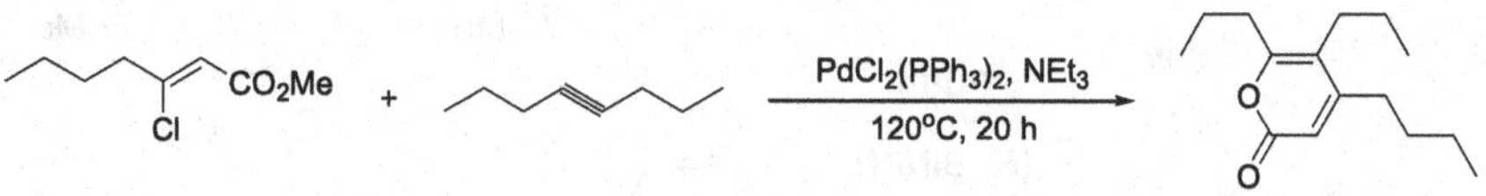

Scheme 59. Palladium-catalyzed cycloaddition of alkynes for six-membered *O*-heterocycles

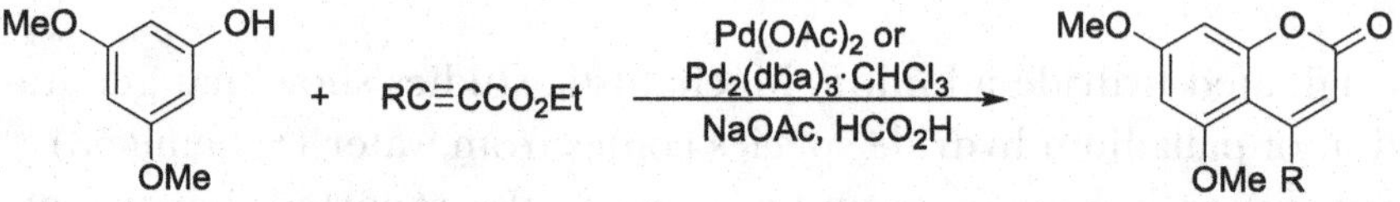

Scheme 60. Palladium-catalyzed cycloaddition of alkynes for six-membered *O*-heterocycles

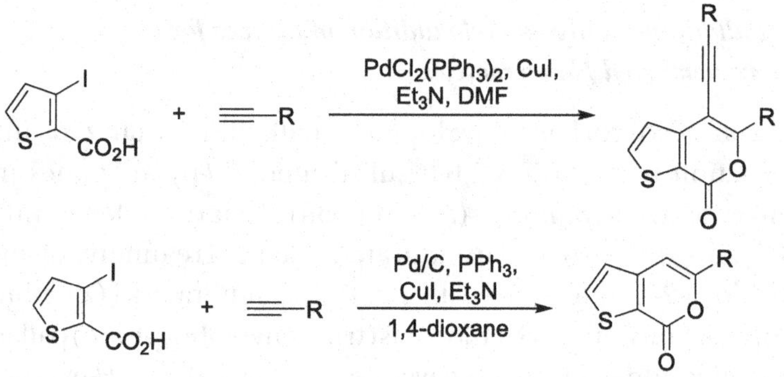

Scheme 61. Palladium-catalyzed cycloaddition of alkynes for six-membered *O*-heterocycles

the use of $PdCl_2(PPh_3)_2$ as a catalyst source and was found to be quite general and highly regioselective, while 5-substituted thieno[2,3-*c*] pyran-7-ones was obtained by using Pd/C-mediated coupling/cyclization of 3-iodothiophene-2-carboxylic acid with terminal alkynes.

3.3.3 *Palladium-catalyzed cycloaddition of allenes for six-membered N-heterocycles*

Ma and co-workers reported a Pd(II)-catalyzed cyclizative coupling reaction of 2,3- or 3,4-allenols with allylic halides for the preparation

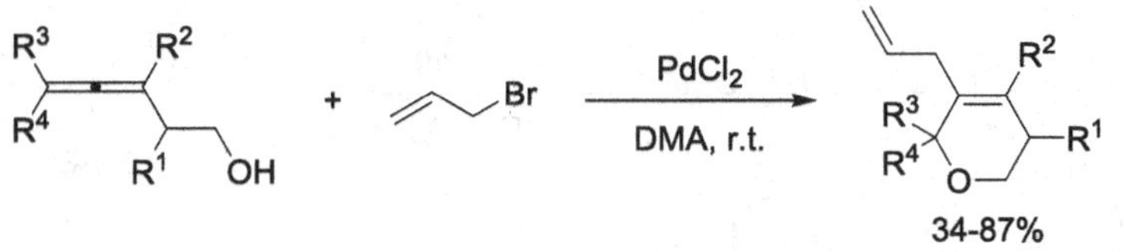

Scheme 62. Palladium-catalyzed cycloaddition of allenes for six-membered *O*-heterocycles

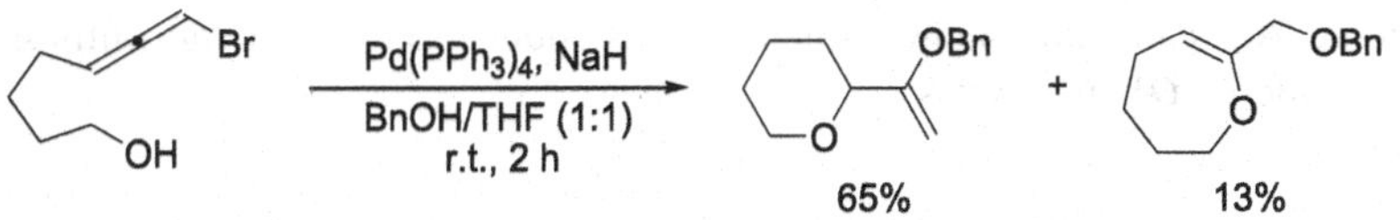

Scheme 63. Palladium-catalyzed cycloaddition of allenes for six-membered *O*-heterocycles

of polysubstituted 2,5-dihydrofurans or 5,6-dihydro-2*H*-pyrans in DMA at room temperature with moderate to good yields in the absence of base (Scheme 62).[71]

Tanaka and co-workers described a palladium(0)-catalyzed synthesis of six-membered *O*-heterocycles using bromoallenes as an allyl di-cation equivalent in the presence of alcohol. In many cases, this reaction proceeds in high regio and stereoselectivity, and affords desired six-membered or medium rings in good to high yields (Scheme 63).[72]

3.3.4 *Palladium-catalyzed hetero-Diels–Alder reactions for six-membered O-heterocycles*

Mikami and co-workers developed a powerful strategy of asymmetric activation for the use of *tropos* ligands without enantiomeric resolution or asymmetric synthesis (Scheme 64).[73] The metal complex with the *tropos* BIPHEP ligand and Palladium-catalyzed hetero-Diels–Alder reactions for the synthesis six-membered O-heterocycles (DABN) activator can establish, *in situ* asymmetric catalysis of carbon–carbon bond-forming reactions, higher enantioselectivity and

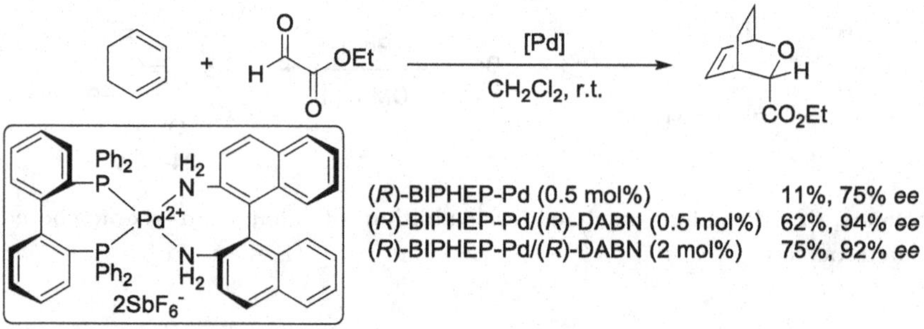

Scheme 64. Palladium-catalyzed hetero-Diels–Alder reactions for the synthesis of six-membered *O*-heterocycles

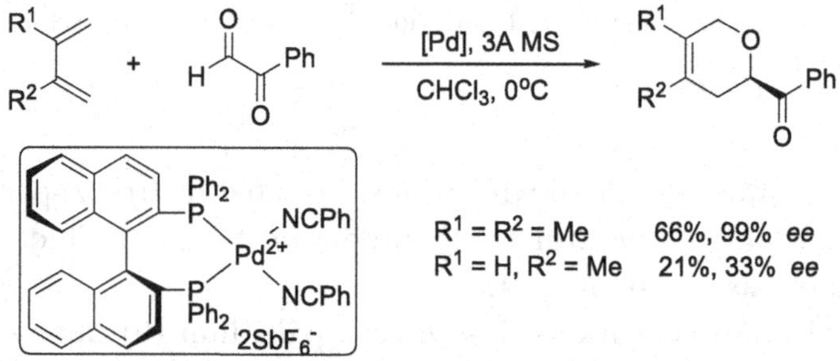

Scheme 65. Hetero-Diels–Alder reaction with phenylglyoxal catalyzed by cationic BINAP-Pd(II)

catalytic efficiency than those attained by the BIPHEP complex without DABN or the *atropos* and racemic BINAP complex with DABN. The catalytic system can prompt hetero-Diels–Alder reactions for the synthesis of six-membered *O*-heterocycles with up to 75% yield and 94% ee.

Oi and co-workers reported a hetero Diels–Alder reaction of non-activated conjugated dienes with arylglyoxals and glyoxylate esters proceeded enantioselectively in the presence of a catalytic amount of cationic chiral BINAP-palladium complexes and 3 Å molecular sieves (Scheme 65).[74] A chiral induction model involving the square-planar palladium complex coordinated with BINAP and a dienophile is believed.

3.4 Palladium-catalyzed annulation reaction for six-membered *O*-heterocycles

Buchwald and co-workers reported a palladium-catalyzed synthesis of aryl ethers involving primary and secondary alcohols. Cyclization of enantiopure alcohols results in cycled ethers without racemization under these reaction conditions with up to 95% yield and 99% ee (Scheme 66).[75]

Boulanger and co-workers applied the Buchwald annulation reaction to the total synthesis of (S)-equol (Scheme 67).[76] The reported route relies on an Evans alkylation to form the stereocenter and an intramolecular Buchwald etherification to generate the chroman ring. The intramolecular Buchwald etherification gained chroman derivatives in 46% yield at 50°C for 22 h.

Li and co-workers reported a mild method for the synthesis of 6*H*-benzo[c]chromenes by palladium-catalyzed annulations of 2-(2-iodophenoxy)-1-arylethanones and 1-(2-iodophenoxy)-propan-2-one with arynes (Scheme 68).[77] This mild route allows formation

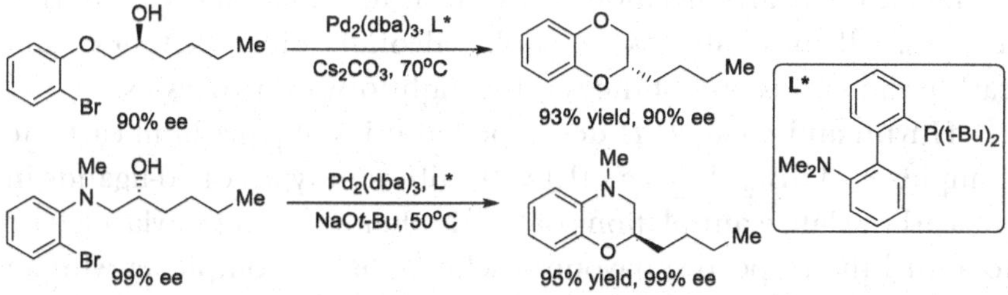

Scheme 66. Pd-catalyzed annulation reaction for six-membered *O*-heterocycles

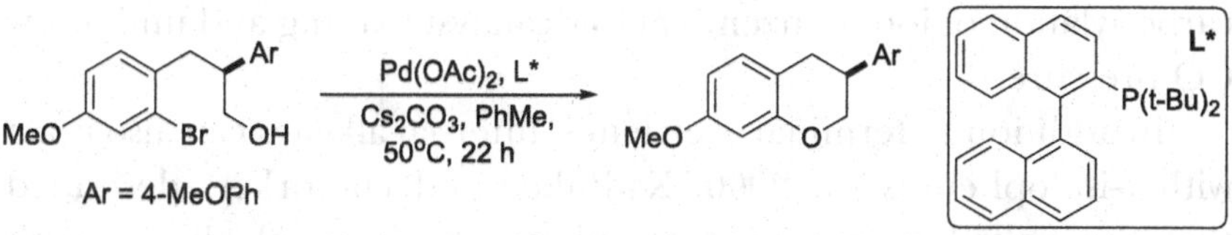

Scheme 67. Pd-catalyzed annulation reaction for six-membered *O*-heterocycles

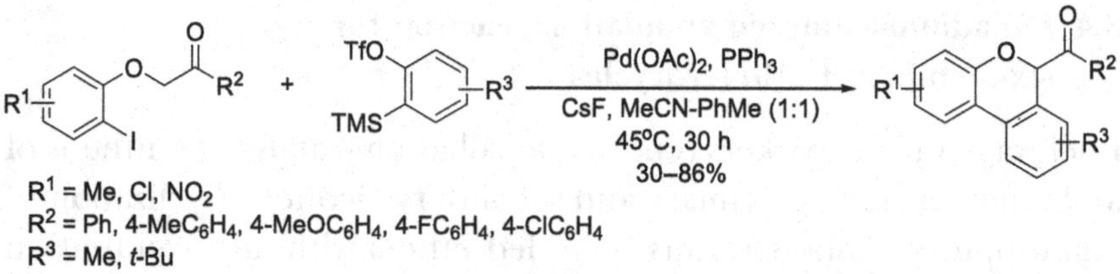
Pd(OAc)2, PPh3
CsF, MeCN-PhMe (1:1)
45°C, 30 h
30–86%
TfO
TMS
R1
R2
R3
O
I
R1 = Me, Cl, NO2
R2 = Ph, 4-MeC6H4, 4-MeOC6H4, 4-FC6H4, 4-ClC6H4
R3 = Me, t-Bu

Scheme 68. Pd-catalyzed annulation reaction for six-membered *O*-heterocycles

of two new carbon–carbon bonds *via* an α-arylation/annulation process in the presence of 2-(trimethylsilyl)phenyl triflate, an aryne precursor.

3.5 Palladium-catalyzed carbonylative synthesis of six-membered *O*-heterocycles

Palladium-catalyzed carbonylation reaction is a general and important methodology for the synthesis of carbonyl containing six-membered *O*-heterocycles. Among all the known procedures, the palladium catalyzed cyclocarbonylation of CO with terminal alkynes, internal alkynes, alkenes, allenes, enynols, alcohol, diols, ketones, and carboxylic acid is one of the most straightforward processes.

Huynh and co-workers developed a series of palladium carbene complexes [$PdBr_2(iPr_2$-bimy)L] with different types of co-ligands in the carbonylative annulations of 2-iodophenol with phenylacetylene to afford the respective flavones (Scheme 69).[78] Complexes with an *N*-phenylimidazole co-ligand provided high yields for the substrates such as aryl or pyridyl acetylenes. The Pd-*N*-heterocyclic carbene (NHC) complex also proved to be an efficient catalyst for the hydroxy-carbonylation of iodobenzenes at low catalyst loading and under low CO pressure.

In addition to terminal acetylenes, internal alkynes can also react with *o*-iodophenols. In 2000, Kadnikov and co-workers described that the palladium-catalyzed annulation of internal alkynes with *o*-iodophenols in the presence of CO resulted in exclusive formation of coumarins (Scheme 70).[79] 2-Iodophenol and 5 equiv of alkyne

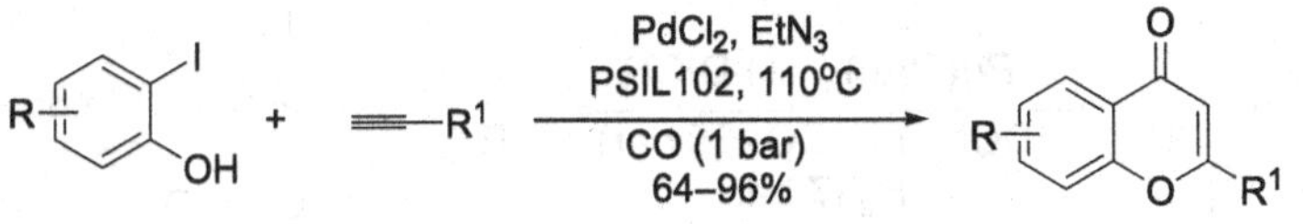

Scheme 69. Pd-catalyzed carbonylation of 2-iodophenol with phenylacetylene

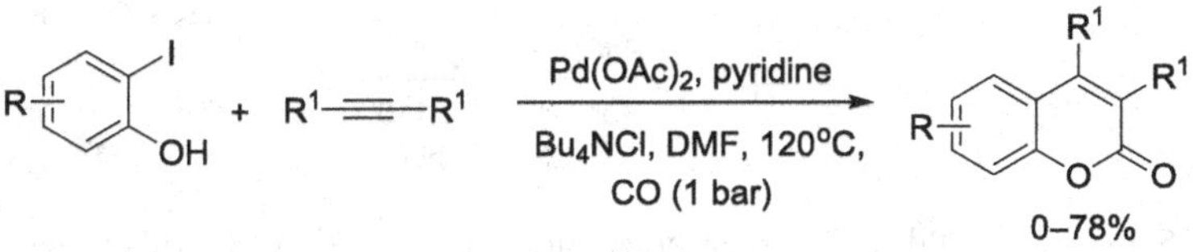

Scheme 70. Pd-catalyzed carbonylation of 2-iodophenol with internal alkynes

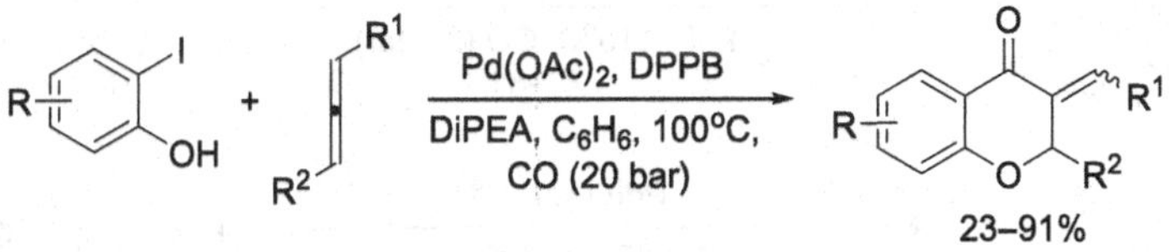

Scheme 71. Pd-catalyzed carbonylation of 2-iodophenol with allenes

were treated with 5 mol% Pd(OAc)$_2$ under 1 atm of CO in the presence of 2 equiv of pyridine, and 1 equiv of n-Bu$_4$NCl in DMF at 120°C. A sterically unhindered pyridine base was essential to achieve high yields, and the achieved chromones is a single isomeric.

Except the palladium-catalyzed carbonylative coupling of o-iodophenols with acetylenes, its coupling with allenes was also observed. Okuro and co-workers developed a regioselective process for the preparation of benzopyranones in fair to high yields (Scheme 71).[80] The Grigg group improved this methodology to be performed for o-iodoanilines at atmospheric CO pressure with Pd(PPh$_3$)$_4$ (5 mol%) as catalyst and K$_2$CO$_3$ as base.[81]

Dong and co-workers described that the cyclocarbonylation of o-isopropenylphenols using Pd(OAc)$_2$ and (+)-DIOP [DIOP = (2,2-dimethyl-1,3-dioxolane-4,5-diylbismethylene)bisdiphenyl-phosphine] as a chiral catalyst afforded 3,4-dihydro-4-methylcoumarins in the

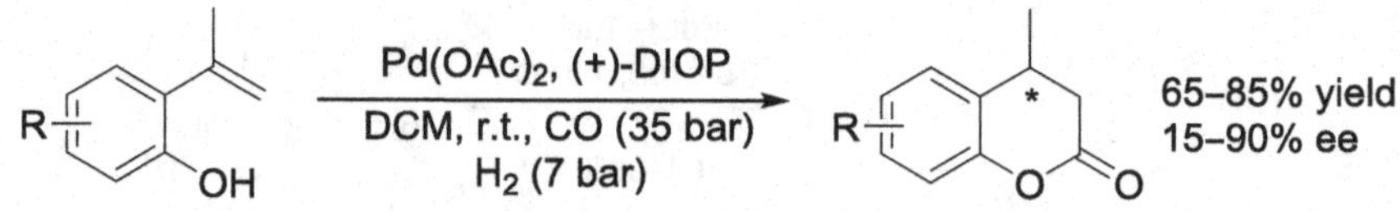

Scheme 72. Pd-catalyzed cyclocarbonylation of *o*-isopropenylphenols

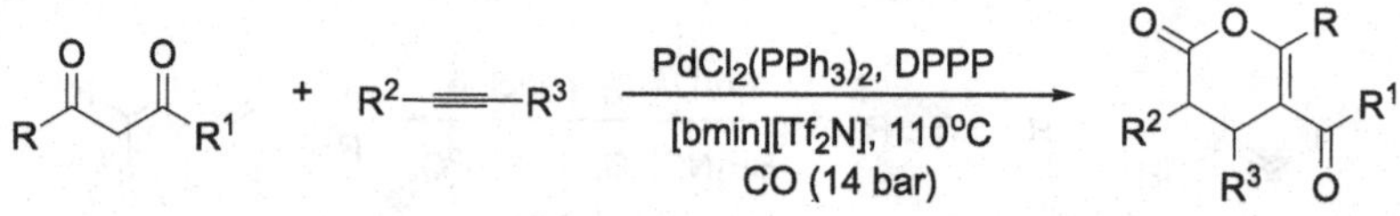

Scheme 73. Pd-catalyzed carbonylations of alkynes and 1,3-diketones

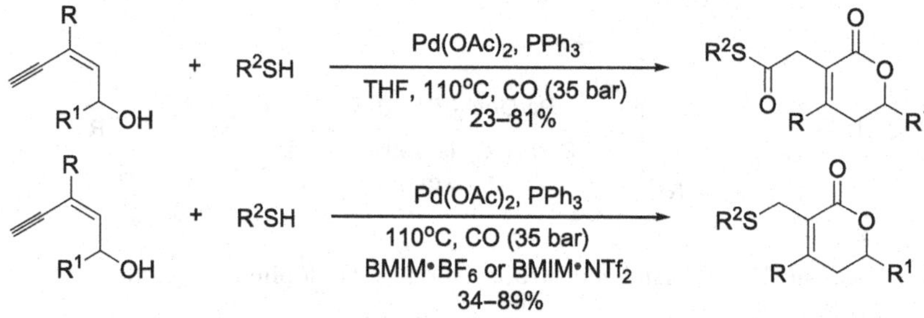

Scheme 74. Pd-catalyzed carbonylation of enynols with thiols

presence of CO (35 psi) and H_2 (7 psi) with 60–85% yield and up to 90% ee (Scheme 72).[82]

Li and co-workers developed an efficient ionic liquid-based protocol for the synthesis of highly substituted endocyclic enol lactones *via* carbonylations of alkynes and 1,3-diketones. The reactions proceeded in excellent regioselectivity and reasonably good yields (Scheme 73).[83] The catalyst system could be recycled five times with only modest loss of catalytic activity.

Cao and co-workers developed a palladium-catalyzed double carbonylation and cyclization reaction of enynols with thiols to form thioester-containing six-membered lactone rings with excellent selectivity and moderate to good yields (Scheme 74a).[84] They also conducted the reaction in ionic liquids, however, only monocarbonylated six-membered lactones were yielded (Scheme 74b).[85]

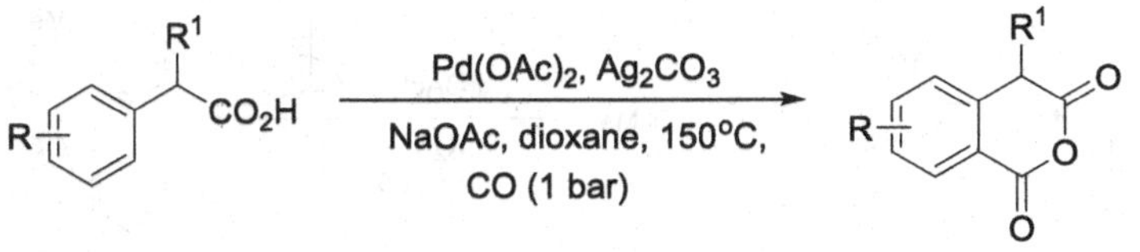

Scheme 75. Pd-catalyzed carbonylation of aromatic carboxylic acids

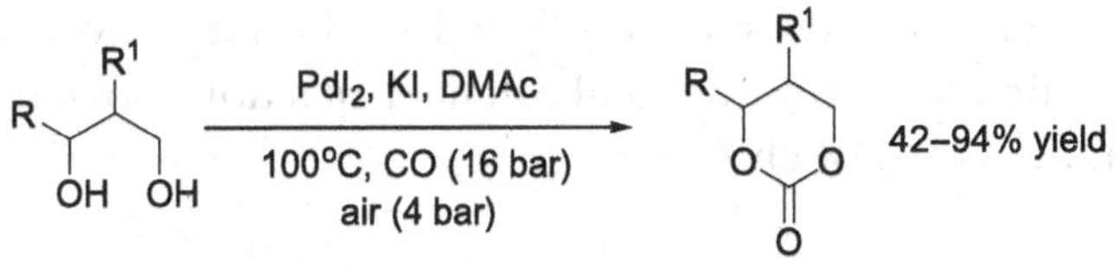

Scheme 76. Pd-catalyzed oxidative carbonylation of 1,3-diols

In 2008, Yu and co-workers developed a selective palladium-catalyzed C–H activation/carbonylation of aromatic carboxylic acids in the presence of $Pd(OAc)_2$, benzoic and phenylacetic acid derivatives were converted into *ortho*-substituted dicarboxylic acids in good yields (Scheme 75).[86]

Gabriele and co-workers developed an efficient method for the oxidative carbonylation of 1,3-diols to six-membered cyclic carbonates (Scheme 76).[87] 1, 3-diols underwent an oxidative carbonylation process in the presence of PdI_2-based system and provided six-membered cyclic carbonates in 66–74% yields. No oxidative addition of the active palladium complex to a C–X bond takes place in the reaction. Instead, activation of carbon monoxide followed by nucleophilic attack of the alcohol and subsequent reductive elimination afforded the cyclic carbonates.

3.6 Norbornene mediated synthesis of six-membered *O*-heterocycles

Catellani and co-workers described an effective Pd-catalyzed norbornene mediated strategy for the synthesis of 6*H*-dibenzopyran in 71% yield (Scheme 77).[88] Although the most substituents on the aryl bromide are the electron-withdrawing ones for the kinds of reaction,

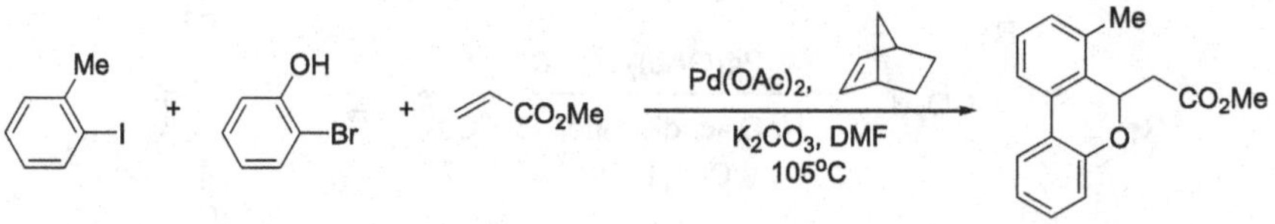

Scheme 77. Norbornene mediated synthesis of six-membered *O*-heterocycles

o-bromophenol reacted successfully led to the formation of a condensed cyclic compound by final Michael reaction, possibly due to the favorable effect of chelation.

3.7 C–H activation-based methodology

Baudoin and co-workers reported a new methodology for the synthesis of fused chromene derivatives in good yields and diastereoselectivities through double C–H arylations mediated by a single palladium/phosphine catalyst. Both double intermolecular/intramolecular and intramolecular/intramolecular C–C couplings were performed successfully (Scheme 78).[89]

Yamamoto and co-workers reported that phenols containing electron-donating groups with propiolic acids in the presence of Pd(II) species in acidic solvents give the corresponding coumarin derivatives (Scheme 79).[90] The same reaction has also been found to occur using a Pd(0) species such as $Pd_2(dba)_3$ in formic acid.

Ackermann and co-workers reported a Palladium-catalyzed intramolecular dehydrogenative direct arylations of 1,2,3-triazoles under ambient pressure of air for the synthesis of annulated phenanthrenes (Scheme 80).[91]

Peng and co-workers developed a one-step protocol for the preparation of xanthones *via* Pd-catalyzed annulations of 1,2-dibromoarenes and salicylaldehydes (Scheme 81).[92] The success of the reaction heavily relies on the careful selection of proper Pd-catalyst, solvent and base.

Yu and co-workers developed a palladium-catalyzed carbonylation of phenethyl alcohols, which gives access to the corresponding saturated isochromanones with moderate to excellent yields (Scheme 82).[93]

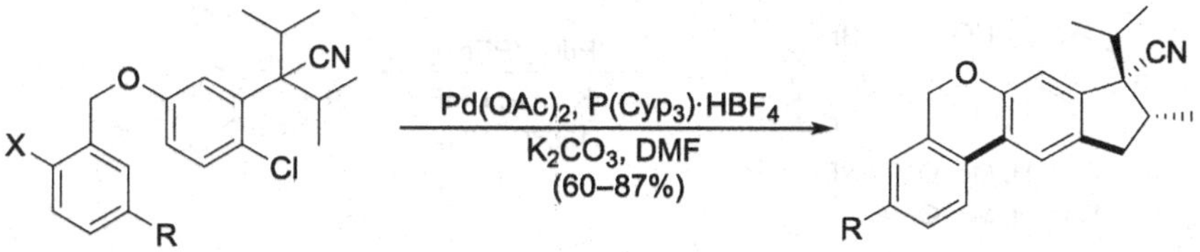

Scheme 78. Pd-catalyzed synthesis of fused chromene derivatives

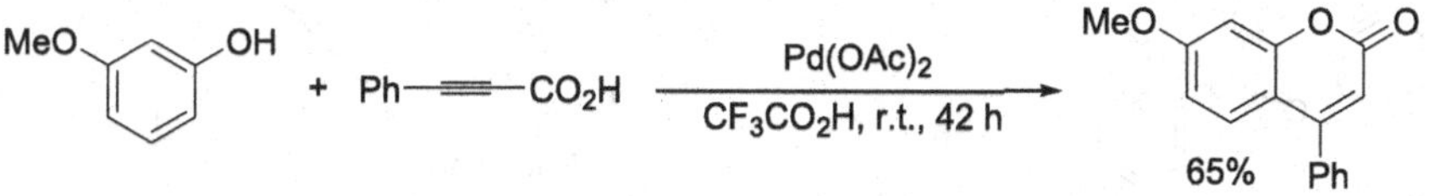

Scheme 79. Pd-catalyzed coupling reaction of phenols with propiolic acids

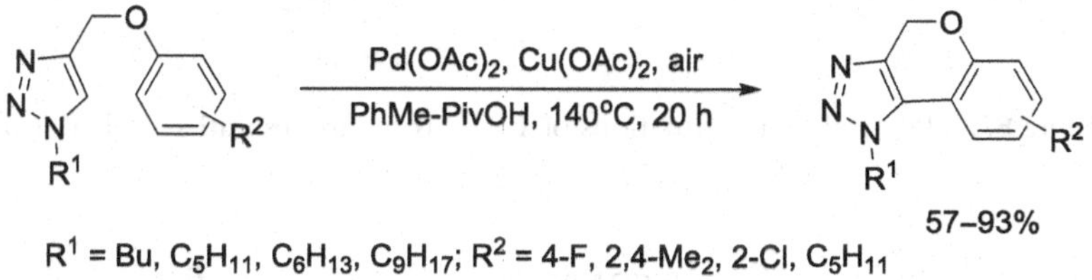

R^1 = Bu, C_5H_{11}, C_6H_{13}, C_9H_{17}; R^2 = 4-F, 2,4-Me$_2$, 2-Cl, C_5H_{11}

Scheme 80. Pd-catalyzed intramolecular dehydrogenative direct arylations of 1,2,3-triazoles

The desired products were achieved *via* C–H activation using amino acid ligands and over stoichiometric amounts of silver acetate.

4. Miscellaneous (Six-Membered Heterocycles Containing S, P, Si, etc.)

4.1 Ring-expansion reaction for other six-membered heterocycles

In 1991, Oshima and co-workers reported a palladium catalyzed ring-expansion reaction of silacyclobutanes with alkynes and allenes for the synthesis of silacycles (Scheme 83).[94] Recently, Zhang and co-workers synthesized six-membered silacycle odorants Si-artemone,

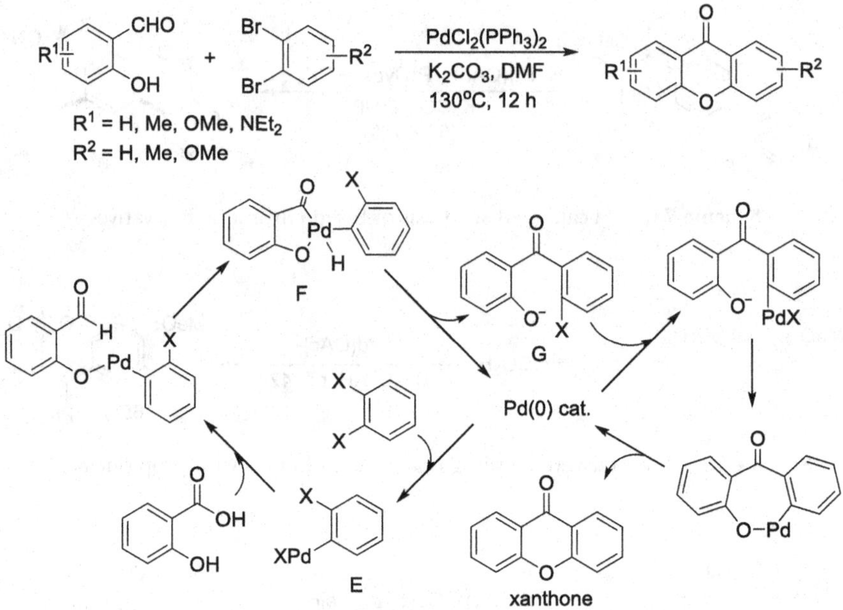

R^1 = H, Me, OMe, NEt$_2$
R^2 = H, Me, OMe

Scheme 81. Pd-catalyzed annulations of 1,2-dibromoarenes and salicylaldehydes

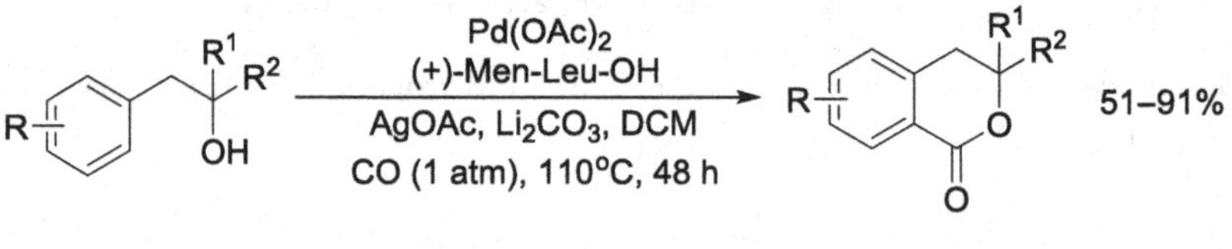

51–91%

Scheme 82. Pd-catalyzed C–H activation of phenethyl alcohols

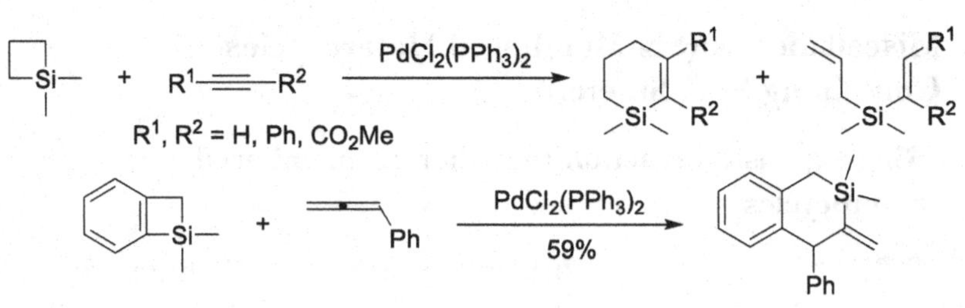

R^1, R^2 = H, Ph, CO$_2$Me

59%

Scheme 83. Pd-catalyzed ring-expansion reaction of silacyclobutanes for the synthesis of six-membered heterocycles

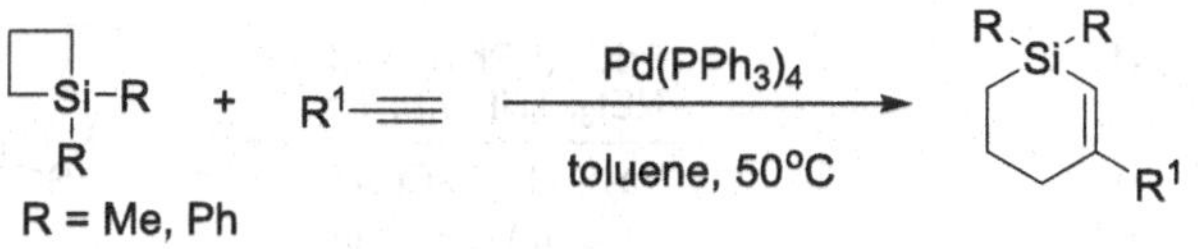

Scheme 84. Pd-catalyzed ring-expansion reaction of silacyclobutanes

Si-β-dynascone, and Si-herbac in high efficiency by insertion reactions of terminal alkynes with silacyclobutane (Scheme 84).[95] Their carbon counterparts artemone, β-dynascone, and herbac demand much more laborious synthetic procedures. The authors also researched structure–odor relationships between silacycles and their counterpart carbocycles.

4.2 Heck-type cyclization

Müller and co-workers described a microwave-assisted coupling-addition-S_NAr (CASNAR) sequence to construct 4H-thiopyran-4-ones starting from readily available (het)aroyl chlorides, alkynes, and sodium sulfide nonahydrate in a consecutive one-pot three component reaction (Scheme 85).[96] A whole family of annelated 4H-thiopyran-4-ones as the core structural unit was readily synthesized in good yields.

Harmata and co-workers reported a new approach to 1,2-benzothiazines possessing a sulfoximine functional group (Scheme 86).[97] The reaction of S-2-bromophenyl-S-methyl-sulfoximine with terminal alkynes in the presence of a palladium catalyst resulted in the formation of 1,2-benzothiazines and 1,2-benzoisothiazoles. A preference for the former was seen with alkylalkynes, while the latter were preferentially formed with alkynylarenes. Pal and co-workers also reported a practical method of constructing a thiazine ring fused with benzene under mild reaction conditions *via* Pd/C-mediated C–C coupling followed by iodocyclization strategy.[98]

Larock and co-workers described a palladium catalyzed annulation reaction of certain hindered propargylic alcohols with (2-iodophenyl)acetonitrile to afford 1,3-benzoxazine derivatives instead of the anticipated 2-aminonaphthalenes (Scheme 87).[99] This reaction

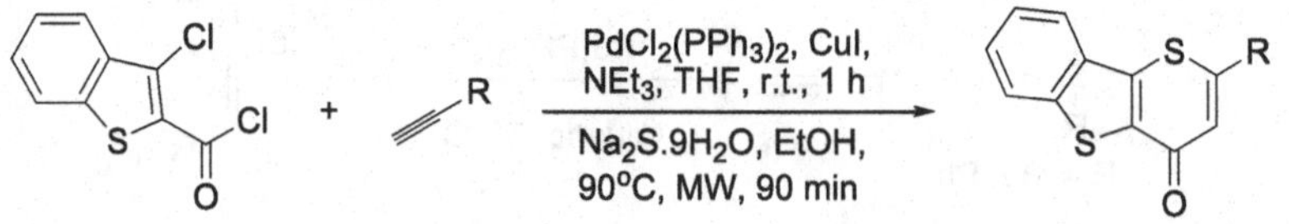

Scheme 85. Pd-catalyzed microwave-assisted coupling-addition-S_NAr sequence

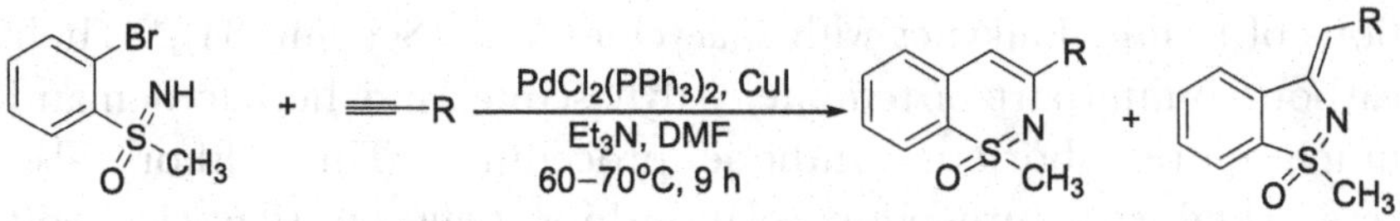

Scheme 86. Pd-catalyzed synthesis of 1,2-benzothiazines

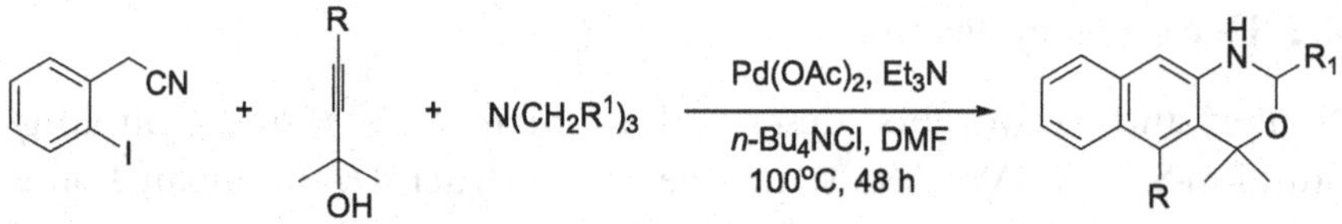

Scheme 87. Pd-catalyzed annulation of propargylic alcohols with (2-iodophenyl) acetonitrile

apparently proceeds with involvement of the trialkylamine base present in the reaction, which transfers one of its alkyl groups to the final product. The reaction mechanism mentioned by authors showed that the synthesis of 1,3-benzoxazine derivatives involves the formation of the expected 2-amino-3-(1-hydroxyalkyl) naphthalenes, followed by their condensation with an iminium ion species formed from the trialkylamine base.

4.3 Cycloaddition reactions for other six-membered heterocycles

Ding and co-workers reported a novel $PdCl_2(CH_3CN)_2$-catalyzed cyclization reaction of o-(1-alkynyl)phenylphosphonamide monoethyl esters to phosphaisoquinolin-1-ones with high regioselectivity and good yields. The present reaction is the first example of intramolecular addition of P–NH to substituted alkynes, which provides

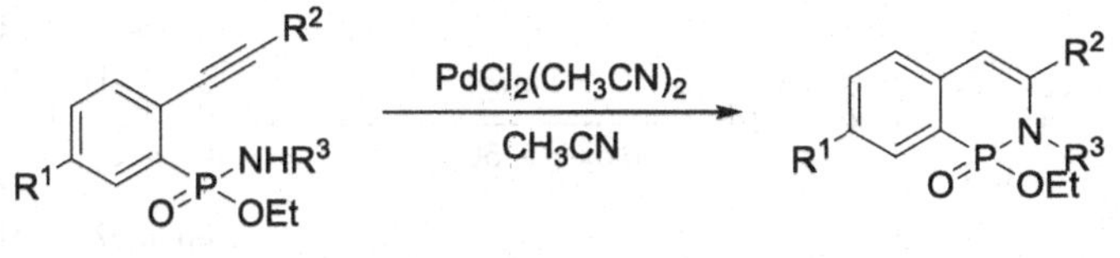

Scheme 88. Pd-catalyzed synthesis of phosphaisoquinolin-1-ones

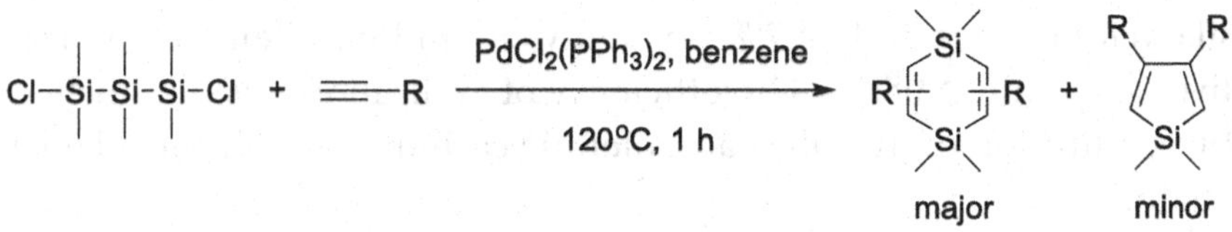

Scheme 89. Pd-catalyzed synthesis of 1,4-disilacyclohexa-2,5-dienes

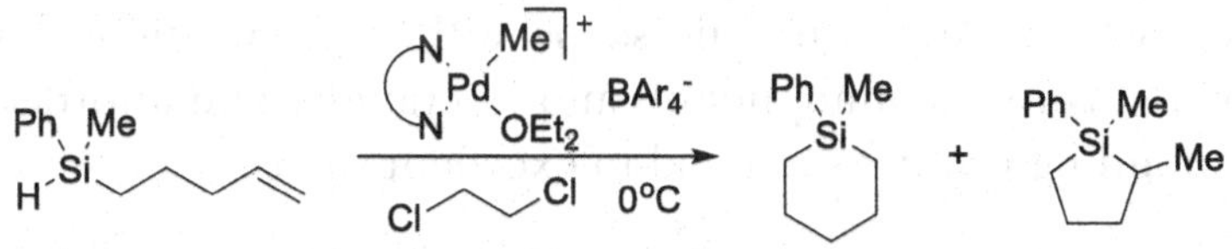

Scheme 90. Pd-catalyzed intramolecular hydrosilylation of alkenylsilanes

a valuable way to synthesize novel phosphorus heterocycles with potential bioactivities (Scheme 88).[100]

Tanaka and co-workers described an unexpected formation of 1,4-disilacyclohexa-2,5-dienes in the palladium-catalyzed reactions of $Cl(SiMe_2)_3Cl$ with acetylenes at 120–140°C in 20–82% yields and its application to polymer synthesis (Scheme 89).[101] Terminal acetylenes exhibit higher reactivities than internal acetylenes in the reaction. $PdCl_2L_2$ catalysts with L = $P(aryl)_3$ or $AsPh_3$ are efficient, while those with L = $P(alkyl)_3$ or $PhCN$ are not effective. Use of $X(SiMe_2)_3X$ (X = F, OMe) in place of $Cl(SiMe_2)_3Cl$ produces very little 1,4-disilacyclohexa-2,5-diene.

Widenhoefer and co-workers reported a palladium-catalyzed intramolecular hydrosilylation of alkenylsilanes for the selective formation of silacyclohexanes (Scheme 90).[102] The cationic palladium complex $(phen)Pd(Me)(OEt_2)^+BAr_4^-$ catalyzed the intramolecular hydrosilylation of 4-pentenyl silanes and 5-hexenylsilanes to form

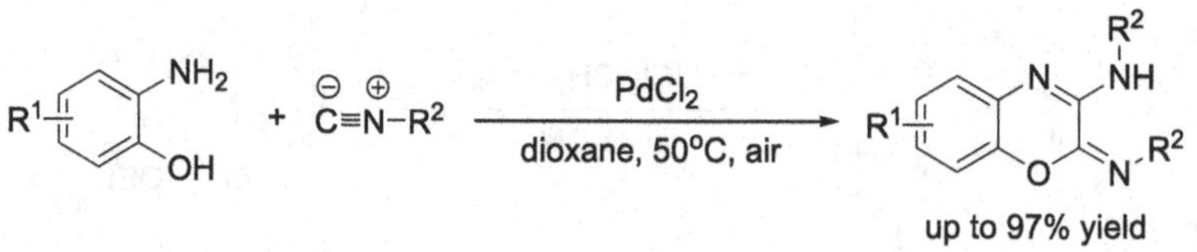

J. Mao & W. Bao

Scheme 91. Pd-catalyzed aerobic oxidation of *o*-aminophenols and isocyanides

silacyclohexanes in 43–87% isolated yield and excellent regioselectivity (typically ≥ 98:1). The efficiency of cyclization was diminished by substitution on the alkyl and olefinic carbon atoms of the alkenyl chain.

Jiang and co-workers developed a Pd-catalyzed aerobic oxidation of *o*-aminophenols and isocyanides for the preparation of 3-aminobenzoxazines in an air atmosphere (Scheme 91).[103] This method was applicable to a large synthetic scope with wide functional group compatibility for *o*-aminophenols and isocyanides and afforded the corresponding products in good to excellent yields.

4.4 Annulation reations for other six-membered heterocycles

Jiang and co-workers described a novel method for the synthesis of substituted 1,4-benzothiazine derivates *via* a Pd-catalyzed coupling reaction (Scheme 92).[104] The important feature of this method is using stable $Na_2S_2O_3$ salt as sulfurating reagent which makes it free from foul-smelling thiols, and render this protocol attractive for studies on metal catalyzed C–S bond formation by employing those inexpensive readily available reagents.

4.5 Palladium-catalyzed carbonylative synthesis of other six-membered heterocycles

Beller and co-workers reported a general and efficient palladium-catalyzed carbonylative synthesis of 2-aryloxazines from aryl bromides in moderate to good yields (Scheme 93).[105] Both electron-donating- and electron-withdrawing group-substituted aryl bromides were reacted and provided the corresponding products in good yield. Heterocycles containing aryl bromides were also reacted with the aminochloride to afford the desired products in good yields.

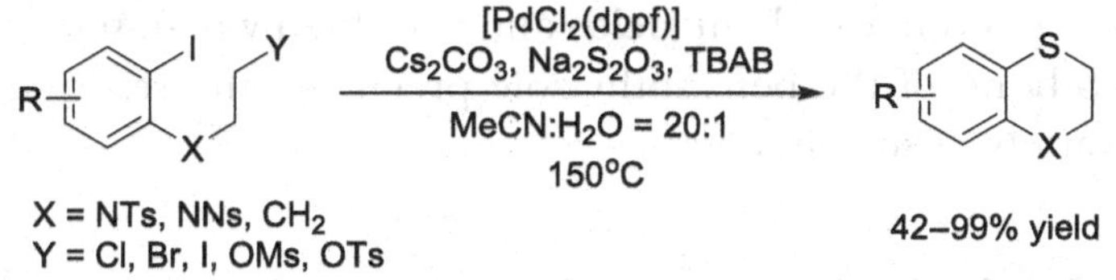

Scheme 92. Pd-catalyzed synthesis of substituted 1,4-benzothiazine derivates

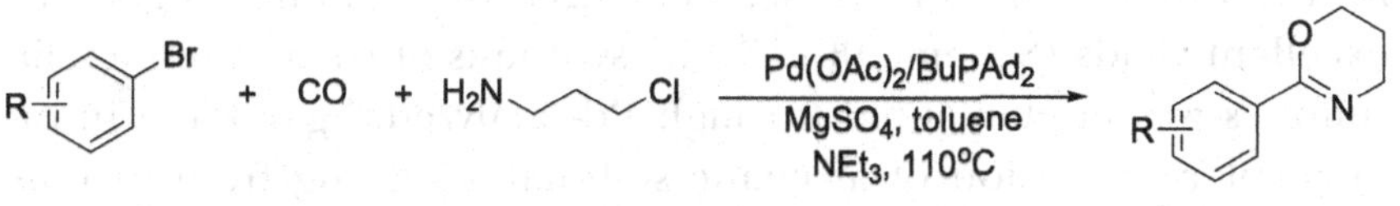

X = NTs, NNs, CH$_2$
Y = Cl, Br, I, OMs, OTs

[PdCl$_2$(dppf)]
Cs$_2$CO$_3$, Na$_2$S$_2$O$_3$, TBAB
MeCN:H$_2$O = 20:1
150°C

42–99% yield

Scheme 93. Pd-catalyzed carbonylative synthesis of 2-aryloxazines

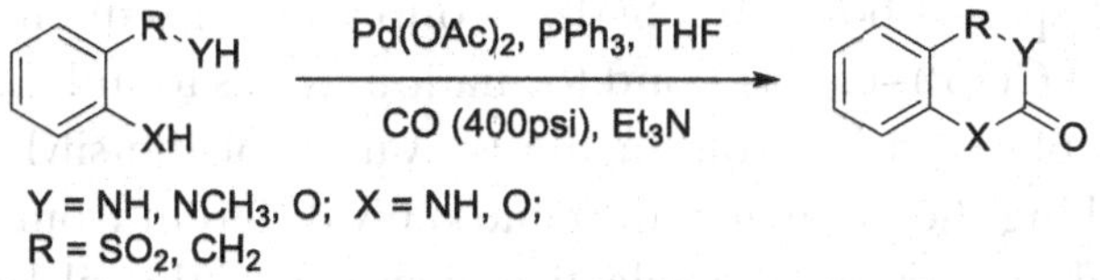

Pd(OAc)$_2$/BuPAd$_2$
MgSO$_4$, toluene
NEt$_3$, 110°C

Scheme 94. Pd-catalyzed cyclocarbonylation reaction

Y = NH, NCH$_3$, O; X = NH, O;
R = SO$_2$, CH$_2$

Pd(OAc)$_2$, PPh$_3$, THF
CO (400psi), Et$_3$N

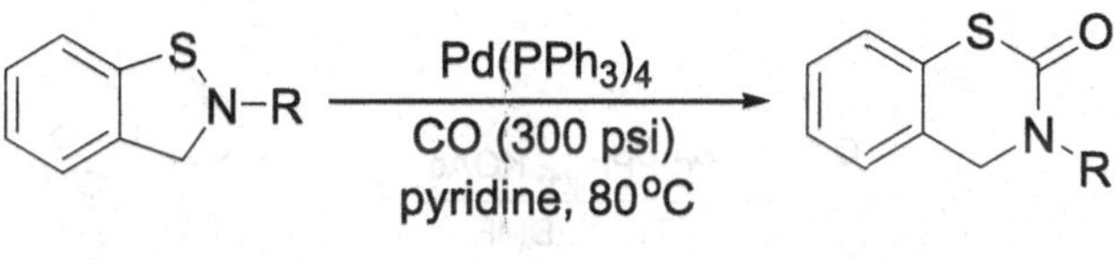

Scheme 95. Pd-catalyzed synthesis of 3-substituted-3,4-dihydro-2H-1,3-benzothiazin-2-ones

Pd(PPh$_3$)$_4$
CO (300 psi)
pyridine, 80°C

Troisi and co-workers described an efficient and simple synthetic procedure for the preparation of benzo-fused six-membered heterocycles by palladium-catalyzed cyclocarbonylation reaction (Scheme 94).[106] Such an easy procedure can be applied to the synthesis of products having great pharmaceutical and agrochemical interest in good yields.

Alper and co-workers reported a novel strategy for the synthesis of 3-substituted-3,4- dihydro-2H-1,3-benzothiazin-2-ones *via* the palladium-catalyzed carbonylation reaction of 2-substituted-2,3-dihydro-1,2-benzisothiazoles (Scheme 95).[107] This carbonylative insertion process occurs in good to excellent yields and with highly regioselectivity at the N–S bond of the benzisothiazole precursor, the reaction tolerates a variety of substituents.

4.6 C–H activation-based methodology

Majumdar and co-workers described a short and efficient method for the synthesis of aromatic six-membered ring sultones in good to excellent yields (Scheme 96).[108] The synthesis of tri- and tetracyclic sultones was implemented through Pd-catalyzed, ligand-free intramolecular cyclization of aromatic sulfonates starting from various bromo phenols and naphthols.

Xi and co-workers reported a Pd-catalyzed selective cleavage of the silyl $C(sp^3)$–H bond in a $SiMe_3$ group and consequent intramolecular silyl $C(sp^3)$–$C(sp^2)$ bond formation. It was found that the silicon atom played a key role in the activation of the silyl $C(sp^3)$–H bond, making the $SiMe_3$ group remarkably different from the CMe_3 group and prompting the selective cleavage of the silyl $C(sp^3)$–H bond in a trialkylsilyl group to construct silacycles (Scheme 97).[109]

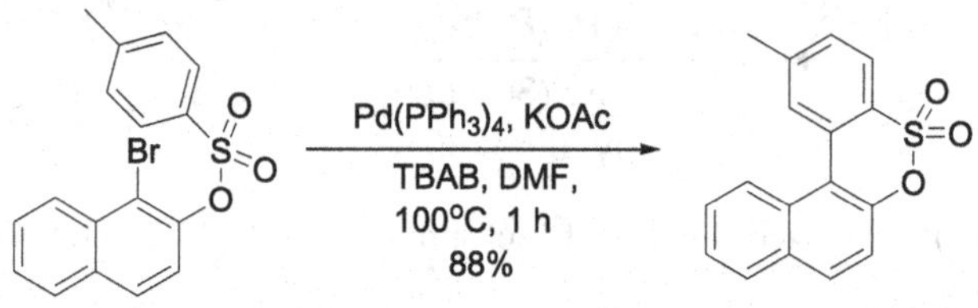

Scheme 96.　Pd-catalyzed synthesis of aromatic six-membered ring sultones

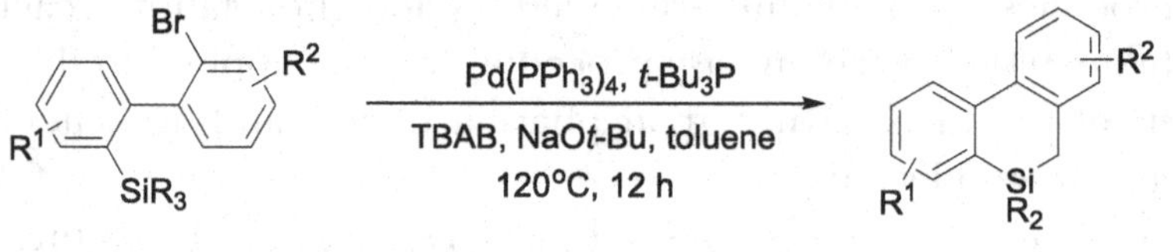

Scheme 97.　Pd-catalyzed intramolecular silyl $C(sp^3)$–$C(sp^2)$ bond formation

References

1. (a) F. Diederich and P. J. Stang, 1998. *Metal-catalyzed Cross-coupling Reactions*. Wiley-VCH, Weinhein, Germany. (b) V. Farina, V. Krishnamurthy and W. J. Scott, 1998. *The Stille Reaction*. Wiley, New York, NY. (c) J. Tsuji, 1999. *Perspectives in Organopalladium Chemistry for the 21st Century*. Elsevier, Lausanne, Switzerland. (d) J. Tsuji, 1995. *Palladium Reagents and Catalysts: Innovations in Organic Synthesis*. Wiley, Chichester, UK (e) L. S. Hegedus, 1999. *Transition Metals in the Synthesis of Complex Organic Molecules*, 2nd Edition. University Science Books, Mill Valley, USA. (f) J.-L. Malleron, J.-C. Fiaud and J.-Y. Legros, 1997. *Handbook of Palladium-catalyzed Organic Reactions*. Academic Press, San Diego, USA.

2. (a) J. P. Wolfe and J. S. Thomas, 2005. *Curr. Org. Chem.*, 9, 625. (b) I. Nakamura and Y. Yamamoto, 2004. *Chem. Rev.*, 104, 2127. (c) G. Zeni and R. C. Larock, 2004. *Chem. Rev.*, 104, 2285. (d) M. Rubin, A. W. Sromek and V. Gevorgyan, 2003. *Synlett*, 2265.

3. (a) G. Dai and R. C. Larock, 2001. *Org. Lett.*, 3, 4035. (b) G. Dai and R. C. Larock, 2003. *J. Org. Chem.*, 68, 920.

4. (a) K. R. Roesch and R. C. Larock, 1998. *J. Org. Chem.*, 63, 5306. (b) K. R. Roesch, H. Zhang and R. C. Larock, 2001. *J. Org. Chem.*, 66, 8042.

5. C. Chowdhury, K. Brahma, S. Mukherjee and A. K. Sasmal, 2010. *Tetrahedron Lett.*, 51, 2859.

6. Z. Lu and S. S. Stahl, 2012. *Org. Lett.*, 14, 1234.

7. X. Wang, Z. Wu, X. Zhu, C. Ye, F. Jiang and W. Zhang, 2013. *Chin. J. Chem.*, 31, 132.

8. E. Merişor and U. Beifuss, 2007. *Tetrahedron Lett.*, 48, 8383.

9. L. M. Lutete, I. Kadota and Y. Yamamoto, 2004. *J. Am. Chem. Soc.*, 126, 1622.

10. M. Narsireddy and Y. Yamamoto, 2008. *J. Org. Chem.*, 73, 9698.

11. H. Ohno, A. Okano, S. Kosaka, K. Tsukamoto, M. Ohata, K. Ishihara, H. Maeda, T. Tanaka and N. Fujii, 2008. *Org. Lett.*, 10, 1171.

12. G. B. Bajracharya, Z. Huo and Y. Yamamoto, 2005. *J. Org. Chem.* 70, 4883.

13. S. Ma, F. Yu and W. Gao, 2003. *J. Org. Chem.*, 68, 5943.

14. K. C. Majumdar, S. Chakravorty and K. Ray, 2008. *Synthesis*, 18, 2991.

15. S. Nandi and J. K. Ray, 2009. *Tetrahedron Lett.*, 50, 6993.

16. E. M. Beccalli, E. Borsini, S. Brenna, S. Galli, M. Rigamonti and G. Broggini, 2010. *Chem. Eur. J.*, 16, 1670.

17. D. A. Goff and R. N. Zuckermann, 1995. *J. Org. Chem.*, 60, 5748.

18. R. B. Bambal and R. D.W. Kemmitt, 1989. *J. Organomet. Chem.*, 362, C18.

19. R. Shintani and T. Hayashi, 2006. *J. Am. Chem. Soc.*, 128, 6330.

20. (a) B. M. Trost and D. M. T. Chan, 1979. *J. Am. Chem. Soc.*, 101, 6429. (b) B. M. Trost and D. M. T. Chan, 1983. *J. Am. Chem. Soc.*, 105, 2315.

21. N. Kaur and D. Kishore, 2014. *Syn. Comm.*, 44, 1173.

22. M. Hatano and K. Mikami, 2003. *J. Am. Chem. Soc.*, 125, 4704.

23. S.-C. Zhao, K.-G. Ji, L. Lu, T. He, A.-X. Zhou, R.-L. Yan, S. Ali, X.-Y. Liu and Y.-M. Liang, 2012. *J. Org. Chem.*, 77, 2763.

24. M. Li and D. J. Dixon, 2010. *Org. Lett.*, 12, 3784.

25. C.-H. Küchenthal and W. Maison, 2010. *Synthesis*, 5, 719.

26. (a) K. R. Roesch and R. C. Larock, 1999. *Org. Lett.*, 1, 553. (b) K. R. Roesch and R. C. Larock, 2002. *J. Org. Chem.*, 67, 86. (c) K. R. Roesch and R. C. Larock, 1998. *J. Org. Chem.*, 63, 5306. (d) K. R. Roesch, H. Zhang and R. C. Larock, 2001. *J. Org. Chem.*, 66, 8042.

27. G. Lu and H. C. Malinakova, 2004. *J. Org. Chem.*, 69, 4701.

28. J. S. Nakhla and J. P. Wolfe, 2007. *Org. Lett.*, 9, 3279.

29. T. Tsuritani, Y. Yamamoto, M. Kawasaki and T. Mase, 2009. *Org. Lett.*, 11, 1043.

30. (a) D. Solé, L. Vallverdú, X. Solans, M. Font-Bardia and J. Bonjoch, 2003. *J. Am. Chem. Soc.*, 125, 1587. (b) I. Fernandez, D. Solé and M. A. Sierra, 2011. *J. Org. Chem.*, 76, 1592.

31. G. Qiu, Y. He and J. Wu, 2012. *Chem. Commun.*, 48, 3836.

32. G. Qiu, G. Liu, S. Pu and J. Wu, 2012. *Chem. Commun.*, 48, 2903.

33. G. Qiu, Y. Lu and J. Wu, 2013. *Org. Biomol. Chem.*, 11, 798.

34. B. Liu, Y. Li, H. Jiang, M. Yin and H. Huang, 2012. *Adv. Synth. Catal.*, 354, 2288.

35. V. Tyagi, S. Khan, A. Giri, H. M. Gauniyal, B. Sridhar and P. M. S. Chauhan, 2012. *Org. Lett.*, 14, 3126.

36. G. V. Baelen, S. Kuijer, L. Rýček, S. Sergeyev, E. Janssen, F. J. J. de Kanter, B. U. W. Maes, E. Ruijter and R. V. A. Orru, 2011. *Chem. Eur. J.*, 17, 15039.

37. N. Haddad, J. Tan and V. Farina, 2006. *J. Org. Chem.*, 71, 5031.

38. F. Ye and H. Alper, 2007. *J. Org. Chem.*, 72, 3218.
39. (a) C. Larksarp and H. Alper, 2000. *J. Org. Chem.*, 65, 2773. (b) C. Larksarp and H. Alper, 1999. *J. Org. Chem.*, 64, 9194.
40. B. Ma, Y. Wang, J. Peng and Q. Zhu, 2011. *J. Org. Chem.*, 76, 6362.
41. (a) J. G. Knight, S. W. Ainge, A. M. Harm, S. J. Harwood, H. I. Maughan, D. R. Armour, D. M. Hollinshead and A. A. Jaxa-Chamiec, 2000. *J. Am. Chem. Soc.*, 122, 2944. (b) J. G. Knight, I. M. Lawson and C. N. Johnson, 2006. *Synthesis*, 2, 227.
42. (a) G. Chouhan and H. Alper, 2008. *Org. Lett.*, 10, 4987. (b) G. Chouhan and H. Alper, 2009. *J. Org. Chem.*, 74, 6181.
43. Z. Zheng and H. Alper, 2008. *Org. Lett.*, 10, 4903.
44. (a) G. Dai and R. C. Larock, 2002. *Org. Lett.*, 4, 193. (b) G. Dai and R. C. Larock, 2002. *J. Org. Chem.*, 67, 7042.
45. G. Abbiati, A. Arcadi, V. Canevari, L. Capezzuto and E. Rossi, 2005. *J. Org. Chem.*, 70, 6454.
46. T. Tsujihara, T. Shinohara, K. Takenaka, S. Takizawa, K. Onitsuka, M. Hatanaka and H. Sasai, 2009. *J. Org. Chem.*, 74, 9274.
47. O. Ferraccioli, D. Carenzi, O. Rombola and M. Catellani, 2004. *Org. lett.*, 6, 4759.
48. M. M. Sun, H. D. Wu, J. N. Zheng and W. L. Bao, 2012. *Adv. Synth. Catal.*, 354, 835.
49. (a) R. Faust, P. J. Garratt, R. Jones and L.-K. Yeh, 2000. *J. Med. Chem.*, 43, 1050. (b) D. Alberico, M. E. Scott and M. Lautens, 2007. *Chem. Rev.*, 107, 174.
50. G. Cuny, M. Bois-Choussy and J. Zhu, 2003. *Angew. Chem.*, 115, 4922.
51. Á. Balázs, A. Hetényi, Z. Szakonyi, R. Sillanpää and F. Fülöp, 2009. *Chem. Eur. J.*, 15, 7376.
52. Y. Zheng, G. Yu, J. Wu and W.-M. Dai, 2010. *Synlett*, 1075.
53. G.-W. Wang, T.-T. Yuan and D.-D. Li, 2011. *Angew. Chem. Int. Ed.*, 50, 1380.
54. J. Li, Y. He, S. Luo, J. Lei, J. Wang, Z. Xie and Q. Zhu, 2015. *J. Org. Chem.*, 80, 2223.
55. Z. J. Wang, J. G. Yang, F. Yang and W. L. Bao, 2010. *Org. Lett.*, 3034.
56. S.-S. Zhang, J.-Q. Wu, X. Liu and H. Wang, 2015. *ACS Catal.*, 5, 210.
57. Y. Wang and Q. Zhu, 2012. *Adv. Synth. Catal.*, 354, 1902.
58. (a) C. Chowdhury and N. G. Kundu, 1996. *Chem. Commun.*, 9, 1067. (b) C. Chowdhury, G. Chaudhuri, S. Guha, A. K. Mukherjee and

N. G. Kundu, 1998. *J. Org. Chem.*, 63, 1863. (c) W. Wang, J. Xue, T. Tian, J. Zhang, L. Wei, J. Shao, Z. Xie and Y. Li, 2013. *Org. Lett.*, 15, 2402.

59. C. V. Ramana, B. Induvadana, B. Srinivas, K. Yadagiri, M. N. Deshmukh and R. G. Gonnade, 2009. *Tetrahedron*, 65, 9819.

60. J. Uenishi and M. Ohmi, 2005. *Angew. Chem. Int. Ed.*, 44, 2756.

61. B. M. Trost, M. R. Machacek and H. C. Tsui, 2005. *J. Am. Chem. Soc.*, 127, 7014.

62. (a) L. Wang and P. L, D. Menche, 2010. *Angew. Chem. Int. Ed.*, 49, 9270. (b) L. Wang and D. Menche, 2012. *J. Org. Chem.*, 77, 10811.

63. B. Alcaide, P. Almendros, T. M. del Campo, M. T. Quiros, E. Soriano and J. L. Marco-Contelles, 2013. *Chem. Eur. J.*, 19, 14233.

64. M. Oberholzer, R. Gerber and C. M. Frech, 2012. *Adv. Synth. Catal.*, 354, 627.

65. R. Jana, S. Samanta and J. K. Ray, 2008. *Tetrahedron Lett.*, 49, 851.

66. K. Mikami and M. Hatano, 2004. *Proc. Natl. Acad. Sci. U. S. A.*, 101, 5767.

67. Y. Yamamoto, S. Kuwabara, Y. Ando, H. Nagata, H. Nishiyama and K. Itoh, 2004. *J. Org. Chem.*, 69, 6697.

68. R. Hua and M. Tanaka, 2001. *New J. Chem.*, 25, 179.

69. B. M. Trost, F. D. Toste, K. Greenman, 2003. *J. Am. Chem. Soc.*, 125, 4518.

70. S. Raju, V. R. Batchu, N. K. Swamy, R. V. Dev, B. R. Sreekanth, J. M. Babu, K. Vyas, P. R. Kumar, K. Mukkanti, P. Annamalai and M. Pal, 2006. *Tetrahedron*, 62, 9554.

71. S. Ma, W. Gao, 2002. *J. Org. Chem.*, 67, 6104.

72. H. Ohno, H. Hamaguchi, M. Ohata, S. Kosaka and T. Tanaka, 2004. *J. Am. Chem. Soc.*, 126, 8744.

73. (a) K. Mikami, K. Aikawa and Y. Yusa, 2002. *Org. Lett.*, 4, 95. (b) K. Mikami, K. Aikawa, Y. Yusa and M. Hatano, 2002. *Org. Lett.*, 4, 91.

74. S. Oi, E. Terada, K. Ohuchi, T. Kato, Y. Tachibana and Y. Inoue, 1999. *J. Org. Chem.*, 64, 8660.

75. K. E. Torraca, S.-I. Kuwabe and S. L. Buchwald, 2000. *J. Am. Chem. Soc.*, 122, 12907.

76. J. M. Heemstra, S. A. Kerrigan, D. R. Doerge, W. G. Helferich and W. A. Boulanger, 2006. *Org. Lett.*, 8, 5441.

77. R.-J. Li, S.-F. Pi, Y. Liang, J.-Q. Wang, R.-J. Song, G.-X. Chen and J.-H. Li, 2010. *Chem. Commun.*, 46, 8183.

78. L. Xue, L. Shi, Y. Han, C. Xia, H. V. Huynh and F. Li, 2011. *Dalton Trans.*, 40, 7632.
79. (a) D. V. Kadnikov, R. C. Larock, 2000. *Org. Lett.*, 2, 3643. (b) D. V. Kadnikov and R. C. Larock, 2003. *J. Org. Chem.*, 68, 9423.
80. K. Okuro and H. Alper, 1997. *J. Org. Chem.*, 62, 1566.
81. R. Grigg, A. Liu, D. Shaw, S. Suganthan, D. E. Woodall and G. Yoganathan, 2000. *Tetrahedron Lett.*, 41, 7125.
82. C. Dong and H. Alper, 2004. *J. Org. Chem.*, 69, 5011.
83. Y. Li, Z. Yu and H. Alper, 2007. *Org. Lett.*, 9, 1647.
84. H. Cao, W.-J. Xiao and H. Alper, 2006. *Adv. Synth. Catal.*, 348, 1807.
85. H. Cao, W.-J. Xiao and H. Alper, 2007. *J. Org. Chem.*, 72, 8562.
86. R. Giri and J.-Q. Yu, 2008. *J. Am. Chem. Soc.*, 130, 14082.
87. (a) B. Gabriele, R. Mancuso, G. Salerno, G. Ruffolo, M. Costa and A. Dibenedetto, 2009. *Tetrahedron Lett.*, 50, 7330. (b) B. Gabriele, R. Mancuso, G. Salerno, L. Veltri, M. Costa and A. Dibenedetto, 2011. *ChemSusChem*, 4, 1778.
88. F. Faccini, E. Motti and M. Catellani, 2004. *J. Am. Chem. Soc.*, 126, 78.
89. C. Pierre and O. Baudoin, 2011. *Org. Lett.*, 13, 1816.
90. M. Kotani, K. Yamamoto, J. Oyamada, Y. Fujiwara and T. Kitamura, 2004. *Synthesis*, 9, 1466.
91. L. Ackermann, R. Jeyachandran, H. K. Potukuchi, P. Novak and L. Buttner, 2010. *Org. Lett.*, 12, 2056.
92. S. Wang, K. Xie, Z. Tan, X. An, X. Zhaou, C.-C. Guo and Z. Peng, 2009. *Chem. Commun.*, 42, 6469.
93. Y. Lu, D. Leow, X. Wang, K. M. Engel and J.-Q. Yu, 2011. *Chem. Sci.*, 2, 967.
94. Y. Takeyama, K. Nozaki, K. Matsumoto, K. Oshima and K. Utimoto, 1991. *Bull. Chem. Soc. Jpn.*, 64, 1461.
95. J. Liu, Q. Zhang, P. Li, Z. Qu, S. Sun, Y. Ma, D. Su, Y. Zong and J. Zhang, 2014. *Eur. J. Inorg. Chem.*, 21, 3435.
96. B. Willy, W. Frank and T. J. J. Müller, 2010. *Org. Biomol. Chem.*, 8, 90.
97. M. Harmata, K.-O. Rayanil, M. G. Gomes, P. Zheng, N. L. Calkins, S.-Y. Kim, Y. Fan, V. Bumbu, D. R. Lee, S. Wacharasindhu and X. Hong, 2005. *Org. Lett.*, 7, 143.
98. D. K. Barange, V. R. Batchu, D. Gorja, V. R. Pattabiraman, L. K. Tatini, J. M. Babu and M. Pal, 2007. *Tetrahedron*, 63, 1775.
99. Q. Tian, A. A. Pletnev and R. C. Larock, 2003. *J. Org. Chem.*, 68, 339.
100. W. Tang and Y.-X. Ding, 2006. *J. Org. Chem.*, 71, 8489.

101. Y. Tanaka, H. Yamashita and M. Tanaka, 1995. *Organomet.*, 14, 530.
102. R. A. Widenhoefer, B. Krzyzanowska and G. Webb-Wood, 1998. *Organometal.*, 17, 5124.
103. B. Liu, M. Yin, H. Gao, W. Wu and H. Jiang, 2013. *J. Org. Chem.*, 78, 3009.
104. Z. Qiao, H. Liu, X. Xiao, Y. Fu, J. Wei, Y. Li, X. Jiang, 2013. *Org. Lett.*, 15, 2594.
105. X.-F. Wu, H. Neumann, S. Neumann and M. Beller, 2012. *Chem. Eur. J.*, 18, 13619.
106. L. Troisi, C. Granito, S. Perrone and F. Rosato, 2011. *Tetrahedron Lett.*, 52, 4330.
107. G. Rescourio and H. Alper, 2008. *J. Org. Chem.*, 73, 1612.
108. K. C. Majumdar, S. Mondal and D. Ghosh, 2009. *Tetrahedron Lett.*, 50, 4781.
109. Y. Liang, W. Geng, J. Wei, K. Ouyang and Z. Xi, 2012. *Org. Biomol. Chem.*, 10, 1537.

CHAPTER 7

Pd-Catalyzed Synthesis of Sulfur- and Phosphorus-Containing Heterocycles *via* C–H Activation

Weiping Su

Fujian Institute of Research on the Structure of Matter,
Chinese Academy of Sciences
wpsu@fjirsm.ac.cn

1. Introduction

The construction of heterocycles constitutes one of the most active research areas in modern organic synthesis because such structure units are ubiquitous in numerous natural products, biologically active compounds, pharmaceutical, and agrochemical molecules.[1] Over the past decades, the C–H functionalization has been intensively applied to the synthesis of various heterocycles, providing a step- and atom-economical approach to heterocycles as it obviates the need to pre-functionalize the starting material and often reduces the amount of undesired waste.[2] Compared with nitrogen- and oxygen-containing heterocycles (see Chapters 2 and 3 of this book), there have been fewer examples of C–S, C–P, C–Ge, or C–Si bond formation for synthesis of heterocycles *via* C–H functionalization, among which palladium catalyst was proved to be an excellent catalyst in many C–S and C–P formation reactions while C–Si[3] and C–Ge[4] formation mostly catalyzed by rhodium catalyst. This chapter will focus specifically on the construction of sulfur- and phosphorus-containing heterocycles through palladium-catalyzed C–H functionalization reactions during the past decades. Those reactions are categorized and present based on the types of C-heteroatom bonds formed.

2. Synthesis of Sulfur-Containing Heterocycles

Although palladium-catalyzed C–H functionalization has been widely studied for C–C bond-forming reaction, its utility for C–S bond-forming reaction[5] was realized only until recently presumably because of the catalyst poisoning by sulfur.[6] Substantial progress towards this goal was made successfully in 2008 by Inamoto and co-workers, who achieved the direct synthesis of substituted benzo[b]thiophenes 1 via the cyclization of thioenols by using 10 mol% PdCl$_2$ or PdCl$_2$(cod) as the catalyst (Scheme 1).[7] The use of dimethyl sulfoxide (DMSO) as the solvent was crucial for high conversion and the reoxidants were not necessary in this process. Electronically diverse 4,4′- and 3,3′- substituted thioenols generally gave the corresponding 2,3-diarylbenzo[b] thiophenes 1 in good yields. The established method could also be applied to the synthesis of 2-alkoxycarbonylbenzo[b]thiophenes from corresponding thioenols. The authors proposed that the

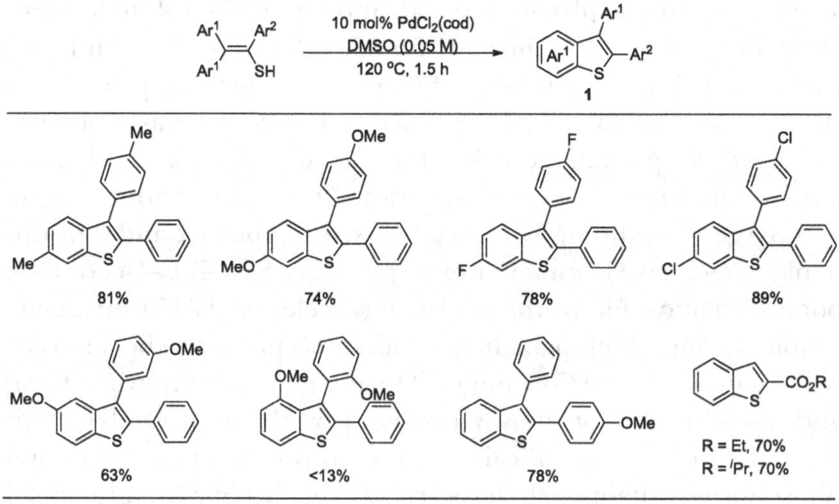

Scheme 1.   The synthesis of benzo[b]thiophenes from thioenols

reaction proceeded *via* the formation of disulfide intermediates from corresponding thioenols, and those disulfides could undergo oxidative addition to the palladium catalyst, so oxidant was not required in this process. Although the detailed mechanism remains unclear at present, it illustrates the possibility of direct C–H functionalization for C–S bond-forming reaction.

In the same year, the report from the same group described a similar intramolecular cyclization of thiobenzanilides to yield 2-substituted benzothiazoles **2** *via* the palladium-catalyzed C–H functionalization/ C–S bond formation process (Scheme 2).[8] The reactions were carried out in DMSO–NMP(1:1) by using 10 mol% palladium(II) as the catalyst and 50 mol% CuI in combination with 2.0 equiv Bu_4NBr as the oxidants. The addition of Bu_4NBr was proved to be crucial for efficient transformation as only less than 2% yield was obtained if it was removed from the catalyst system. The protocol was applicable to an array of thiobenzanilides with either electron-donating or electron-withdrawing groups. Various functional groups such as methoxyl, nitro, even sensitive alkoxycarbonyl, cyano, and halogen atoms including iodine were tolerated under the established conditions.

In 2010, the same group developed an improved procedure for palladium-catalyzed synthesis of 2-substituted benzothiazoles *via*

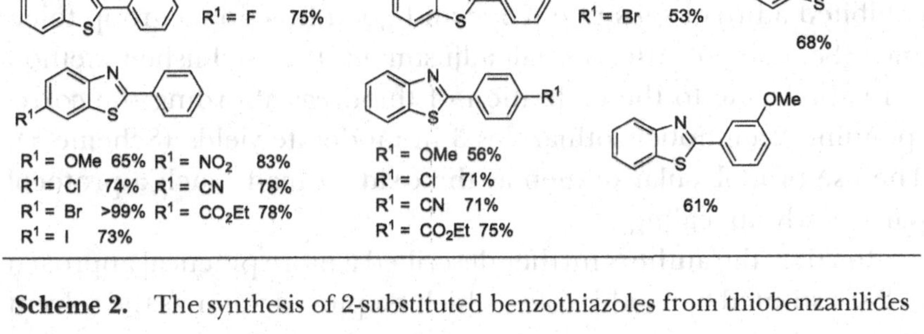

Scheme 2. The synthesis of 2-substituted benzothiazoles from thiobenzanilides

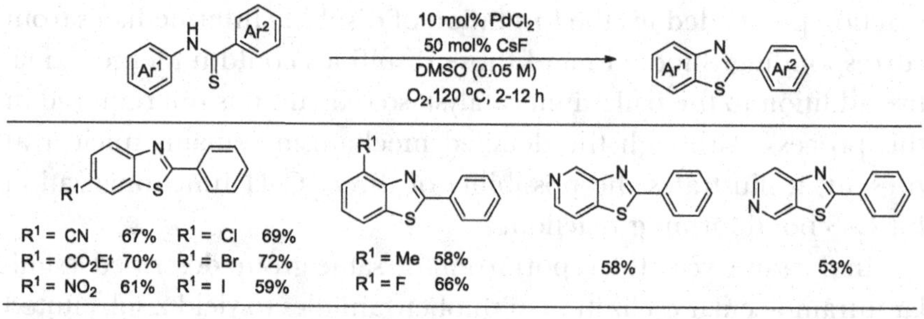

Scheme 3. The synthesis of 2-substituted benzothiazoles from thiobenzanilides under oxygen

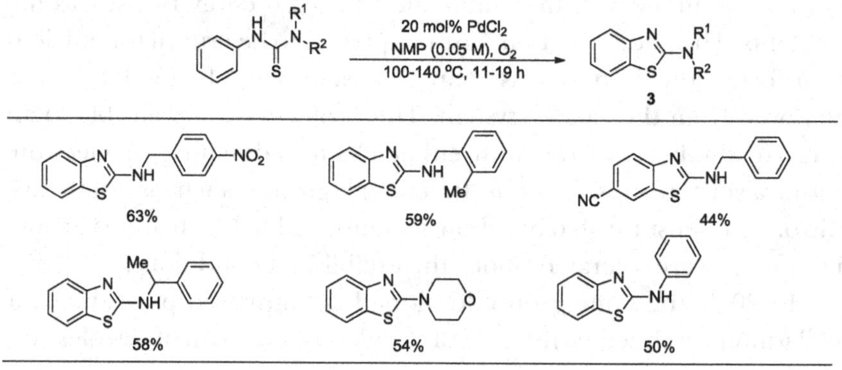

Scheme 4. The synthesis of 2-aminobenzothiazoles from thioureas

cyclization of thiobenzanilides, in which molecular oxygen was used as the oxidant.[9] The introduction of a basic additive such as Caesium flouride (CsF) was found to be the key to success. This protocol also exhibited a broad substrate scope and good functional group tolerance (Scheme 3). After a small adjustment, the established method were applicable to the cyclization of thioureas affording the corresponding 2-aminobenzothiazoles **3** in moderate yields (Scheme 4). The use of molecular oxygen as the oxidant made such a protocol particularly appealing.

In 2012, the authors further described a more practical approach to 2-substituted benzothiazoles, which employed water as the solvent

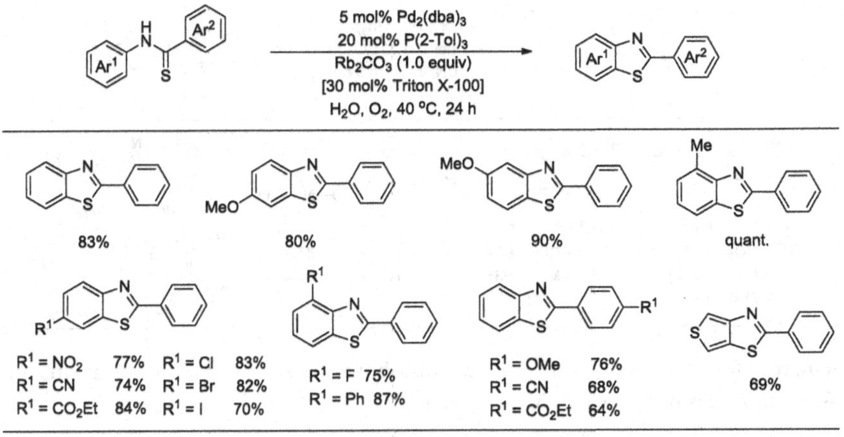

Scheme 5. The synthesis of 2-substituted benzothiazoles *via* cyclization of thiobenzanilides in water

(Scheme 5).[10] The reaction was carried out at 40°C using 10 mol% $PdCl_2$ as the catalyst, 20 mol% $P(2\text{-}Tol)_3$ as the ligand, and oxygen as the oxidant. Substrates with an electron-donating methoxy or methyl group on the benzene ring afforded the desired products in good yields. For substrates possessing an electron-withdrawing group, the addition of a surfactant such as Triton X-100 was found to be beneficial for obtaining a better yield.

Meanwhile, Batey and co-workers described a catalytic $Pd(0)/MnO_2$ system for the synthesis of 2-aminobenzothiazoles **4** *via* intramolecular cyclization of N-arylthioureas using 3 mol% $Pd(PPh_3)_4$ as the catalyst and 10 mol% MnO_2 as the co-oxidant under an oxygen atmosphere at 80°C.[11] Replacing the catalyst $Pd(PPh_3)_4$ with 1 mol% $Pd(dba)_3$ in combination with 8 mol% PPh_3 was also effective for the desired reaction while no product was obtained in the absence of PPh_3, which indicated that the $Pd(0)/phosphine$ based catalyst system was very important for this transformation. A range of N-arylthioureas bearing substitution at the ortho, *meta*, and *para* positions of benzene ring underwent desired cyclization to form products **4** in good yields (Scheme 6). The reaction was sensitive to steric effects as di- or tri-substituted N-arylthioureas generally resoluted in low conversions. A primary kinetic isotope effects (KIE) of 5.9 excluded

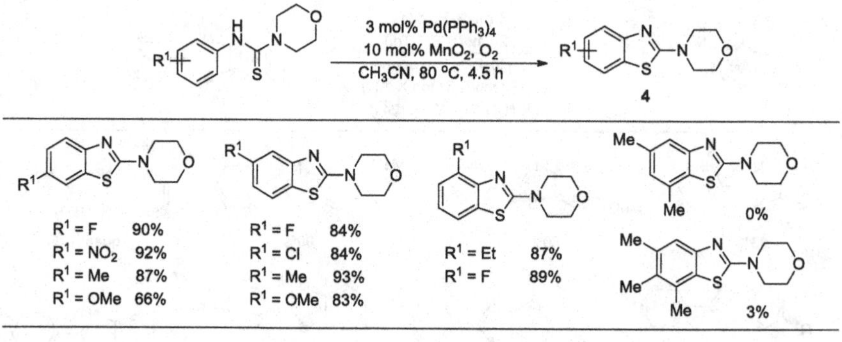

Scheme 6. The synthesis of 2-aminobenzothiazoles *via* cyclization of N-arylthioureas using Pd(0)/Mno$_2$ system

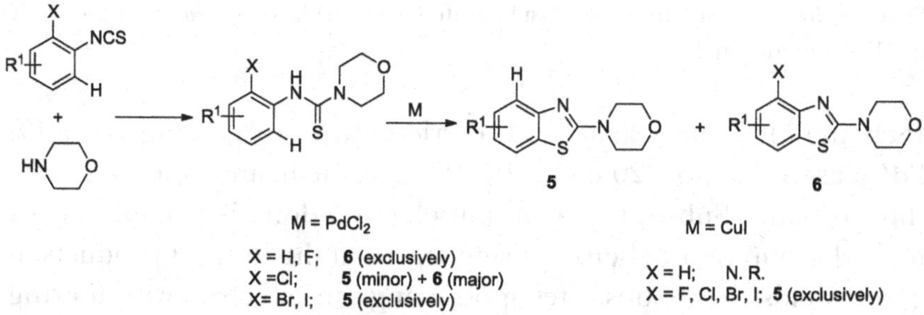

Scheme 7. Regioselective cyclization of N-arylthioureas to synthesize 2-aminobenzothiazoles by Pd(II) and Cu(I)

the electrophilic palladation mechanism. The authors proposed that the C–H bond activation occurs *via* a σ-bond metathesis, wherein proton abstraction was promoted by an anionic peroxo/peroxide-Pd complex in this process.

In 2012, Patel and co-workers demonstrated a similar approach to 2-aminobenzothiazoles by using PdCl$_2$ as the catalyst and K$_2$CO$_3$ as the base in dimethylformamide (DMF) under an open atmosphere.[12] The regioselective intramolecular C–S bond formation was observed for 2-halothioureas using either Pd or Cu as the catalyst (Scheme 7). Because of the presence of halogen atom, competition occurred between C–H or C–X activation. It seems that the selectivity of the reaction was controlled by the metal catalyst and substituents on the aryl thioureas. In most cases, for substrates with a less reactive 2-halo groups such as fluoro, palladium prefer a C–H activation path over dehalogenation, while copper catalyst often promoted the C–X bond cleavage. For 2-bromo and 2-iodo arylthioureas, both palladium and copper catalyst afforded the dehalogenated products while 2-chloro substituted arylthioureas underwent the reaction *via* either of the paths giving a mixture of dehydrogenated/dehalogenated products.

In 2012, Zhang and co-workers reported the synthesis of sugar-based benzothiazoles **7** *via* the Pd(II)-catalyzed intramolecular cyclization of glycosyl thioureas.[13] The reaction was carried out in DMSO in the presence of 10 mol% Pd(cod)Cl$_2$ as the catalyst and 2.0 equiv Bu$_4$NBr as the additive under oxygen atmosphere. Substrates containing either an electron donating or withdrawing group all underwent the reaction smoothly to afford the desired sugar-based benzothiazoles **7** in good yields (Scheme 8). The presence of pivaloyl carbonyl group played a significant role for the reaction as non-carbohydrate thiourea derivatives and non-pivaloylated glycosyl thiourea substrate all failed to undergo the reaction under standard conditions. The authors performed theoretical calculations to obtain insight into the mechanism of the reaction, and it is proposed that the generation of intermediate **A**, in which palladium coordinates to both S atom of thiourea and O atom of one of the neighboring pivaloyl carbonyl groups, and the presence of intramolecular

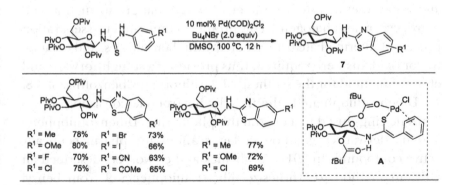

Scheme 8. The synthesis of sugar-based benzothiazoles *via* chemoselective intramolecular C–S coupling

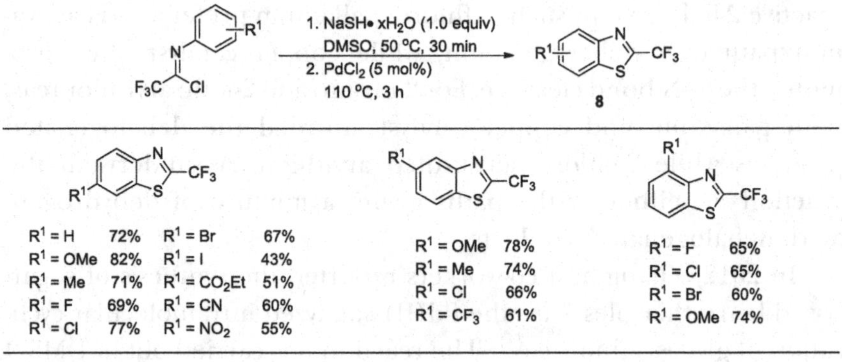
1. NaSH• xH2O (1.0 equiv)
DMSO, 50 °C, 30 min
2. PdCl2 (5 mol%)
110 °C, 3 h
8
R1 = H 72%
R1 = OMe 82%
R1 = Me 71%
R1 = F 69%
R1 = Cl 77%
R1 = Br 67%
R1 = I 43%
R1 = CO2Et 51%
R1 = CN 60%
R1 = NO2 55%
R1 = OMe 78%
R1 = Me 74%
R1 = Cl 66%
R1 = CF3 61%
R1 = F 65%
R1 = Cl 65%
R1 = Br 60%
R1 = OMe 74%

Scheme 9. The cascade synthesis of 2-trifluoroethylbenzothiazoles

hydrogen bonding between 2-O-pivaloyl carbonyl group and amide N–H group of the thiourea moiety, may play important roles in dictating reactivity and selectivity of the reaction.

In 2010, Wu and co-workers developed a cascade synthesis of 2-trifluoroethylbenzothiazoles **8** by treatment of trifluoromethylimidoyl chlorides with sodium hydrosulfide hydrate using $PdCl_2$ as the catalyst under air atmosphere (Scheme 9).[14] An optimal reaction condition was established by heating trifluoromethylimidoyl chlorides with sodium hydrosulfide in DMSO at 50°C for half an hour, followed by addition of 5 mol% $PdCl_2$ and heating the mixture at 110°C for 3 h. The reaction featured wide range of substrate scope and good tolerance to a number of functional group such as carbonyl, cycan, and halogens. This protocol eliminated the separation process of intermediate trifluoromethylthiobenzamides, and no additive or oxidants were required, thus providing a straightforward and efficient method for the synthesis of 2-trifluoroethylbenzothiazoles **8.**

Dibenzothiophene is the tricyclic heterocycle consisting of two benzene rings fused to a central thiophene ring. Dibenzothiophene and its derivatives compose a key scaffold for pharmaceutically active compound. In 2011, Antibchick and co-workers described an efficient method to synthesize dibenzothiophenes **9** from benzyl phenyl sulfoxide precursors, which were catalyzed by 15 mol% $PdCl_2$

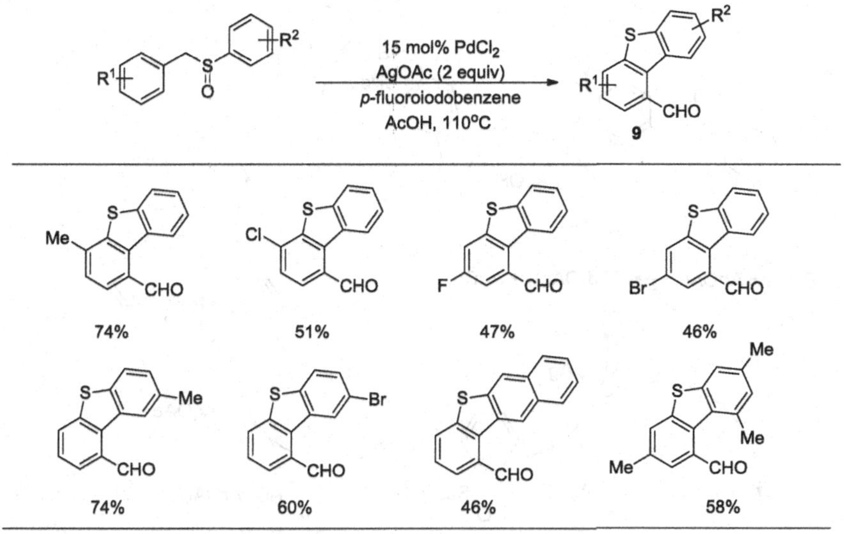

Scheme 10.   The synthesis of dibenzothiophenes from benzyl phenyl sulfoxides

in AcOH in the presence of 2.0 equiv AgOAc and 2.0 equiv *p*-fluoro-iodobenzene.[15] Under the optimum reaction conditions, a variety of benzyl phenyl sulfoxides with either electron-donating or -withdrawing groups all underwent the reaction smoothly affording the desired products in good yields (Scheme 10). A set of mechanistic experiments were carried out to gain insight into the reaction mechanism, it is proposed that the reaction proceeded *via* an initially Pd(II)-catalyzed sulfoxides-directed C–H/C–H dehydrogenation coupling leading to the formation of cyclic sulfoxide intermediate **10**, which undertakes a Pummerer rearrangement providing a mercapto-aldehyde **11**. In the next cycle, **11** undergoes a cyclization by Pd(II)-catalyzed C–H functionalization/intramolecular C–S bond formation giving the desired product (Scheme 11).

In the same year, Duan and co-workers described the synthesis of dibenzothiophenes **12** *via* the Pd(II)-catalyzed ring rearrange-ment of bromothiophenes with alkynes, in which both activation of C–H bond and C–S bond of bromothiophene were involved to form a new C–S bond between the cleaved sulfur moiety and its neighboring

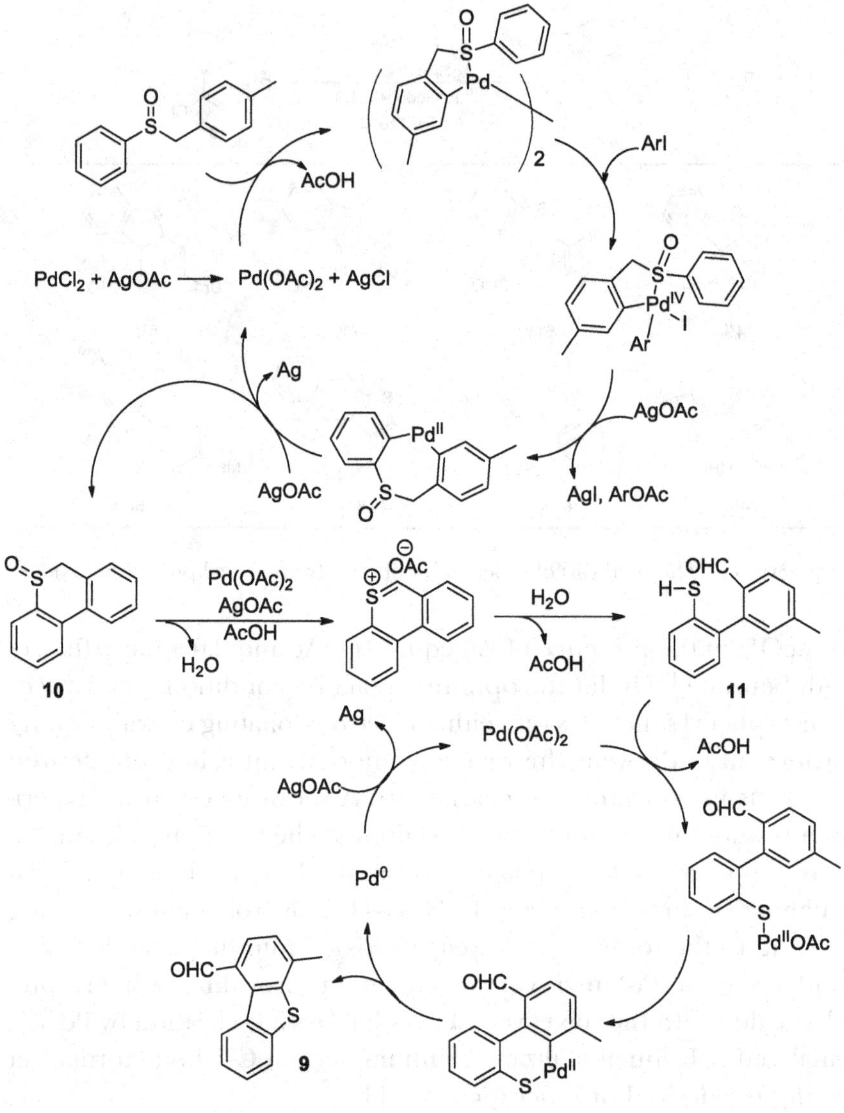

Scheme 11. The possible mechanism for synthesis of dibenzothiophenes from benzyl phenyl sulfoxides

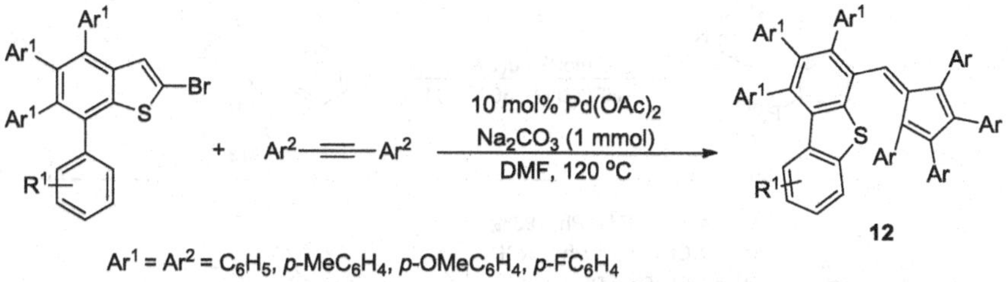

$Ar^1 = Ar^2 = C_6H_5$, p-MeC_6H_4, p-$OMeC_6H_4$, p-FC_6H_4

Scheme 12. The ring rearrangement of bromothiophenes with alkynes to prepare dibenzothiophenes

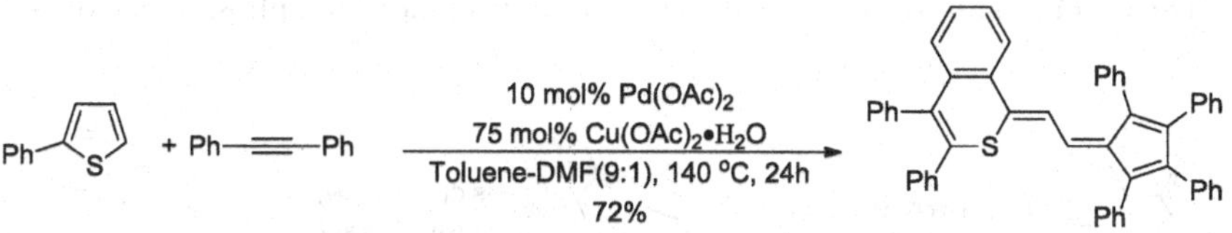

Scheme 13. The ring rearrangement of simple thiophenes with alkynes to prepare sulfur-containing compounds

phenyl group (Scheme 12).[16] The reaction was carried out in DMF at 120°C in the presence of 10 mol% Pd(OAc)$_2$ and 2.0 equiv Na$_2$CO$_3$. Under the established reaction conditions, electronically diverse bromothiophenes as well as alkynes gave the desired products **12** in good yields. Soon after, the same group developed a new palladium catalyzed reaction of simple thiophenes with alkynes through a domino type C–H and C–S bond activation, which eliminated the requirement of α–C–Br (Scheme 13).[17] It was found that the amount of the salt used significantly, impact the efficiency and selectivity of the reaction.

3. Synthesis of Phosphorus-Containing Heterocycles

The first example of the construction of phosphorus-containing heterocycle *via* C–H activation was reported by Takai and co-workers in 2011, who described the palladium-catalyzed intramolecular dehydrogenative cyclization of secondary hydrophosphine oxides with a biphenyl group. A variety of phospholes **13** was synthesized

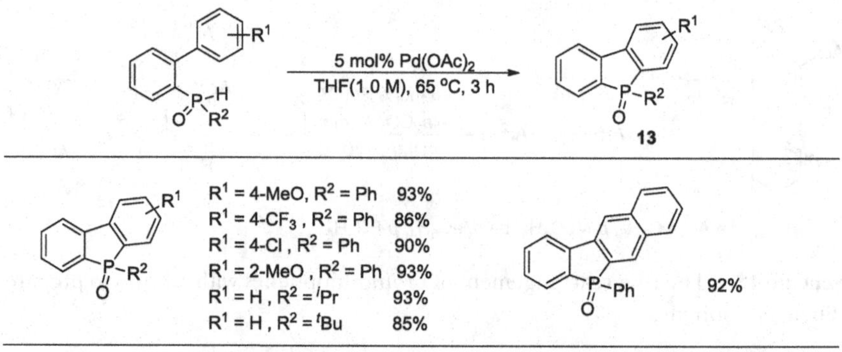

Scheme 14. The synthesis of dibenzophosphole oxides from phosphine oxides

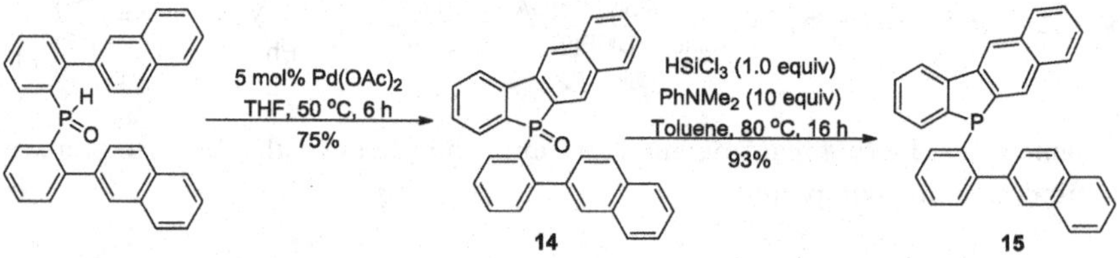

Scheme 15. The synthesis of phosphine oxides with a chiral phosphorus center

using 5 mol% Pd(OAc)$_2$ as the catalyst (Scheme 14).[18] This transformation proceeded via successive P–H and C–H bond cleavage producing only H$_2$ as a side product. A catalytic cycle between Pd(0) and Pd(II) was involved in this reaction, additional oxidant was not necessary because the resulting Pd(0) catalyst was oxidized to Pd(II) by P–H bond via oxidative addition. A primary KIE of 2.3 indicated that the C–H bond activation was involved in the rate-determining step. The authors further applied this protocol into the synthesis of a phosphine oxide with a chiral phosphorus center **14**, which could easily produce corresponding chiral phosphine derivatives **15** by reduction (Scheme 15).[19]

In view of the widespread availability and stability of triarylphosphines, Chatani and co-workers subsequently realized the synthesis of phosphole derivatives from triarylphosphines via simultaneous cleavage of C–H and C–P bonds (Scheme 16).[20] The reaction was

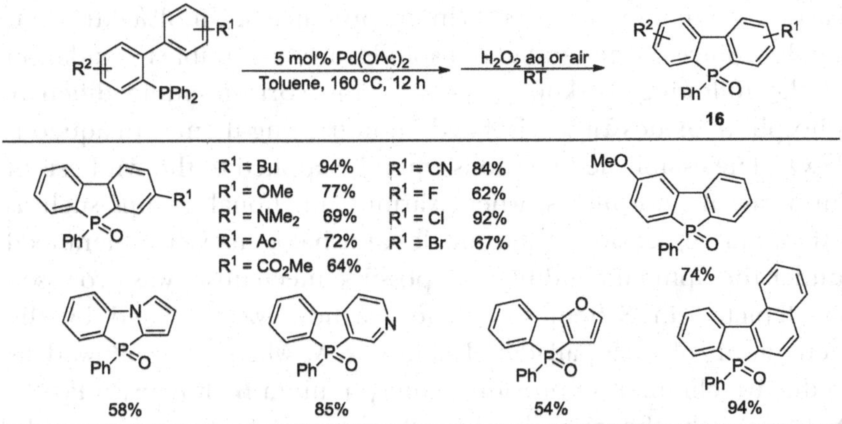

Scheme 16. The synthesis of phosphole derivatives from triarylphosphines

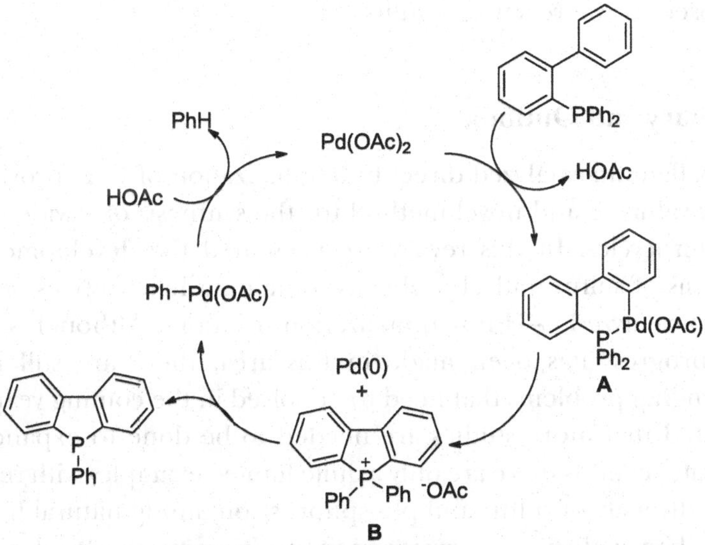

Scheme 17. The possible mechanism for synthesis of phosphole derivatives from triarylphosphines

carried out in toluene at 160°C in the presence of 5 mol% $Pd(OAc)_2$ catalyst. Since the generated phospholes are susceptible to oxidation at the following workup procedure, the corresponding dibenzo-phosphine oxides **16** was isolated upon treating them with aqueous H_2O_2. The established methods could be applied to the synthesis of an array of phospholes, where various functional groups such as ether, amine, ketone, ester, nitrile, and halogens were toleranced under the optimal conditions. A possible mechanism was proposed as depicted in Scheme 17. The reaction was initiated by the generation of cyclopalladated complex **A**, which was followed by reductive elimination providing phosphonium **B** along with Pd(0). Subsequently, the phosphonium **B** immediately undergoes oxidative addition to Pd(0) leading to the formation of the products phospholes and PhPd(OAc) *via* cleavage of C–P bond. Finally, PhPd(OAc) is protonated by AcOH to regenerate $Pd(OAc)_2$ catalyst. Several experiments were performed and the proposal was supposed by the reaction results.

Summary and Outlook

The palladium-catalyzed direct functionalization of C–H provides a highly valuable and novel method for the synthesis of a wide range of heterocycles. In this review, we presented the development in synthesis of sulfur and phosphorus-containing heterocycles *via* pal-ladium-catalyzed C–H functionalization reactions. Although significant progress has been made in this area, there are still many challenging problems that need to be solved in the coming year. It is expected that more studies are needed to be done to expand the substrate scope as there are only a quite limited examples with respect to the diversity of sulfur and phosphorus-containing natural hetero-cycles. The exploring of new reaction involving intramolecular dehy-drogenative cyclization represents a promising future as oxidant was not required and only H_2 was released as side product. The continu-ous efforts towards this field will provide more economical and diverse approach to heterocycles.

References

1. (a) A. F. Pozharskii, A. T. Soldatenkov and A. R. Katritzky, *Heterocycles in Life and Society: An Introduction to Heterocyclic Chemistry, Biochemistry and Applications*, 2nd Edition, *John Wiley & Sons*, Ltd., 2011. (b) L. D. Quin and J. A. Tyrell, *Fundamentals of Heterocyclic Chemistry: Introduction in Nature and in the Synthesis of Pharmaceuticals*, John Wiley & Sons, Ltd., 2010. (c) Rajiv Dua, Suman Shrivastava, S. K. Sonwane and S. K. Srivastava, 2011. *Advan. Biol. Res.*, 5, 120.

2. (a) Y. Liu, J. Kim and J. Chae, 2014. *Curr. Org. Chem.*, 18, 2049. (b) M. Zhang, A.-Q. Zhang and Y. Peng, 2013. *J. Organomet. Chem.*, 723, 224. (c) E. M. Beccalli, G. Broggini, A. Fasana and M. Rigamonti, 2011. *J. Organomet. Chem.*, 696, 277. (d) M. Zhang, 2009. *Adv. Synth. Catal.*, 351, 2243. (e) H. M. L. Davies and M. S. Long, 2005. *Angew. Chem. Int. Ed.*, 44, 3518.

3. (a) T. Ureshino, T. Yoshida, Y. Kuninobu and K. Takai, 2010. *J. Am. Chem. Soc.*, 132, 14324. (b) Y. Kuninobu, T. Nakahara, H. Takeshima and K. Takai, 2013. *Org. Lett.*, 15, 426. (c) Y. Kuninobu, K. Yamauchi, N. Tamura, T. Seiki and K. Takai, 2013. *Angew. Chem. Int. Ed.*, 52, 1520. (d) T. Shibata, T. Shizuno and T. Sasaki, 2015. *Chem. Commun.*, 51, 7802. (e) Q.-W. Zhang, K. An, L.-C. Liu, Y. Yue and W. He, 2015. *Angew. Chem. Int. Ed.*, 54, 6918.

4. M. Murai, K. Matsumoto, R. Okada and K. Takai, 2014. *Org. Lett.*, 16, 6492.

5. C. Shen, P. Zhang, Q. Sun, S. Bai, T. S. A. Hor and X. Liu, 2015. *Chem. Soc. Rev.*, 44, 291.

6. L. L. Hegedus and R. W. McCabe, 1984. Marcel Dekker, New York.

7. K. Inamoto, Y. Arai, K. Hiroya and T. Doi, 2008. *Chem. Commun.*, 5529.

8. K. Inamoto, C. Hasegawa, K. Hiroya and T. Doi, 2008. *Org. Lett.*, 10, 5147.

9. K. Inamoto, C. Hasegawa, J. Kawasaki, K. Hiroya and T. Doi, 2010. *Adv. Synth. Catal.*, 352, 2643.

10. Y. Kondo, K. Inamoto and K. Nozawa, 2012. *Synlett*, 23, 1678.

11. L. L. Joyce and R. A. Batey, 2009. *Org. Lett.*, 11, 2792.

12. S. K. Sahoo, A. Banerjee, S. Chakraborty and B. K. Patel, 2012. *ACS Catal.*, 2, 544.

13. C. Shen, H. Xia, H. Yan, X. Chen, S. Ranjit, X. Xie, D. Tan, R. Lee, Y. Yang, B. Xing, K.-W. Huang, P. Zhang and X. Liu, 2012. *Chem. Sci.*, 3, 2388.

14. J. Zhu, Z. Chen, H. Xie, S. Li and Y. Wu, 2010. *Org. Lett.*, 12, 2434.
15. R. Samanta and A. P. Antonchick, 2011. *Angew. Chem. Int. Ed.*, 50, 5217.
16. H. Huang, J. Li, C. Lescop and Z. Duan, 2011. *Org. Lett.*, 13, 5252.
17. J. Li, H. Huang, W. Liang, Q. Gao and Z. Duan, 2013. *Org. Lett.*, 15, 282.
18. Y. Kuninobu, T. Yoshida and K. Takai, 2011. *J. Org. Chem.*, 76, 7370.
19. K. Takai, Y. Kuninobu and K. Origuchi, 2012. *Heterocycles*, 85, 3029.
20. K. Baba, M. Tobisu and N. Chatani, 2013. *Angew. Chem. Int. Ed.*, 52, 11892.

CHAPTER 8

Pd-Catalyzed Heterocycle Synthesis in Ionic Liquids

Jianxiao Li, Huanfeng Jiang

School of Chemistry and Chemical Engineering, South China University of Technology, Guangzhou 510640, China
jianghf@scut.edu.cn

Heterocyclic and fused heterocyclic compounds are ubiquitously found in natural products and biologically interesting molecules, and many currently marketed drugs hold heterocycles as their core structure. In this chapter, recent advances on Pd-catalyzed synthesis of heterocycles in ionic liquids (ILs) are reviewed. In palladium catalysis, ILs with different cations and anions are investigated as an alternative recyclable and environmentally benign reaction medium, and a variety of heterocyclic compounds including cyclic ketals, quinolones, quinolinones, isoindolinones, and lactones are conveniently constructed. Compared to the traditional methods, these new approaches have many advantages, such as environmentally friendly synthetic procedure, easy product and catalyst separation, recyclable medium, which make them have the potential applications in industry.

Keywords: Pd-Catalyzed, Synthesis, Heterocycles, Ionic Liquids

1. Introduction

Nowadays, the development of environmentally friendly synthetic procedures has become a major concern throughout the chemical industry, due to continuing depletion of natural resources and growing environmental awareness.[1-6] One of the major efforts in contemporary academic research is the search for replacements to the environmentally damaging organic solvents, especially those which are volatile and difficult to contain. Most notably, ILs have attracted considerable interest as environmentally benign reaction media because of their fascinating and intriguing properties.[7-12] They offer an alternative and ecologically sound medium compared to the conventional organic solvents due to their negligible vapor pressure, ease of handling and potential for recycling. Moreover, their high compatibility with transition metal catalysts and limited miscibility with common solvents, enables easy product and catalyst separation with the retention of the stabilized catalyst in the ionic phase.[13,14] Consequently, the ILs have been successfully used as an alternative recyclable and environmentally benign reaction medium for chemical processes in the past few years.

Given the above mentioned advantages, the applications of ILs as excellent reaction or process media have been widely studied in the past decade, and a great deal of elegant works have been done.[15,16] In recent years, there have been a number of excellent reviews concerning catalysis in ILs, including biocatalysts in ILs,[17-19] two-phase catalysis in ILs,[20] asymmetric catalysis in ILs,[21,22] nanoparticle catalysis in ILs,[23] supported IL catalysis,[24] etc. In this chapter, recent advances on Pd-catalyzed heterocycle synthesis in ILs will be summarized.

2. Applications in the Synthesis of Heterocycles

2.1 Synthesis of cyclic ketals

Cyclic ketals and its derivatives are widely used in synthetic organic chemistry. Many products that contain the subunit of indane exhibit useful and diverse biological activity. These compounds are applications found in pharmaceuticals and protease inhibitors.[25-27] Hallberg and Larhed were the first to report the highly regioselective arylation of hydroxyalkyl vinyl ethers to synthesize indane derivatives, using aryl triflates in dimethylformamide (DMF). When aryl halides were used, this reaction required the addition of bidentate ligands

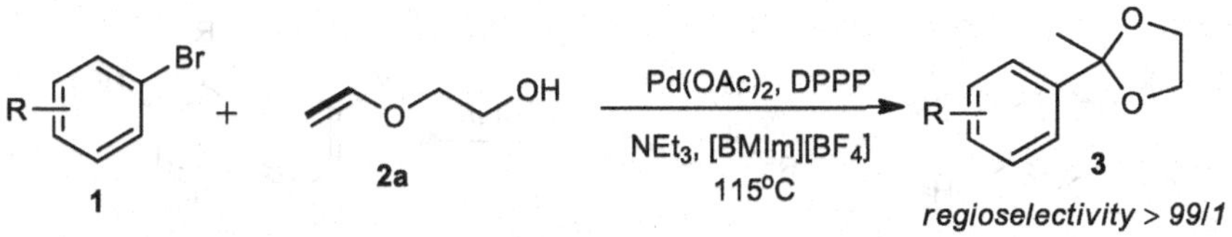

Scheme 1. Regioselective arylation/cyclization of hydroxy vinyl ether **2a**

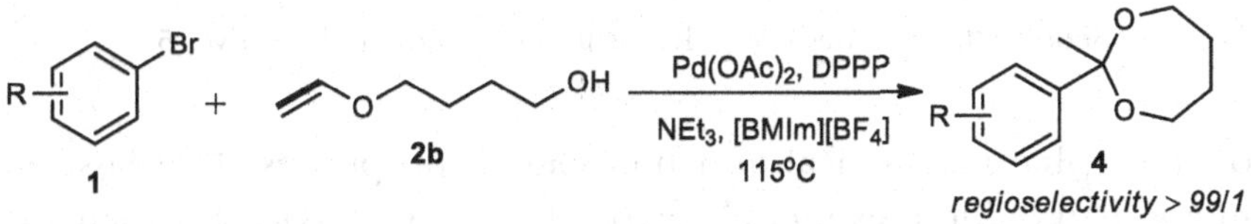

Scheme 2. Regioselective arylation/cyclization of 1,4-butanediol vinyl ether **2b**

and stoichiometric amounts of silver or thallium salts as halide scavengers to slow arylation process.[28] A drawback to the chemistry is that triflates are base-sensitive, thermally labile, and are rarely commercially available, as well as the inorganic additives silver or thallium salts are costly and toxic. Xiao and co-workers demonstrated the synthesis of indane derivatives by tandem process from commercially available aryl bromides (**1**) and hydroxyalkyl vinyl ether (**2a**) using $Pd(OAc)_2$ as a catalyst in [BMIM][BF_4] (Scheme 1).[29] All the reactions were finished with >99/1 regioselectivity in favor of the branched products in good to excellent yields. Particularly, no halide scavenger is required, and the reaction appears to be faster. It is also interesting that the IL [BMIM][BF_4] allows the internal arylation to give the cyclization product **3** without the addition of an acid. Furthermore, the hydroxyalkyl vinyl ether (**2b**) could also be arylated with high regioselectivity, and afford the seven-membered cyclic ketals **4** (Scheme 2).

2.2 Synthesis of quinolines and quinolinones

The quinoline skeleton attracting significant synthetic interest has emerged as an important class of nitrogenated heterocycles because

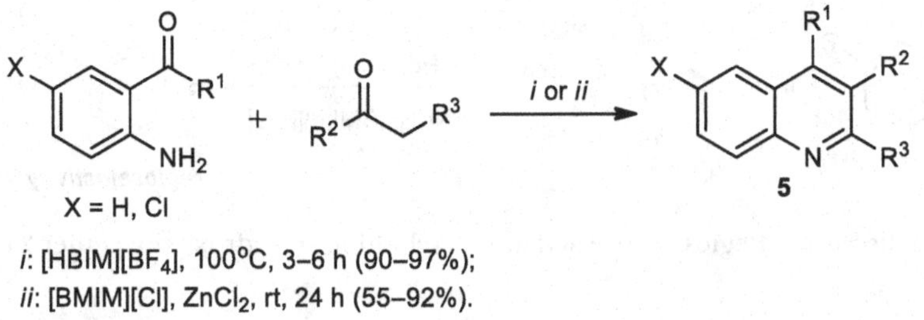

X = H, Cl

i: [HBIM][BF$_4$], 100°C, 3–6 h (90–97%);
ii: [BMIM][Cl], ZnCl$_2$, rt, 24 h (55–92%).

Scheme 3.   Synthesis of polysubstituted quinoline derivatives **5**

of their pharmacological and therapeutic properties.[30–32] Classical method for quinolines and related polyheterocycles is the Friedlander quinoline synthesis,[33–35] in which strong acids or bases are usually employed as the catalyst. The reaction can also occur without a catalyst at high temperature. However, the harsh reaction conditions are incompatible with either acid- or base-sensitive groups, which will reduce the synthetic scope of this reaction. In this scenario, the use of ILs has gained considerable importance. Srinivasan and co-workers innovated the use of IL [HBIM][BF$_4$] as reaction media for the synthesis of quinoline derivatives **5** (Scheme 3).[36] Later, the same reaction was investigated by Perumal and Karthikeyan using a [BMIM][Cl]/ ZnCl$_2$ melt (1:2 molar ratio), which can act as both a solvent and catalyst on account of its high polarity and Lewis acidity.[37] The authors mentioned that the polarity and the large electrochemical window of the IL may contribute to the observed regiospecificity.[36]

Palladium-catalyzed domino hydroarylation/cyclization reaction represents a useful tool for the construction of functionalized quinoline derivatives in [BMIM][BF$_4$].[38] Under mild conditions, the reaction can be successfully carried out starting from internal alkynes (**6**) and aryl iodides (Scheme 4). Compared to the traditional methods, the new approach has many advantages, including higher yield and shorter reaction time. As described, the solution containing the catalytically active palladium species can be used at least six times with essentially no loss of activity.

Quinolinone derivatives, an important class of fused heterocycles, have also attracted much attention due to their potential biological and pharmaceutical activities.[39–41] Cyclocarbonylation reactions catalyzed

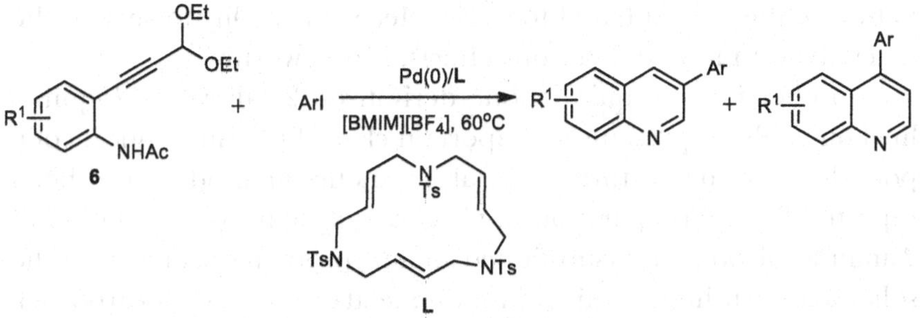

Scheme 4. Synthesis of substituted quinoline derivatives from **6**

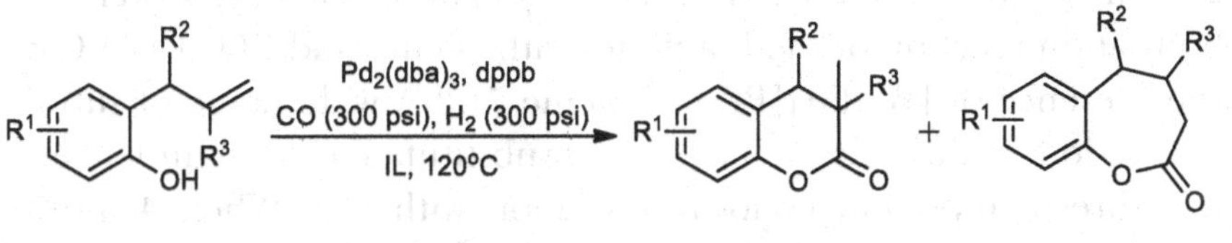

ILs:			
[BMIM][PF$_6$]:		15%	77%
[BMIM][BF$_4$]:		60%	22%
[BMIM][NTf$_2$]:		39%	58%
[BMIM][Cl]:		0%	0%

Scheme 5. Cyclocarbonylation of 2-allylphenol in various ILs

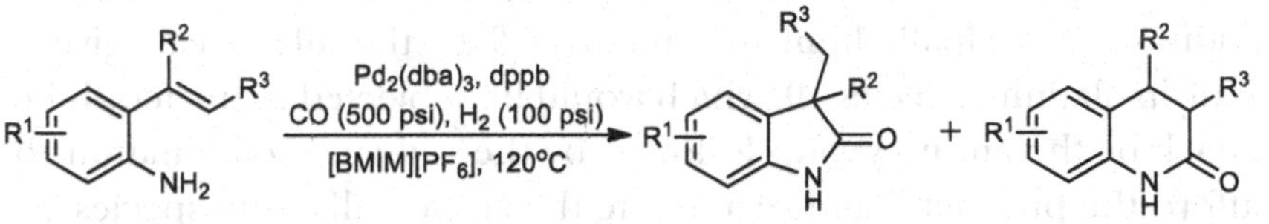

Scheme 6. Cyclocarbonylation of 2-aminostyrenes in [BMIM][PF$_6$]

by transition metal complexes have attracted considerable interest as a method to prepare these heterocyclic compounds.[42–44] Despite the significance, most of these processes often suffer from multistep procedures, low yields, or the need for a large amount of catalyst. Recently, Alper reported that the cyclocarbonylation of unsaturated phenols and anilines, catalyzed by Pd$_2$(dba)$_3$·CHCl$_3$ in IL [BMIM][PF$_6$], provided lactones (Scheme 5) or lactams (Scheme 6) in moderate to good yields. Different ring sizes were favored, subject to the nature of the reactant, and to steric effects, the ability to recycle the reactants with little deleterious effects, is noteworthy.[45]

Another kind of quinolinone derivatives, 2,3-dihydro-1*H*-quinolin-4-ones, also represents an important class of pharmaceutical compounds.[46] To date, three typical synthetic methods have been reported for the preparation of these ring systems: (i) cyclization of 2-aminochalcones or 3-(substituted anilino)propionic acid[47–48]; (ii) the solid-phase synthesis[49]; (iii) transition metal-catalyzed cyclocarbonylation in classical organic solvents.[50] Given the previous research results in ILs, Alper and co-workers developed the synthetic method for 2,3-dihydro-1*H*-quinolin-4-ones *via* a palladium-catalyzed cyclocarbonylation reaction of *o*-iodoanilines with allenes and CO at 120°C in the presence of [BMIM][PF$_6$] (Scheme 7).[51] A wide range of substituted *o*-iodoanilines, as well as several substituted allenes, underwent this three-component cyclocarbonylation with CO. When 4 mol% 1,4-Bis(diphenylphosphino)butane (DPPB) was used, as expected, the system of IL, palladium catalyst, and DPPB could be recovered and reused 4–6 times without a significant decrease in the yield of **7**.

A possible mechanism for the cyclocarbonylation of *o*-iodoanilines with allenes is outlined in Scheme 8. The oxidative addition of the palladium species to an iodoaniline generates **8**, which undergoes CO insertion to form the acylpalladium intermediate **9**. The addition of acylpalladium intermediate **9** to the allene can give a π-allylpalladium species **10**, which would be subjected to nucleophilic attack by the amino group followed by the reductive elimination to afford the product **7** and regenerate the active palladium species.

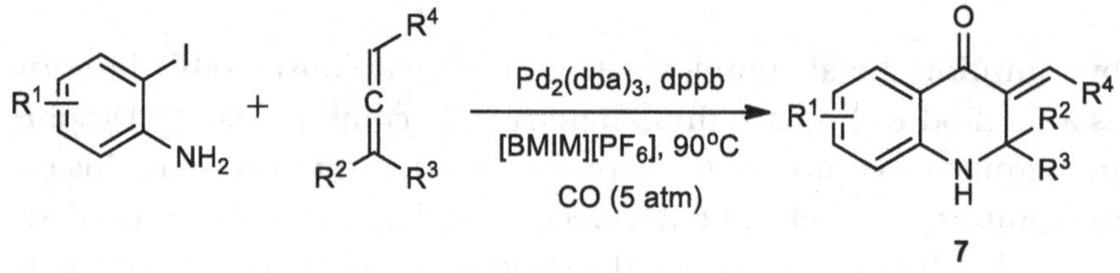

Scheme 7.   Cyclocarbonylation reaction of *o*-iodoanilines and allenes

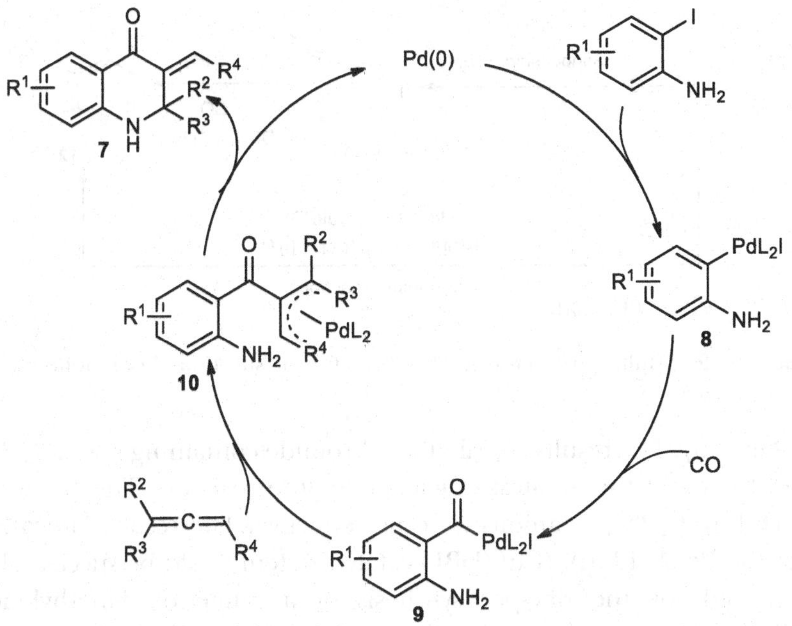

Scheme 8. Plausible reaction mechanism

2.3 Synthesis of isoindolinones

Recently, the combination of task-specific ILs as versatile and novel reaction media, with transition metal complexes as catalysts, has resulted in some effective and easily separable catalytic systems that were successfully used for carbonylation reactions.[52–54] Most ILs research has been conducted in nitrogen-based solvents.[55] Although beneficial in many cases, ammonium ILs have been shown to degrade under strong base and sonication conditions. For these reasons, researchers began to focus on developing processes in phosphonium salt-based ionic liquids (PSILs). PSILs are also non-volatile, economical, and available on an industrial scale.

Alper and co-workers described two efficient approaches for the synthesis of isoindolin-1-one derivatives **12** in PSILs.[56] The reactions were conducted under 1 atm of CO, 110°C, with $PdCl_2(PPh_3)_2$ as the catalyst and 1,8-diazabicyclo[5.4.0]undec-7-ene (DBU) as the base

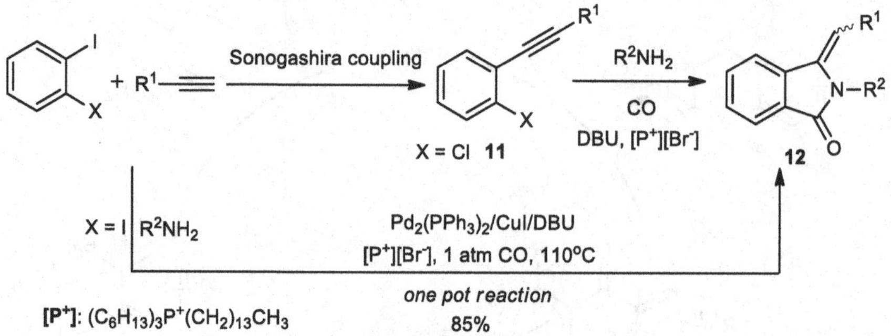

[P⁺]: $(C_6H_{13})_3P^+(CH_2)_{13}CH_3$

Scheme 9. Multistep and one-pot synthesis for the substituted isoindolinone

(Scheme 9). The results revealed that bromide containing media ([P⁺] [Br⁻]) showed the greatest efficiency for the reaction. The ILs with [NTf$_2^-$] and [PF$_6^-$] as anions were not effective. The results indicated that the $PdCl_2(PPh_3)_2$/CuI/DBU catalyst system could be successfully employed for the one-pot synthesis of a range of 3-methylene-isolindolin-1-ones. All processes were highly stereoselective, and gave the Z-isomer as the main product. When different arene dihalides were employed, the 1,2-diiodobenzene was efficiently transferred to the isoindolin-1-one derivative in 85% yield, whereas the reaction stopped at the first step with 1-chloro-2-iodobenzene as the substrate, and afforded the Sonogashira coupling product **11** as the only isolated product.

2.4 Synthesis of lactones

Transition metal-catalyzed enyne cyclization is a powerful reaction for the assembly of heterocyclic structures, which are important scaffolds in some bioactive compounds. In 2008, Jiang and co-workers demonstrated the Pd-catalyzed cyclizations of 1,6-enynes for the synthesis of α-methylene-γ-butyrolactones and lactams in imidazolium-type IL by the protonolysis of sp³–C–Pd bond (Scheme 10).[57] By contrast, in conventional organic solvents, protonolysis is usually restricted to the sp³–C–Pd bond adjacent to an electron-withdrawing group. In addition, with the conventional solvents such as CH_3CN, CH_2Cl_2, and 1,4-dioxane as solvents, there was no product **13** detected.

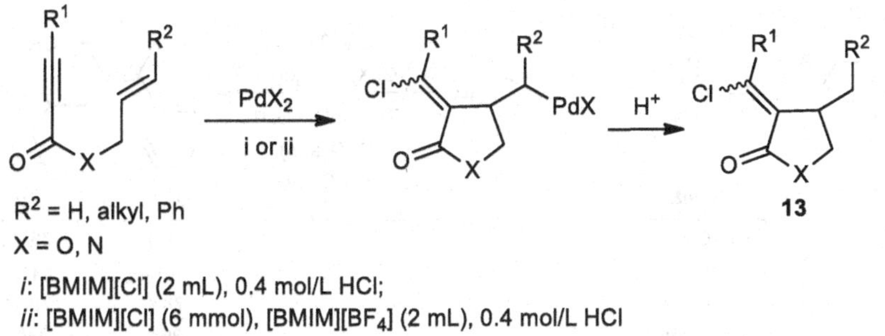

R^2 = H, alkyl, Ph

X = O, N

13

i: [BMIM][Cl] (2 mL), 0.4 mol/L HCl;
ii: [BMIM][Cl] (6 mmol), [BMIM][BF$_4$] (2 mL), 0.4 mol/L HCl

Scheme 10. Cyclization of enyne substrates in ILs

It was observed that imidazolium-type ILs played an important role in the reaction both as a ligand for the palladium catalyst and as a solvent. The reaction system of the Pd catalyst as well as the IL can be recycled and reused for three times with a gradual decrease in the yield of **13** (93%, 90%, and 88%).

The γ-lactone skeleton is quite important in organic synthesis and has also been found as a substructure in numerous bioactive natural products and potential pharmaceutically interesting compounds.[58] Transition metal-catalyzed reactions have emerged as a powerful tool for the construction of γ-lactones in conventional organic solvents.[59–61] Motived by the increasing requirements for sustainable chemistry, green and atom-economic synthesis have attracted great attention. In 2014, Jiang and co-workers developed the first example of palladium-catalyzed intermolecular carbonylation of alkynes with homoallylic alcohols in [C_2O_2mim]X to construct γ-lactones with high regio and stereo-selectivity (Z/E ≥ 98/2).[62] Various conventional solvents were examined, such as CH_3CN, tetrahydrofuran (THF), 1,4-dioxane, and DMF, which were found to significantly decrease the yields and stereoselectivities (Scheme 11).

A postulated mechanism is depicted in Scheme 12. Pd complex is initially formed *in situ* in ILs, and vinylpalladium intermediate **14** is formed by *trans*-chloropalladation of the alkyne in a polar solvent system in the presence of excess chloride ions. Then, intermediate **14** undergoes alkene insertion, and the vinylpalladium species simultaneously coordinates to both oxygen atoms of OR^2 as well as the

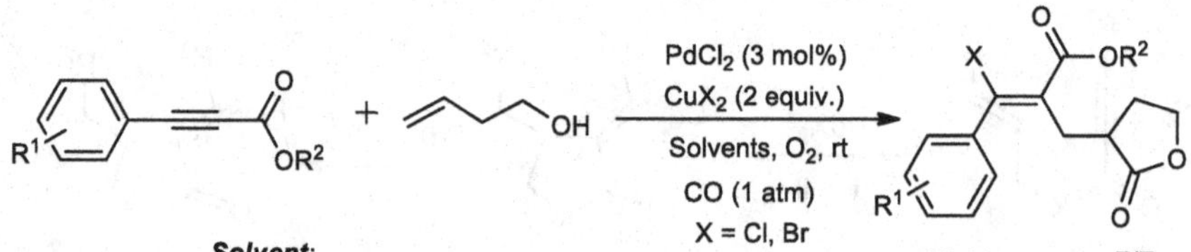

Scheme 11. Solvent effects of the carbonylation of alkynes

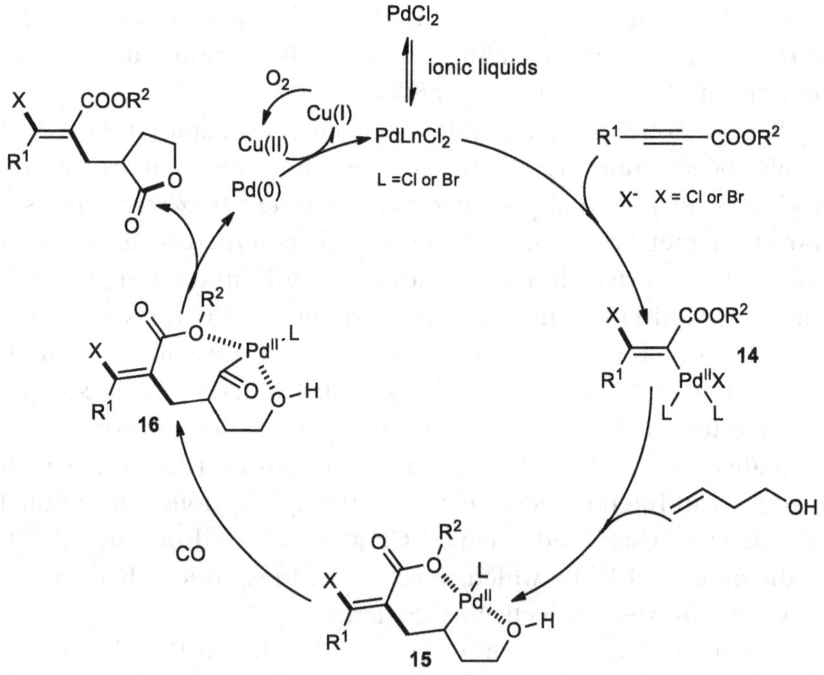

Solvent:	Yield	Z/E
CH₃CN	80%	78/22
DMF	18%	54/16
THF	74%	79/21
[C₂O₂mim]Cl	86%	98/2

Scheme 12. Proposed mechanism of the carbonylation

hydroxyl to generate a Pd–alkyl intermediate **15**. Subsequently, migratory insertion of CO into the palladium-carbon σ-bond is done to produce intermediate **16**. Finally, a reductive elimination gives the target product.

Just after, using a similar catalytic system $PdCl_2/AgNO_3$ with $[C_2O_2mim]Cl$ under open atmosphere, Jiang and co-workers reported the first example of a palladium-catalyzed intermolecular cascade annulation for the synthesis of functionalized β- and γ-lactones with high regio and stereoselectivity in ILs. The cascade annulation of substituted bromoalkynes **17** with pent-4-enoic acid **18** in $[C_2O_2mim]$ Cl afforded the corresponding γ-lactones **19** in good yields (69–83%) with high stereoselectivities (Z/E $\geq$ 98/2) (Scheme 13).[63] Furthermore, the reaction was found to be applicable to chloroalkynes, although the reaction was relatively sluggish.

Using $[C_2OHmim]Cl$ (0.5 ml), and CH_3CN (0.1 ml) as co-solvents, Jiang and co-workers have also extended the method to alkynoate **20** and but-3-enoic acid **21**, preparing a variety of substituted β-lactones **22** under similar reaction conditions (Scheme 14).[63] The traditional strategies for preparing β-lactones, in some cases, suffer from certain limitations such as multiple steps, harsh reaction conditions or low yields.

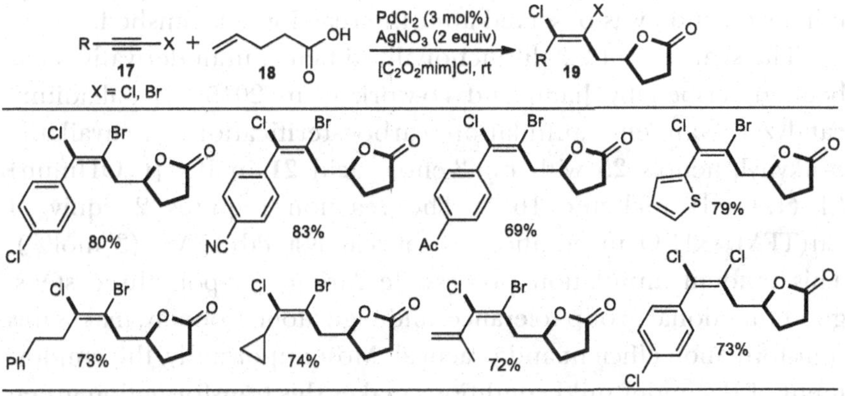

Scheme 13. Palladium-catalyzed cascade annulation for the synthesis of γ-lactones

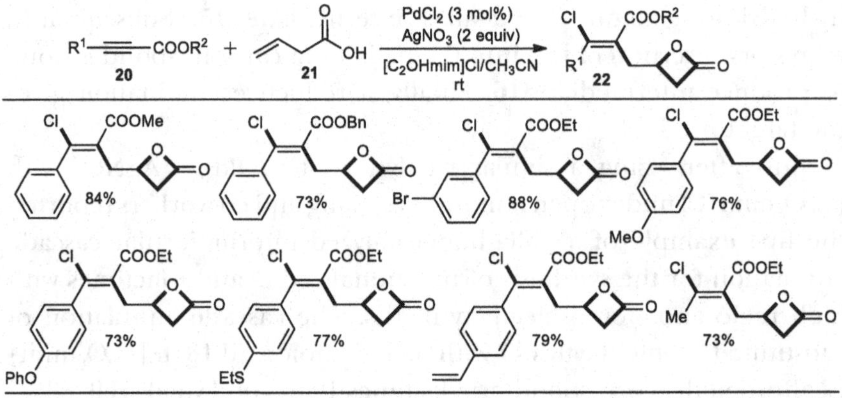

Scheme 14. Palladium-catalyzed cascade annulation for the synthesis of β-lactones

Also, chloropalladation triggering intermolecular carboesterification of alkynes with enoic acid in ILs was proposed (Scheme 15). A palladium complex is initially formed *in situ* in ILs, and the vinylpalladium intermediate 23 is formed by *trans*-chloropalladation of the alkyne in the presence of excess chloride ions. Subsequently, 23 could undergo alkene insertion. Simultaneously, the vinylpalladium species coordinates to the oxygen atoms of the hydroxy group to generate the palladium/alkyl intermediate 24. Finally, a reductive elimination gives the target product. It should be noted that a silver mirror reaction was observed after the reaction was finished.

The synthesis of 2,3-difunctionalized benzofuran derivatives has been described by Jiang and co-workers in 2015 *via* palladium-catalyzed tandem annulation/carboesterification of available 2-alkynylphenols 25 with but-3-enoic acid 21 in ILs [C$_2$OHmim] Cl (1.0 ml) (Scheme 16).[64] The reaction requires 2 equiv of Cu(TFA)$_2$·xH$_2$O in addition to the catalyst Pd(TFA)$_2$ (3 mol%). This tandem annulation process, featuring one-pot, three steps, good functional group tolerance, and high atom economy, makes this transformation efficient and practical. Most importantly, the employment of ILs under mild conditions makes this transformation green and environmentally friendly. Furthermore, this protocol should provide an ecofriendly method to functionalized benzofurans.

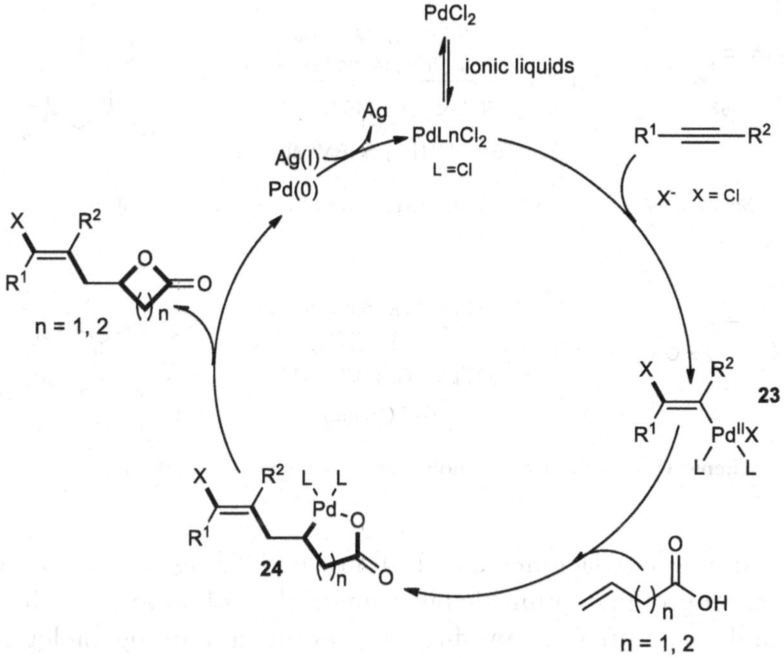

Scheme 15. Proposed mechanism for the synthesis of β- or γ-lactones

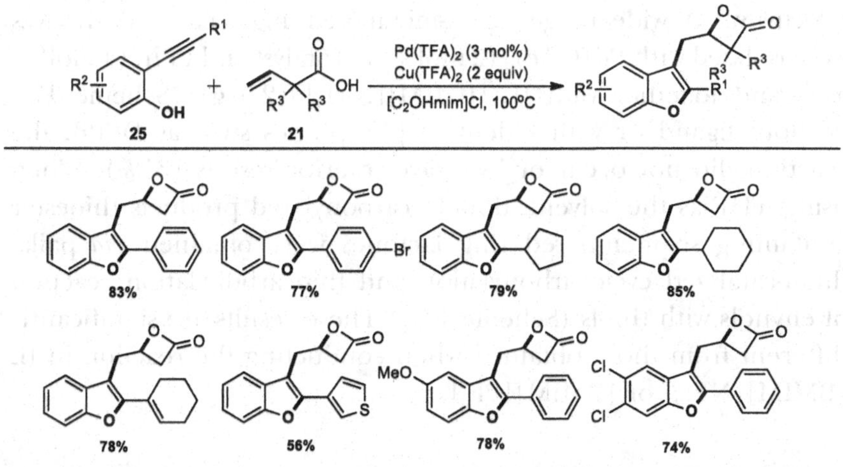

Scheme 16. Tandem annulation for the synthesis of functionalized benzofuran derivatives

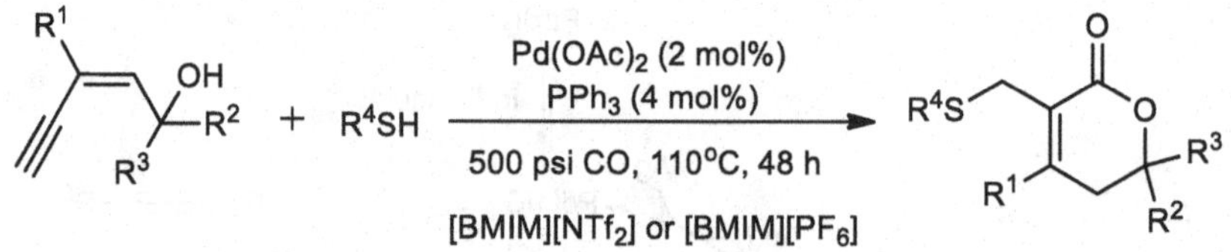

Scheme 17. Cyclocarbonylation reaction of enynols and thiols in ILs

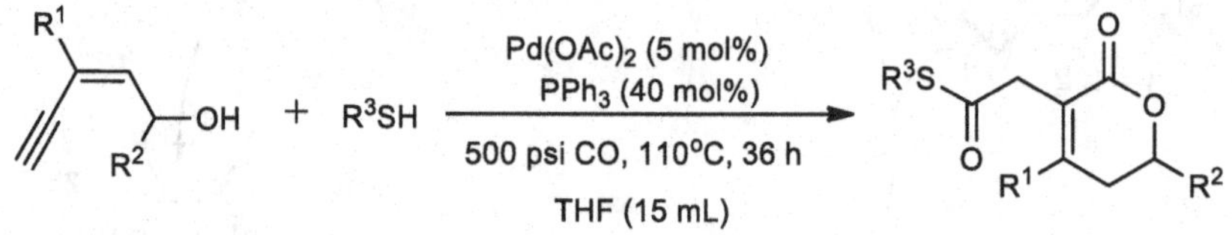

Scheme 18. Cyclocarbonylation reaction of enynols and thiols in THF

Unsaturated lactones and lactones containing a thioester or thioether group are valuable building blocks and useful subunits in natural and unnatural products possessing interesting biological activities.[65,66] Xiao and Alper demonstrated $Pd(OAc)_2$-catalyzed carbonylation reactions of enynols with thiols in ILs, affording monocarbonylated six-membered-ring lactones in high yields and good selectivity.[67] A wide range of six-membered-ring lactone derivatives were isolated with $Pd(OAc)_2$ (2 mol%) as catalyst and PPh_3 (4 mol%) as ligand together with IL [BMIM][NTf_2] (2.5 g) (Scheme 17). Without ligand or with bidentate phosphines such as DPPB, the reaction did not occur or just gave inferior results (31%). When using THF as the solvent, double carbonylated products thioester containing six-membered ring lactones were obtained *via* palladium-catalyzed cyclocarbonylation and thiocarbonylation reaction of enynols with thiols (Scheme 18).[68] These results are significantly different from those obtained when conducting the reaction in IL [BMIM][NTf_2] or [BMIM][PF_6].

3. Conclusions and Outlook

Over the past decade, ILs have been shown to be promising "green" solvents with several advantages compared to traditional organic solvents. Naturally, ILs can be used as an alternative recyclable and environmentally benign reaction medium for chemical processes. In addition, ILs can also act as a catalyst, catalyst activator, or co-catalyst for a reaction. Applications have also been found in different areas, including gas chromatography as stationary phases, in electrochemistry as solvents and electrolytes, and in pervaporation. In brief, the chemistry of ILs is at an incredibly exciting stage in its development.

On the other hand, although a number of chemical reaction processes in which ILs are used have been established, several urgent questions regarding the fundamental aspects of ILs need to be clarified: (i) the ILs purity issue in the chemical reaction should be emphasized; (ii) the stability of the ILs in particular, those functionalized ILs, should be carefully investigated; (iii) the toxicity and effect on human health of ILs should be further evaluated before a wide and large-scale applications.

Abbreviations

DPPP	Bis(diphenylphosphino)propane
DPPB	1,4-Bis(diphenylphosphino)butane
DBU	1,8-Diazabicyclo[5.4.0]undec-7-ene
[BMIM][Cl]	1-Butyl-3-methylimidazolium chloride
[BMIM][BF$_4$]	1-Butyl-3-methylimidazolium tetrafluoroborate
[BMIM][PF$_6$]	1-Butyl-3-methylimidazolium hexafluorophosphate
[BMIM][NTf$_2$]	1-Butyl-3-methylimidazolium bis(trifluoromethylsulfonyl)imides
[HBIM][BF$_4$]	1-Butyl imidazolium tetrafluoroborate
[C$_2$O$_2$MIM][Cl]	1-Carboxymethyl-3-methylimidazolium chloride
[C$_2$O$_2$MIM][Br]	1-Carboxymethyl-3-methylimidazolium bromide
[C$_2$OHMIM][Cl]	1-Hydroxyethyl-3-methylimidazolium chloride

References

1. P. T. Anastas and M. M. Kirchhoff, 2002. *Acc. Chem. Res.*, 35, 686.

2. B. M. Trost, 2002. *Acc. Chem. Res.*, 35, 695.

3. G. Wulff, 2002. *Chem. Rev.*, 102, 1.

4. I. T. Horváth and P. T. Anastas, 2007. *Chem. Rev.*, 107, 2169.

5. C. Gunanathan and D. Milstein, 2011. *Acc. Chem. Res.*, 44, 588.

6. R. A. Sheldon, 2012. *Chem. Soc. Rev.*, 41, 1437.

7. T. Welton, 1999. *Chem. Rev.*, 99, 2071.

8. J. Dupont, R. F. de Souza and P. A. Z. Suarez, 2002. *Chem. Rev.*, 102, 3667.

9. C. E. Song, 2004. *Chem. Commun.*, (9), 1033.

10. W. Wu, B. Han, H. Gao, Z. Liu, T. Jiang and J. Huang, 2004. *Angew. Chem., Int. Ed.*, 43, 2415.

11. M. A. P. Martins, C. P. Frizzo, D. N. Moreira, N. Zanatta and H. G. Bonacorso, 2008. *Chem. Rev.*, 108, 2015.

12. Q. Zhang, S. Zhang and Y. Deng, 2011. *Green Chem.*, 13, 2619.

13. P. Wasserscheid and W. Keim, 2000. *Angew. Chem., Int. Ed.*, 39, 3772.

14. H. Olivier-Bourbigou and L. Magna, 2002. *J. Mol. Catal. A: Chem.*, 182–183, 419.

15. M. Haumann and A. Riisager, 2008. *Chem. Rev.*, 108, 1474.

16. H. Olivier-Bourbigou, L. Magna and D. Morvan, 2010. *Appl. Catal., A*, 373, 1.

17. R. A. Sheldon, R. M. Lau, M. J. Sorgedrager, F. Rantwijk and K. R. Seddon, 2002. *Green Chem.*, 4, 147.

18. F. Rantwijk and R. A. Sheldon, 2007. *Chem. Rev.*, 107, 2757.

19. Y. Fan and J. Qian, 2010. *J. Mol. Catal. B: Enzym.*, 66, 1.

20. H. Olivier, 1999. *J. Mol. Catal. A: Chem.*, 146, 285.

21. B. Ni and A. D. Headley, 2010. *Chem.-Eur. J.*, 16, 4426.

22. J. Durand, E. Teuma and M. Gómez, 2007. *C. R. Chim.*, 10, 152.

23. P. Migowski and J. Dupont, 2007. *Chem.-Eur. J.*, 13, 32.

24. C. P. Mehnert, 2005. *Chem.-Eur. J.*, 11, 50.

25. J. Hulten, H. O. Anderson, W. Schaal, H. U. Danielson, B. Classon, I. Kvarntrom, A. Karlen, T. Unge, B. Samuelsson and A. Hallberg, 1999. *J. Med. Chem.*, 42, 4054.

26. A. Bengtson, M. Larhed and A. Hallberg, 2002. *J. Org. Chem.*, 67, 5852.

27. A. Arefalk, M. Larhed and A. Hallberg, 2005. *J. Org. Chem.*, 70, 938.

28. M. Larhed and A. Hallberg, 1997. *J. Org. Chem.*, 62, 7858.

29. Z. Hyder, J. Mo and J. Xiao, 2006. *Adv. Synth. Catal.*, 348, 1699.
30. A. R. Katritzky and C. W. Rees, 1984. *Comprehensive Heterocyclic Chemistry*, Vol. 2. Pergamon Press, New York.
31. A. R. Katritzky, C. W. Ress, E. F. V. Scriven, 1996. *Comprehensive Heterocyclic Chemistry II*, Vol. 5. Pergamon Press, New York.
32. P. M. S. Chauhan and S. K. Srivastava, 2001. *Curr. Med. Chem.*, 8, 1535.
33. P. Friedlander, 1882. *Ber.*, 15, 2572.
34. C.-C. Cheng and S.-J. Yan, 1982. In *Organic Reactions*, W. G. Dauben (ed.), Vol. 28. Wiley, New York.
35. A. R. Katritzky, C. W. Ress and E. F. V. Scriven, 1996. *Comprehensive Heterocyclic Chemistry II*, Vol. 2. Pergamon Press, New York.
36. S. S. Palimkar, S. A. Siddiqui, T. Daniel, R. J. Lahoti and K. V. Srinivasan, 2003. *J. Org. Chem.*, 68, 9371.
37. G. Karthikeyan and P. T. Perumal, 2004. *J. Heterocycl. Chem.*, 41, 1039.
38. S. Cacchi, G. Fabrizi, A. Goggiamani, M. Moreno-Mañas and A. Vallribera, 2002. *Tetrahedron Lett.*, 43, 5537.
39. Y. Kitahara, M. Shimizu and A. Kubo, 1990. *Heterocycles*, 31, 2085.
40. M. Venet, D. End and P. Angibaud, 2003. *Curr. Top. Med. Chem.*, 3, 1095.
41. J. M. Kraus, C. L. M. J. Verlinde, M. Karimi, G. I. Lepesheva, M. H. Gelb and F. S. Buckner, 2009. *J. Med. Chem.*, 52, 1639.
42. I. Ojima, M. Tzamarioudaki, Z. Li and R. J. Donovan, 1996. *Chem. Rev.*, 96, 635.
43. G. Zeni and R. C. Larock, 2004. *Chem. Rev.*, 104, 2285.
44. I. Nakamura and Y. Yamamoto, 2004. *Chem. Rev.*, 104, 2127.
45. F. Ye and H. Alper, 2006. *Adv. Synth. Catal.*, 348, 1855.
46. A. R. Katritzky and C. W. Rees, 1984. In *Pyridine and their Benzo Derivatives: Applications, in Comprehensive Heterocyclic Chemistry*, F. S. Yates (ed.), Vol. 2, p. 511. Pergamon Press, Oxford.
47. R. Grigg, A. Liu, D. Shaw, S. Suganthan, D. E. Woodall and G. Yoganathan, 2000. *Tetrahedron Lett.*, 41, 7125.
48. T. Nemoto, T. Fukuda and Y. Hamada, 2006. *Tetrahedron Lett.*, 47, 4365.
49. S. Wendeborn, 2000. *Synlett*, (1), 45.
50. C. Larksarp and H. Alper, 2000. *J. Org. Chem.*, 65, 2773.
51. F. Ye and H. Alper, 2007. *J. Org. Chem.*, 72, 3218.
52. J. H. Davis, 2004. *Chem. Lett.*, 33, 1072.
53. P. Wasserscheid and T. Welton, 2002. *Ionic Liquid in Synthesis*. Wiley-VCH & Co. KGaA, Weinheim, Germany.

54. E. Mizushima, T. Hayashi and M. Tanaka, 2001. *Green Chem.*, 3, 76.

55. P. Wasserscheid and T. Welton, 2003. *Ionic Liquids in Synthesis.* Wiley-VCH, Weinheim, Germany,

56. H. Cao, L. McNamee and H. Alper, 2008. *Org. Lett.*, 10, 5281.

57. S.-R. Yang, H.-F. Jiang, Y.-Q. Li, H.-J. Chen, W. Luo and Y.-B. Xu, 2008. *Tetrahedron*, 64, 2930.

58. R. Bandichhor, B. Nosse and O. Reiser, 2005. *Top. Curr. Chem.*, 243, 43.

59. B. W. Greatrex, M. C. Kimber, D. K. Taylor, G. Fallon and E. R. T. Tiekink, 2002. *J. Org. Chem.*, 67, 5307.

60. G. Liu and X. Lu, 2003. *Tetrahedron Lett.*, 44, 127.

61. M. He, A. Lei and X. Zhang, 2005. *Tetrahedron Lett.*, 46, 1823.

62. J. Li, S. Yang, W. Wu and H. Jiang, 2014. *Chem. Commun.*, 50, 1381.

63. J. Li, W. Yang, S. Yang, L. Huang, W. Wu, Y. Sun and H. Jiang, 2014. *Angew. Chem. Int. Ed.*, 53, 7219.

64. J. Li, Z. Zhu, S. Yang, Z. Zhang, W. Wu and H. Jiang, 2015. *J. Org. Chem.*, 80, 3870.

65. L. A. Collet, M. T. Davis-Coleman and D. E. A. Rivett, 1998. In *Progress in the Chemistry of Organic Natural Products*, W. Herz, H. Falk, G. W. Kirby, E. Moore and Ch. Tamm, (eds.), Vol. 75, p 182. Springer, New York.

66. S. V. Ley, L. R. Cox and G. Meek, 1996. *Chem. Rev.*, 96, 423.

67. H. Cao, W.-J. Xiao and H. Alper, 2007. *J. Org. Chem.*, 72, 8562.

68. H. Cao, W.-J. Xiao and H. Alper, 2006. *Adv. Synth. Catal.*, 348, 1807.

Index

Printed in the United States
By Bookmasters